Advances in Nanofluid Heat Transfer

Advances in Nanofluid Heat Transfer

Edited by

Hafiz Muhammad Ali
Mechanical Engineering Department,
King Fahd University of Petroleum and Minerals,
Dhahran, Saudi Arabia
Interdisciplinary Research Center for Renewable Energy and
Power Systems (IRC-REPS), King Fahd University of Petroleum and
Minerals, Dhahran, Saudi Arabia

ELSEVIER

Elsevier
Radarweg 29, PO Box 211, 1000 AE Amsterdam, Netherlands
The Boulevard, Langford Lane, Kidlington, Oxford OX5 1GB, United Kingdom
50 Hampshire Street, 5th Floor, Cambridge, MA 02139, United States

ISBN: 978-0-323-88656-7

For Information on all Elsevier publications
visit our website at https://www.elsevier.com/books-and-journals

Publisher: Matthew Deans
Acquisitions Editor: Brian Guerin
Editorial Project Manager: Mariana L. Kuhl
Production Project Manager: Manju Paramasivam
Cover Designer: Christian J. Bilbow

Typeset by MPS Limited, Chennai, India

Dedication

Hafiz Muhammad Ali dedicates this work to his beloved sons Muhammad Ibraheem Ali, Muhammad Ismaeel Ali, and Muhammad Ishaac Ali.

Contents

List of contributors xvii
Acknowledgment xxi

1. Experimental correlations for Nusselt number and friction factor of nanofluids 1

L. Syam Sundar, E. Venkata Ramana, Hafiz Muhammad Ali and Antonio C.M. Sousa

1.1 Introduction 1
1.2 Preparation of nanofluids 2
1.2.1 One-step method 3
1.2.2 Two-step method 3
1.3 Experimental methods 3
1.3.1 Nusselt number 4
1.3.2 Friction factor 5
1.3.3 Nondimensional numbers 5
1.4 Nusselt number correlations for single-phase fluid 6
1.4.1 Laminar flow 6
1.4.2 Turbulent flow 7
1.5 Friction factor correlations for single-phase fluids 8
1.5.1 Laminar flow 8
1.5.2 Turbulent flow 8
1.6 Factors influencing the development of correlations 8
1.6.1 Nusselt number 8
1.6.2 Friction factor 9
1.7 Developed corrections for nanofliuids 9
1.7.1 Nusselt number 9
1.7.2 Friction factor 16
1.8 Conclusion 20
References 20

2. Preparation and evaluation of stable nanofluids for heat transfer application 25

Zeyu Liu, Rong Fu and Yan Yuying

Nomenclature 25
2.1 Introduction 26
2.2 Preparation 27
2.2.1 Two-step method 27

2.2.2 One-step method 29
2.3 **Evaluation of nanofluid stability** 30
2.3.1 Zeta potential analysis 31
2.3.2 Sedimentation and centrifugation methods 32
2.3.3 Spectral analysis method 33
2.3.4 3ω method 34
2.3.5 Electron microscopy and light scattering 34
2.3.6 pH measurement 36
2.4 **Stabilization techniques** 36
2.4.1 Physical method 37
2.4.2 Chemical stabilization 38
2.5 **Stability mechanisms** 40
2.5.1 Electrostatic stabilization 41
2.5.2 Steric stabilization 42
2.5.3 Electrosteric stabilization 44
2.6 **Impact of nanofluid stability on thermophysical properties** 45
2.6.1 Effects of stability on density 45
2.6.2 Effects of stability on viscosity 45
2.6.3 Effects of stability on specific heat capacity of nanofluids 48
2.6.4 Effects of stability on thermal conductivity 49
2.7 **Conclusion** 52
References 53

3. Synthesis, characterization, and measurement techniques for the thermophysical properties of nanofluids 59

Arun Kumar Tiwari, Amit Kumar and Zafar Said

3.1 **Introduction** 59
3.2 **Synthesis of nanofluid** 59
3.2.1 One-step method 60
3.2.2 Two-step method 64
3.3 **Characterization of nanofluid** 67
3.3.1 Zeta potential 67
3.3.2 Fourier-transform infrared spectroscopy 70
3.3.3 Transmission electron microscopy 71
3.3.4 X-ray crystallography 74
3.3.5 UV–Vis technique 74
3.4 **Thermophysical properties measurement techniques of nanofluid** 75
3.4.1 Thermal conductivity 76
3.4.2 Viscosity of nanofluids 79
3.4.3 Specific heat 81
3.4.4 Density 83
3.5 **Conclusion** 84
Nomenclature 85
References 86

4. Thermophysical and rheological properties of unitary and hybrid nanofluids 95

Hatice Mercan, Ali Celen and Tolga Taner

4.1 Introduction 95
4.2 Thermophysical properties 96
4.2.1 Thermal conductivity 96
4.2.2 Specific heat capacity 104
4.2.3 Density 112
4.2.4 Viscosity 116
4.3 Conclusion 122
Nomenclature 123
References 124

5. Comparison of physical properties enhancement in various heat transfer nanofluids by MXene 131

Lingenthiran Samylingam, Kumaran Kadirgama, Navid Aslfattahi, M. Samykano and R. Saidur

Nomenclature 131
5.1 Introduction 132
5.2 Methodology 134
5.2.1 Preparation of MXene nanomaterial as an additive to heat transfer nanofluids 134
5.2.2 Preparation of various heat transfer nanofluids by MXene 135
5.2.3 Physical properties 138
5.3 Results and discussion 140
5.3.1 Thermal conductivity of various heat transfer nanofluids by MXene 140
5.3.2 Viscosity analysis of various heat transfer nanofluids by MXene 144
5.3.3 Thermal Stability of various heat transfer nanofluids by MXene 146
5.4 Conclusion 148
Acknowledgment 149
References 149

6. Numerical modeling of nanofluids' flow and heat transfer 151

Tehmina Ambreen, Arslan Saleem and Cheol Woo Park

Nomenclature 151
6.1 Introduction 153
6.2 Heat transfer enhancement mechanism of nanofluid 154
6.2.1 Particle–particle interactions 155
6.2.2 Particle–liquid interactions 159
6.2.3 External forces on particles 161

6.3 **Thermophysical properties of nanofluids** 162
- 6.3.1 Thermal conductivity 162
- 6.3.2 Viscosity 165
- 6.3.3 Specific heat 172
- 6.3.4 Density 173

6.4 **Mathematical models to simulate nanofluids** 173
- 6.4.1 Single-phase model 173
- 6.4.2 Multiphase Eulerian–Eulerian model 178
- 6.4.3 Eulerian model 180
- 6.4.4 Volume of fluid model 181
- 6.4.5 Multiphase Eulerian–Lagrangian model 183

6.5 **Numerical techniques to simulate nanofluid** 186
- 6.5.1 Macroscale techniques 187
- 6.5.2 Microscale techniques 189
- 6.5.3 Mesoscale techniques 191

6.6 **Conclusion** 192

References 193

7. Recent advances in machine learning research for nanofluid heat transfer in renewable energy 203

Zafar Said, Maham Sohail and Arun Kumar Tiwari

Nomenclature 203

7.1 **Introduction** 203
- 7.1.1 Data collection and representation 204
- 7.1.2 Model selection and validation 204
- 7.1.3 Model optimization 205

7.2 **Machine learning techniques** 205
- 7.2.1 Multilayer perception artificial neural network 206
- 7.2.2 Adaptive neuro fuzzy inference system 207
- 7.2.3 Radial basis function network 208
- 7.2.4 Least square support vector machine 209

7.3 **Nanofluid heat transfer and machine learning** 210

7.4 **Machine learning of nanofluids' thermophysical properties and thermal performance** 217

7.5 **Challenges and future opportunities** 223

7.6 **Conclusion** 223

References 223

8. Heat transfer enhancement with nanofluids in automotive 229

Xihui Wang, Kuo Huang, Yan Yuying and Hailin Cen

8.1 **Historical background** 229
- 8.1.1 Applications of computational fluid dynamics in nanofluids studies 229
- 8.1.2 Applications of nanofluid in automotive system 232

8.2 Physical properties 234
8.2.1 Thermal conductivity of nanofluids 234
8.2.2 Viscosity of nanofluids 234
8.3 The fundamental relation for computational fluid dynamics model 239
8.3.1 Macroscopic models 239
8.3.2 Lattice Boltzmann method 240
8.4 Heat transfer enhancement with nanofluids in automotive 243
8.4.1 Utilization in engine coolant 243
8.4.2 Utilization in refrigerant of automotive air conditioning 250
Nomenclature 259
Problems 259
References 260

9. The use of nanofluids in solar desalination of saline water resources as antibacterial agents 265

Roya Mehrkhah, Mohammad Mustafa Ghafurian, Hamid Niazmand, Elaheh K. Goharshadi and Omid Mahian

Nomenclature 265
9.1 Harvesting solar energy by nanofluids 266
9.1.1 Introduction 266
9.1.2 Solar steam generation experiments by the nanofluids 267
9.1.3 Effective parameters on light harvesting capability of the nanofluids 271
9.1.4 Effect of NFs on heat localization 278
9.1.5 Comparison of economic performance of NFs 278
9.1.6 Photocatalytic experiments by NFs 279
9.2 Antibacterial activity of some NFs 282
9.2.1 Introduction 282
9.2.2 Effect of NFs on prevention of *Escherichia coli* proliferation 284
9.2.3 Mechanisms of antibacterial activity of NFs 285
9.2.4 Effective parameters on antibacterial activity of NFs 289
9.2.5 Rate of antibacterial activity of NPs 294
9.2.6 The effect of NFs on the bacteria morphology 294
9.3 Conclusion 296
References 296

10. Application of nanofluids in combustion engines with focusing on improving heat transfer process 303

Mohammad Hemmat Esfe, Saeed Esfandeh and Mohammad Hassan Kamyab

Nomenclature 303
10.1 Introduction 305
10.1.1 History of heat transfer in combustion engines 306

10.2 Parameters affecting the heat transfer of combustion engines 307
10.2.1 Engine size and dimension 307
10.2.2 Engine speed 307
10.2.3 Engine load 308
10.2.4 Inlet air temperature 309
10.2.5 Coolant temperature 309
10.2.6 Engine materials 309
10.2.7 Friction coefficient 310
10.2.8 Heat transfer coefficient 311
10.2.9 Conduction heat transfer coefficient 313
10.3 Type of lubricants 316
10.3.1 Lubrication of combustion engines 316
10.3.2 Basic engine oils and challenges 319
10.4 Using nanoparticles in internal combustion engines 320
10.4.1 Adding nanoparticles to engine oils (nano oil) 322
10.4.2 Adding nanoparticles to fuel (nanofuel) 327
10.4.3 Effects of nanoparticles additives on friction 327
10.4.4 Effects of nanoparticles additives on engines power 329
10.4.5 Effects of nanoparticle additives on the cooling cycle of combustion engines 331
10.5 Conclusion on threats and opportunities of applying nanoscience in combustion engines 333
References 334

11. Applications of nanofluids in solar energy collectors focusing on solar stills 341

Saeed Esfandeh, Mohammad Hemmat Esfe and Mohammad Hassan Kamyab

Nomenclature 341
11.1 History of solar energy collectors 342
11.2 Classification of Solar energy collectors 344
11.2.1 Classification based on solar energy application 344
11.2.2 Classification based on concentration level ability 345
11.2.3 Classification based on tracking or nontracking solar collectors 345
11.2.4 Passive and active solar systems 346
11.2.5 Main types of solar collectors 347
11.3 Effective parameters on solar still performance 356
11.3.1 Various design parameters 356
11.3.2 Climatic parameters 359
11.4 Application of nanofluids in solar stills 361
11.4.1 Solar stills without nanofluids 361
11.4.2 Solar stills with nanofluids 362

11.5 Most applied nanoparticles in solar stills 363
11.6 Challenges of nanofluid application in solar collectors 364
References 369

12. Utilization of nanofluids (mono and hybrid) in parabolic trough solar collector: a comparative analysis 375
Muhammad Sajid Khan, Mi Yan and Hafiz Muhammad Ali

12.1 Introduction 375
12.2 System description and thermodynamic modeling 380
12.2.1 Energy and exergy analysis of parabolic trough collector 383
12.2.2 Heat transfer analysis 385
12.2.3 Hybrid and mono nanofluids specifications 386
12.2.4 Performance evaluation criteria 388
12.2.5 Validation of parabolic trough collector model 389
12.3 Results and discussion 389
12.4 Conclusion 396
Nomenclature 397
Acknowledgment 399
References 399

13. Electronics thermal management applying heat pipes and pulsating heat pipes 403
Roger R. Riehl, Jesús Esarte San Martin and Juncal Estella

13.1 Introduction 403
13.2 Design parameters 405
13.3 Heat pipes 405
13.3.1 Heat pipe designs 412
13.3.2 Application of heat pipes 414
13.4 Pulsating heat pipes 418
13.4.1 Application of pulsating heat pipes 421
13.5 Nanofluids capabilities and models 424
13.6 Nanofluids in heat transfer systems: pros and cons 430
13.6.1 Types of nanoparticles used in nanofluids 430
13.6.2 Nanofluid preparation 432
13.6.3 Nanofluid stability strategies 433
13.6.4 Nanofluid stability mechanisms 435
13.6.5 Nanofluid stability evaluation 436
13.6.6 Nanofluids after operation cycles 437
13.6.7 Nanofluids applications 438
13.7 Concluding remarks 440
Nomenclature 440
References 442

14. Role of nanofluids in microchannel heat sinks 447

Ali Turab Jafry, Sohail Malik, Naseem Abbas, Uzair Sajjad, Hafiz Muhammad Ali and Chi-Chuan Wang

Nomenclature 447
14.1 Introduction 448
14.2 Key characteristics of nanofluids 449
14.2.1 Thermophysical properties 449
14.2.2 Effect of nanoparticle concentration 454
14.2.3 Effect of nanoparticle size 455
14.2.4 Effect of nanoparticle shape 456
14.2.5 Effect of nanoparticle thermal conductivity 456
14.2.6 Effect of base fluid 457
14.2.7 Effect of nanofluid temperature 457
14.2.8 Effect of preparation technique 458
14.2.9 Summary 459
14.3 Flow of nanofluids in microchannels 459
14.4 Thermal performance of nanofluids in microchannels 461
14.5 Entropy analysis of nanofluid-based microchannel heat sinks 464
14.6 Geometry effect of microchannels 467
14.6.1 Changing shape of channels 467
14.6.2 Adding geometrical inclusions/exclusions in the cross section 467
14.6.3 Varying path of the fluid 467
14.7 Future advances and challenges 468
14.8 Conclusions 472
References 473

15. Nanofluids for enhanced performance of building thermal energy systems 479

Naseem Abbas, Muhammad Bilal Awan, Mohsin Ali Badshah, Uzair Sajjad and Waseem Raza

15.1 Introduction 479
15.2 Overview of domain knowledge related to nanofluids 480
15.3 Role of nanofluids in efficiency enhancement of building energy systems 483
15.3.1 Nanofluids for performance enhancement of photovoltaic thermal systems 483
15.3.2 Nanofluids for performance enhancement of heating, ventilation, and air conditioning systems 487
15.3.3 Nanofluids for performance enhancement of thermal storage systems 489
15.4 Barriers 494
15.5 Conclusions 495
References 496

16. Ionic nanofluids: preparation, characteristics, heat transfer mechanism, and thermal applications 503

Tayyab Raza Shah, Chao Zhou, Hafiz Muhammad Rizwan, Muhammad Abdullah, Asad Iqbal, Awais Awan and Hafiz Muhammad Ali

Abbreviations 503
16.1 Introduction 504
16.2 Preparation methods 506
16.3 Characteristics 511
16.3.1 Thermal conductivity 511
16.3.2 Viscosity 512
16.3.3 Specific heat 512
16.3.4 Density 512
16.4 Heat transfer mechanism 513
16.5 Thermal applications 526
16.6 Future prospects and challenges 526
16.7 Conclusions 532
References 532

17. Hybrid nanofluids towards advancement in nanofluids for heat sink 537

Furqan Jamil, Taha Baig, Hafiz Muhammad Ali, Muhammad Shehryar, Shahab Khushnood and Faisal Hassan

Nomenclature 537
17.1 Introduction 538
17.2 Preparation of hybrid nanofluids 539
17.2.1 Two-step method 539
17.2.2 One-step method 540
17.3 Various hybrid nanofluids used in different heat sinks 541
17.3.1 Thermal management of electronics using other fluids 545
17.4 Conclusion 551
References 552

Index 557

List of contributors

Naseem **Abbas** Department of Mechanical Engineering, Sejong University, Seoul, South Korea

Muhammad **Abdullah** Mechanical Engineering Department, Quaid-E-Azam College of Engineering and Technology, Sahiwal, Pakistan

Hafiz Muhammad **Ali** Mechanical Engineering Department, King Fahd University of Petroleum and Minerals, Dhahran, Saudi Arabia; Interdisciplinary Research Center for Renewable Energy and Power Systems (IRC-REPS), King Fahd University of Petroleum and Minerals, Dhahran, Saudi Arabia

Tehmina **Ambreen** School of Mechanical Engineering, Kyungpook National University, Bukgu Daegu, South Korea

Navid **Aslfattahi** Department of Fluid Mechanics and Thermodynamics, Faculty of Mechanical Engineering, Czech Technical University in Prague, Prague, Czech Republic

Awais **Awan** Mechanical Engineering Department, Quaid-E-Azam College of Engineering and Technology, Sahiwal, Pakistan

Muhammad Bilal **Awan** Department of Mechanical Engineering, University of Central Punjab, Lahore, Pakistan

Mohsin Ali **Badshah** Department of Chemical Engineering and Materials Science, The Henry Samueli School of Engineering, University of California, Irvine, CA, United States

Taha **Baig** Mechanical Engineering Department, University of Wah, Wah Cantonment, Pakistan; Mechanical Engineering Department, University of Engineering and Technology, Taxila, Pakistan

Ali **Celen** Department of Mechanical Engineering, Faculty of Engineering, Erzincan Binali Yıldırım University, Erzincan, Turkey

Hailin **Cen** School of Mechanical and Automotive Engineering, South China University of Technology, GuangZhou, P.R. China

Saeed **Esfandeh** Department of Mechanical Engineering, Jundi-Shapur University of Technology, Dezful, Iran; Department of Mechanical Engineering, Imam Hossein University, Tehran, Iran

Juncal **Estella** NAITEC—Centro Tecnológico en Automoción y Mecatrónica, Pamplona, Spain

Rong **Fu** Institute of Engineering Thermophysics, Chinese Academy of Sciences, Beijing, P.R. China

Mohammad Mustafa **Ghafurian** Department of Mechanical Engineering, Bozorgmehr University of Qaenat, Qaen, Iran; Center for Nanotechnology in Renewable Energies, Ferdowsi University of Mashhad, Mashhad, Iran

Elaheh K. **Goharshadi** Department of Chemistry, Faculty of Science, Ferdowsi University of Mashhad, Mashhad, Iran

Faisal **Hassan** Mechanical Engineering Department, University of Engineering and Technology, Taxila, Pakistan

Mohammad **Hemmat Esfe** Department of Mechanical Engineering, Imam Hossein University, Tehran, Iran

Kuo **Huang** Faculty of Engineering, The University of Nottingham, Nottingham, United Kingdom

Asad **Iqbal** Mechanical Engineering Department, Quaid-E-Azam College of Engineering and Technology, Sahiwal, Pakistan

Ali Turab **Jafry** Faculty of Mechanical Engineering, Ghulam Ishaq Khan Institute of Engineering Sciences and Technology, Topi, Pakistan

Furqan **Jamil** Mechanical Engineering Department, University of Wah, Wah Cantonment, Pakistan

Kumaran **Kadirgama** Faculty of Mechanical Engineering, Universiti Malaysia Pahang, Pekan, Malaysia; Automotive Engineering Centre, Universiti Malaysia Pahang, Pekan, Malaysia

Mohammad Hassan **Kamyab** Department of Mechanical Engineering, Imam Hossein University, Tehran, Iran; Department of Mechanical Engineering, University of Kashan, Kashan, Iran

Muhammad Sajid **Khan** Institute of Energy and Power Engineering, College of Mechanical Engineering, Zhejiang University of Technology, Hangzhou, P.R. China; Department of Mechanical Engineering, Mirpur University of Science & Technology (MUST), Mirpur, AJK, Pakistan

Shahab **Khushnood** Mechanical Engineering Department, University of Wah, Wah Cantonment, Pakistan

Amit **Kumar** Mechanical Engineering Department, Institute of Engineering & Technology, Dr. A.P.J. Abdul Kalam Technical University, Lucknow, Uttar Pradesh, India

Zeyu **Liu** Marine Engineering College, Dalian Maritime University, Dalian, Liaoning, P.R. China; Faculty of Engineering, The University of Nottingham, Nottingham, Nottinghamshire, United Kingdom

Omid **Mahian** Center for Nanotechnology in Renewable Energies, Ferdowsi University of Mashhad, Mashhad, Iran; School of Chemical Engineering and Technology, Xi'an Jiaotong University, Xi'an, Shaanxi, P.R. China

Sohail **Malik** Faculty of Mechanical Engineering, Ghulam Ishaq Khan Institute of Engineering Sciences and Technology, Topi, Pakistan

Jesús Esarte San **Martin** NAITEC—Centro Tecnológico en Automoción y Mecatrónica, Pamplona, Spain

Roya **Mehrkhah** Department of Chemistry, Faculty of Science, Ferdowsi University of Mashhad, Mashhad, Iran

Hatice **Mercan** Department of Mechatronics Engineering, Yildiz Technical University, Istanbul, Turkey

Hamid **Niazmand** Center for Nanotechnology in Renewable Energies, Ferdowsi University of Mashhad, Mashhad, Iran

Cheol Woo **Park** School of Mechanical Engineering, Kyungpook National University, Bukgu Daegu, South Korea

E. Venkata **Ramana** Department of Physics, University of Aveiro, Aveiro, Portugal

Waseem **Raza** Department of Mechanical Engineering, Jeju National University, Jeju, South Korea

Roger R. **Riehl** GamaTech Thermal Solutions, São José dos Campos, Brazil

Hafiz Muhammad **Rizwan** Faculty of Mechanical Engineering, GIK Institute of Engineering Sciences and Technology, Topi, Pakistan

Zafar **Said** U.S.-Pakistan Center for Advanced Studies in Energy (USPCAS-E), National University of Sciences and Technology (NUST), Islamabad, Pakistan; Department of Sustainable and Renewable Energy Engineering, University of Sharjah, Sharjah, United Arab Emirates

R. **Saidur** Research Center for Nano-Materials and Energy Technology (RCNMET), School of Engineering and Technology, Sunway University, Selangor Darul Ehsan, Malaysia; Department of Engineering, Lancaster University, Lancaster, United Kingdom

Uzair **Sajjad** Department of Energy and Refrigerating Air-Conditioning Engineering, National Taipei University of Technology, Taipei, Taiwan

Arslan **Saleem** School of Engineering, Cardiff University, Cardiff, Wales, United Kingdom

M. **Samykano** Faculty of Mechanical Engineering, Universiti Malaysia Pahang, Pekan, Malaysia

Lingenthiran **Samylingam** Research Center for Nano-Materials and Energy Technology (RCNMET), School of Engineering and Technology, Sunway University, Selangor Darul Ehsan, Malaysia

Tayyab Raza **Shah** College of Engineering, Peking University, Beijing, P.R. China

Muhammad **Shehryar** Mechanical Engineering Department, University of Engineering and Technology, Taxila, Pakistan

Maham **Sohail** Department of Sustainable and Renewable Energy Engineering, University of Sharjah, Sharjah, United Arab Emirates

Antonio C.M. **Sousa** Department of Mechanical Engineering, Prince Mohammad Bin Fahd University, Al-Khobar, Saudi Arabia

L. Syam **Sundar** Department of Mechanical Engineering, Prince Mohammad Bin Fahd University, Al-Khobar, Saudi Arabia

Tolga **Taner** Department of Motor Vehicles and Transportation Technology, Aksaray University, Aksaray, Turkey

Arun Kumar **Tiwari** Mechanical Engineering Department, Institute of Engineering and Technology, Dr. A.P.J. Abdul Kalam Technical University, Lucknow, Uttar Pradesh, India

Chi-Chuan **Wang** Department of Mechanical Engineering, National Yang Ming Chiao Tung University, Hsinchu, Taiwan

Xihui **Wang** School of Mechanical and Automotive Engineering, South China University of Technology, GuangZhou, P.R. China

Mi **Yan** Institute of Energy and Power Engineering, College of Mechanical Engineering, Zhejiang University of Technology, Hangzhou, P.R. China

Yan **Yuying** Faculty of Engineering, The University of Nottingham, Nottingham, Nottinghamshire, United Kingdom

Chao **Zhou** College of Engineering, Peking University, Beijing, P.R. China

Acknowledgment

Hafiz Muhammad Ali acknowledges the support received from the King Fahd University of Petroleum and Minerals, Dhahran, 31261, Kingdom of Saudi Arabia.

Chapter 1

Experimental correlations for Nusselt number and friction factor of nanofluids

L. Syam Sundar[1], E. Venkata Ramana[2], Hafiz Muhammad Ali[3,4] and Antonio C.M. Sousa[1]

[1]*Department of Mechanical Engineering, Prince Mohammad Bin Fahd University, Al-Khobar, Saudi Arabia,* [2]*Department of Physics, University of Aveiro, Aveiro, Portugal,* [3]*Mechanical Engineering Department, King Fahd University of Petroleum and Minerals, Dhahran, Saudi Arabia,* [4]*Interdisciplinary Research Center for Renewable Energy and Power Systems (IRC-REPS), King Fahd University of Petroleum and Minerals, Dhahran, Saudi Arabia*

1.1 Introduction

Heat transfer augmentation is necessary for each and every industry. Heat transfer augmentation can be achieved by using active techniques and passive techniques. An active method of heat transfer enhancement is possible by using mechanical stirring, surface vibrations, electrostatic fields, jet impingement, spray, etc. [1–3]. To achieve the active method of heat transfer enhancement it needs extra devices, which means that it requires extra power. So, this type of heat transfer enhancement is not advisable, because it needs more power. A passive method of heat transfer enhancement is possible by providing surface texturing [4], extended surfaces [5], and flow turbulence [6,7] and adding additives to the fluid [8,9]. Just adding additives (metal or nonmetal particles) to the base fluid may get heat transfer enhancement, since the additives have more thermal conductivity than the fluids. This type of heat transfer enhancement does not require any extra accessories. The initial research work was related to the dispersion of micron-size $\left(1 \times 10^{-6}\mathrm{m}\right)$ particles in the fluid and it gives the enhanced heat transfer but faced the problem of particle agglomeration in the base fluid [10].

Due to the development in nanotechnology, the microsize particles are further reduced to nanosize $\left(1 \times 10^{-9}\mathrm{m}\right)$. The dilution of nanosized particles with some treatment on the particle surface can avoid particle agglomeration in the base fluid. Diluted nanosized particles in the base fluid are called

Advances in Nanofluid Heat Transfer. DOI: https://doi.org/10.1016/B978-0-323-88656-7.00002-7

nanofluid, and were first invented by Choi [11]. Improved thermal conductivity was observed for nanofluids compared with the base fluid. The problem of particle clogging, sedimentation, and erosion is minimized with the use of nanofluids while they flow in a tube or minichannel.

Discussion has been going on for more than 20 years but the mechanism behind the improvement of heat transfer for these nanofluids is still doubtful [12–16]. However, these nanofluids possess augmented thermal conductivity, which is the key parameter for heat transfer intensification. Many theories have been available in the open literature to know about the thermal conductivity augmentation of these nanofluids [17,18].

The diluted nanoparticles Brownian motion, liquid–particle interface, microconvection play a significant vital role in thermal conductivity augmentation [19]. The influence of Brownian motion of nanoparticles causes the heat transfer augmentation of nanofluids [20]. Based on the static model of Maxwell, particle conduction is the mode of heat transfer augmentation, but based on the dynamic model, the particle–particle and particle–liquid interactions is the mode for heat transfer enhancement.

When using these nanofluids in heat exchanger devices, heat transfer is important. Several studies are available in the literature for heat transfer characteristics of various nanofluids. Since the single-phase correlations for Nusselt number and friction factor are not valid for nanofluids, the authors have developed their own correlations based on the use of the nanofluids. Until now there is no standard Nusselt number and friction factor correlation applicable for all type of nanofluids. Developing a single correlation for all the nanofluids is difficult because the heat transfer augmentation depends on several parameters, such as particle size, shape, concentration, geometry, and operating conditions.

This chapter covers the Nusselt number and friction correlations for different nanofluids in various geometrical and operating conditions. These developed equations are helpful to identify the necessity of developing another kind of equation for forthcoming research.

1.2 Preparation of nanofluids

The preparation of nanofluids is necessary before they are used in any equipment. There are two methods to prepare the nanofluids: (1) one-step method and (2) two-step method. Stable nanofluids are important for heat transfer applications. The nanofluids are stable with the use of surfactants. If the nanofluids are not stable, there may be a clogging of larger particles in pumps, pipes, and also fouling in the tubes may occur using these nanofluids in thermal equipments.

1.2.1 One- step method

The one-step method of nanofluids is that the nanoparticles are directly made and diluted in the base fluid. This method is advantageous because the nanoparticles cannot react with oxygen in the atmosphere. Most of the research has been used this one-step method of nanofluid preparation technique. The accessories like dryers and storage are not required for this method. This method also has the advantage of less nanoparticles aggregation in the base fluid, resulting in an increase in nanofluid stability. The laser ablation method and submerged arc nanoparticles synthesis system are the commonly used methods for the preparation of nanofluids by the one-step method.

1.2.2 Two-step method

In the two-step method of nanofluids the nanoparticles, nanotubes, and other nanomaterials are primarily prepared in the form of dry powers by using chemical, physical, and mechanical methods. The prepared nanoparticles are dispersed in the base fluid. This method is the most economical method for the preparation of nanofluids. Nanoparticle agglomeration and sedimentation are the main difficulties in this type of nanofluids preparation. Ultrasonication or the stirring process will decrease the nanoparticle sedimentation in the base fluid. Because of the van der Waals forces, the nanoparticles aggregate very easily in the base fluids. By using the surfactant in the base fluid nanoparticle sedimentation is avoided. Nanoparticle agglomeration will affect the thermophysical properties of the nanofluid. However, this method of nanofluids preparation is most economical and also suitable for large-scale nanofluids preparation. For preparing the oxides-based nanofluids, this method is more suitable. Commonly used methods for the synthesis of nanoparticles are the direct synthesis method, sol–gel method, and chemical precipitation method. Commonly used surfactants are gum Arabic, SDBS, chitosan, polyvinylpyrolidine, and cetyl trimethyl ammonium bromide.

One can use either the one-step method or two-step method for the preparation of nanofluids, but the nanofluids should be stable to avoid the particle agglomeration in the tubes and channel. Therefore for nanofluid application in engineering systems new technology is required to increase the stability of nanofluid.

1.3 Experimental methods

The experimental equations to evaluate the heat transfer coefficient, Nusselt number, and friction factor are given below. Sundar et al. [21] experimental schematic diagram of forced convection equipment is shown in Fig. 1.1.

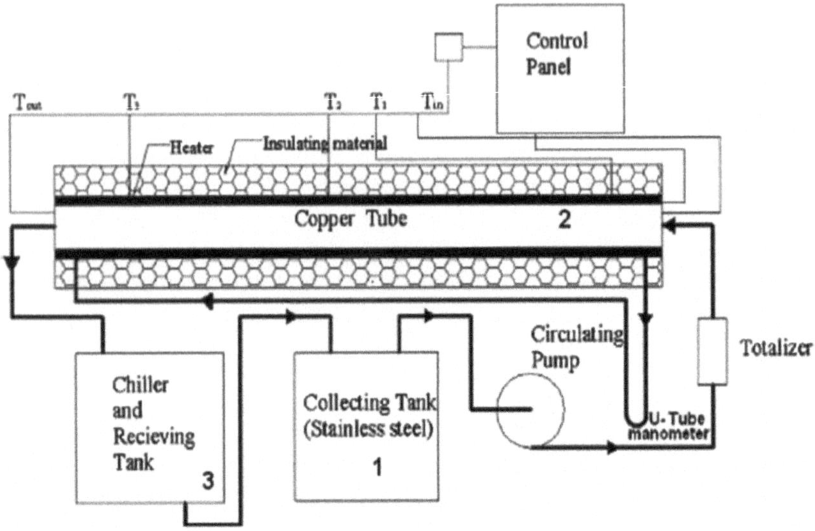

FIGURE 1.1 Schematic diagram of forced convection apparatus [21].

1.3.1 Nusselt number

The Newton's law of cooling equation is used to estimate the heat transfer coefficient.

a. Heat supplied to the test tube,

$$Q_h = V \times I \tag{1.1}$$

b. Heat absorbed by the fluid,

$$Q_a = m^{\cdot} \times C_p \times (T_o - T_i) \tag{1.2}$$

The average $Q_{ag} = \frac{(Q_h + Q_a)}{2}$

c. Heat transfer coefficient,

$$h = \frac{Q_{ag}}{A(T_w - T_m)} \tag{1.3}$$

Where, $A = \pi D_i L; T_w = \frac{T_1 + T_2 + T_3 + T_4 + T_5}{5}; T_m = \frac{T_o + T_i}{2}$

d. Nusselt number,

$$Nu = \frac{h \times D_i}{k} \tag{1.4}$$

e. Prandtl number,

$$Pr = \frac{\mu \times C_P}{k} \tag{1.5}$$

1.3.2 Friction factor

The fluid pressure drop for both water and nanofluids between inlet and outlet is measured through the pressure gauge. When the fluid is flowing inside the tube, the resistance between the fluid and tube wall provides the pressure drop.

f. Pressure drop,

$$\Delta P = \rho_{nf} g H \tag{1.6}$$

The terms in Eq. (1.6) are density of the fluid (ρ) (kg/m^3), acceleration due to gravity (g) (m/s^2), and manometric fluid height (H) (m). The manometric fluid height is converted into equivalent working fluid height for pressure drop calculations.

The pumping power of water and nanofluids can be given as:

g. Pumping power

$$= \left(\frac{\dot{m}}{\rho_{nf}}\right) \times \Delta P \tag{1.7}$$

From the pressure drop, the friction factor of water and nanofluid is estimated.

h. Friction factor,

$$f = \frac{(\Delta P)}{\left(\frac{L}{d_i}\right)\left(\frac{\rho v^2}{2}\right)} \tag{1.8}$$

where $\dot{m} = \rho A v \Rightarrow v = \frac{\dot{m}}{\rho A}$

1.3.3 Nondimensional numbers

The nondimensional numbers such as Reynolds number, Nusselt number, Prandtl number, and Peclet number are generally used to develop correlations.

i. Reynolds number,

$$Re = \frac{4\dot{m}}{\pi D \mu} \tag{1.9}$$

j. Prandtl number,

$$Pr = \frac{\mu C_p}{k} \tag{1.10}$$

k. Nusselt number,

$$Nu = \frac{hD}{k} \tag{1.11}$$

l. Peclet number,

$$Pe_d = \frac{u_m d_p}{\alpha} \tag{1.12}$$

The thermal diffusivity is: $\alpha = \frac{k_{nf}}{(\rho C_p)_{nf}}$

1.4 Nusselt number correlations for single-phase fluid

Different correlations are available to estimate the Nusselt number for the case of single-phase fluids flow in a tube. The fluid flow may be defined as laminar and turbulent flow. When the Reynolds number is less than or equal to 2300, the flow is called laminar flow; when the Reynolds number is greater than 2300, the flow is called as turbulent flow. The generally used correlations are compiled below.

1.4.1 Laminar flow

a. Shah [22]:

$$Nu = 1.953\left(RePr\frac{D}{x}\right)^{1/3} \quad \left(RePr\frac{D}{x}\right) \geq 33.3 \tag{1.13}$$

b. Churchill and Ugasi [23]

$$Nu^{10} = Nu^{10} + \left[\exp\left(\frac{2200-(Re/365)}{Nu_{lc}^2} + \frac{1}{Nu_t^2}\right)\right]^{-5} \tag{1.14}$$

Nu_l = Laminar Nusselt number
Nu_{lc} = Nusselt number at critical Reynolds number of 2200
Nu_t = Turbulent Nusselt number

c. Tam and Ghajar [24]

$$Nu = 0.0233 \quad Re^{0.8}Pr^{0.385}\left(\frac{x}{D}\right)^{-0.0054}\left(\frac{\mu_b}{\mu_w}\right)^{0.14} \tag{1.15}$$

$$3 \leq \frac{x}{D} \leq 192; 7000 \leq Re \leq 49,000; 4 \leq Pr \leq 34; 1.1 \leq \frac{\mu_b}{\mu_w} \leq 1.7$$

d. Sider-Tate [25]

$$Nu = 0.027Re^{0.8}Pr^{0.3}\left(\frac{\mu_b}{\mu_w}\right)^{0.14} \tag{1.16}$$

$$700 \le Re \le 10,000; 0.7 \le Pr \le 16; 1.1 \le \frac{L}{d} \ge 1.7$$

1.4.2 Turbulent flow

e. Dittus-Boelter [26]

$$Nu = 0.023Re^{0.23}Pr^{0.4} \tag{1.17}$$

f. Gnelinski's [27]

$$Nu = 0.012\left(Re^{0.87} - 280\right)Pr^{0.4} \tag{1.18}$$

$$1.5 \le Pr \le 500; 3000 \le Re \le 10^5$$

Another Gnelinski [27] equation is,

$$Nu = \frac{(f/8)(Re - 1000)Pr}{1 + 12.7(f/8)^{1/2}\left(Pr^{2/3} - 1\right)} \tag{1.19}$$

$$0.5 \le Pr \le 2 \times 10^3; 3000 \le Re \le 10^6$$

g. Petukhov [28]

$$Nu = \frac{(f/8)(Re - 1000)Pr}{1.07 + 12.7(f/8)^{1/2}\left(Pr^{2/3} - 1\right)} \tag{1.20}$$

$$0.5 \le Pr \le 2 \times 10^3; 10^4 \le Re \le 5 \times 10^6$$

h. Notter-Sleicher [29]

$$Nu = 5 + 0.015Re^{0.856}Pr^{0.347} \tag{1.21}$$

$$0 < Pr < 10^4; 10^4 < Re < 10^5$$

1.5 Friction factor correlations for single-phase fluids

The friction factor is estimated based on the pressure drop, and velocity of the fluid. Various correlations are available to predict the friction factor for the case of single-phase fluid flow in a tube. The fluid flow may be defined as laminar or turbulent flow.

1.5.1 Laminar flow

a. Hagen-Poiseuille [30]

$$f = \frac{64}{Re} \tag{1.22}$$

1.5.2 Turbulent flow

a. Moody [31]

$$f = 0.046 \quad Re^{-0.2} \tag{1.23}$$

$$2 \times 10^4 < Re < 10^6$$

b. Blasius [32]

$$f = 0.079Re^{-0.25} \tag{1.24}$$

$$3 \times 10^3 < Re < 2 \times 10^4$$

c. Petukhov [28]

$$f = (0.790 \quad ln(Re) - 1.64)^{-2} \tag{1.25}$$

$$3000 < Re < 5 \times 10^6$$

d. Filonenko [33] for smooth tubes

$$f = 0.25(0.790\ln(Re) - 1.64)^{-2} \tag{1.26}$$

$$3000 < Re < 5 \times 10^6$$

1.6 Factors influencing the development of correlations

1.6.1 Nusselt number

The Nusselt number of nanofluid is a function of Reynolds number, Prandtl number, and particle volume loadings.

$$Nu = f(Re, Pr, \phi) \tag{1.27a}$$

By releasing the functionality limit, the Eq. (1.27a) becomes Eq. (1.27b)

$$Nu = C \quad Re^a Pr^b \phi^c \tag{1.27b}$$

where "C" is a constant; "a", "b", and "c" are the exponents.

1.6.2 Friction factor

The friction factor is a function of Reynolds number and particle volume loadings.

$$f = f(Re, \phi) \tag{1.28a}$$

By releasing the function, the Eq. (1.28a) becomes Eq. (1.28b)

$$f = C \quad Re^a \phi^b \tag{1.28b}$$

where "C" is a constant; "a" and "b" are exponents.

1.7 Developed corrections for nanofliuids

1.7.1 Nusselt number

Based on the experimental data of nanofluids, many researchers have developed Nusselt number correlations. Fig. 1.2 indicates the experimental Nusselt number of Al_2O_3/water at different Reynolds numbers and particle loadings [21]. The Nusselt number of 0.5% volume concentration of Al_2O_3/

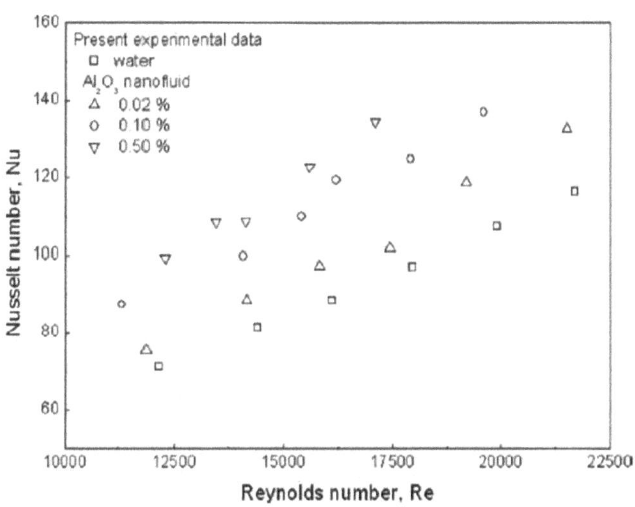

FIGURE 1.2 Nusselt number of Al_2O_3/water at different Reynolds numbers and particle loadings [21].

water nanofluid is enhanced by 22.76% and 30.30% at Reynolds number of 10,000 and 22,000, compared with water data, respectively.

The experimental heat transfer coefficient and Nusselt number of Ni/water nanofluids at various Reynolds numbers and particle loadings are shown in Fig. 1.3A and 1.3B [34]. The Nusselt number is enhanced to 10% and 9.5% at 0.02% vol. of Ni/water nanofluid at Reynolds numbers of 3000 and 22,000, respectively; while for the 0.6% vol. of Ni/water nanofluid it is approximately enhanced to 29.12% and 39.18% at the same Reynolds numbers compared to water.

The proposed Nusselt number correlations of various researchers are mentioned below. Anoop et al. [35] developed a Nusselt number for Al_2O_3/water nanofluid flow in a tube and they also noticed that the heat transfer enhancement is about 25% and 11% for 45 and 150 nm size nanoparticles at a weight concentration (ψ) of 4%.

$$Nu = 4.36 + \left[a + x_+^{-b}(1 + \psi^c)exp^{(dx_+)}\right]\left[1 + e\left(\frac{d_p}{d_{ref}}\right)^{-f}\right] \tag{1.29}$$

$$a = 6.219 \times 10^{-3}, b = 1.1522, c = 0.1533, d = 2.5228, e = 0.57825$$

$$f = 0.2183, d_{ref} = 100 \quad nm, d_p = \text{diameter of the particle}, \quad x_+ = \frac{x}{RePrD}$$

$$50 < \frac{x}{D} < 200, 500 < Re < 2000, 0 \leq \psi \leq 4\%$$

Li and Xuan [36] developed a Nusselt number correlation for Cu/water nanofluid and also revealed heat transfer enhancement of 60% at $\phi = 2\%$ of Cu/water nanofluid.

$$Nu = 0.4328 \quad \left(1 + 11.258 \quad \phi^{0.754} \quad Pe_d^{0.218}\right)Re^{0.333}Pr^{0.4} \tag{1.30}$$

$$800 < Re < 4000; 0 < \phi < 2\%$$

Yang et al. [37] proposed a Nusselt number correlation for graphite/commercial automatic transmission fluid-based nanofluid and also observed 22% of heat transfer enhancement at 2.5 wt.% and at Re of 120.

$$Nu = aRe^bPr^{1/3}\left(\frac{D}{L}\right)^{1/3}\left(\frac{\mu_w}{\mu_b}\right)^{1/3} \tag{1.31}$$

$$5 < Re < 120; 0 < \psi < 2.5\%$$

where "a" and "b" are the constant and exponent, respectively.

Similarly, Rea et al. [38] developed a regression equation based on the experimental data of alumina/water nanofluid and zirconia/water and

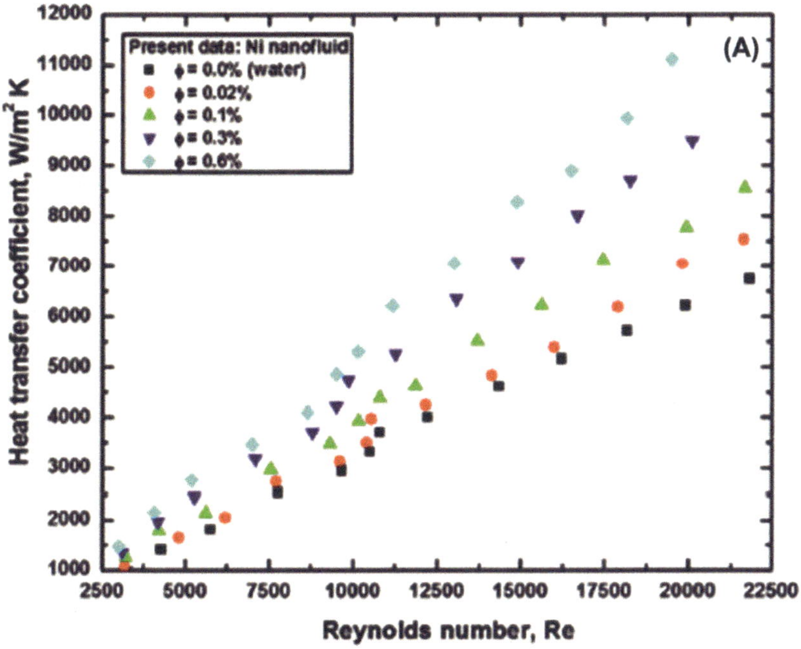

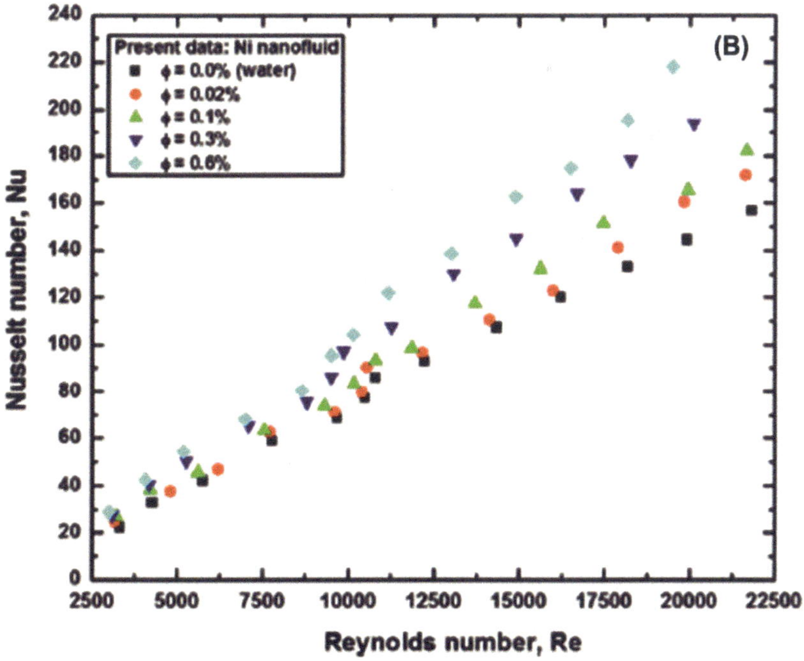

FIGURE 1.3 Ni/water nanofluid at different Reynolds numbers and particle loadings: (A) heat transfer coefficient and (B) Nusselt number [34].

achieved heat transfer enhancements of 27% and 3% at $\phi = 6\%$ and $\phi = 1.32\%$, respectively.

$$Nu = 1.619\left(x^{+}\right)^{1/3} \tag{1.32}$$

$$x^{+} < 0.01; x^{+} = \frac{2(x/D)}{Re \quad Pr}$$

$431 < Re < 2000; 0 < \phi < 6\%$ For Al_2O_3/water nanofluid

$140 < Re < 362;\ 0 < \phi < 3\%$ For ZrO_2/water nanofluid

Based on the experimental data of Al_2O_3/water and TiO_2/water nanofluid, Pak and Cho [39] have developed Nusselt number correlation.

$$Nu = 0.021 Re^{0.8} Pr^{0.4} \tag{1.33}$$

$$10^4 < Re < 10^5; 6.54 < Pr < 12.33; 0 < \phi < 3\%$$

Turbulent convective heat transfer for Cu/water nanofluid flow in a tube has been estimated by Xuan and Li [40] experimentally and they also proposed a Nusselt number correlation.

$$Nu = 0.0059\left(1.0 + 7.6286 \quad \phi^{0.8} Pe_d^{0.001}\right) \quad Re^{0.9238} Pr^{0.4} \tag{1.34}$$

$$1 \times 10^4 < Re < 2.5 \times 10^5; 0 < \phi < 2\%$$

From the experimental data of TiO_2/water nanofluid, Duangthongsuk and Wongwises [41] have noticed heat transfer coefficient enhancement of 26% at $\phi = 1\%$ and also observed 14% enhancement for $\phi = 2\%$ and proposed a correlation.

$$Nu = 0.074 \quad Re^{0.707} Pr^{0.385} \phi^{0.074} \tag{1.35}$$

$$3000 < Re < 18000; 0 < \phi < 2\%$$

Numerical study was performed by Maiga et al. [42] for heat transfer of water/Al_2O_3 and ethylene glycol/Al_2O_3 nanofluids and they presented a correlation under laminar flow.

$$Nu = 0.086 \quad Re^{0.55} Pr^{0.5} (\text{Constant wall heat flux}) \tag{1.36}$$

$$Nu = 0.28 \quad Re^{0.35} Pr^{0.36} (\text{Constant wall temperature}) \tag{1.37}$$

$$Re \leq 1000; 6 \leq Pr \leq 753; 0 < \phi < 10\%$$

Maiga et al. [43] also proposed a correlation for Nusselt number from the data of water/Al_2O_3 and ethylene glycol/Al_2O_3 mixtures under turbulent flow.

$$Nu = 0.085 \quad Re^{0.71} Pr^{0.35} \tag{1.38}$$

$$10^4 \leq Re \leq 5 \times 10^5; 6.6 \leq Pr \leq 13.9; 0 < \phi < 10\%$$

Sajadi and Kazemi [44] used TiO_2/water nanofluid in a tube and revealed an increase of h of 22% and a pressure drop increase of 25% at $\phi = 0.25\%$ under turbulent flow (Fig. 1.4).

$$Nu = 0.067 \quad Re^{0.71} Pr^{0.35} + 0.0005 \quad Re \tag{1.39}$$

$$5000 < Re < 30,000; 0 < \phi < 0.25\%$$

Sundar et al. [45] also estimated the Nusselt number of Fe_3O_4/water nanofluid flow in a tube—the data are shown in Fig. 1.5—and they observed that the Nusselt number enhancement of 0.6% volume concentration of Fe_3O_4/water nanofluid is 20.99% and 30.96% at Reynolds number of 3000 and 22,000, respectively, compared to water, and they proposed a Nusselt number correlation.

$$Nu = 0.02171 \quad Re^{0.8} Pr^{0.5} (1 + \phi)^{0.5181} \tag{1.40}$$

$$3000 < Re < 22,000; 0 < \phi < 0.6\%; 3.72 < Re < 6.50$$

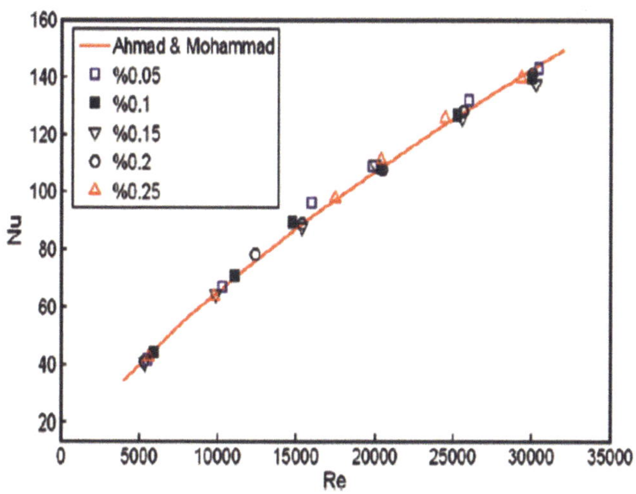

FIGURE 1.4 Nusselt number of TiO_2/water at different Reynolds numbers and particle loadings [44].

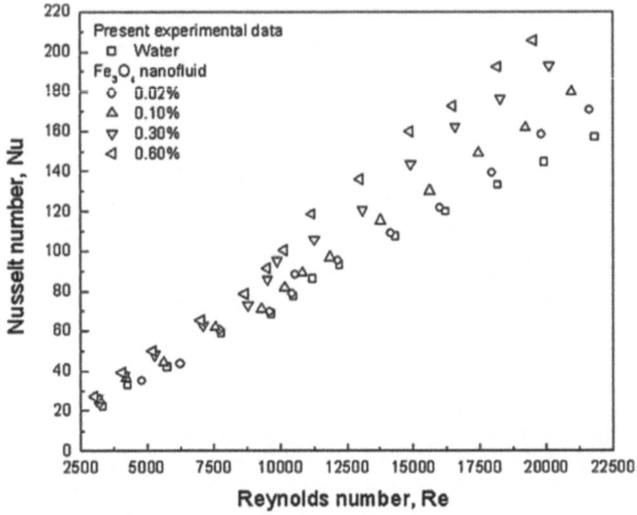

FIGURE 1.5 Nusselt number of Fe_3O_4/water at different Reynolds numbers and particle loadings [45].

Buongiorno [46] proposed an alternative explanation for the abnormal heat transfer coefficient increment by considering viscosity within the boundary layer.

$$Nu = \frac{(f/8)(Re - 1000)Pr}{1 + \delta_v^+ (f/8)^{1/2}(Pr^{2/3} - 1)} \tag{1.41}$$

δ_v^+ = Thickness, that is, 15.5
f = friction factor correlation for turbulent flow
$5000 < Re < 65,000; 0 < \phi < 3.6\%$ For Al_2O_3 nanofluid
$5000 < Re < 65,000; 0 < \phi < 0.9\%$ For ZrO_2 nanofluid

Based on the silver/water nanofluid, Asirvatham et al. [47] proposed a Nusselt number correlation.

$$Nu = 0.0023 \quad Re^{0.8}Pr^{0.3}(0.617\phi - 0.135)Re^{(0.445\phi - 0.37)}Pr^{(1.081\phi - 1.305)} \tag{1.42}$$

$$900 < Re < 12,100; 0 < \phi < 0.9\%$$

The 60:40% of EG/W based Al_2O_3, CuO, SiO_2 nanofluids have been used by Vajjha and Das [48] and presented a Nusselt number equation.

$$Nu = 0.065 \quad (Re^{0.8} - 60.22)(1 + 0.0169 \quad \phi^{0.3})Pr^{0.542} \tag{1.43}$$

$$R^2 = 0.97; 3000 < Re < 16,000$$

$0 < \phi < 0.06\%$ For CuO and SiO_2 nanofluid
$0 < \phi < 0.1\%$ For Al_2O_3 nanofluid

Yang et al. [37] developed a Nusselt number equation for graphite/water nanofluid in laminar flow.

$$Nu = 1.86Re^{0.333} \quad Pr^{0.333}\left(\frac{D}{L}\right)^{0.333}\left(\frac{\mu_b}{\mu_w}\right)^{0.14} \tag{1.44}$$

Moraveji et al. [49] proposed a correlation for Nusselt number based on the single-phase model of Al_2O_3/water nanofluid in the developing region.

$$Nu = 2.03 \quad Re^{0.293} \quad Pr^{0.6}\phi^{0.06}\left(\frac{x}{D_h}\right)^{-0.37} \tag{1.45}$$

$$500 < Re < 2500$$

Moraveji et al. [50] developed a correlation based on the experimental data of non-Newtonian Al_2O_3/XAS (Xanthan Aqueous Solution) nanofluid.

$$Nu = 2.14 \quad \left(Re_{nf}.D_h/x\right)^{0.025} \quad Pr^{0.333}\phi^{0.004}\left(\frac{\mu_{nf}}{\mu_{bf}}\right)^{0.1927} \tag{1.46}$$

Keshavarz et al. [51] developed a correlation for Al_2O_3/water nanofluid under laminar flow.

$$Nu = 0.716 \quad Re^{0.314}Pr^{0.6}\phi^{0.3} \tag{1.47}$$

Hejazian et al. [52] presented an equation for TiO_2/water nanofluid based on the numerical analysis.

$$Nu = 0.0218 \quad Re^{1.0037}Pr^{0.5}(1+\phi)^{154.6471} \tag{1.48}$$

$$4800 \le Re \le 30500; 5.5 \le Pr \le 5.59; 0 \le \phi \le 0.25\%$$

Hejazian et al. [53] developed Nusselt number correlations based on their numerical study on forced convection of alumina/water nanofluid.

$$Nu = 0.02178 \quad Re^{0.811}Pr^{0.266}(1+\phi)^{1.140} \tag{1.49}$$

Minea [54] presented a Nusselt number expression of Al_2O_3/water nanofluid using the Brinkman number.

$$Nu = Re^{0.62}Pr^{0.33}(0.058 + 0.3\phi)Br^{-0.27} \tag{1.50}$$

where $Br = \mu_{nf}w_{nf}/QD_h$

Moghadassi et al. [55] presented a correlation for Al_2O_3/water nanofluid based on the numerical models.

$$Nu = 0.119 \quad Re^{0.593}Pr^{0.33}(1+\phi)^{69.23} \tag{1.51}$$

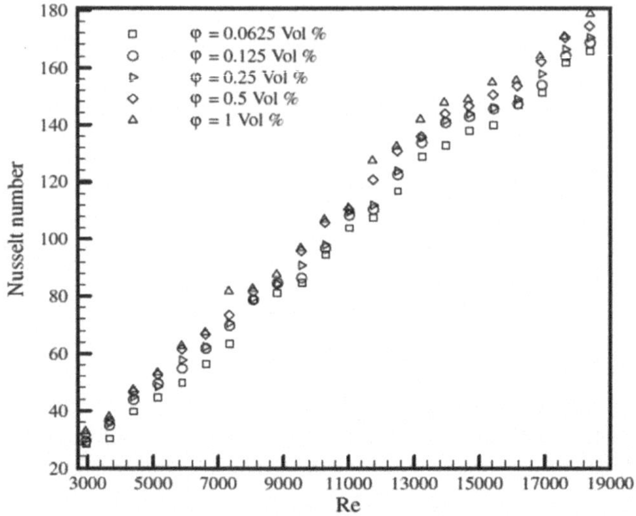

FIGURE 1.6 Nusselt number of MgO/water at different Reynolds numbers and particle loadings [56].

Fig. 1.6 represents the experimental data of Hemmat Esfe [56] who showed Nusselt number enhancement of MgO/water nanofluid flow in a tube at particle concentrations of 0.0625%, 0.125%, 0.25%, 0.5%, and 1%.

Iyahraja et al. [57] conducted the heat transfer experiments for silver/water nanofluids flow in a tube under laminar flow conditions and they observed Nusselt number ratio increased up to 1.128 for 0.1% volume fraction at the Reynolds number of 2300 and they also proposed a Nusselt number correlation.

$$Nu = 0.1279 \quad Re^{0.592} Pr^{0.333} (1+\phi)^{77.13} \tag{1.52}$$

$$Re < 2300; \phi < 0.1\%$$

Heris et al. [58] have conducted experiments for Al_2O_3/water nanofluid flow inside a circular tube under laminar flow, they have observed that, at 2.5% vol. of nanofluid, at Peclet number from 2560 to 6000, the ratio increases from 1.08 to 1.22. Also for Peclet numbers 2500 and 6000 as the nanofluid concentration changes from 0.2% to 2.5% the ratio increases from 1.01 to 1.09 and 1.07 to 1.22, respectively, and their data are presented in Fig. 1.7.

1.7.2 Friction factor

The experimental data of nanofluids are used to develop a friction factor correlaton. Available correlations are presented here. From a study of Fe_3O_4/

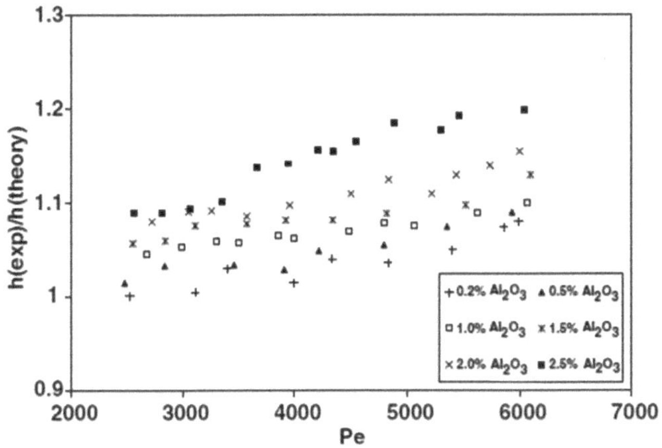

FIGURE 1.7 The heat transfer coefficient ratio of Al_2O_3/water nanofluid vs. Peclet number at different volume concentrations [58].

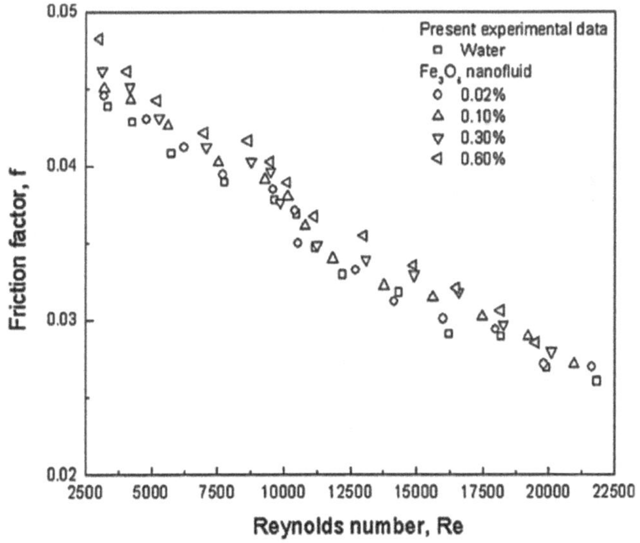

FIGURE 1.8 Friction factor of Fe_3O_4/water at different Reynolds numbers and particle loadings [45].

water nanofluid, Sundar et al. [45] proposed a friction factor correlation under turbulent flow and the data are shown in Fig. 1.8. They have observed the enhancement of friction factor in a plain tube with 0.6% vol. of Fe_3O_4

nanofluid when compared to water is 1.09 times and 1.10 times for Reynolds number of 3000 and 22,000 respectively.

$$f = 0.3491 Re^{-0.25} \quad (1+\phi)^{0.1517} \tag{1.53}$$

$$3000 < Re < 22,000; 0 < \phi < 0.6\%$$

Vajjha and Das [94] experimentally investigated Al_2O_3, CuO, and SiO_2 nanofluid in a tube under turbulent flow condition. They observed an increase of 10.0% pressure drop of Al_2O_3 nanofluid at a Reynolds number of 6700.

$$f = 0.3164 \quad Re^{-0.25} \left(\frac{\rho_{nf}}{\rho_{bf}}\right)^{0.707} \left(\frac{\mu_{nf}}{\mu_{bf}}\right)^{0.108} \tag{1.54}$$

$$4000 < Re < 16,000; 0 < \phi < 0.1\%$$

Hejazian et al. [53] developed a friction factor correlation for Al_2O_3/water nanofluid based on the numerical study.

$$f = 0.14332 \quad Re^{-0.1615}(1+\phi)^{0.1709} \tag{1.55}$$

Moghadassi et al. [55] a performed similar study on the laminar forced convection of Al_2O_3/water nanofluid based on the numerical models.

$$f = 103.46 \quad Re^{-1.10}(1+\phi)^{99.71} \tag{1.56}$$

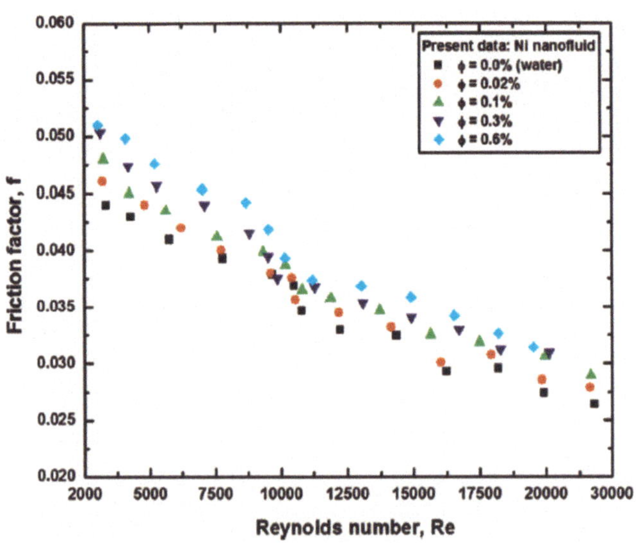

FIGURE 1.9 Friction factor of Ni/water at different Reynolds numbers and particle loadings [45].

Sundar et al. [34] developed a friction factor correlation for Ni/water nanofluid based on an experimental study, which is shown in Fig. 1.9. They noted that the friction factor of Ni/water nanofluid is increased to 19.12% at 0.6% vol. and at a Reynolds number of 22,000 compared to water.

$$f = 0.35662 \quad Re^{-0.25}(1+\phi)^{0.23759} \tag{1.57}$$

$$3000 < Re < 22,000; 0 < \phi < 0.6\%; 3.72 < Pr < 6.37$$

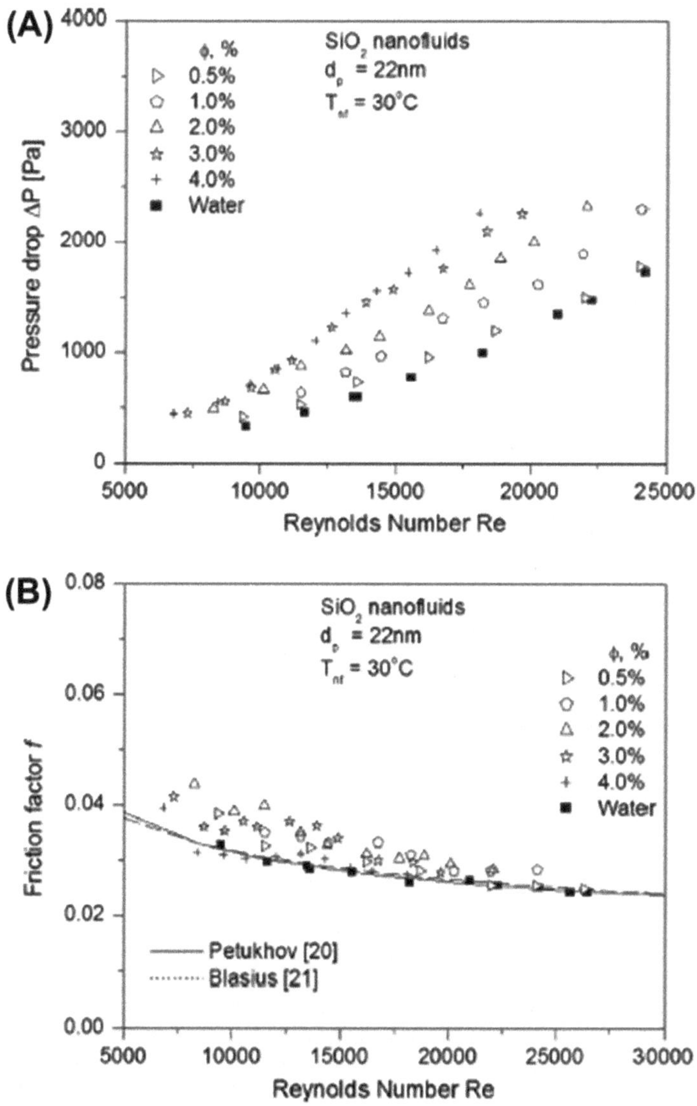

FIGURE 1.10 (A) Pressure drop of SiO_2 nanofluid and (B) friction factor of SiO_2/water nanofluid [59].

Iyahraja et al. [57] conducted the friction factor experiments for silver/water nanofluids flow in a tube under laminar flow conditions and their data are mentioned in Fig. 1.10. They found that the friction factor increases by 4.05%, 6.8%, and 10.9%, respectively, for the particle volume concentrations of 0.01%, 0.05%, and 0.1%. Their friction factor correlation is shown below.

$$f = 71.529 \quad Re^{-1.0085}(1+\phi)^{104.373} \tag{1.58}$$

$$Re < 2300; \phi < 0.1\%$$

Azmi et al. [59] conducted heat transfer and friction factor experiments for SiO_2/water nanofluid flow in a tube up to 4% particle loadings at constant heat flux boundary condition. In the Reynolds number range of 5000–27,000, they have observed Nusselt number and friction factor at 3.0% nanofluid particle loading is greater than the values of water by 32.7% and 17.1% and their pressure drop and friction factor data are shown in Fig. 1.10A and B.

1.8 Conclusion

This chapter focuses on the available Nusselt number and friction correlations of various kinds of nanofluids. The chapter also explains the developed correlations under various flow and boundary conditions. The heat transfer and pressure drop characteristics of nanofluids are dependent on several factors, such as shape, size, material, base fluid, temperature, particle loading, geometry, etc. Major research reports indicate that the heat transfer augmentation is dependent on the particle loading. The majority of the developed Nusselt number and friction factor correlations are dependent on the Reynolds number, Prandtl number, and particle loading; some of the correlations include the Peclet number, Stefan number, and viscosity terms.

Despite the heat transfer augmentation with nanofluids and potential applications in various sectors, a deficiency is the lack of a theoretical approach for the heat transfer mechanism and the poor stability of the nanoparticles in the base fluid. There need to be further investigations to establish a detailed heat transfer mechanism for nanofluids. A proper and standard nanoparticle synthesis method is required so that the stable nanofluids are achieved. Moreover, it is necessary to develop a common Nusselt number and friction factor correlation that is applicable for all type of nanofluids for ease of use.

References

[1] L. Cheng, T. Luan, W. Du, M. Xu, Heat transfer enhancement by flow-induced vibration in heat exchangers, Int. Comm. Heat. Mass. Transf. 52 (2009) 1053–1057.

[2] A. Hosseinian, A.H.M. Isfahani, Experimental study of heat transfer enhancement due to the surface vibrations in a flexible double pipe heat exchanger, Heat. Mass. Transf. 54 (2018) 1113–1120.

[3] S. Gururatana, X. Li, Heat transfer enhancement of small scale heat sinks using vibrating pin fin, Am. J. Appl. Sci. 10 (8) (2013) 801–809.

[4] N. Zheng, P. Liu, F. Shan, Z. Liu, W. Liu, Heat transfer enhancement in a novel internally grooved tube by generating longitudinal swirl flows with multi-vortexes, Appl. Therm. Eng. 95 (2016) 421–432.

[5] S. Baskaya, M. Sivrioglu, M. Ozek, Parametric study of natural convection heat transfer from horizontal rectangular fin arrays, Int. J. Therm. Sci. 39 (2000) 797–805.

[6] N. Piriyarungrod, S. Eiamsa-ard, C. Thianpong, M. Pimsarn, K. Nanan, Heat transfer enhancement by tapered twisted tape inserts, Chem. Eng. Process.: Process. Intensif. 96 (2015) 62–71.

[7] L.S. Sundar, K.V. Sharma, Experimental investigation of heat transfer and friction factor characteristics in a circular tube with longitudinal strip inserts, J. Enhanc. Heat. Transf. 15 (2008) 325–333.

[8] A.S. Ahuja, Augmentation of heat transport in laminar flow of polystyrene suspension, I. Experimental and results, J. Appl. Phys. 46 (8) (1975) 3408–3416.

[9] A.E. Bergles, Recent development in convective heat transfers augmentation, Appl. Mech. Rev. 26 (1973) 675–682.

[10] A.S. Ahuja, Augmentation of heat transport in laminar flow of polystyrene suspensions. II. Analysis of the data, J. Appl. Phys. 46 (1975) 3417.

[11] S.U.S. Choi, Enhancing thermal conductivity of fluid with nanoparticles, in: D.A. Siginer, H.P. Wang (Eds.), Developments and Applications of Non-Newtonian Flows, FED-V.231/MD-V.66, ASME, New York, 1995, pp. 99–105.

[12] D. Wen, G. Lin, S. Vafaei, K. Zhang, Review of nanofluids for heat transfer applications, Particuology 7 (2009) 141–150.

[13] L. Godson, B. Raja, D. Mohan Lal, S. Wongwises, Enhancement of heat transfer using nanofluids-an overview, Renew. Sustain. Energy Rev. 14 (2010) 629–641.

[14] R.B. Ganvir, P.V. Walke, V.M. Kriplani, Heat transfer characteristics in nanofluid–a review, Renew. Sustain. Energy Rev. 75 (2017) 451–460.

[15] M.M. Tawfik, Experimental studies of nanofluid thermal conductivity enhancement and applications: a review, Renew. Sustain. Energy Rev. 75 (2017) 1239–1253.

[16] M. Awais, N. Ullah, J. Ahmad, F. Sikandar, M.M. Ehsan, S. Salehin, et al., Heat transfer and pressure drop performance of nanofluid: a state-of- the-art review, Int. J. Thermofluids 9 (2021) 100065.

[17] S. Lee, S.U.S. Choi, S. Li, J.A. Eastman, Measuring thermal conductivity of fluids containing oxide nanoparticles, J. Heat. Transf. 121 (1999) 280–289.

[18] S.U.S. Choi, Z.G. Zhang, W. Yu, F.E. Lockwood, E.A. Grulke, Anomalous thermal conductivity enhancement in nanotube suspensions, Appl. Phys. Lett. 79 (14) (2001) 2252–2254.

[19] P. Keblinski, R. Phillpot, S.U.S. Choi, J.A. Eastman, Mechanisms of heat flow in suspensions of nano-sized particles (nano fluids), Int. J. Heat. Mass. Transf. 45 (2002) 855–863.

[20] X. Wang, X. Xu, S.U.S. Choi, Thermal conductivity of nanoparticle-fluid mixture, J. Thermophys. Heat. Transf. 13 (1999) 474–480.

[21] L.S. Sundar, K.V. Sharma, Turbulent heat transfer and friction factor of Al_2O_3 nanofluid in circular tube with twisted tape inserts, Int. J. Heat. Mass. Transf. 53 (2010) 1409–1416.

[22] R.K. Shah, Thermal entry length solutions for the circular tube and parallel plates, Proc. Third Natl Heat. Mass. Transf. Conf. 1 (1975) 11–75. Indian Institute of Technology, Bombay.

[23] S.W. Churchill, R.A. Usagi, General expression for the correlation of rates of transfer and other phenomena, AICHE 18 (1972) 1121–1128.
[24] L.M. Tam, A.J. Ghajar, Transitional heat transfer in plain horizontal tubes, Heat. Transf. Eng. 27 (2006) 23–38.
[25] E.N. Seider, G.E. Tate, Heat transfer and pressure drop of liquids in tubes, Ind. Eng. Chem. 28 (1936) 1429–1436.
[26] F.W. Dittus, L.M.K. Boelter, Heat transfer in automobile radiators of the tubular type, Univ. Calif. Publ. Eng. 11 (1930) 443–461.
[27] V. Gnielinski, New equations for heat and mass transfer in turbulent pipe and channel flow, Int. Chem. Eng. 16 (1976) 359–368.
[28] B.S. Petukhov, Heat transfer and friction in turbulent pipe flow with variable physical properties, in: J.P. Hartnett, T.F. Irvine (Eds.), Advances in Heat Transfer, Academic Press, New York, 1970, pp. 504–564.
[29] R.H. Notter, C.A. Sleicher, A solution to the turbulent Graetz problem—III. Fully developed and entry region heat transfer rates, Chem. Eng. Sci. 27 (1972) 2073–2093.
[30] P.S. Sutera, R. Skalak, The history of Poiseuille's law, Annu. Rev. Fluid Mech. 25 (1993) 1–19.
[31] L.F. Moody, Friction factors for pipe flow, Trans. ASME 66 (1944) 671–684.
[32] H. Blasius, Grenzschichten in Flussigkeiten mit kleiner Reibung (German), Z. Math. Phys. 56 (1908) 1–37.
[33] G.K. Filonenko, Hydraulic resistance in pipes, Teploenergetika 1, 1954, pp. 40–44. Russia.
[34] L.S. Sundar, M.K. Singh, I. Bidkin, A.C.M. Sousa, Experimental investigations in heat transfer and friction factor of magnetic Ni nanofluid flowing in a tube, Int. J. Heat. Mass. Transf. 70 (2014) 224–234.
[35] K.B. Anoop, T. Sundararajan, S.K. Das, Effect of particle size on the convective heat transfer in nanofluid in the developing region, Int. J. Heat. Mass. Transf. 52 (2009) 2189–2195.
[36] Q. Li, Y. Xuan, Convective heat transfer and flow characteristics of Cu–water nanofluid, Sci. China Ser. E: Technol. Sci. 45 (2002) 408–416.
[37] Y. Yang, Z.G. Zhang, E.A. Grulke, W.B. Anderson, G. Wu, Heat transfer properties of nanoparticle-in-fluid dispersions (nanofluids) in laminar flow, Int. J. Heat. Mass. Transf. 48 (2005) 1107–1116.
[38] U. Rea, T. McKrell, L.W. Hu, J. Buongiorno, Laminar convective heat transfer and viscous pressure loss of alumina/water and zirconia/water nanofluids, Int. J. Heat. Mass. Transf. 52 (2009) 2042–2048.
[39] B.C. Pak, Y.I. Cho, Hydrodynamic and heat transfer study of dispersed fluids with submicron metallic oxide particles, Exp. Heat. Transf. 11 (1998) 151–170.
[40] Y. Xuan, Q. Li, Investigation on convective heat transfer and flow features of nanofluids, J. Heat. Transf. 125 (2003) 151–155.
[41] W. Duangthongsuk, S. Wongwises, An experimental study on the heat transfer performance and pressure drop of TiO_2–water nanofluids flowing under a turbulent flow regime, Int. J. Heat. Mass. Transf. 53 (2010) 334–344.
[42] S.E.B. Maiga, S.J. Palm, C.T. Nguyen, G. Roy, N. Galanis, Heat transfer enhancement by using nanofluids in forced convection flows, Int. J. Heat. Fluid Flow. 36 (2005) 530–546.
[43] S.E.B. Maiga, C.T. Nguyen, N. Galanis, G. Roy, T. Mare, M. Coqueux, Heat transfer enhancement in turbulent tube flow using Al_2O_3 nanoparticle suspension, Int. J. Numer. Methods Heat. Fluid Flow. 16 (2006) 275–292.

[44] A.R. Sajadi, M.H. Kazemi, Investigation of turbulent convective heat transfer and pressure drop of TiO_2/water nanofluid in circular tube, Int. Comm. Heat. Mass. Transf. 38 (2011) 1474–1478.

[45] L.S. Sundar, M.T. Naik, K.V. Sharma, M.K. Singh, T. Ch, Siva Reddy, Experimental investigation of forced convection heat transfer and friction factor in a tube with Fe_3O_4 magnetic nanofluid, Exp. Therm. Fluid Sci. 37 (2012) 65–71.

[46] J. Buongiorno, Convective transport in nanofluids, J. Heat. Transf. 128 (2006) 240–250.

[47] L.G. Asirvathama, B. Rajab, D.M. Lal, S. Wongwises, Convective heat transfer of nanofluids with correlations, Particuology 9 (2011) 626–631.

[48] R.S. Vajjha, D.K. Das, D.P. Kulkarni, Development of new correlations for convective heat transfer and friction factor in turbulent regime for nanofluids, Int. J. Heat. Mass. Transf. 53 (2010) 4607–4618.

[49] M.K. Moraveji, M. Darabi, S.M.H. Haddad, R. Davarnejad, Modeling of convective heat transfer of a nanofluid in the developing region of tube flow with computational fluid dynamics, Int. Comm. Heat. Mass. Transf. 38 (2011) 1291–1295.

[50] M.K. Moraveji, S.M.H. Haddad, M. Darabi, Modeling of forced convective heat transfer of a non-Newtonian nanofluid in the horizontal tube under constant heat flux with computational fluid dynamics, Int. Comm. Heat. Mass. Transf. 39 (2012) 995–999.

[51] K.M. Moraveji, E. Esmaeili, Comparison between single-phase and two phases CFD modeling of laminar forced convection flow of nanofluids in a circular tube under constant heat flux, Int. Comm. Heat. Mass. Transf. 39 (2012) 1297–1302.

[52] M. Hejazian, M.K. Moraveji, A comparative analysis of single and two-phase models of turbulent convective heat transfer in a tube for TiO_2 nanofluid with CFD, Numer. Heat. Transf. Part. A Appl. 63 (2013) 795–806.

[53] M. Hejazian, M.K. Moraveji, A. Beheshti, Comparative study of Euler and mixture models for turbulent flow of Al_2O_3 nanofluid inside a horizontal tube, Int. Comm. Heat. Mass. Transf. 52 (2014) 152–158.

[54] A.A. Minea, A study on Brinkman number variation on water based nanofluid heat transfer in partially heated tubes, Mech. Res. Commun. 73 (2016) 7–11.

[55] A. Moghadassi, E. Ghomi, F. Parvizian, A numerical study of water based Al_2O_3 and Al_2O_3-Cu hybrid nanofluid effect on forced convective heat transfer, Int. J. Therm. Sci. 92 (2015) 50–57.

[56] M.H. Esfe, S. Saedodin, M. Mahmoodi, Experimental studies on the convective heat transfer performance and thermophysical properties of MgO–water nanofluid under turbulent flow, Exp. Therm. Fluid Sci. 52 (2014) 68–78.

[57] S. Iyahraja, J. Selwin Rajadurai, B. Siva Subramanian, S. Siva Subramanian, P. Sivasankar, S. Subash, Investigation on convective heat transfer and friction factor of silver–water nanofluid under laminar flow—an experimental study, Heat. Mass. Transf. (2019). Available from: https://doi.org/10.1007/s00231-019-02640-y.

[58] S.Z. Heris, M. Nasr Esfahany, S. Gh. Etemad, Experimental investigation of convective heat transfer of Al_2O_3/water nanofluid in circular tube, Int. J. Heat and Fluid Flow 28 (2007) 203–210.

[59] W.H. Azmi, K.V. Sharma, P.K. Sarma, R. Mamat, S. Anuar, V. Dharma Rao, Experimental determination of turbulent forced convection heat transfer and friction factor with SiO_2 nanofluid, Exp. Thermal and Fluid Sci. 51 (2013) 103–111.

Chapter 2

Preparation and evaluation of stable nanofluids for heat transfer application

Zeyu Liu[1,2], Rong Fu[3] and Yan Yuying[2]

[1]Marine Engineering College, Dalian Maritime University, Dalian, Liaoning, P.R. China, [2]Faculty of Engineering, The University of Nottingham, Nottingham, Nottinghamshire, United Kingdom, [3]Institute of Engineering Thermophysics, Chinese Academy of Sciences, Beijing, P.R. China

Nomenclature

A_1	Particle Hamaker constants (-)
A_2	Medium Hamaker constants (-)
A_λ	Absorbance (-)
C_{bl}	Specific heat capacity of base liquid (J/(kg • K))
C_{nf}	Specific heat capacity of nanofluids (J/(kg • K))
C_p	Specific heat capacity of nanoparticles (J/(kg • K))
D_p	Diameter of nanoparticle (m)
G_A	Gibbs free energy of attraction (J)
G_{OS}	Gibbs free energy of osmotic repulsion (J)
G_{ST}	Gibbs free energy of steric repulsion (J)
G_T	Total Gibbs free energy (J)
G_{VR}	Gibbs free energy of volume restraint interaction (J)
I_0	Intensity of incident laser light (W/m^2)
R_H	Hydrodynamic radius (m)
R_a	Radius of gyration (m)
T_λ	Transmittance (-)
V_A	van der Waals attractive potential (V)
V_R	Electrostatic repulsive potential (V)
V_T	Total potential (V)
V_t	Settlement velocity (m/s)
k_B	Boltzmann constant (-)
k_{bl}	Thermal conductivity of base liquid (W/(m • K))
k_{nf}	Thermal conductivity of nanofluids (W/(m • K))
k_p	Thermal conductivity of nanoparticles (W/(m • K))

Advances in Nanofluid Heat Transfer. DOI: https://doi.org/10.1016/B978-0-323-88656-7.00013-1

k_r	Ratio of k_{nf} to k_{bl} (-)
r_a	Radius of aggregates (m)
r_p	Radius of particles (m)
ρ_0	Density of base liquid (kg/m^3)
ρ_{nf}	Density of nanofluids (kg/m^3)
ρ_p	Density of nanoparticle (kg/m^3)
φ_a	Volume fraction of aggregates (-)
φ_m	Volume fraction of densely packed particles (-)
r	Radius of nanoparticles (m)
ε	Permittivity (F/m)
ε_0	Vacuum permittivity (F/m)
ζ	Zeta potential (mV)
A	Hamaker constant (-)
I	Intensity of laser beam (W/m^2)
T	Absolute temperature (K)
X	Distance between specific location of tube and rotation in the centrifuge (m)
c	Particle concentration (kg/m^3)
k	Thermal conductivity ($W/(m \bullet K)$)
l	Optical path length (m)
n	Empirical shape factor (-)
α	Absorptivity (-)
μ	Viscosity ($N \bullet s/m^2$)
φ	Volume fraction (-)
ω	Angular velocity of the centrifuge (s^{-1})

2.1 Introduction

Heat transfer is becoming a core challenge with an increasing degree of integration and boosting of novel technology [1] that is encouraging researchers to explore new generation working fluids. In this progress, nanofluids present great potential for heat transfer performance due to their excellent thermal physical and rheology properties. Many studies have reported the heat transfer enhancement of nanofluids [2,3]. For instance, Xuan and Li [4] found a 78% enhancement in thermal conductivity in Cu nanofluid. Yu et al. [5] experimentally studied the thermal conductivity of Fe_3O_4-kerosene nanofluid, which showed an increase of 34%. Both studies also tried to decrease frictional loss and pumping power for nanofluids. Additionally, some authors also found drag reduction of nanofluids [6–8]. Nakhchi et al. [9] found that water-based copper nanofluid reduced the friction factor by 23.6% in a circular pipe.

The performance and functionality of nanofluids are closely dependent on their stability. Therefore the long-term nanofluid stability should be investigated prior to practical application for high heat flux components. The collision of nanoparticles in the suspension leads to agglomeration that causes nanofluid destabilization. The high surface activity is a dominating factor for the agglomeration tendency of nanoparticle suspension. Moreover, the precipitation phenomenon is an important obstacle to the industrial production of nanofluids. In microfluidics, the cluster negates the beneficial properties of nanofluids and leads to clogging of the microchannel. According to the study of Lin et al. [10], nanofluid instability affects both thermal and mechanical performance, resulting in the inefficiency of heat transfer performance. For hybrid nanofluids, nanofluid instability is a big challenge for optimizing the combination of different types of nanoparticles, base liquid, and surfactant. Unexpected changes in thermophysical properties are attributed to the precipitations of nanofluids. In addition, agglomeration also affects the exergy due to the increase of viscosity and pumping cost. Therefore instability must be eliminated for industrial applications.

However, limited understanding of the anomalous behavior of nanofluids makes nanoparticle suspensions easily obtain poor characterizations. Besides, operational conditions (such as temperature, concentration, shearing rate, etc.) in practical applications can also adversely affect nanofluid stability. Therefore it is necessary to fully understand nanofluid stability from different perspectives, such as preparation, evaluation, stabilization techniques, etc. To the best of our knowledge, there are limited reports on the factors influencing nanofluid stability and their influences on thermophysical properties. The previous studies have concentrated on identifying and explaining influencing factors that affect nanofluid stability. A lack of understanding of theoretical models and mechanisms impedes the effective control of nanofluid stability. Instead of overemphasizing the thermophysical aspects of nanofluids, this chapter aims to review the current research gap and developments of nanofluid stability that include preparation, evaluation of nanofluid stability, stabilization techniques, stability mechanisms, and effects on thermal properties. This chapter expects to improve the wide availability of nanofluids from the viewpoint of stability.

2.2 Preparation

Nanofluids are prepared by dispersing nanosize particles into base liquid. Synthesizing well-dispersed nanofluids is the prerequisite for the commercial application of nanofluids. Surfactants and surface modifications are two strategies to improve nanofluid stability. The two-step method and one-step method are the two main techniques to prepare nanofluids. Fig. 2.1 summarizes different preparation methods for various types of nanofluids. In addition, the following section introduces synthesis methods from the perspective of nanofluid stability.

2.2.1 Two-step method

For the two-step method, nanoparticles or nanomaterials are commercially available nanosize powders that are distributed in the base liquid by mechanical agitation or ultrasonic agitation (Fig. 2.2). According to the previous literature, 95% of researchers prepared nanofluids by the two-step method due

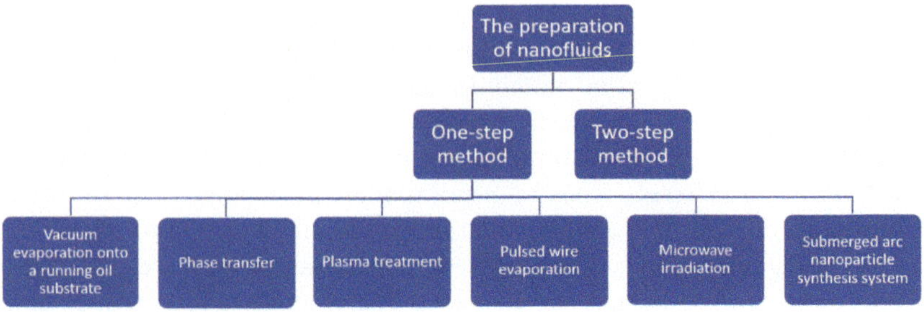

FIGURE 2.1 Various synthesis methods of nanofluids.

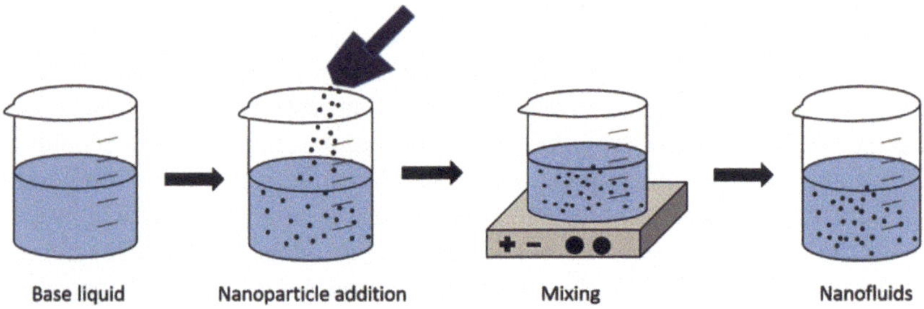

FIGURE 2.2 The two-step synthesis process of nanofluids.

to their simplicity without sophisticated apparatus. For example, Lee et al. [11] and Eastman et al. [12] both selected the two-step technique for the preparation of Al nanofluids. Xuan et al. [2] synthesized transformer oil-based and water-based Cu nanofluids, respectively. The two-step method shows physical friendly potentials that can prepare relatively stable nanofluids without any surfactants. Kim et al. [13] dispersed ethylene glycol (EG)-based CuO nanofluid without stabilizers. Besides, some studies pointed out that two-step preparation was more suitable for oxide nanoparticles than metallic nanoparticles [14].

Instability is the main drawback of the two-step technique. Nanoparticles prepared by this technique are prone to aggregation due to surface activity and high surface area. There are two common methods to keep the prepared nanofluids stable, namely ultrasonication and stirring. Unfortunately, these physical methods cannot effectively stabilize the nanofluid for a long time. Surfactants are selected to control particle aggregations, which include anionic [sodium dodecyl sulfate (SDS)] [15], nonionic (Span 80, Tween 20), cationic [cetyltrimethylammonium bromide (CTAB)], and polymers [polyvinylpyrrolidone (PVP)]. However, surfactant degradation needs to be concerned about under high-temperature conditions.

In summary, the key advantage of the two-step method is that it is an ideal method for large-scale and cost-effective production. This is because synthesis techniques of nanosize powders have already been in use for industrial mass production; while the disadvantages of these techniques are as follows: first, nanofluids prepared with this technique tend to agglomerate. Second, particle shape and size cannot be controlled during the preparation. Thirdly, the purchased nanoparticles are prone to agglomeration in both storage and drying stages.

2.2.2 One-step method

Compared to the two-step method, the one-step method is developed for more stable nanofluids. This method is a combination of nanoparticle production with nanofluid synthesis simultaneously (Fig. 2.3). This synthesis method can be undertaken by both physical and chemical techniques without intermediate drying steps (powder drying, dispersion, etc.).

In terms of one-step physical synthesis, Akoh et al. [16] synthesized oil-based magnetic nanofluids (Fe, Co) by a direct evaporation method, which is called vacuum evaporation onto a running oil substrate. Similarly, Eastman et al. [12] condensed Cu into nanoparticles with VEROS by contacting with flowing low vapor pressure generated by EG. However, VEROS cannot separate large amounts of nanoparticles from base liquids. Submerged arc nanoparticle synthesis system (SANSS) is a physical technique that can control the nanoparticle shape by minimising nanoparticles. Lo et al. [17] synthesized CuO and Cu nanoparticles in dielectric liquids by SANSS. Their TEM analysis showed that the formation of nanoparticles had a different shape and uniform size distribution. However, it is difficult to synthesize nanofluids at a large scale by this method.

One-step chemical methods improve the scale of production to a certain extent. Zhu et al. [18] reported microwave irradiation to prepared Cu nanofluid by $CuSO_4 \cdot 5H_2O$ with $NaH_2PO_2 \cdot H_2O$. Ayyappan et al. [19] used PVP as a stabilizer to synthesize mineral oil-based silver nanofluid in an ethanol medium. They found nanofluid stability with stabilizer was investigated at

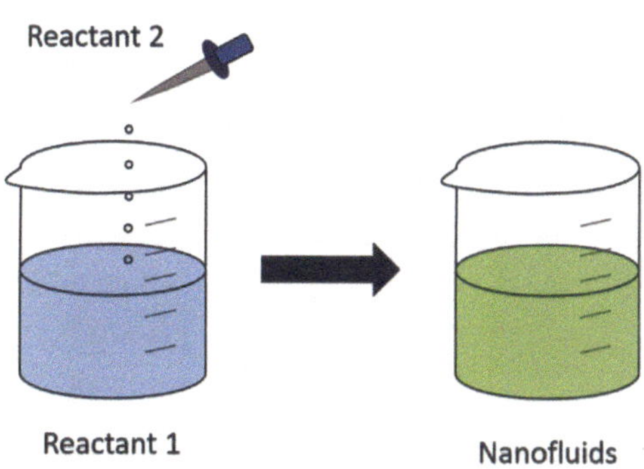

FIGURE 2.3 The one-step synthesis process of nanofluids.

high temperature for 2 weeks without any sign of sedimentation. Phase transfer is another effective chemical method, Yu et al. [20] coated functional groups (-COOH, -OH, and C = O) onto the surface of graphene oxide nanoparticles by transferring functional groups in oxygen. Graphene oxide nanoparticles can be easily dispersed in the base liquid due to the repulsive force provided by functional groups. Munkhbayar et al. [21] prepared Ag-MWCNTs and Ag nanofluids by pulsed wire evaporation with a long and thin wire under 300 V. Subsequently, supplied voltage induced heat to evaporate within milliseconds, and it then turned into plasma with the evaluated temperature.

In summary, there are three factors improving nanofluid stability in a one-step method: the addition of a steric stabilizer, decreasing particle size, and better dispersibility. The main strength of the one-step technique is that it can control the shape and size of nanoparticles in the process of preparation. However, low quantity and high cost are still the prime problems that hinder the application of this method.

2.3 Evaluation of nanofluid stability

Preparation is the prerequisite of nanofluid stability, but either one-step or two-step methods cannot directly synthesisz well-dispersed nanofluids. The effective evaluation of nanofluid stability is significant to implement the industrial application of nanofluids. Fig. 2.4 summarizes the main methods to evaluate nanofluid stability. A comprehensive review of these evaluation techniques is introduced in the following sections.

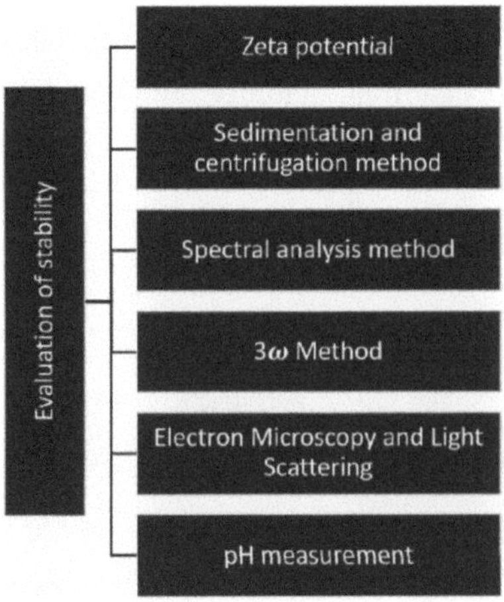

FIGURE 2.4 Evaluation methods of nanofluid stability.

2.3.1 Zeta potential analysis

Zeta potential is defined as electric potential in the interfacial double layer, which is between the stationary layer of fluid attached to the dispersed particles and the dispersion medium. Subsequently, electric surface charges are generated around nanoparticles that form an electrostatic field affecting the ions in the solution. Both thermal motion of ions and electrostatic field can screen the electric surface charge when the net surface charge is equal to the resultant electric charge in magnitude, but opposite in polarity.

Zeta potential is an effective parameter to evaluate nanofluid stability in the synthesis process of nanofluids. For example, Kim et al. [22] analyzed the effect of NaOH and HCl in small quantities on zeta potential. Experimental results found that the isoelectric point (IEP), an electric neutralization, was at a pH of 3. Leong et al. [23] investigated the effect of zeta potential on different modified nanofluids with various concentrations. They found that nanofluids with surfactants exhibited high zeta potential values. Additionally, Li et al. [24] also studied the effects of different concentrations, pH values, and dispersants on copper nanofluid.

To obtain an accurate relationship between zeta potential and nanofluid stability, Zetasizer Nano is the most reliable method. It indicates the precise potential difference between base liquid and dispersion in nanofluids. The repulsive force is not only highly dependent on the zeta potential but also related to dispersion. According to previous investigations, researchers classified the degree of stability with zeta potential as shown Table 2.1, which considered a nanofluid to be stable when zeta potential is beyond ± 30 mV.

Due to the complicated operation of Zetasizer Nano, alternative devices are also used for zeta potential analysis. For example, Li et al. [25] conducted PALS 190 plus to analyze the stability of SiC nanofluid. Nikkhah et al. [26] analyzed the zeta potential of aqueous-based copper oxide nanofluid by dynamic light scattering (DLS).

TABLE 2.1 Stability of nanofluids at different zeta potentials.

Zeta potential (mV)	Stability degree
0	Slight to no stability
15	Less stability with light settling
30	Moderate stability
45	Decent stability with possible settling
60	Excellent stability with lesser possibility of settling

2.3.2 Sedimentation and centrifugation methods

The sedimentation method, also known as the photographic technique, is a common technique to assess nanofluid stability. This method is based on the gravity precipitation of particles at the bottom of the container. Nanofluids stay stable with constant particle size and concentration. The sedimentation method does not need sophisticated equipment under stationary conditions. However, it takes a long period compared to other evaluation techniques. Therefore sedimentation photograph is a method to assist the sedimentation method. As shown in Fig. 2.5, it takes 8 months to verify nanofluid stability [7].

Researchers are using this sedimentation technique to evaluate colloidal stability. Zhu et al. [27] captured photographs of graphite nanofluids during the preparation period and measured the stability of the prepared nanofluids by sedimentation method. Khaleduzzama et al. [28] used this method in the microsize technology to verify nanofluid stability by observing 30 days with no sign of aggregation. Sedimentation is a time-consuming method. Therefore the centrifugation method, as an alternative method, shortens the sedimentation process and accelerates the valuation of colloidal stability. Singh et al. [29] estimated the stability of silver nanofluids by the centrifugation method (3000 rpm) for 10 hours. Their results showed there was no sedimentation that was as same as the 1-month observation. Besides, the centrifugation method is also used to identify the function of the stabilizer (PVP) and modifications by Sing et al. [29]. For the spherical, smooth nanoparticles, the final settlement velocity in the centrifuge can be expressed by Stokes law as follows:

$$V_t = \frac{D_p^2(\rho_p - \rho_0)\omega^2 X}{18\mu} \tag{2.1}$$

FIGURE 2.5 Evaluation of nanofluid stability by observation method [7].

where $\rho_0, \rho_p, D_p, X, \mu$, and ω are the density of the base liquid, the density of the nanoparticle, particle diameter, the distance between the specific location of the test tube and the axis of rotation in the centrifuge, the viscosity of base liquid, and angular velocity of centrifuge, respectively. The expression shows that four main parameters affect nanofluid stability, which are the viscosity of the base liquid, particle size, density difference of nanoparticle and base liquid. Besides, lowering the particle size and higher settling velocity can lead to particle clustering. In summary, the combination of centrifugation with the sedimentation method offers a quantitative evaluation of colloidal stability by measuring the height of sedimentation and settling time.

2.3.3 Spectral analysis method

UV–Vis spectrophotometer measurement, as a reliable technique for spectral analysis, can measure the concentration of nanofluids quantitatively. It is applicable when the extinction bands range from 190 to 1100 nm wavelength. Nanofluid stability can be evaluated by testing absorption peak characteristics over time. The relationship between absorbance (A_λ) and concentration is given as Eq. (2.2),

$$A_\lambda = \log_{10}\left(\frac{I_0}{I}\right) = \alpha \times l \times c \tag{2.2}$$

where I_0, I, α, l, and c are the intensity of incident laser light, the intensity of laser beam, absorptivity, optical path length, and particle concentration, respectively. Based on Eq. (2.2), the decrease of absorbance indicates the decrease of concentration due to sedimentation. However, it is difficult to evaluate dark color, high concentration nanofluids by the absorbance measurement [30]. Similarly, transmittance measurement can also evaluate stability. Transmittance (T_λ) is dependent on absorbance (A_λ), and transmittance and can be expressed as follows,

$$T_\lambda = I/I_0 \tag{2.3}$$

Table 2.2 exhibits the stability of Cu-Al LDH nanofluids by transmittance. The decrease in transmittance indicates the instability of the nanofluid. The experimental result exhibits instability at a higher concentration of nanofluids.

Jiang et al. [32] were the pioneers in determining nanofluid stability by absorbency. They analyzed both refluxed and unrefluxed CNTs by UV–Vis Spectrograph. The results showed that nanofluids were stable for 80 days. In addition, Yang et al. [33] used the transmittance method to measure nanorefrigerants. The transmittance decreased with the increase of concentration, and stability decreased with the increasing transmittance. Furthermore, Li et al. [24] combined absorbency and zeta potential with sedimentation to evaluate the stability of water-based copper nanofluids under dispersant types, concentrations, and pH conditions.

TABLE 2.2 Transmittance measurement of Cu-Al LDH nanofluid under various concentrations and time durations [31].

Sample	Concentration (vol.%)	% Transmittance		
		Day 0	Day 2	Day 3
Water	0	100	100	100
Cu-Al LDH nanofluid	1	77.21	81.70	85.22
	2	72.78	77.78	81.72
	4	30.03	36.34	42.34

2.3.4 3ω method

3ω method detects suspension stability by measuring thermal conductivity growth. This method is suitable for a wider range of volume fractions. Karthikeyan et al. [34] measured an 87.5% drop in the thermal conductivity of CuO nanofluid because of agglomeration and nanoparticle sedimentation. This is because particle clustering decreases both effective thermal interaction area and the ratio of surface to volume. On the other hand, Oh et al. [35] analyzed the increase of thermal conductivity in both EG and water-based nanofluids. This increase is also attributed to moderate particle agglomeration. They found that there was a peak value of particle agglomeration and thermal conductivity drops when the clustering is beyond the peak value. However, the evaluation of colloidal stability by the 3ω method is still very limited.

2.3.5 Electron microscopy and light scattering

Electron microscopy and light scattering techniques can not only measure the size distributions of particles but also analyze particle aggregation. For the electron microscopy method, there are two main optical microscopies to visualize high-resolution images of particle clusters of dried nanofluids samples: scanning electron microscope (SEM) and transmission electron microscopy (TEM). Both SEM and TEM can detect and analyze the microstructure and the distribution of nanoparticles.

Specifically, uniform particle distribution means well-dispersed nanofluids, while clusters and aggregations in the test sample indicate unstable nanofluids. Many scholars used electron microscopy to analyze particle aggregation, since TEM and SEM are two significant indicators of nanofluid characterizations. For instance, Chakraborty et al. [36] used TEM to study the effect of polymer addition and surfactant on the dispersion of water-based TiO_2 by TEM measurement. The measurement exhibits a good agreement with the result of zeta potential and sedimentation measurements. Fu et al. [7] observed well-dispersed nanoparticles within 200 nm by TEM measurement. TEM images show that few are overlapping in nanoparticle distribution, for example, Fig. 2.6. In terms of SEM, it utilizes scattered light to measure the particle diameter due to Brownian motion. Li et al. [24] investigated particle distribution and microstructure of 0.2 vol.% EG SiC nanoparticles by SEM. The SEM measurement reveals agglomerates in a microscopic dimension, but there was no obvious precipitation in nanofluids.

For DLS, it can calculate various particle sizes in higher concentrations. According to light scattering theory, the diffusion coefficient can be expressed based on intensity fluctuation by the Stokes equation as follows,

$$R_H = \frac{k_B T}{6\pi\mu D} \tag{2.4}$$

where D, T, R_H, k_B, and μ are translational diffusion coefficient, absolute temperature, hydrodynamic radius, Boltzmann constant, and viscosity, respectively. The agglomeration tendency can be indicated by nanoparticle size at various time intervals over a long period. DLS can monitor changes in nanoparticle size over time, which makes DLS an effective method to detect the clustering rate of nanofluids with time (Fig. 2.7). For instance, Kole et al. [37] determined that the cluster size of CuO nanoparticles increased nine times compared to the original size of nanoparticles due to nanofluid instability. Sadeghi et al. [38] also investigated the relationship between sonication time and the stability of Al_2O_3 nanofluids by checking increasing particle size with concentrations.

SAXS is another nanoparticle characterization tool that can detect nanofluid stability by the deviation of particle scattering function because of

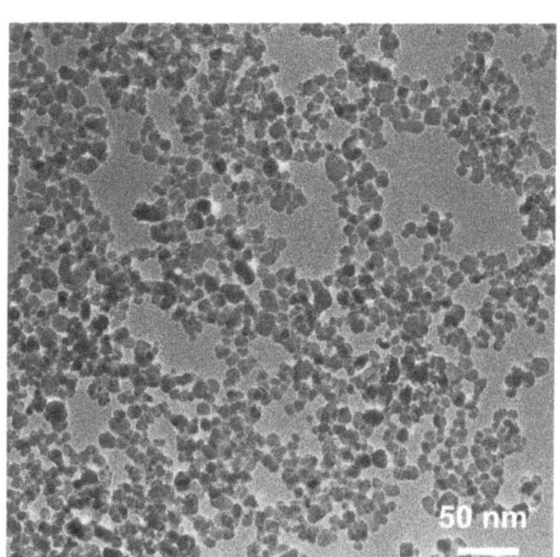

FIGURE 2.6 TEM measurement of Fe_3O_4 nanoparticles [7].

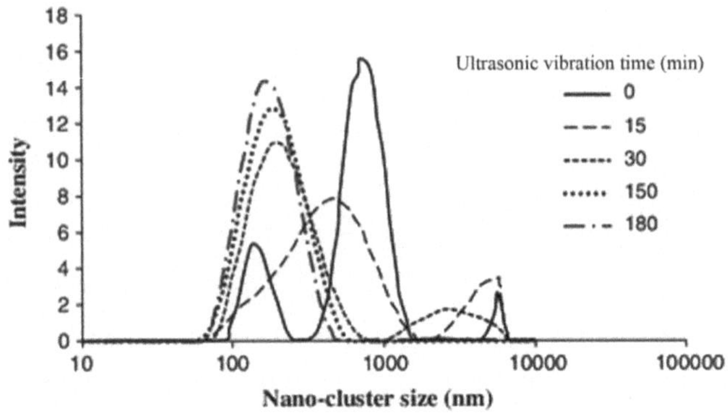

FIGURE 2.7 Measurement of nanocluster sizes by DLS [39].

aggregated scattering with the low momentum transfer. Rieker et al. [40] combined TEM with SAXS to investigate the distribution of particle size. Hoffmann et al. [41] also investigated nanoparticle size in the water-based silicon carbide nanofluids by both DLS and SAXS. However, much more research reported electron microscopy and light scattering than that of SAXS [17,18,42–44].

2.3.6 pH measurement

pH measurement is an effective and simple technique to analyze nanofluid stability. It can improve nanofluid stability by increasing repulsive forces. different nanofluids have their pH value ranges. Chandrasekar et al. [15] and Kamatchi et al. [45], respectively, measured stable Al_2O_3 and rGO nanofluids at the pH of 5 and 7.56–8.69, which are far away from their IEPs. Suitable pH values for water-based graphite, alumina, and copper nanofluids are 2, 8, and 9.5, respectively. Qi et al. [46] used transmittance to analyze the relationship between pH values and nanofluid stability. They selected four samples of Al_2O_3–water and TiO_2–water at pH values of 7, 8, 9, and 10, respectively. They found that there was better stability when the pH value of the target nanofluids is at 8 since they had the lowest transmittance over time at a pH of 8.

2.4 Stabilization techniques

Physical and chemical methods are two categories to improve nanofluid stability, as shown in Fig. 2.8. Physical methods generally provide high energy to break down nanoparticle clusters. Chemical methods stabilize nanoparticles by pH control, surface modifications, additions, etc. In this section, both methods are introduced.

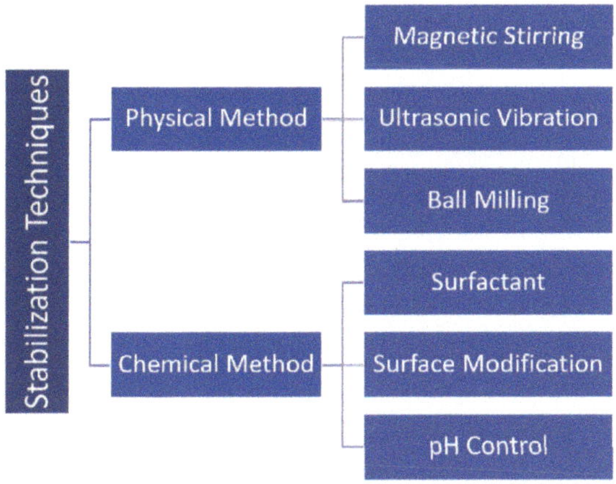

FIGURE 2.8 Techniques to optimize nanofluid stability.

2.4.1 Physical method

2.4.1.1 Mechanical stirring

Mechanical stirring is a basic mixing technique during the preparation of nanofluids. Under the magnetic field, the magnetic bead rotates at the bottom to break large clusters. Kusure et al. [47] synthesized CuO and Al_2O_3 nanofluids by mechanical stirring and then modified the addition of SDS by ultrasonic vibration. To synthesize high-quality nanofluids, stirring speed should be controlled in a reasonable range. This is because high-speed stirring affects the mass fraction of nanofluids by bubble formation. Therefore it is important to optimize stirring speed during the preparation of nanofluids. For instance, Hu et al. [48] controlled stirring speed within the range from 800 to 1200 rpm to avoid foam forming. Besides, a combination of high-pressure homogenizer, ultrasonication with surfactant is also a solution to disperse CuO–water nanofluids.

2.4.1.2 Ultrasonic vibration

Ultrasonic vibration is a more effective mechanical method for cluster size reduction and stability. There are two forms of sonication to implement ultrasonic vibration, namely ultrasonic bath, and probe sonicator. Probe sonicator exhibits better performance in lower the size of clusters and breaking particle clusters compared to ultrasonic bath. Chen and Wen [49] synthesized water-based gold nanofluids with sonication times ranging from 10 to 60 min. They reported that the particle size continuously decreased until sonication time increased to 45 min. However, excessive ultrasonication can damage the structure of nanofluids and even affects the thermophysical properties of nanofluids.

Amrollahi et al. [50] found that excessive ultrasonication could damage the structure of CNT/EG-based nanoparticles and shorten the length of CNT. Therefore the duration and degree of sonication need to be optimized to prevent such defects [51]. Additionally, ultrasonication also affects the thermophysical properties of MWCNT due to the increasing temperature [51]. To solve this problem, ultrasonication equipment with autotuning mode and programmable pulse mode is used to adjust energy delivered to nanofluids. The autotuning mode can adjust electrical efficiency by monitoring the frequency change due to the variation of load.

2.4.1.3 Ball milling

Ball milling is a rarely reported technique and sometimes combines with other stabilization methods to obtain well-dispersed nanofluids. To prepare aqueous MWCNT nanofluids, Munkhbayar et al. [21] combined planetary ball mill and ultrasonication. In their studies, the main function of planetary ball milling was to decrease the average size of the cluster and improve nanofluid stability. Similarly, Boopathy et al. [52] prepared Cu and Al nanofluids by combining ultrasonication and mechanical stirring with wet milling simultaneously.

2.4.2 Chemical stabilization

Chemical stabilization, including adding surfactants, modifying surfaces, and changing pH values, is also an effective method to stabilize nanofluid. The details are introduced in the following sections.

2.4.2.1 Surfactant

Surfactants are chemical compounds that attach to nanoparticle surfaces to decrease the surface tension of the liquid. In this way, surfactants can slow down the accumulation of nanoparticles. There are several kinds of surfactants utilized in nanofluids, including SDS [15], oleic acid [3,8], PVP [29], etc. Changing nanofluid properties is the main reason to improve the dispersibility of nanoparticles by surfactants. Both electrostatic repulsive forces and hydrophobic surface forces contribute to breaking down agglomerations.

Many studies are focusing on adding surfactants to suppress nanoparticle sedimentation. Leong et al. [53] reported that surfactants can improve nanofluid stability. In their report, they found that nanofluids with surfactants [gum Arabic, PVP, and CTAB] showed a maximum enhancement of 26% in thermal conductivity. Similarly, Al-Waeli et al. [54] compared SDS, CTAB, and SDBS surfactants. They found that CTAB exhibited better nanofluids stability for over 80 days.

Although adding surfactant is a feasible method to optimize the dispersibility of nanoparticles, there are still two main disadvantages in the preparation of nanofluids with surfactants. First, surfactants are prone to produce foams at high temperatures. Second, attaching surfactant molecules increases the nanoparticle thermal resistance between base liquid and nanoparticles, which hinders the thermal conductivity. Therefore Munkhbayar et al. [21] studied surfactant-free synthesis of Ag-MWCNT hybrid nanofluids. They found that nanofluids can be stable and dispersed for a long duration when there is no surfactant present in the nanofluids. Besides, Yu et al. [55] proposed that additional thermal resistance is likely the potential reason for the decrease in thermal conductivity.

2.4.2.2 Surface modification

Surface modification is an effective surfactant-free technique that functionalizes the surface of nanoparticles. The injection of functional nanoparticles into base liquids is a potential method for long-term nanofluid stability. Yang et al. [33] prepared nanofluid by modifying silanes onto the surface of SiO_2 nanoparticles in the solution. They found that there was no deposition layer formed on the heated surface after a pool boiling process. In addition, Hwang et al. [56] used a mechanochemical reaction to modify hydrophilic functional groups. This modification exhibited high stability and thermal conductivity with low viscosity, which showed great potential in thermal applications. In addition, hydroxyl groups, as a wet mechanochemical reaction, had been popularly applied to the treated CNT surfaces. To optimize the stability of aqueous ZnO nanofluids, Tang et al. [57] modified the ZnO nanoparticle surface with poly methacrylic (PMAA) acid as a modification. PMAA interacts with hydroxyl groups that are attached to the ZnO surface. This interaction eventually produces zinc methacrylate which improves the stability of ZnO nanofluids. Some special nanoparticles are difficult to modify directly. For instance, the diamond nanoparticle can be modified effectively by plasma treatment. Polar groups are coated onto diamond nanoparticle surface by plasma by mixing oxygen and methane.

2.4.2.3 pH control of nanofluids

The control of pH can directly relate to the electrokinetic properties of nanofluids which are strong indicators for nanofluid stability. This is because repulsive forces stabilize well-dispersed nanosuspensions by high surface charge density. For instance, Xie et al. [58] achieved well-dispersed CNTs by acid treatment that produced hydroxyl groups converting a hydrophobic nature to a hydrophilic nature. When zeta potential is at IEP, nanoparticles lack repulsive forces in base liquids and nanoparticles are prone to coagulate [59]. Therefore it is crucial to remain a high level of repulsive forces to improve nanofluid stability [60–62]. Wang and Li [63] further investigated the relationship between pH value and the degree of stability. They studied water-based Cu and Al_2O_3 nanofluids at different pH values. Cu and Al_2O_3

nanofluids exhibited stability until pH values were over 7.6 and 7.5–8.9, respectively. The stability of Cu and Al_2O_3 nanofluids is attributed to the increase of electrostatic repulsion among nanoparticles. Li et al. [64] found that the best stability for water-based copper nanofluid is at pH of 9.5. However, the study has proved that the pH value of stable nanofluids should be maintained around 7. This is because acidic or alkaline solutions may erode heat transfer devices, especially at high temperature. Therefore nanofluids with the best stability at neutral pH values will be potential working media in industrial applications.

2.5 Stability mechanisms

The colloidal stability indicates the degree of nanoparticle aggregation. This indicator is determined by the probability of cohesion and frequency of collisions during the collision. Academia generally accepted DVLO theory to explain the mechanism of nanofluid stability. There are five assumptions for DVLO theory [65]: (1) homogeneous colloidal suspensions; (2) dilute particle concentration; (3) neglected gravity and buoyancy forces; (4) two main forces acting on suspended particles, namely van der Waals attractive and electrostatic forces; and (5) ion distribution in nanofluid is related to Brownian motion, electrostatic force, and entropy induced dispersion. Based on the total potential, DLVO theory describes the degree of nanofluid stability that can be expressed as,

$$V_T = V_R + V_A \tag{2.5}$$

where V_R and V_A are electrostatic repulsive potential and van der Waals attractive potential, respectively. Nanofluid stability can be reflected by the sum of V_R and V_A that can describe the separation distance between nanoparticles. When the attractive force dominates over the repulsive force, nanoparticles tend to aggregate which makes nanofluids unstable. When the repulsive force is sufficient, the suspensions will exist to ensure stability. On one hand, Hamaker constant (A) is highly related to van der Waals attractive potential (V_A). Hamaker constant is highly related to the dielectric property of base liquid and refractive index, which can be expressed as,

$$A = \left(A_1^{\frac{1}{2}} - A_2^{\frac{1}{2}}\right)^2 \tag{2.6}$$

where A_1 and A_2 refer to particle and medium Hamaker constants, respectively. According to [66], nanoparticles with a high Hamaker constant can produce high van der Waals attractive potential [67] that, consequently, are more likely to aggregate. On the other hand, in terms of V_R, the surface potential is highly related to zeta potential (ζ). The relationship can be expressed as,

$$V_R = 2\pi\varepsilon\varepsilon_0 r\zeta^2 e^{-kx} \tag{2.7}$$

Even though there are sufficient energy barriers, external conditions, like elevated temperature, can trigger particles to collide with each other to breach the energy barrier. In this condition, the attractive force dominates and pulls particles together forming clusters. Moreover, when particle loading is high, while zeta potential is relatively low, repulsive potential produces a secondary minimum with the decreasing separation distance. Thanks to the secondary minimum, the weakly bonded can be broken up by physical techniques. In other words, it is desirable that repulsive force dominates between nanoparticles due to less agglomeration. Therefore high repulsive force is essential for obtaining long-term nanofluid stability. There are three dispersion methods developed for the synthesis of long-term nanofluids, including electrostatic, steric, and electrosteric stabilization. All three methods can intensify the repulsive forces that can counteract the van der Waals attraction.

2.5.1 Electrostatic stabilization

Electrostatic stabilization occurs by introducing charges onto the nanoparticle surfaces. The surrounding particles create an electrical double layer that forms a coulombic repulsion force between the nanoclusters. Positive charge density around negative charges forms an electrostatic repulsive force that makes nanoparticles dispersed in the base liquid. The nanoparticle surroundings can be divided into three layers:

1. Surface charge contains charges attached to the nanoparticle.
2. The Stern layer contains counterions that are firmly absorbed onto the surface of the nanoparticle by the electrostatic repulsive force.
3. The diffuse layer contains positive and negative ions that are adsorbed around the particle surface by the electrostatic repulsive force.

Electrostatic stabilization can be achieved when the electrostatic repulsive force can be counterbalanced by van der Waals attractive force (Fig. 2.9).

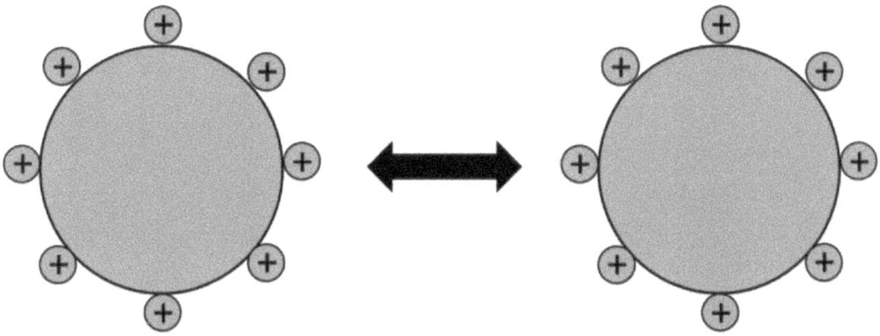

FIGURE 2.9 Schematic diagram of electrostatic stabilization between nanoparticles.

It has been proven that electrostatic stabilization is the most effective method to stabilize metal and metal oxide particles [68].

For electrostatic stabilization, the nanoparticle surface exhibits the highest electrical potential in the electric double layer (EDL). The electric potential gradually decreases as one escapes from the nanoparticle surface until potential reaches zero at the boundary of EDL.

As shown in Fig. 2.10, zeta potential (ς), as an evaluation method of nanofluids, is defined as the potential of the slipping plane. Electrostatic stabilization aims to promote potential barriers among nanoparticles to prevent strongly bonded clusters. A high potential barrier can stabilize nanofluids.

There are generally four forms of charged surfactants attached to the nanofluid suspension for improved electrostatic stability, including cation (CTAB, CTAC), anion (SDS, SDBS), nonion (oleic acid, sodium cholate), and amphoteric ion (lecithin, hydroxysultaine, sodium lauroamphoacetate). Zwitterionic and amphoteric surfactants are the surfactants that contain anionic and cationic hydrophilic groups. They produce anions and cations due to the acidity and alkalinity of nanofluids.

Although electrostatic stabilization is the most effective method, some limitations should be noted prior to modification: (1) electrostatic stabilization only applies to dilute suspension; (2) this technique cannot be applied to electrolyte sensitive systems and saline environment; and (3) this method can only be applied to water-based suspension.

2.5.2 Steric stabilization

Steric stabilization is obtained by grafting large molecules including nonionic surfactants and polymers to the nanoparticle surface (Fig. 2.11). Steric stabilization is an obstruction among nanoparticles that can prevent the van

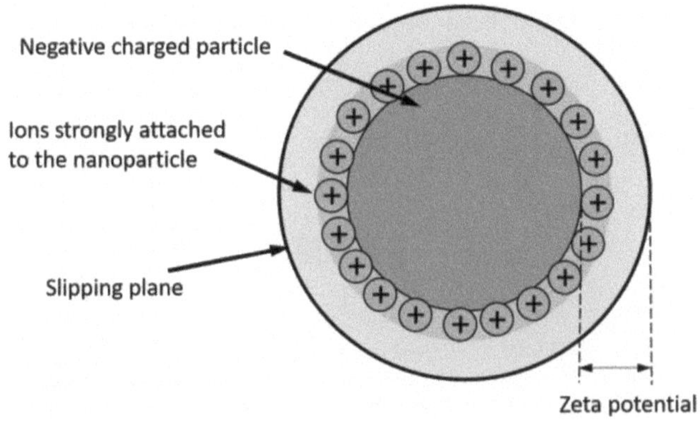

FIGURE 2.10 Schematic diagram of zeta potential concept.

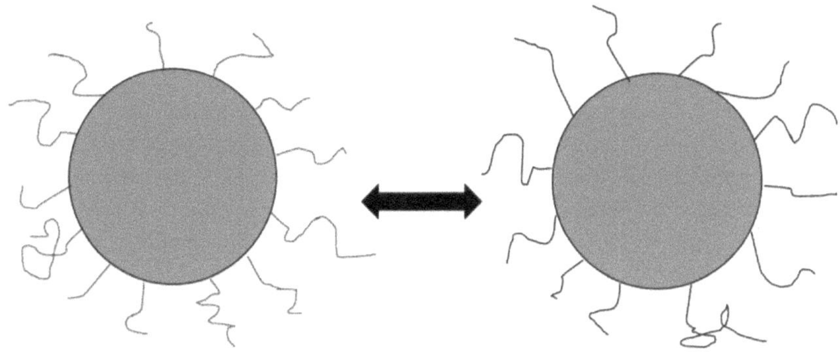

FIGURE 2.11 Steric stabilization of nanoparticles.

der Waals attraction. It is essential to obtain strong binding between the polymeric chain and nanoparticles for steric stabilization. The longer the polymer chains, the better the freedom of polymer chains and shielding of the particle core when nanoparticles are closer to each other.

For nonionic amphiphilic polymer, steric stabilization contains two parts, namely the anchor part and polymeric tail. The anchor part attaches to the surface of the nanoparticle. Meanwhile, the polymeric tail forms a protective layer against agglomeration, consequently, it stops nanoparticles from attaching to solvent.

There are two dominating parameters for steric stabilization, including strong enthalpic interaction, density, and length of polymeric tail. In terms of thermodynamic stabilization scheme, steric stabilization is calculated by total Gibbs free energy G_T,

$$G_T = G_{ST} + G_A = G_{OS} + G_{VR} + G_A \tag{2.8}$$

G_{ST}, G_A, G_{OS}, and G_{VR} are Gibbs free energy of steric repulsion, attraction, osmotic repulsion, and volume restraint interaction, respectively. G_{OS} refers to the increase of osmotic pressure that surrounds nanoparticles because of the increasing concentration of solvated polymeric tail. Moreover, due to this effect, when solvent molecules overlap with the polymer layer, they can produce a strong repulsion that can stop nanoparticle agglomeration. The sign of G_{OS} is highly dependent on the solubility of the stabilizer in the dispersion medium. When G_{OS} is positive, the solvent is stable. Moreover, mixing interaction prompts nanoparticle repulsion and colloidal stability. When G_{OS} is negative, it can be termed as a poor solvent, and mixing interaction is attractive, which makes nanofluids destabilization. When G_{VR} is positive, it is favorable for repulsion. Due to the compression and overlapping of the polymeric chain, volume restraint interaction decreases interacting polymer volume, which leads to a decline of entropy. Therefore G_{OS} and G_{VR} contribute to repulsive free energy. In terms of G_T, it exhibits steric stabilization when $\Delta G > 0$. For the stable solvent, $G_T > 0$, $G_{ST} = G_{OS} + G_{VR} > G_A$; while for the unstable solvent, $G_T < 0, G_{OS} + G_{VR} < G_A$.

Compared to electrostatic stabilization, steric stabilization can stabilize high concentration nanofluids. However, the attachment cannot change the surface potential, so that steric stabilization cannot be measured quantitatively by the zeta potential technique. Therefore transmittance, sedimentation, and DLS techniques are effective methods to quantify nanofluid stability.

2.5.3 Electrosteric stabilization

Electrosteric stabilization is a combination of electrostatic stabilization and steric stabilization. Both stabilization mechanisms contribute to nanofluid stabilization when two modified particles are applied in one solution. Electrosteric stabilizers are often ionic polymers with the anchor part and stabilize ionic polymeric chain that can provide functions of electrostatic and steric stabilization. The commonly used electrosteric stabilizers are polystyrene sulfonic acid, polyvinyl sulfonic acid, poly (methacrylic acid), etc. The ionizable group is dissociated to form a charged polymer. Polyelectrolytes, ionizable groups (carboxylic, sulfonic acid group, etc.), are often used for electrosteric stabilization. To firmly attach polymer onto the surface of nanofluids, nanoparticles and polyelectrolyte need to have opposite charges. Polyampholytes oppose both cationic and anionic groups that offer electrosteric stabilization. Ionic liquids can obtain electrosteric stabilization because of their strong chemisorption and high charge density. Tetrafluoroborate, pyridinium, hexafluorophosphate, and imidazolium are commonly used ionic liquids to stabilize nanoparticles [69].

Kong et al. [70] electrosterically prepared aqueous cement particles by adding nonionic polymer and strong anionic polyelectrolyte. The experimental results show that the particle distribution exhibits poor performance when the stabilizer is MFS. However, when stabilization exhibits superior when both MFS and HPMC are applied simultaneously. Zhang et al. [71] delivered hydrophobic Fe2O3 and TiO2 nanoparticles into solution by ligand exchange process under high-temperature conditions. Polyelectrolytes replaced the oleic acid ligands. This is because polyelectrolytes possess anchoring groups that support the attachment of polymer chains to nanoparticles. Additionally, activation energy can be transferred to stimulate the ligand exchange at high temperatures. The water-based nanofluids synthesized by the electrosteric method show good stability owing to electrostatic and steric stabilization. Moreover, for high concentration, Sen et al. [72] coated 50 wt.% TiO_2 nanofluids with 3-trihydroxysiyl-1-propenasulfonic acid. The surface-modified TiO2 nanofluids showed good dispersion stability and high fluidity.

2.6 Impact of nanofluid stability on thermophysical properties

The thermophysical properties of nanofluids are highly dependent on morphological and colloidal characterizations of nanoparticles. Characterizations of suspensions and colloids change with the degree of nanofluid stability. In other words, for large-scale nanofluid production, the impact of nanofluid stability is the main bottleneck on the thermophysical properties of nanofluids. In this section, different thermophysical properties are reviewed.

2.6.1 Effects of stability on density

The conventional expression of nanofluid density can be expressed in Eq. (2.9), where ρ_{nf}, ρ_0, ρ_p, and φ refer to density of nanofluids, base fluid, solid particles, and volume fraction, respectively.

$$\rho_{nf} = (1 - \varphi)\rho_0 + \varphi \rho_p \tag{2.9}$$

However, Eq. (2.9) cannot describe the influence of stability on density. There are limited pieces of literature investigating density based on stability. Halefadl et al. [73] reported that the density of nanofluids is dependent on surfactants in nanofluids. The density value increases with a greater quantity of surfactant in nanofluids. In their study, they studied the density of aqueous CNTs with SDBS as a stabilizer under a low concentration with temperatures of 20°C, 30°C, and 40°C, respectively. They compared the theoretical results with experimental results and proposed a modified density-based model that can explain the impact of surfactants. Based on their derived model, it is evident that existing theories regarding nanofluids density lack consideration of surfactants. Sandhu et al. [74] also verified the effect of surfactants on nanofluids density. In their research, they remained MWCNT−double distilled water nanofluids with no sign of agglomeration for 36 days when the ratio of water to the addition of EG reached 1:1. Their results also revealed that the nanofluids density increased with EG and reached the maximum value at the ratio of 1:1. There is no existing model that can express the density of nanofluids in the function of stability.

2.6.2 Effects of stability on viscosity

Viscosity indicates the rheological nanofluid characteristics. Currently, many models can express the concentrated-dependent viscosity of nanofluids. When particles are ideally dispersed at low particle concentrations, the Einstein equation [75] can predict the relative viscosity as follows,

$$\mu_r = 1 + 2.5\varphi \tag{2.10}$$

where μ_r is relative viscosity of nanofluids and φ is concentration. The Einstein model can apply to low volume concentration $\varphi < 1\%$. The Brickman model [76] is an extension of the Einstein model, which is used for moderate volume concentrations,

$$\mu_r = \frac{1}{(1-\varphi)^{2.5}} \tag{2.11}$$

When $\varphi \geq 1\%$, hydrodynamic interactions among nanoparticles become more and more important. The Batchelor model [77] takes Brownian motion and particles' interactions into consideration, which is suggested to be useful at $\varphi \sim 1\%$.

$$\mu_r = 1 + 2.5\varphi + 6.5\varphi^2 \tag{2.12}$$

Krieger and Dougherty [78] proposed a semiempirical relationship that covers the full range of particle volume fraction,

$$\mu_r = \left(1 - \frac{\varphi_a}{\varphi_m}\right)^{-2.5\varphi_m} \tag{2.13}$$

where φ_a is the volume fraction of aggregates, and φ_m is the volume fraction of densely packed particles. To take aggregations into consideration, Chen et al. [79] improved Krieger and Dougherty's model as,

$$\mu_r = \left(1 - \frac{\varphi_a}{\varphi_m}\right)^{-2.5\varphi_m}$$

with

$$\varphi_a = \varphi\left(\frac{r_a}{r_p}\right)^{3-d_f} \tag{2.14}$$

where r_a is the radius of aggregates, r_p is the radius of particles, and d_f is the fractal dimension of the aggregates which is around 1.8. Fig. 2.12 compares experimental data and calculated results by Eq. (2.13). The model fits best when $\frac{r_a}{r_p} = 3$. Prasher et al. [80] found that Eq. (2.13) can be reduced to a linear relationship, when φ_a is small,

$$\mu_r = 1 + 2.5\varphi_a \tag{2.15}$$

where φ_a is the slope. φ_a is equal to φ with ideally dispersed particles, which means the Krieger and Dougherty model is equal to the Einstein model.

Additionally, the experimental study in viscosity reveals more influencing factors. Nanofluids can exhibit either Newtonian or non-Newtonian behaviors. According to recent studies, different rheological behaviors depend on particle size, concentration, stabilizer, temperature, and shear rate. Irrespective of nanoparticle types, nanofluid viscosity increases with the increasing nanofluid concentration [31,37,81], and decreases with elevated temperature [81].

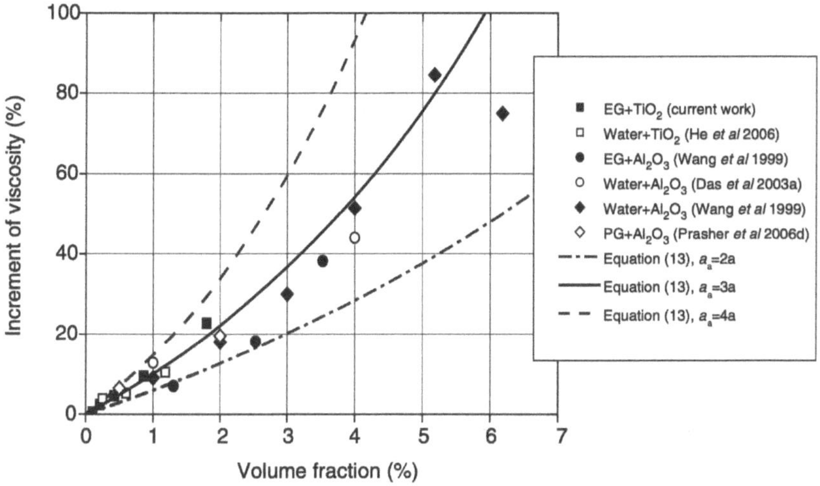

FIGURE 2.12 Comparison of experimental and numerical models of viscosities spherical particle nanofluids [79].

Furthermore, many investigations show that sonication plays an important role in nanofluid viscosity. Duan et al. [82] investigated the viscosity of CTAB (cetyltrimethylammonium ammonium bromide) modified water-based Al_2O_3 nanofluids at 1%–5% concentration. They evaluated nanofluid stability and homogeneous dispersion by ultrasonication. They studied the effect of shear rate and volume fraction on viscosity. Al_2O_3 nanofluids exhibited non-Newtonian behavior when measured after 2 weeks of synthesis, while nanofluids show Newtonian behavior after resonication. Haghighi et al. [83] studied the viscosity of distilled water-based CeO_2 (1–2 wt.%) in EG and alumina (9 wt.% fractions) nanofluids as a function of sonication at 14, 45, and 75 rpm, respectively. Their experimental results showed viscosity decreased with speeds due to their non-Newtonian behavior. On the contrary, some studies reported that water-based Al_2O_3 nanofluids exhibited good stability and kept almost unchanged viscosity with shear rate and irrelevant to time. The Newtonian-like behavior makes alumina–water nanofluids promising in heat transfer applications.

Additionally, the relative viscosity data were compared with theoretical values. The increment of relative viscosity can be observed with the increasing nanoparticle loading, which is very different from theoretical predictions. They stated that such an abnormal difference is attributed to the aggregation properties of nanofluids. To make the theoretical values fit relative viscosity data, Gaganpreet et al. [84] considered particle size as an important influencing factor when analyzing the relative viscosity of aqueous TiO_2 and Al_2O_3 nanofluids. Moreover, a novel viscosity model was derived by the *K-D* equation, which explains the aggregate size of nanoparticles. Based on this model, the viscosity of nanofluids increases rapidly with the increasing aggregate size. This abnormal increment is impartial to the volume fraction of nanoparticles.

2.6.3 Effects of stability on specific heat capacity of nanofluids

Generally, there are two commonly used models to describe φ-dependent C_{nf}. Similar to the mixing theory, the first model [85] can be calculated as,

$$C_{nf} = \varphi C_p + (1 - \varphi)C_{bl} \tag{2.16}$$

where subscripts $nf, p, \text{and} bl$ refer to the nanofluid, nanoparticle, and base liquid, respectively. According to the classical and statistical mechanisms, the second model is called the thermal equilibrium model [86], which can be calculated as,

$$C_{nf} = \frac{\varphi \rho_n C_p + (1 - \varphi)\rho_{bl} C_{bl}}{\varphi \rho_n + (1 - \varphi)\rho_{bl}} \tag{2.17}$$

where ρ_n is the density of particles and ρ_{bl} is the base liquid. Eq. (2.17) considers the effect of density, which is more fit than Eq. (2.16).

There are still limited scientific reports available regarding the influence of stability on specific heat capacity. Experimental measurements show that nanofluids contain lower specific heat than that of base liquids. Sabiha et al. [87] investigated specific heat capacity of SWCNT with SDS as a surfactant at 0.05–0.25 vol.% fractions as shown in Fig. 2.13A. The results showed that stable SWCNT nanofluids possessed lower heat capacity than water. Heat capacity decreases with the increasing volume fractions as well as elevated temperature. However, their experimental results are very different from theoretical results (Fig. 2.13B). The studies on the mechanism of heat capacity of nanofluids are still limited. More systematic investigations are expected in specific heat capacity of nanofluids.

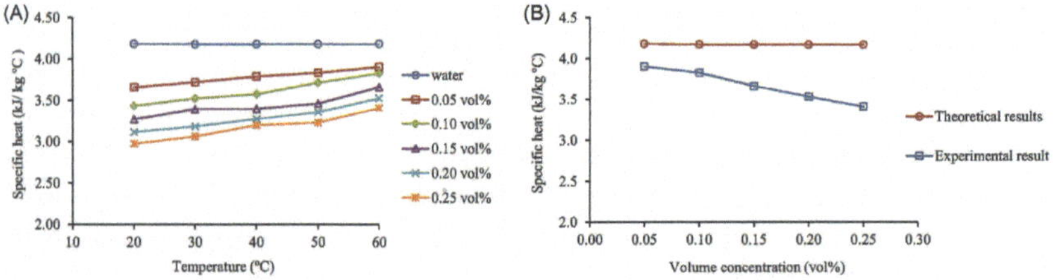

FIGURE 2.13 (A) Effect of temperature on specific heat capacity at different concentrations. (B) Comparison of theoretical and experimental results [87].

2.6.4 Effects of stability on thermal conductivity

Thermal conductivity enhancement is in great demand for the thermophysical properties of nanofluids. In terms of numerical models, most models are derived from the Hamilton–Crosser model. Nanofluid, as a two-phase liquid, is expected to have common features with solid–liquid mixtures. Therefore the Hamilton–Crosser model is described as,

$$k_n = k_{bl}\left[\frac{k_p + (n-1)k_{bl} + (n-1)\varphi(k_p - k_{bl})}{k_p + (n-1)k_{bl} - \varphi(k_p - k_{bl})}\right] \tag{2.18}$$

where subscripts nf, bl, and P stand for nanofluid, base liquid, and particle; n is an empirical shape factor. When particles are spherical, n equals 3, which is the same as the Maxwell model,

$$k_{nf} = k_{bl}\left[\frac{k_p + 2k_{bl} + 2\varphi(k_p - k_{bl})}{k_p + 2k_{bl} - \varphi(k_p - k_{bl})}\right] \tag{2.19}$$

The Maxwell model is suggested to be satisfactory for spherical particles at low concentrations. Bruggeman et al. [88] proposed a model for a full range of particle concentrations.

$$\varphi\left(\frac{k_p - k_r}{k_p + k_r}\right) + (1-\varphi)\left(\frac{k_{bf} - k_r}{k_{bf} + 2k_{nf}}\right) = 0 \tag{2.20}$$

where k_r is the ratio of k_{nf} to k_{bl}. It is noticeable that at low volume fractions, the Maxwell model and Bruggeman model generate similar results [89].

Lots of evidence demonstrate the disagreement between the experimental data and values calculated by the above models. Studies have shown that thermal conductivity is highly dependent on nanofluid stability [90]. In other words, the underlying mechanism of stability may be an important influencing factor for the thermal conductivity enhancement of nanofluids. Generally, most investigations have shown that even with a small volume fraction, stable nanofluids are higher than destabilized nanofluids in thermal conductivity. Researchers [14] have reported the effect of stabilizers and stabilization techniques on thermal conductivity. However, experimental results from different groups showed the different relationships between stable conditions and thermal conductivity. Additionally, the potential effect of nanoparticle aggregation on thermal conductivity showed that chemicals play an important role in thermal conductivity enhancement.

Dispersant/surfactant can not only stabilize nanoparticles but also enhance thermal conductivity. Yu et al. [91] studied the stability of paraffin-based Cu nanofluid. Compared to samples without dispersants, they found dispersed samples exhibit more stability than samples without dispersants. On the other hand, researchers discussed the optimum degree of aggregation for thermal conductivity enhancement. Prasher et al. [80] and Wen et al. [14] found that the degree of aggregation contradicts the concept of homogeneous dispersion. Additionally, Wen et al. [14] reported a relationship between the degree of aggregation and thermal conductivity, which can be divided into three phases (Fig. 2.14). There is a maximum thermal conductivity at an optimized aggregation rather than at a well-dispersed condition.

Additionally, Prasher et al. [80] further analyzed the potential mechanisms behind the relationship between aggregation degree and thermal conductivity. Compared to homogenization concepts of colloids, they suggested that the radius of gyration of clusters along with fractal dimensions are the main influencing factors for the accurate evaluation of thermal conductivity enhancement. Moreover, they found that there were two main kinds of nanoparticle clusters, including the linear chain and the side chain (Fig. 2.15). Linear chains are the main part to accelerate heat conduction. Similarly, Timofeeva et al.'s work [42] supports Prashant et al.'s idea [80]. They found that aggregation could increase thermal conductivity before a larger aggregate caused sedimentation. Therefore it is essential to keep a rate of clustering that can be controlled by stabilization techniques as introduced in Section 2.4, including ultrasonication, stabilizer addition, pH control, etc.

Furthermore, Bordi et al. found a similar phenomenon in electrical conductivity. They investigated the influence of aggregation kinetics on the electrical conductivity of nanofluids. Aggregation is considered as a time-dependent phenomenon. Although nanofluids initially exhibit well-dispersed behavior, they will start forming aggregations that can improve thermal conductivity. In this process, aggregated particles are regarded as "new particles" with an effective radius of Ra that improves the conductivity of nanofluid. However, excessive aggregation decreases this enhancement. Li et al. [92] also found the three-level phenomenon. They attempt to control the stability by the combination of pH and surfactant as shown in Fig. 2.16.

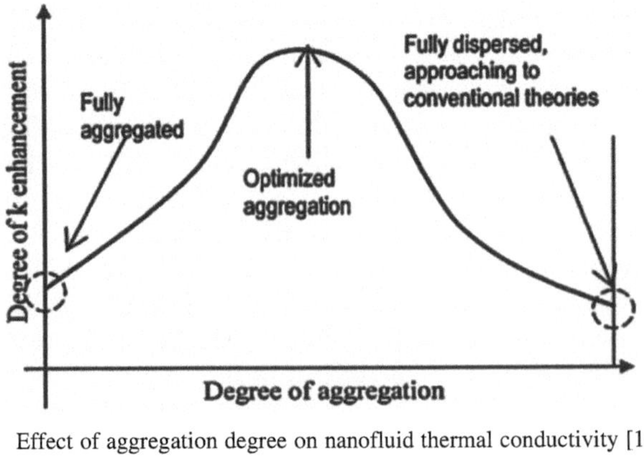

FIGURE 2.14 Effect of aggregation degree on nanofluid thermal conductivity [14].

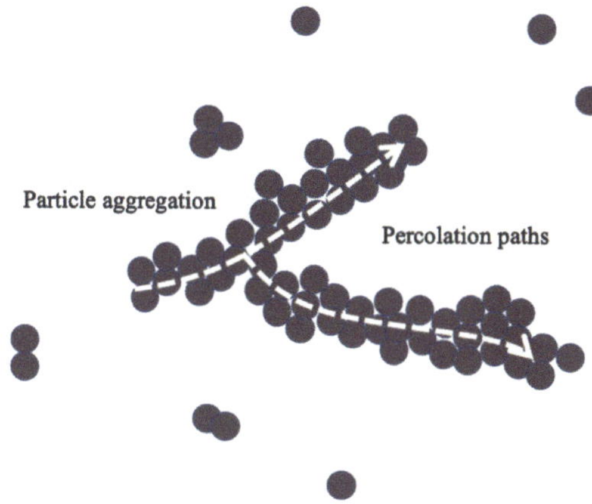

FIGURE 2.15 schematic diagram of nanoparticle aggregate structures (linear chains and side chains).

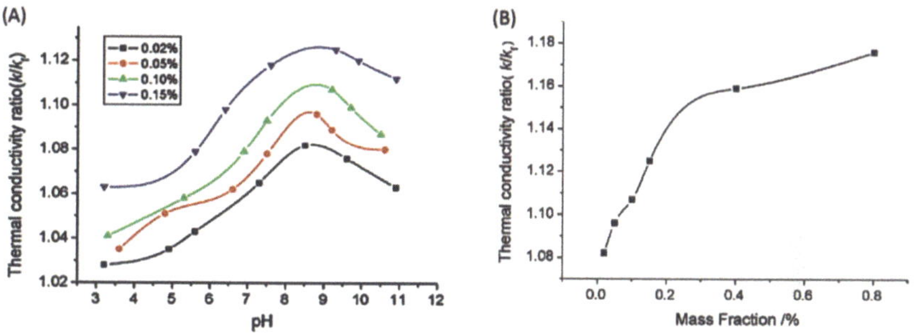

FIGURE 2.16 (A) Effect of pH values on thermal conductivity of water-based Cu nanofluids at different concentrations. (B) Effect of mass fractions on thermal conductivity of water-based Cu nanofluids [92].

For pH values, they controlled pH at 3−9.5. When the pH value is away from IEP, hydration around nanoparticles improves the mobility of nanoparticles that leads to the increasing microconvection. Therefore increasing zeta potential makes better stability. The addition of SDBS becomes the maximum thermal conductivity at a pH between 8.5 and 9.5. However, the thermal conductivity decreases when the pH value is beyond 9.5. Therefore they advise the optimal combination of surfactant and pH values for both stability and heat transfer.

Generally, there are two main potential mechanisms for the thermal conductivity enhancement of nanofluids. First, Brownian motion is responsible for microconvection that leads to thermal conductivity enhancement. Second, aggregation condition is the other important factor for thermal conductivity enhancement.

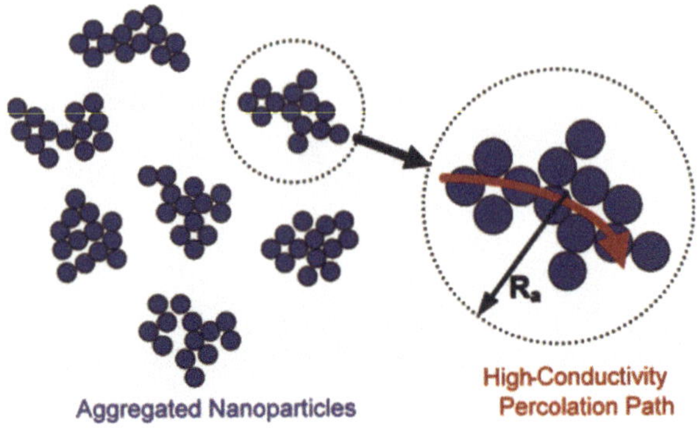

FIGURE 2.17 Schematic diagram of well-dispersed aggregates that are characterized by the radius of gyration (Ra) [95].

Furthermore, some works [93,94] demonstrate the relationship between Brownian motion and aggregation. As shown in Fig. 2.17, the decreasing particle size shortens the interaction distance between nanoparticles, which makes the attractive van der Waals force play a dominant role. In terms of aggregation, although the cluster hinders Brownian motion, the linear chain-like structure makes a shortcut for heat convection as well. Current existing models can indicate the aggregation model and aggregation kinetics model, respectively. However, there is no compatible model that includes two factors simultaneously.

2.7 Conclusion

Nanofluid stability is a prerequisite for improving heat transfer performance in thermal applications. The effectiveness of nanofluids reduces over time due to their instability. This chapter reviewed synthesis methods, stability evaluation methods, stabilization techniques, relevant mechanisms, effects of thermophysical properties on nanofluid stability, etc. For the preparation, the two-step method is the mainstream technique due to the economy and simplicity, but stability is the main problem. For the evaluation of nanofluid stability, zeta potential and spectral analysis are quantitative methods that can evaluate the degree of stability. In terms of the stabilization technique, physical methods can break large aggregations, while chemical methods can stable nanofluids by controlling the interaction between particles. For the mechanisms of stability, electrostatic stabilization and steric stabilization are the main potential mechanisms. Although mechanisms of stability are still controversial, researchers prefer to use electrosteric stabilization, which is a combination of electrostatic stabilization and steric stabilization. Study of the effects of stability on thermophysical properties is still limited. Two topics need to be paid more attention. First, pH values and concentrations of nanofluids are potential factors that affect the rheology property of nanofluids. Second, moderate aggregation can increase the thermal conductivity of nanofluids due to microconduction enhancement. Overall, stability needs more systematic investigations for the development of nanofluids in heat transfer. We hope this chapter can highlight nanofluid stability and the challenges related to industrial feasibility in wider applications.

References

[1] N. Kumar, S.S. Sonawane, Experimental study of Fe2O3/water and Fe2O3/ethylene glycol nanofluid heat transfer enhancement in a shell and tube heat exchanger, Int. Commun. Heat. Mass. Transf. 78 (2016) 277–284.

[2] Y. Xuan, Q. Li, Heat transfer enhancement of nanofluids, Int. J. Heat. Fluid Flow. 21 (1) (2000) 58–64.

[3] Z. Liu, et al., Enhancement of solar energy collection with magnetic nanofluids, Therm. Sci. Eng. Prog. 8 (2018) 130–135.

[4] Q. Li, Y. Xuan, Convective heat transfer and flow characteristics of Cu-water nanofluid, Sci. China Ser. E: Technol. Sci. 45 (4) (2002) 408–416.

[5] W. Yu, et al., Enhancement of thermal conductivity of kerosene-based Fe3O4 nanofluids prepared via phase-transfer method, Colloids Surf. A: Physicochem. Eng. Asp. 355 (1–3) (2010) 109–113.

[6] A. Pouranfard, D. Mowla, F. Esmaeilzadeh, An experimental study of drag reduction by nanofluids through horizontal pipe turbulent flow of a Newtonian liquid, J. Ind. Eng. Chem. 20 (2) (2014) 633–637.

[7] R. Fu, et al., Experimental investigation of turbulent forced heat transfer of Fe3O4 ethylene glycol–water nanofluid with highly disaggregated particles, Therm. Sci. Eng. Prog. 10 (2019) 1–9.

[8] R. Fu, et al., Experimental investigation of the influence of particle disaggregation on shear thinning of Fe3O4 ethylene glycol-water nanofluid, J. Nanofluids 7 (4) (2018) 613–619.

[9] M.E. Nakhchi, J.A. Esfahani, Cu-water nanofluid flow and heat transfer in a heat exchanger tube equipped with cross-cut twisted tape, Powder Technol. 339 (2018) 985–994.

[10] K.C. Lin, A. Violi, Natural convection heat transfer of nanofluids in a vertical cavity: effects of non-uniform particle diameter and temperature on thermal conductivity, Int. J. Heat. Fluid Flow. 31 (2) (2010) 236–245.

[11] S. Lee, et al., Measuring thermal conductivity of fluids containing oxide nanoparticles, Journal of Heat Transfer 121 (2) (1999) 280–289.

[12] J.A. Eastman, et al., Enhanced Thermal Conductivity through the Development of Nanofluids, Argonne National Lab., IL, 1996.

[13] K.-Y. Kwak, C.-Y. Kim, Viscosity and thermal conductivity of copper oxide nanofluid dispersed in ethylene glycol, Korea-Aust. Rheol. J. 17 (2) (2005) 35–40.

[14] D. Wen, et al., Review of nanofluids for heat transfer applications, Particuology 7 (2) (2009) 141–150.

[15] M. Chandrasekar, S. Suresh, A.C. Bose, Experimental investigations and theoretical determination of thermal conductivity and viscosity of Al2O3/water nanofluid, Exp. Therm. Fluid Sci. 34 (2) (2010) 210–216.

[16] H. Akoh, et al., Magnetic properties of ferromagnetic ultrafine particles prepared by vacuum evaporation on running oil substrate, J. Cryst. Growth 45 (1978) 495–500.

[17] C.-H. Lo, et al., Fabrication of copper oxide nanofluid using submerged arc nanoparticle synthesis system (SANSS), J. Nanopart. Res. 7 (2) (2005) 313–320.

[18] H.-t Zhu, Y.-s Lin, Y.-s Yin, A novel one-step chemical method for preparation of copper nanofluids, J. Colloid Interface Sci. 277 (1) (2004) 100–103.

[19] S. Ayyappan, et al., Nanoparticles of Ag, Au, Pd, and Cu produced by alcohol reduction of the salts, J. Mater. Res. 12 (2) (1997) 398–401.

[20] W. Yu, et al., Highly efficient method for preparing homogeneous and stable colloids containing graphene oxide, Nanoscale Res. Lett. 6 (1) (2011) 47.

[21] B. Munkhbayar, et al., Surfactant-free dispersion of silver nanoparticles into MWCNT-aqueous nanofluids prepared by one-step technique and their thermal characteristics, Ceram. Int. 39 (6) (2013) 6415–6425.

[22] H.J. Kim, I.C. Bang, J. Onoe, Characteristic stability of bare Au-water nanofluids fabricated by pulsed laser ablation in liquids, Opt. Lasers Eng. 47 (5) (2009) 532–538.

[23] Y.-K. Leong, B. Ong, Critical zeta potential and the Hamaker constant of oxides in water, Powder Technol. 134 (3) (2003) 249–254.

[24] X. Li, D. Zhu, X. Wang, Evaluation on dispersion behavior of the aqueous copper nanosuspensions, J. Colloid Interface Sci. 310 (2) (2007) 456–463.

[25] W. Li, C. Zou, Experimental investigation of stability and thermo-physical properties of functionalized β-CD-TiO2-Ag nanofluids for antifreeze, Powder Technol. 340 (2018) 290–298.

[26] V. Nikkhah, M. Sarafraz, F. Hormozi, Application of spherical copper oxide (II) water nano-fluid as a potential coolant in a boiling annular heat exchanger, Chem. Biochem. Eng. Q. 29 (3) (2015) 405–415.

[27] H. Zhu, et al., Preparation and thermal conductivity of suspensions of graphite nanoparticles, Carbon 45 (1) (2007) 226–228.

[28] S. Khaleduzzaman, et al., Stability of Al2O3-water nanofluid for electronics cooling system, Procedia Eng. 105 (2015) 406–411.

[29] A.K. Singh, V.S. Raykar, Microwave synthesis of silver nanofluids with polyvinylpyrrolidone (PVP) and their transport properties, Colloid Polym. Sci. 286 (14) (2008) 1667–1673.

[30] D.F. Swinehart, The beer-lambert law, J. Chem. Educ. 39 (7) (1962) 333.

[31] S. Chakraborty, et al., Synthesis of Cu-Al LDH nanofluid and its application in spray cooling heat transfer of a hot steel plate, Powder Technol. 335 (2018) 285–300.

[32] L. Jiang, L. Gao, J. Sun, Production of aqueous colloidal dispersions of carbon nanotubes, J. Colloid Interface Sci. 260 (1) (2003) 89–94.

[33] X. Yang, Z.-h Liu, A kind of nanofluid consisting of surface-functionalized nanoparticles, Nanoscale Res. Lett. 5 (8) (2010) 1324–1328.

[34] N. Karthikeyan, J. Philip, B. Raj, Effect of clustering on the thermal conductivity of nanofluids, Mater. Chem. Phys. 109 (1) (2008) 50–55.

[35] D.-W. Oh, et al., Thermal conductivity measurement and sedimentation detection of aluminum oxide nanofluids by using the 3ω method, Int. J. Heat. Fluid Flow. 29 (5) (2008) 1456–1461.

[36] S. Chakraborty, et al., Effect of surfactant on thermo-physical properties and spray cooling heat transfer performance of Cu-Zn-Al LDH nanofluid, Appl. Clay Sci. 168 (2019) 43–55.

[37] M. Kole, T. Dey, Effect of aggregation on the viscosity of copper oxide–gear oil nanofluids, Int. J. Therm. Sci. 50 (9) (2011) 1741–1747.
[38] R. Sadeghi, et al., Investigation of alumina nanofluid stability by UV–vis spectrum, Microfluid. Nanofluid. 18 (5) (2015) 1023–1030.
[39] R. Sadeghi, et al., Investigation of alumina nanofluid stability by UV-vis spectrum, Microfluid. Nanofluid. 18 (5–6) (2015) 1023–1030.
[40] T.P. Rieker, S. Misono, F. Ehrburger-Dolle, Small-angle x-ray scattering from carbon blacks: crossover between the fractal and Porod regimes, Langmuir 15 (4) (1999) 914–917.
[41] A.O.C. Gomez, A.R.K. Hoffmann, E.P. Bandarra, Experimental evaluation of CNT nanofluids in single-phase flow, Int. J. Heat. Mass. Transf. 86 (2015) 277–287.
[42] E.V. Timofeeva, et al., Thermal conductivity and particle agglomeration in alumina nanofluids: experiment and theory, Phys. Rev. E 76 (6) (2007) 061203.
[43] A. Mishra, S. Ram, G. Ghosh, Dynamic light scattering and optical absorption in biological nanofluids of gold nanoparticles in poly(vinyl pyrrolidone) molecules, J. Phys. Chem. C. 113 (17) (2009) 6976–6982.
[44] M.S. Saterlie, et al., Surfactant effects on dispersion characteristics of copper-based nanofluids: a dynamic light scattering study, Chem. Mater. 24 (17) (2012) 3299–3306.
[45] R. Kamatchi, S. Venkatachalapathy, B.A. Srinivas, Synthesis, stability, transport properties, and surface wettability of reduced graphene oxide/water nanofluids, Int. J. Therm. Sci. 97 (2015) 17–25.
[46] N. Zhao, et al., Experimental study on influences of cylindrical grooves on thermal efficiency, exergy efficiency and entropy generation of CPU cooled by nanofluids, Int. J. Heat. Mass. Transf. 135 (2019) 16–32.
[47] S.E.B. Maïga, et al., Heat transfer enhancement by using nanofluids in forced convection flows, Int. J. Heat. Fluid Flow. 26 (4) (2005) 530–546.
[48] Y.Y. Hu, H.J. Liu, H. Fujii, Improving the mechanical properties of 2219-T6 aluminum alloy joints by ultrasonic vibrations during friction stir welding, J. Mater. Process. Technol. 271 (2019) 75–84.
[49] H.-J. Chen, D. Wen, Ultrasonic-aided fabrication of gold nanofluids, Nanoscale Res. Lett. 6 (1) (2011) 1–8.
[50] A. Amrollahi, A. Hamidi, A. Rashidi, The effects of temperature, volume fraction and vibration time on the thermo-physical properties of a carbon nanotube suspension (carbon nanofluid), Nanotechnology 19 (31) (2008) 315701.
[51] A. Shahsavar, et al., An experimental study on the effect of ultrasonication on thermal conductivity of ferrofluid loaded with carbon nanotubes, Thermochim. Acta 617 (2015) 102–110.
[52] J. Boopathy, et al., Preparation of nano fluids by mechanical method, AIP Conference Proceedings, American Institute of Physics, 2012.
[53] K.Y. Leong, et al., Performance investigation of an automotive car radiator operated with nanofluid-based coolants (nanofluid as a coolant in a radiator), Appl. Therm. Eng. 30 (17–18) (2010) 2685–2692.
[54] A.H.A. Al-Waeli, et al., Evaluation and analysis of nanofluid and surfactant impact on photovoltaic-thermal systems, Case Stud. Therm. Eng. 13 (2019).
[55] W. Yu, H. Xie, A review on nanofluids: preparation, stability mechanisms, and applications, J. Nanomat. 2012 (2012) 1–17.
[56] Y.-j Hwang, et al., Stability and thermal conductivity characteristics of nanofluids, Thermochim. Acta 455 (1–2) (2007) 70–74.

[57] E. Tang, G. Cheng, X. Ma, Preparation of nano-ZnO/PMMA composite particles via grafting of the copolymer onto the surface of zinc oxide nanoparticles, Powder Technol. 161 (3) (2006) 209–214.
[58] H.Q. Xie, et al., Nanofluids containing multiwalled carbon nanotubes and their enhanced thermal conductivities, J. Appl. Phys. 94 (8) (2003) 4967–4971.
[59] N. Bizmark, M.A. Ioannidis, Effects of ionic strength on the colloidal stability and interfacial assembly of hydrophobic ethyl cellulose nanoparticles, Langmuir 31 (34) (2015) 9282–9289.
[60] S.U. Ilyas, R. Pendyala, N. Marneni, Stability of nanofluids, Engineering Applications of Nanotechnology, Springer, 2017, pp. 1–31.
[61] A.R.I. Ali, B. Salam, A review on nanofluid: preparation, stability, thermophysical properties, heat transfer characteristics and application, SN Appl. Sci. 2 (10) (2020).
[62] M. Khairul, et al., Effects of surfactant on stability and thermo-physical properties of metal oxide nanofluids, Int. J. Heat. Mass. Transf. 98 (2016) 778–787.
[63] H. Wang, et al., The preparation of Cu-coated Al2O3 composite powders by electroless plating, Ceram. Int. 37 (7) (2011) 2181–2184.
[64] D. Li, et al., Preparation of well-dispersed silver nanoparticles for oil-based nanofluids, Ind. Eng. Chem. Res. 49 (4) (2010) 1697–1702.
[65] B.W. Ninham, On progress in forces since the DLVO theory, Adv. Colloid Interface Sci. 83 (1–3) (1999) 1–17.
[66] C. Rodriguez-Abreu, Nanocolloids: some basic concepts and principles of their stabilization, Nanocolloids (2016) 1.
[67] E.M. Hotze, T. Phenrat, G.V. Lowry, Nanoparticle aggregation: challenges to understanding transport and reactivity in the environment, J. Environ. Qual. 39 (6) (2010) 1909–1924.
[68] H. Vatanparast, et al., The role of electrostatic repulsion on increasing surface activity of anionic surfactants in the presence of hydrophilic silica nanoparticles, Sci. Rep. 8 (1) (2018) 1–11.
[69] E. Piacenza, A. Presentato, R.J. Turner, Stability of biogenic metal (loid) nanomaterials related to the colloidal stabilization theory of chemical nanostructures, Crit. Rev. Biotechnol. 38 (8) (2018) 1137–1156.
[70] H.J. Kong, S.G. Bike, V.C. Li, Electrosteric stabilization of concentrated cement suspensions imparted by a strong anionic polyelectrolyte and a non-ionic polymer, Cem. Concr. Res. 36 (5) (2006) 842–850.
[71] R. Zhang, et al., Hierarchical Fe2O3 nanorods/TiO2 nanosheets heterostructure: growth mechanism, enhanced visible-light photocatalytic and photoelectrochemical performances, Appl. Surf. Sci. 475 (2019) 380–388.
[72] S. Sen, et al., Surface modification approach to TiO2 nanofluids with high particle concentration, low viscosity, and electrochemical activity, ACS Appl. Mater. Interfaces 7 (37) (2015) 20538–20547.
[73] S. Halelfadl, T. Maré, P. Estellé, Efficiency of carbon nanotubes water based nanofluids as coolants, Exp. Therm. Fluid Sci. 53 (2014) 104–110.
[74] H. Sandhu, D. Gangacharyulu, An experimental study on stability and some thermophysical properties of multiwalled carbon nanotubes with water–ethylene glycol mixtures, Part. Sci. Technol. 35 (5) (2017) 547–554.
[75] A. Einstein, Eine neue bestimmung der moleküldimensionen, ETH Zurich, 1905.
[76] H. Brinkman, The viscosity of concentrated suspensions and solutions, J. Chem. Phys. 20 (4) (1952) 571. 571.

[77] G. Batchelor, The effect of Brownian motion on the bulk stress in a suspension of spherical particles, J. Fluid Mech. 83 (1) (1977) 97–117.
[78] I.M. Krieger, T.J. Dougherty, A mechanism for non-Newtonian flow in suspensions of rigid spheres, Trans. Soc. Rheol. 3 (1) (1959) 137–152.
[79] H. Chen, Y. Ding, C. Tan, Rheological behaviour of nanofluids, N. J. Phys. 9 (10) (2007) 367.
[80] R. Prasher, et al., Effect of aggregation on thermal conduction in colloidal nanofluids, Appl. Phys. Lett. 89 (14) (2006).
[81] A.K. Tiwari, P. Ghosh, J. Sarkar, Heat transfer and pressure drop characteristics of CeO2/water nanofluid in plate heat exchanger, Appl. Therm. Eng. 57 (1–2) (2013) 24–32.
[82] F. Duan, D. Kwek, A. Crivoi, Viscosity affected by nanoparticle aggregation in Al2O 3-water nanofluids, Nanoscale Res. Lett. 6 (1) (2011) 248.
[83] E. Haghighi, et al., Shelf stability of nanofluids and its effect on thermal conductivity and viscosity, Meas. Sci. Technol. 24 (10) (2013) 105301.
[84] G.M. Moldoveanu, et al., Experimental study on viscosity of stabilized Al2O3, TiO2 nanofluids and their hybrid, Thermochim. Acta 659 (2018) 203–212.
[85] B.C. Pak, Y.I. Cho, Hydrodynamic and heat transfer study of dispersed fluids with submicron metallic oxide particles, Exp. Heat. Transf. Int. J. 11 (2) (1998) 151–170.
[86] I.M. Shahrul, et al., A comparative review on the specific heat of nanofluids for energy perspective, Renew. Sustain. Energy Rev. 38 (2014) 88–98.
[87] M. Sabiha, et al., Experimental investigation on thermo physical properties of single walled carbon nanotube nanofluids, Int. J. Heat. Mass. Transf. 93 (2016) 862–871.
[88] D. Bruggeman, Berechnung verschiedener physikalischer Konstanten von heterogenen Substanzen. II. Dielektrizitätskonstanten und Leitfähigkeiten von Vielkristallen der nichtregulären Systeme, Annal. Phys. 417 (7) (1936) 645–672.
[89] K. Khanafer, K. Vafai, A critical synthesis of thermophysical characteristics of nanofluids, Int. J. Heat. Mass. Transf. 54 (19–20) (2011) 4410–4428.
[90] A. Nasiri, et al., Effect of CNT structures on thermal conductivity and stability of nanofluid, Int. J. Heat. Mass. Transf. 55 (5–6) (2012) 1529–1535.
[91] W. Yu, H.Q. Xie, X.P. Wang, Enhanced thermal conductivity of liquid paraffin based nanofluids containing copper nanoparticles, J. Dispers. Sci. Technol. 32 (7) (2011) 948–951.
[92] X.F. Li, et al., Thermal conductivity enhancement dependent pH and chemical surfactant for Cu-H2O nanofluids, Thermochim. Acta 469 (1–2) (2008) 98–103.
[93] J. Hong, D. Kim, Effects of aggregation on the thermal conductivity of alumina/water nanofluids, Thermochim. Acta 542 (2012) 28–32.
[94] J.L. Baldwin, B.A. Dempsey, Effects of Brownian motion and structured water on aggregation of charged particles, Colloids Surf. A-Physicochem. Eng. Asp. 177 (2–3) (2001) 111–122.
[95] R. Prasher, P.E. Phelan, P. Bhattacharya, Effect of aggregation kinetics on the thermal conductivity of nanoscale colloidal solutions (nanofluid), Nano Lett. 6 (7) (2006) 1529–1534.

Chapter 3

Synthesis, characterization, and measurement techniques for the thermophysical properties of nanofluids

Arun Kumar Tiwari[1], Amit Kumar[1] and Zafar Said[2]

[1]Mechanical Engineering Department, Institute of Engineering & Technology, Dr. A.P.J. Abdul Kalam Technical University, Lucknow, Uttar Pradesh, India, [2]Department of Sustainable and Renewable Energy Engineering, University of Sharjah, Sharjah, United Arab Emirates

3.1 Introduction

Nanofluids are the colloidal suspension of nanosize particles (i.e., Ag, CuO, Al_2O_3, ZnO, BN, CNTs, SiC, MgO, TiO_2, Co_3O_4, CeO_2, etc.) into the base fluid (i.e., water, ethylene glycol, acetone, palm oil, vegetable oil, etc.). Due to recent advancements in technology in different industries like air conditioning, heating, automobiles, thermal power plants, etc., intensifying heat generation demands a heat transfer medium with a greater heat transfer rate than the conventional one. Thus the foremost objective of the synthesis and characterization of nanofluids is to attain a better heat transfer rate by enhancing the thermophysical properties of nanofluids [1–4]. Different synthesis, characterization, and measurement techniques have been used to prepare, analyze, and measure the thermophysical properties of nanofluids.

3.2 Synthesis of nanofluid

The stability and thermophysical properties of nanofluids play a great role in choosing the suitable synthesis method. The most common synthesis methods of nanofluid are the one-step and two-step method (classified in Fig. 3.1).

Advances in Nanofluid Heat Transfer. DOI: https://doi.org/10.1016/B978-0-323-88656-7.00012-X

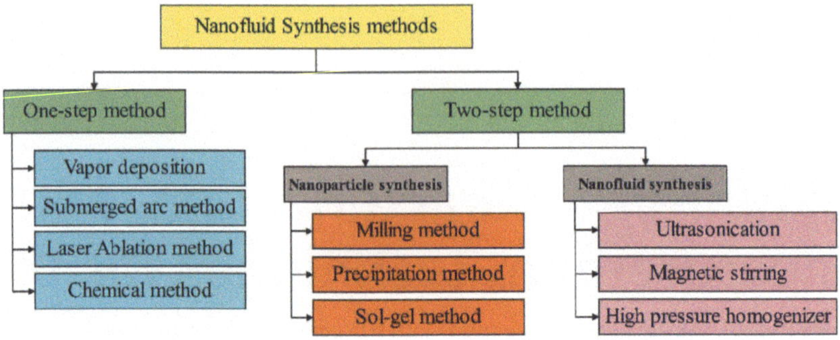

FIGURE 3.1 Classification of nanofluid synthesis method.

3.2.1 One-step method

It is a process of simultaneously manufacturing and dispersing nanosized particles into the base fluid [5]. In this method, different processes like drying, storage, transportations, and dispersion of nanoparticles are avoided to minimize the chances of nanoparticle agglomeration, hence increasing the stability of the nanofluids [6]. Synthesizing nanofluids using a one-step method can be performed either by physical methods (like direct evaporation and condensation techniques, submerged arc nanoparticle synthesis, pulsed wire evaporation, and laser ablation in liquids) or chemical methods [7–13]. The one-step method, being a slow process, can be used in a limited quantity of nanofluid production, and only low-pressure fluids can be synthesized [14].

3.2.1.1 Vapor deposition

The vapor deposition method was developed by Choi and Eastman [15] in 2001, and it is one of the most used one-step methods until now. In this method, a fine film of base fluid is generated against the inner surface of a container through centrifugal action by rotating a disk attached to a container. The crucible is heated and vaporized to lower pressure in this container. After the settling down of the evaporated material vapor into the thin film of the rotating base fluid, the nanofluid is achieved. This is an appropriate method for metallic nanofluids preparation. Due to the high boiling temperature of TiO_2, this process cannot be employed to synthesize the TiO_2 nanofluid. However, using an electric heater, high temperatures can be attained [16].

Pulsed wire evaporation is an advanced vapor deposition technique used to synthesize the nanofluids for high thermal characteristics. Lee et al. [17] used the pulsed wire evaporation technique to synthesize TiO_2/EG nanofluid. This apparatus has four main components: a DC power supply source of high voltage, a high voltage gap switch, a capacitor, and an evaporation or condensation chamber. A Ti (99.99% purity) wire of 100 mm length and

0.5 mm diameter is fed into the reaction chamber. A high-voltage pulse of 25 kV supplied through Ti wire evaporates it and converts the wire into plasma. This plasma is condensed into small size particles by cooling through the argon–oxygen gas mixture. These synthesized small particles are directly connected to the EG container to form TiO_2/EG nanofluid. Wire explosion numbers can control nanofluid volume concentration.

3.2.1.2 Submerged arc method

The synthesis process of nanofluid using the submerged arc method is demonstrated in Fig. 3.2. The system comprises an input heating source, a vacuum chamber, a parameter control module, an isothermal unit, and a pressure control unit. A desired pure metal bar (as an electrode) is immersed into the dielectric liquid tank of this system. A high-temperature electric arc of 6000°C to 12,000°C is generated through the heating source to melt the metals. The metal bar is melted and evaporated in the arc generation region. The nearby cooling region also vaporizes quickly due to the very high temperatures. The metal vapors expand their volume and produce an inertia force near to the dielectric liquid. The highly pressurized vapors of liquid can quickly remove the evaporated metal from the arc zone by immediate cooling to the limit excessive growth of particles. The nanoparticles developed from this technique are suspended in the base fluid uniformly and follow the three transformation stages, that is, nucleation, growth, and condensation. The dielectric liquid is kept at a lower steady temperature as the evaporated metal condenses immediately in the dielectric liquid with the submerged arc simultaneously. In the meantime, as the steady submerged arc is generated, the vacuum chamber pressure and temperature of the dielectric liquid is to be controlled at the required level such that the vaporized metal gas can be efficiently converted into particles at the nanometric scale through particle nucleation and grain growth. The vaporized metallic powders are then quickly quenched using a cooling system, resulting in nucleation and formation of crystalline nanoparticles. Thus the nanoparticles are prepared and dissolved uniformly in a dielectric liquid without aggregating unwanted particles.

Chung et al. [18] analyzed the submerged arc method's process parameters and prepared the Cu/water–EG nanofluids of 20 nm particle size. Chih-Hung et al. [19] synthesized the silver nanofluid of the average size of 6–25 nm and 0.5% volume concentration using this method (Fig. 3.2). Kao et al. [20] synthesized the Cu_2O/DOT3 brake oil and got the thermal conductivity 1.6 times greater than the base fluid.

3.2.1.3 Laser ablation method

In this method, the nanoparticles from the metal surface are suspended into the base fluid by directing a high-intensity laser beam over the metal

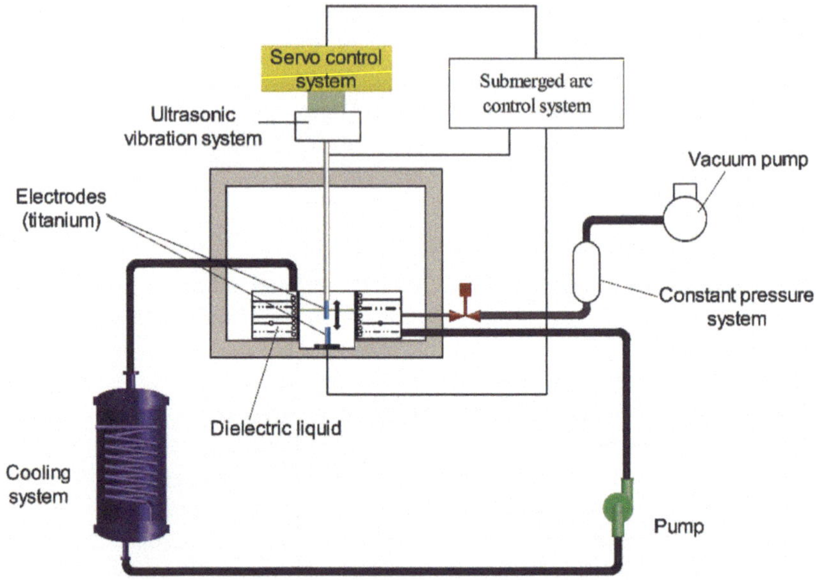

FIGURE 3.2 Schematic diagram of submerged arc method of nanofluid synthesis. Copyright Elsevier (LIC# 5100100522656) [19].

surface sink in the base fluid. Kim et al. [21] synthesized Au/water nanofluid using this process as shown in Fig. 3.3. A gold plate 3 mm thick and 10 mm in diameter is positioned at the base of a glass container having 15 mL pure water. An Nd:YAG laser was used to form the laser beam of 6 mm diameter, 0.8 mrad divergence, and 300 nm focal length using a biconvex lens for a time period of 1–18 hours. Khamliche et al. [22] synthesized Cu/EG nanofluid using this method. A copper plate of 3 mm thickness and 10 mm diameter with 99% purity is placed at the bottom of a glass container having 20 mL ethylene glycol. An Nd:YAG laser was used to produce the laser beam of 300 nm focal length using a convex lens over the copper plate. Riahi et al. [23] synthesized Al_2O_3/water nanofluid, while Phuoc et al. [10] synthesized Ag/deionized water nanofluid using this process.

3.2.1.4 Chemical method

The chemical methods for nanofluid synthesis are cheaper and faster compared to physical methods. So far, distinct chemical methods for nanofluid synthesis were used by the researchers. Some of those chemical methods are described below:

Garg et al. [24] synthesized the Cu/water nanofluid using a chemical reduction method. Initially, they dissolved 2 gm of copper acetate in 50 mL of DI water with continuous magnetic stirring. Further, 0.5 g NaOH was

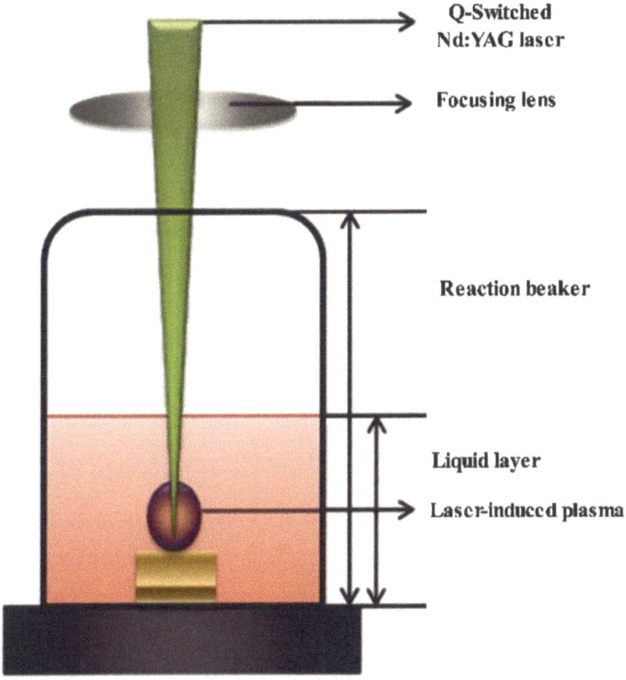

FIGURE 3.3 Schematic diagram of Laser ablation method. Copyright Elsevier (LIC# 5100100784976) [21].

dissolved into the solution and cooled up to 5°C temperature. After this, 3 mL of hydrazine was added as a drop-by-drop solution. To allow a complete reduction process, the solution was stirred for up to 12 hours. The Cu nanoparticles were separated by the centrifugal rotating of the solution. The separated Cu nanoparticles were then washed with acetone and dried at 50°C in vacuum conditions to remove the water. The Cu nanoparticles obtained from the above process are dispersed into EG in a glovebox under an argon environment.

Yu et al. [25] synthesized Cu/water and Cu/EG nanofluids with the chemical reduction method. For the nanofluid synthesis, a particular amount of polyvinyl pyrrolidone (PVP) and ascorbic acid were mixed in aqueous $CuSO_4$ (200 mL of 0.2 mmol/L) solution under continuous mechanical stirring at 80°C for a certain period. The colloid solution was subsequently removed from the oil bath and cooled to ambient temperature. To separate nanoparticles from the suspension, the colloidal was adulterated by ethanol, and the solution was centrifuged at 8000 rpm for 15 min. To eliminate surfactant, the particles were resuspended into ethanol and centrifuged three times. The precipitates were then vacuum dried for 12 hours to obtain the nanoparticles.

Wei et al. [26] used the chemical solution method for Cu_2O/water nanofluid synthesis. The following chemical reaction takes place during the synthesis process:

$$CuSO_4 + 2NaOH = Cu(OH)_2 + Na_2SO_4 \tag{3.1}$$

$$Cu(OH)_2 \triangleq CuO + \quad H_2O \tag{3.2}$$

$$4CuO + N_2H_4 = 2Cu_2O + N_2 + 2H_2O \tag{3.3}$$

Kumar et al. [27] synthesized Cu/EG nanofluid while using sodium hypophosphite as a reducing agent to obtain copper nanoparticles.

$$2\ NaH_2PO_2 + 3\ H_2O + CuSO_4 \rightarrow 2\ NaH_2PO_3 + H_2SO_4 + 2\ H_2 + Cu^0 \tag{3.4}$$

Zhu et al. [28] synthesized CuO nanofluid using the wet-chemical method. The following chemical reactions were used for nanoparticle synthesis:

$$Cu^{2+} + 2\ NaOH = Cu(OH)_2 + 2\ Na^+ \tag{3.5}$$

$$Cu(OH)_2 \triangleq CuO + H_2O \tag{3.6}$$

To synthesize CuO nanofluid a mixture of 600 mL 0.2 M copper acetate solution and 2 mL of glacial acetic acid is taken in a glass beaker and boiled with continuous magnetic stirring. Subsequently, 30 mL of 8 M NaOH was added to the beaker, and the color changed immediately from blue to black. The reaction was performed for 2 hours while stirring and boiling. The mixture is then centrifuged and cooled to ambient temperature to produce a wet CuO precipitate. To remove the impurities of the ions, the wet precipitate is then washed with DI water. CuO nanofluid of different volume concentrations has been obtained by redispersing the wet precipitate in different amounts of DI water under ultrasonic vibrations of 120 W and 40 Hz.

3.2.2 Two-step method

The two-step method is extensively implemented to synthesize nanofluid. Nanotubes, nanofiber, or nanoparticles are synthesized as a dry powder in the first step using different physical and chemical processes [29]. In the next step, these dry powders are then suspended into the base fluid using different methods like ultrasonic agitation, high-shear mixing, intensive magnetic force agitation, and homogenizing. Since nanoparticles are commercially available on an industrial scale, it is currently the most energy-efficient and economic nanofluid synthesis method. However, nanoparticles tend to aggregate due to their high surface area and surface activity [30].

3.2.2.1 Nanoparticles preparations

Two techniques are widely used to synthesized nanoparticles: a top-down and a bottom-up approach. The top-down approach refers to the mechanical methods of nanoparticle preparation, such as mechanical crushing using the milling process, while the bottom-up approach refers to the chemical methods of nanoparticles preparations [31].

3.2.2.1.1 Milling process

The mechanical method of nanoparticle synthesis uses milling to crush the microparticles into the nanoparticles. Metallic and ceramic nanomaterials are synthesized using high-energy ball mill crushers. These ball mill crushers are equipped with grinding devices composed of wolfram carbide or steel. Gou et al. [32] described the ball mill process to synthesize the nanotubes' powder (<100 nm particle size). In this process, the collision between the tiny rigid balls in a sealed container generates localized high pressure to form the very fine nanoparticles. Additional functional chemical groups are added to the surface of the nanotubes to further improve quality. Factors that affect the dispersion quality include milling time, rotation speed, ball size, and nanotube/ball quantity ratio.

3.2.2.1.2 Precipitation processes

The precipitation method for synthesizing nanoparticles is frequently used to produce metallic oxide and nonmetal oxide nanoparticles. The precipitation method involves the reaction of salts in solvents. A precipitating agent is added to produce the desired particle precipitation, which can be filtered out after postheat processing. In this method, particle shape, size, size distribution, and crystallinity can be determined by reaction speed.

Rajaeiyan et al. [33] synthesized alumina nanoparticles by precipitation method. This process dissolves aluminum trichloride ($AlCl_3$) into 150 mL ethanol and a small amount of DI water to obtain a crystal clear solution. After that 60 mL of NH_3 at a rate of 2.5 mL/min was gradually added to stirred $AlCl_3.6H_2O$ solution to obtain white Al^{3+} cations in the form of $Al(OH)_3$. The precipitation is then filtered in a vacuum system, dried in the oven for 2 hours to 200°C and annealed for 1 hour to 750,1000, and 1250°C. Thus a very fine alumina nanopowder (white color) is obtained.

3.2.2.1.3 Sol–gel method

It is a wet-chemical synthesis method to produce porous nonmaterial, ceramic, and oxide nanoparticles at a relatively low temperature. The term sol refers to the dispersion of small particles (<100 nm size) in water or organic solvent. Behnajady et al. [34] synthesized the TiO_2 nanoparticles using the sol–gel method shown in Fig. 3.4. The process consists of four steps. In an initial step, a titanium electrode is liquefied in the solvent. After that, the mixture is sonicated using an ultrasonic bath sonicator. The hydrolysis process is executed by the gradual addition of DI water into the mixture under constant magnetic stirring. The gel product achieved in the final stage

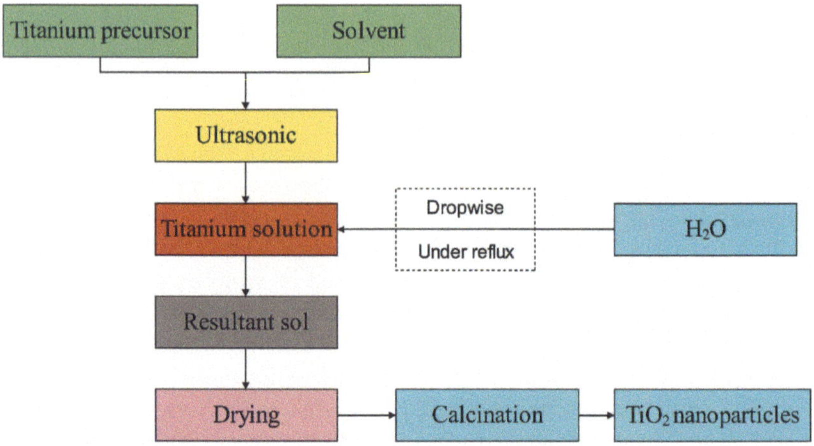

FIGURE 3.4 Schematic diagram of sol–gel synthesis method of TiO_2. Copyright Elsevier (LIC# 5100101081323) [34].

is then dried and calcinated to obtained a crystalline powder. The advantage of the sol–gel method is to regulate the reaction rate by regulating the hydrolysis and condensation by using alkoxides through chemical processes. Leena et al. [35] synthesized the TiO_2 nanoparticles using the sol–gel method. They took a mixture of titanium(IV) isopropoxide (TTIP), ethanol, and distilled water (DW) in the ratio of 1:4:1 and stirred it for 4 hours to obtain a white slurry solution. Nitric acid was used to maintain the optimum pH of the solution. The mixture was heated at 393.5K for 2 hours to evaporate the water and other organic impurities. At last, the dry powder was sintered at 723.15K for 2 hours to get TiO_2 nanoparticles.

3.2.2.2 Nanofluid preparation

Nanofluid preparation comprises the suspension and dispersion of nanoparticles (already synthesized) into the base fluid. The presence of a strong van der Waals force between the nanoparticles and the base liquid makes it very difficult to obtain the homogeneous and stable nanofluid solution. So far, distinct methodologies have been utilized by the researchers to get a stable and homogeneous nanofluid solution. Ultrasonication, magnetic stirring, and high-pressure homogenizer are the most commonly used techniques. To control the volume concentration of nanofluid Tiwari et al. [36] used the following relations

$$\varnothing = \frac{\frac{m_n}{\rho_n}}{\frac{m_n}{\rho_n} + \frac{m_{bf}}{\rho_{bf}}} \tag{3.7}$$

where m and ρ are denoted as the mass and density while subscripts n and bf are used for nanoparticles and base fluid, respectively.

3.2.2.2.1 Ultrasonication method

In the ultrasonication method, the ultrasonic sound wave frequency (> 20 kHz) is applied to the solution for homogeneous dispersion of nanoparticles into the base fluid as it decreases the cluster formation of nanoparticles by disturbing the intermolecular force. Based on the working principle, the ultrasonication process is categorized into two types, that is, direct sonication and indirect sonication. In direct sonication, the ultrasonic tip or probe is directly submerged into the mixture. The ultrasonic bath type is an indirect ultrasonication method. A sample is filled in a container in this method, and the container is immersed in the liquid. The ultrasound waves are transferred to the liquid, which is then transmitted to the sample. The indirect ultrasonic method is not suitable for dispersing dry powder into the high viscosity base fluid. So the direct ultrasonic method is the most effective nanofluid synthesis method. So far, different ultrasonication processes have been utilized by the researchers. Begum et al. [37] prepared the ZnO/DW nanofluid using a direct ultrasonicator (Hielscher UP200Ht) at different operating amplitudes (40% and 100%) and ultrasonication time (0.5, 1, 2, 3, 4 minutes). Chawhan et al. [38] prepared the Ag-doped TiO_2/water nanofluid using the ultrasonication method for better stability.

3.2.2.2.2 Magnetic stirring

In this method, the stirring action is formed with a stirrer. This stir bar is immersed into the mixture and rotates the mixture at a very high speed due to the influence of a rotating magnetic field. This magnetic field is generated by either a rotating electromagnetic assembly or a stationary electromagnetic assembly. Some of the researchers used this technique in combination with the ultrasonication method, while others used it separately [39]. Table 3.1 further summarizes the synthesis and characterization techniques of different nanofluids.

3.3 Characterization of nanofluid

Characterization of nanofluid is the analysis of materials, structure, shape, size, composition, and physical and chemical properties. There are two famous characterization techniques, that is, microscopy method and the spectroscopy method.

3.3.1 Zeta potential

It is the potential difference between the fluid layer absorbed on the nanoparticle surface and bulk phase of liquid [53]. It is a measurement technique to quantify the stability of nanoparticle dispersion in the base fluid or we can say zeta potential is an indicator of the stability of the nanoparticles

TABLE 3.1 Summary of synthesis method and characterization techniques of different nanofluids.

Nanofluid	Method	Technique	Particle size (nm)	Volume concentration (vol%)	Characterizing techniques	Ref.
CuO/DI water	One-step	Submerged arc method	49.1	–	TEM, FESM, XRD, SAED, ESCA	[8]
ZnO/EG	One-step	Pulsed wire evaporation	<100	0.5, 1, 2, 3, 4, 5.5	XRD, TEM	[40]
Cu/water	One-step	Chemical method	–	–	XRD, UV–Vis, FTIR	[27]
Ag/water	One-step	Chemical method	–	–	XRD, UV–Vis, TEM	[41]
Cu-engine oil	One-step	Electrical explosion of wire	–	0.2, 0.5, 1	Zeta potential, TEM	[42]
TiO_2/ethanol	One-step	Laser ablation	–	–	TEM, UV–Vis, FS	[43]
Cu_2O/water	One-step	Chemical solution method	–	–	SEM	[26]
Cu/EG	One-step	Dual plasma method	20–30	–	TEM, SEM, UV–Vis	[7]
TiO_2/poly (vinylpyrrolidone)	One-step	Laser ablation	35	–	XRD, TEM, XPS, UV–Vis	[44]
TiO_2/DI water	One-step	Laser ablation	3	–	XRD, TEM, UV–Vis	[45]

ZnO/water	Two-step	Ultrasonication	20	0.02, 0.04, 0.06, 0.08, 0.1 wt.%	–	[46]
CuO/water, EG, engine oil	Two-step	Magnetic stirrer + ultrasonication	–	–	UV–Vis, SEM, TEM, XRD, DLS	[47]
MWCNT/water	Two-step	Ultrasonication + magnetic stirrer	10–40	0.005, 0.01, 0.05	–	[48]
CeO_2/water	Two-step	Ultrasonic homogenizer	25	0.015, 0.025, 0.035	Zeta potential	[49]
ZnO/water	Two-step	Ultrasonication	–	0.025, 0.05, 0.075, 0.1	XRD, FTIR, FESEM, EDX, UV–Vis	[50]
rGO-Al_2O_3/DI water	Two-step	Ultrasonication	–	0.01–0.1	UV–Vis, TEM, XRD, FTIR	[51]
TiO_2/diathermic oil	Two-step	Magnetic stirrer + ultrasonic homogenizer	10	0.01–0.1	Zeta potential	[52]

suspension in a colloidal solution [54]. This technique determines the surface charge of nanoparticles in the colloidal solution. The charged particles of the surface attract oppositely charged particles to develop a thin film called the Stern layer. The continuous diffused particle in the solution generates an outer layer called the electrical double layer. Zeta potential finds the voltage at the edge of the double-layer [55]. The zeta potential generally varies between −60 and +60 mV. Tiwari et al. [56] measured the zeta potential value for CeO_2-MWCNT/water nanofluid at different surfactant mixtures and different sonication times. Based on the result obtained, they concluded that the zeta potential increases significantly with an increase in the quantity of surfactant into the mixture. However, this increase in zeta potential value is limited to a certain ratio of surfactant to the nanofluid mixture. Further increase in this ratio leads to a decrease in the zeta potential value and hence the stability (as shown in Fig. 3.5). Also, the zeta potential value depends upon the sonication time of the nanofluid. Fig. 3.6 shows that the zeta potential value increases up to a certain sonication time for different kinds of surfactants and decreases with a further increase in sonication time.

3.3.2 Fourier-transform infrared spectroscopy

FTIR system is utilized to collect infrared spectrum data of absorption or transmission. This technique uses the Fourier transformation method to

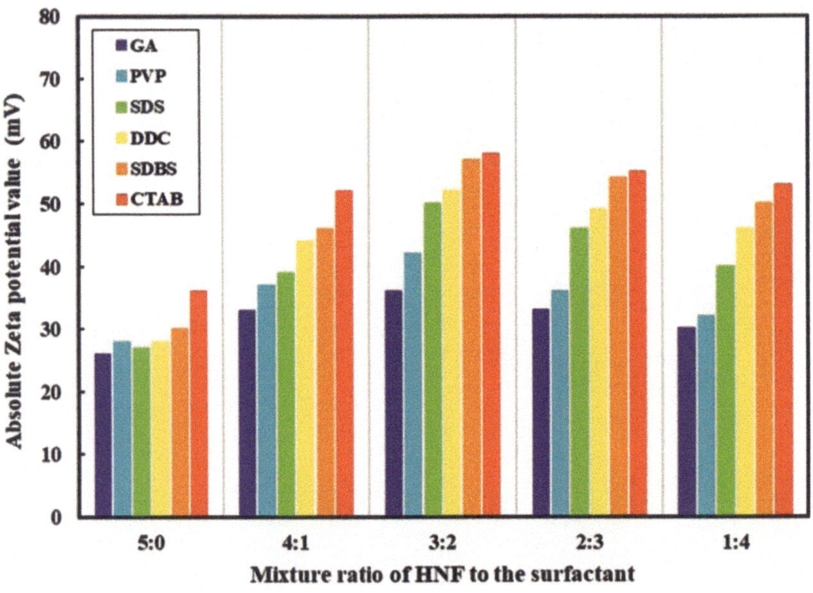

FIGURE 3.5 Effect of mixture ratio of volume concentration to the surfactant on the stability of CeO_2-MWCNT/water. Copyright Elsevier (LIC# 5100101310148) [56].

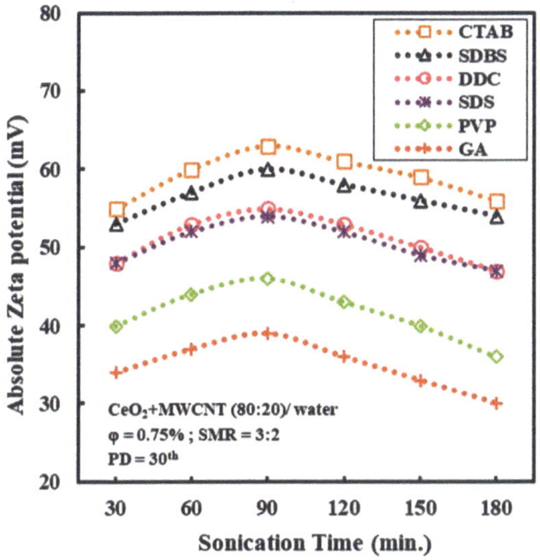

FIGURE 3.6 Effect of Sonication Time on the Absolute Zeta potential of CeO_2-MWCNT/water. Copyright Elsevier (LIC# 5100101485469) [56].

convert the raw data into the actual spectrum. FTIR spectroscopy recognizes the existence of inorganic and organic compounds in the sample [57]. Generally, FTIR spectroscopy collects the spectrum of 4000 to 400 cm^{-1} resolution [58]. Fig. 3.7 shows the schematic representation of FTIR spectroscopy. An FTIR spectroscopy contains the following major components, that is, IR source, beam splitter, fixed and moving mirror, sample cell, detector, decoder and amplifier, and A/D converter. The IR source reaching the detector is transferred to the interferometer. The A/D converter converts the interferometer signal into the digital signal, and the same is transferred to a computer for Fourier transformation [59]. Oliveira et al. [60] performed FTIR analysis to specify the presence of –OH and –COOH chemical groups. It was also used to identify the bond of Ag on the surface of nanodiamond and Di-Ag hybrid nanoparticles. Said et al. [61] characterized the rGO-Co_3O_4 nanofluid using FTIR spectroscopy and found several functional groups. It can be seen in Fig. 3.8 that the two functional groups of rGO are identified as C = C and C = O at a wavenumber of 1623 cm^{-1} and 1726 cm^{-1}, which are related to the formation of –COOH on the rGO nanosheet.

3.3.3 Transmission electron microscopy

TEM is a characterization technique used to measure the nanoparticle shape, size, distribution, and morphology. It is a high-resolution imaging technique that provides the crystallographic image of the sample at the atomic scale.

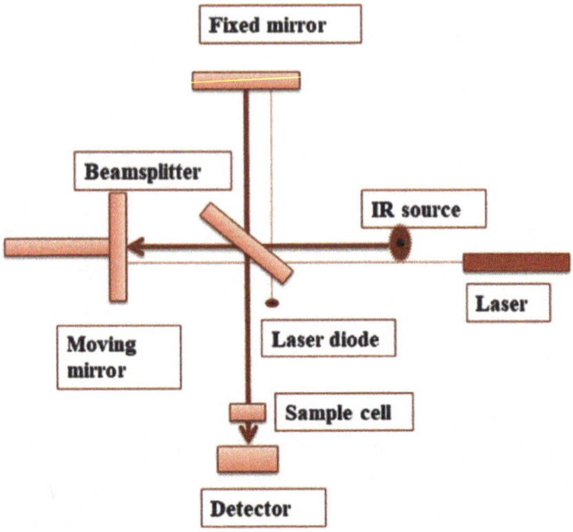

FIGURE 3.7 Schematic diagram of FTIR spectroscopy working principle. Copyright Elsevier (LIC# 5100110251636) [59].

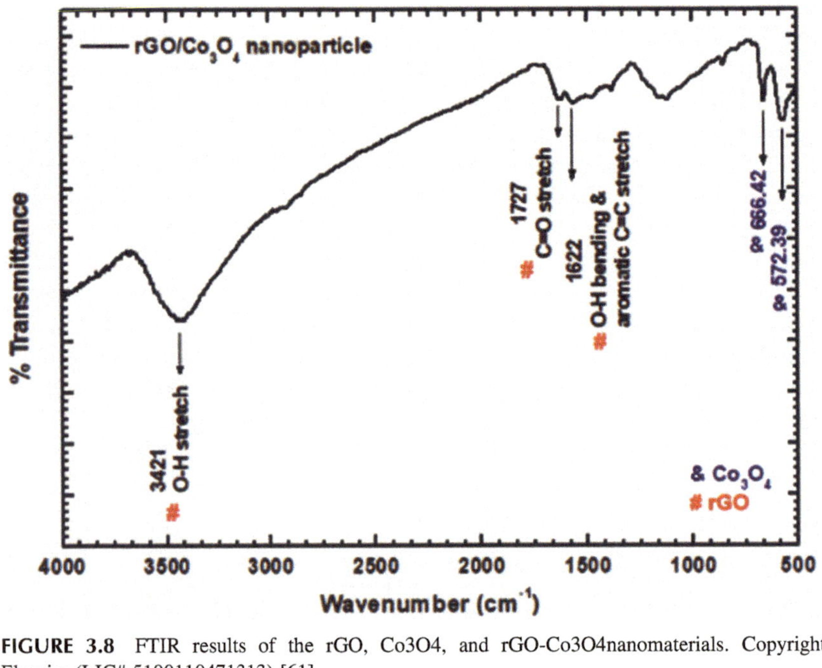

FIGURE 3.8 FTIR results of the rGO, Co3O4, and rGO-Co3O4nanomaterials. Copyright Elsevier (LIC# 5100110471313) [61].

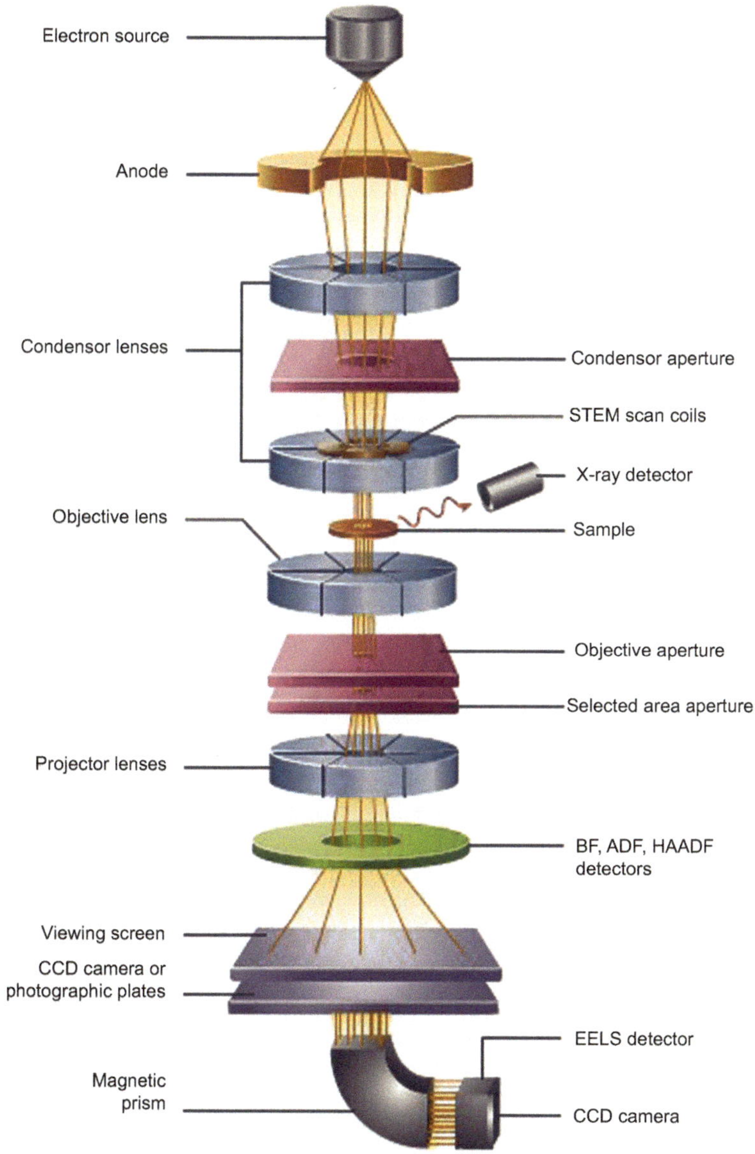

FIGURE 3.9 Schematic representation of TEM. Copyright Elsevier (LIC# 5100110644575) [62].

TEM (Fig. 3.9) uses a particle beam of the electrons to visualize the nanofluid sample for generating highly magnified images. In TEM microscopic technique, a beam of electrons is transmitted to the nanofluid sample. The beam of electrons transmitted through the sample generates an image while interacting with the sample. This image is then magnified and focused on an imaging device like a fluorescent screen or detected by a sensor like a charged couple device camera. Salehi et al. [41] investigated the shape and size of silver nanofluid using the TEM technique. Hung et al. [8] characterized the CuO using the TEM technique and found that the CuO nanoparticles prepared by the one-step method were of 20 nm average width and 80 nm length.

3.3.4 X-ray crystallography

X-ray crystallography (XRD) is a simple, adjustable, and nondestructive method to identify crystalline materials. The XRD method is also used to determine the element proportion in a mixture sample. Along with this crystallinity, deviation of a particular element from its ideal.

composition and its structural state can also be derived from the data analysis. A part of the X-ray beam incident upon the atomic planes gets transmitted while the remaining part is reflected, absorbed, scattered, and diffracted by the sample.

The XRD principle is based upon Bragg's law

$$n\lambda = 2d \sin \theta \tag{3.8}$$

where λ is denoted as the wavelength of X-ray, n is the order of diffraction, d is the distance between the sequential parallel plane of a crystal, and θ is the complementary angle of the X-ray incidence.

Oliveira et al. [60] analyzed the diffraction pattern of the Ag sample through the XRD technique (shown in Fig. 3.10). They observed four diffraction peaks at the following coordinates (38.14°, 111), (44.2°, 200), (64.5°, 220), (77.4°, 311) at 2θ versus intensity graph of Ag sample. The corresponding peaks of nanodiamond and Ag are considered as the peak for hybrid Di-Ag structures.

3.3.5 UV–Vis technique

UV–Vis spectroscopy is a commonly used characterization technique of nanofluids to determine their properties. It measures nanoparticles' extinction (scattering and absorption) properties by passing the light through the sample. The nanoparticle possesses exclusive optical properties that are sensitive to the size, shape, concentration, agglomeration, and refractive index close to the surface of the nanoparticles. This makes UV–Vis spectroscopy a vital tool for identifying, characterizing, and studying nanomaterials [63]. UV–Vis spectroscopy is generally applicable for the UV (185–400 nm) and visible (400–700 nm) range of the electromagnetic spectrum. Metallic nanoparticles like iron and selenium are generally characterized by the UV–Vis method as their nanofluid mixture changes color after the synthesis [64].

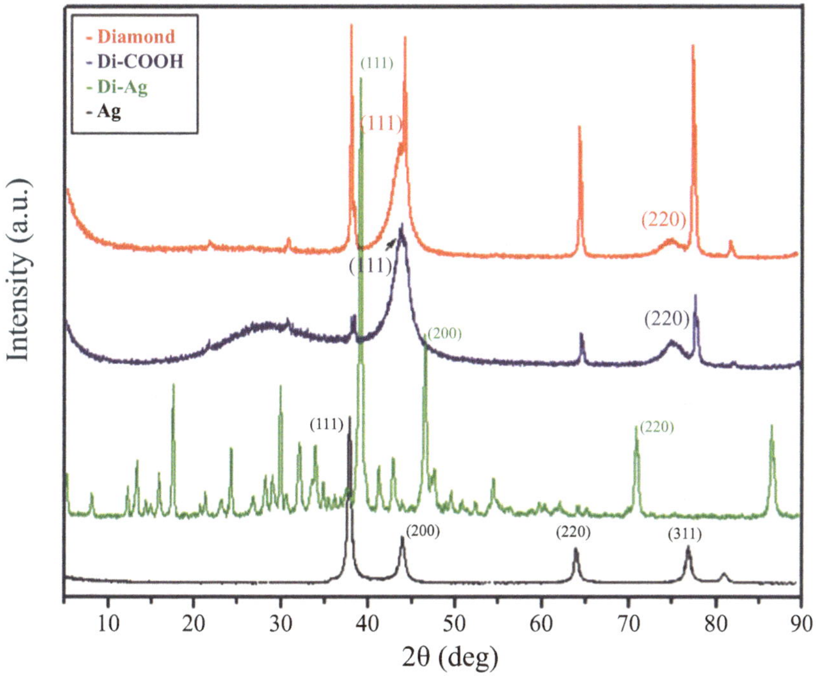

FIGURE 3.10 XDR pattern of nanodiamond, Di-COOH, Di-Ag and Ag nanocomposites. Copyright Elsevier (LIC# 5100110828855) [60].

Kimpton et al. [65] characterized the silver nanofluid using UV–Vis spectroscopy. Agarwal et al. [47] characterized the CuO nanofluid by UV–Vis spectroscopy and recorded the absorption spectrum of copper nanoparticles. Zeroual et al. [12] characterized the Ag/EG nanofluid using UV–Vis spectroscopy. UV3100 Selecta spectrophotometer was used for UV–Visible analysis in a spectra range from 300–600 nm. Fig. 3.11 shows that a sharper surface plasmon surface (SPR) was observed for the lowest copolymer concentration, which means that the highest copolymer concentrations induce larger Ag nanoparticles with a broader distribution size; hence copolymer favors the nanoparticle formation.

3.4 Thermophysical properties measurement techniques of nanofluid

The thermophysical properties have a significant influence on nanofluid's heat transfer phenomena. Various experimental studies have been accomplished to enhance thermophysical properties, including viscosity, thermal conductivity, density, and specific heat. So far, many good review studies

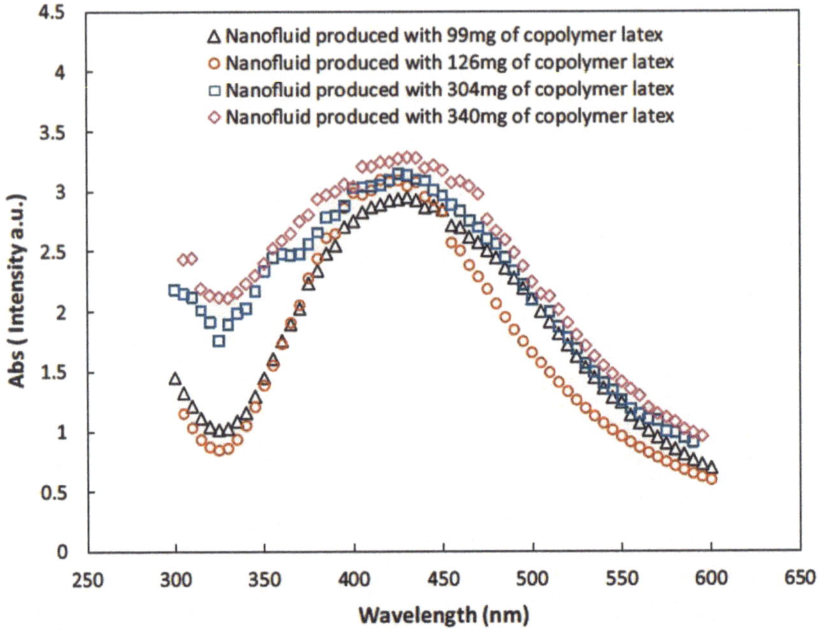

FIGURE 3.11 UV–Vis spectra of EGS-based nanofluids for different initial copolymer latex concentrations. Copyright Elsevier (LIC# 5100110988279) [12].

[66–71] have been published on the enhancement and measurement techniques of thermophysical properties of nanofluids.

3.4.1 Thermal conductivity

The thermal conductivity of a nanofluid is known as its transport property. It transports heat energy by conduction in the nanofluids [72]. Different techniques exist for measuring the thermal conductivity of nanofluids, that is, analyzing thermal constants, transient hot-wire (THW) method, cylindrical cell method, temperature oscillation technique, and 3ω method.

3.4.1.1 Transient hot-wire method

THW method is an extensively applied method for thermal conductivity measurement Fig. 3.12 [73]. The following assumptions are to be considered to calculate thermal conductivity using an ideal model.

- Infinite long hot wire placed in isotropic fluid vertically.
- The hot wire is considered to be of infinite thermal conductivity and zero specific heat.

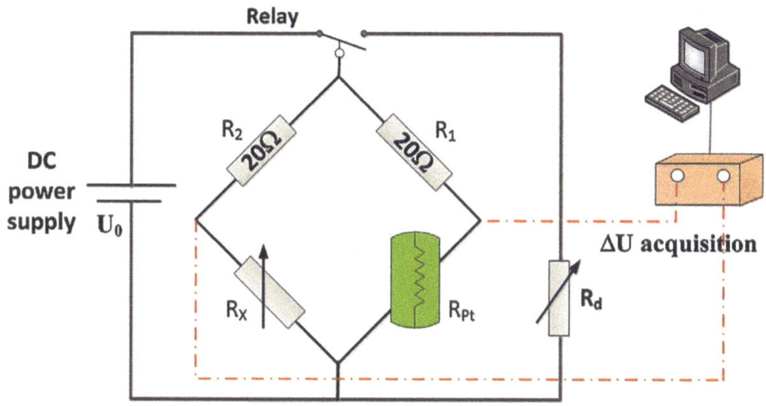

FIGURE 3.12 The schematic diagram of the transient hot-wire method. Copyright Elsevier (LIC# 5100111454806) [73].

As a first step, the system is held in thermodynamic balance at T_0. Next, there is a steady flow of heat in the hot wire.

Then the temperature rise at a radial position r can be written as:

$$\Delta T(r,t) = T(r,t) - T_0 \tag{3.9}$$

The Fourier heat conduction equation is given as:

$$\frac{\partial \Delta T}{\partial r} = \alpha \frac{1}{r}\frac{\partial}{\partial r}\left(r\frac{\partial \Delta T}{\partial r}\right) \tag{3.10}$$

The following initial and boundary conditions as to be followed:

$$T(r,t) = T_0, \quad t = 0 \tag{3.11}$$

$$\lim_{r\to\infty} r\frac{\partial T}{\partial r} = -\frac{q}{2\pi K} \quad t > 0 \tag{3.12}$$

$$\lim_{r\to\infty} T(r,t) = T_0 \quad t \geq 0 \tag{3.13}$$

The thermophysical properties are to be assumed constant for the fluid then Eq. (3.9) and Eq. (3.10) can be given as:

$$T(r,t) = -\frac{q}{4\pi K}E_1\left(\frac{r^2}{4at}\right) \tag{3.14}$$

where $E(x)$ demonstrates the exponential integral function such that the temperature distribution of the hot wire can be given as:

$$T(r,t) - T_0 = \frac{q}{4\pi K}\left[\ln\left(\frac{4at}{r_0^2 C}\right) + \left(\frac{r_0^2}{4at}\right) + \ldots\right] \tag{3.15}$$

Consider $\frac{r_0^2}{4at} \rightarrow 0$ and higher-order terms of Eq. (3.15) can be neglected and it can be written as:

$$\Delta T(r,t) = \frac{q}{4\pi K}\ln\left(\frac{4at}{r_0^2 C}\right) \tag{3.16}$$

Then the nanofluid thermal conductivity can be written as:

$$K = \frac{q}{4\pi}\frac{d\ln t}{d\Delta T(t)} \tag{3.17}$$

3.4.1.2 3ω method

This is another popular method for thermal conductivity measurement similar to the hot wire method. In this method, a sinusoidal current with an angular frequency of ω passes through a thin-layer metallic heater (acts as a thermometer) patterned on the solid substrate of interest. Amplifying the input current frequency twice, during the sinusoidal heating, the temperature of the metal also oscillates at twice the angular frequency (2ω) [74]. As the metal heater's electrical resistance is a linear function of temperature, the temperature oscillation can be measured indirectly by measuring the associated 3ω voltage through the metallic heater. The amplitude and phase of the temperature oscillation can be utilized to measure the thermal conductivity of nanofluid [74]. The following relation of temperature oscillation and heat generation rate measured the thermal conductivity [75].

$$\Delta T = \frac{P}{l\pi k}\int_0^{\infty}\frac{\sin^2 kb}{(kb)^2(k^2+q^2)^{1/2}}dk \tag{3.18}$$

where q is denoted as the magnitude of heat transmission, and p/l is denoted as the amplitude of the power per unit length generated at a frequency of 2ω.

3.4.1.3 Thermal constants analyzer technique

The transient plane source (TPS) method is utilized for the measurement of nanofluids thermal conductivity using the thermal constant analyzer technique. The TPS method uses the Fourier heat conduction law as a fundamental principle to measure thermal conductivity, as does the transient hot-wire (THW) method [76]. This technique has the following advantages:

- Fast thermal conductivity measurement and no need for sample preparation.
- Can be utilized for the thermal conductivity measurement of an extensive range of nanofluid samples (0.02 to 200 W/m-K).

The setup consists of a vessel, thermometer, a constant temperature bath, and a thermal constant analyzer [77]. The thermal constant analyzer probe is submerged vertically in the container containing the nanofluid. The container is placed in a constant temperature bath, and the thermometer is dipped into the container to measure the temperature of the nanofluid. Nanofluid thermal conductivity is calculated by measuring probe resistance. Li et al. [78] used the following correlation to calculate the thermal conductivity.

$$\Delta T(\tau) = \frac{1}{\alpha}\left[\frac{R_p(\tau)}{R_0} - 1\right] \tag{3.19}$$

where α is denoted as the temperature coefficient of the electric resistance, R_0 is denoted as electric resistance at $\tau = 0$, and τ is denoted as electrification time.

$$\tau = \sqrt{\frac{tR}{r_p^2}} \tag{3.20}$$

where t is denoted as the measuring time, R is denoted as nanofluid thermal diffusivity, and r_p is denoted as probe radius.

3.4.2 Viscosity of nanofluids

The viscosity of nanofluids primarily depends on temperature and volume concentration. It is one of the important transport properties that determine the rheological properties of the nanofluid [70,79–81]. It can be categorized into two types, that is, dynamic viscosity and kinematic viscosity. The rheological properties of the nanofluid should be assessed to examine whether the colloidal suspension renders the nanofluid Newtonian or non-Newtonian, while measuring the viscosity of the nanofluids. The following different types of viscometers are used to measure the viscosity of nanofluids.

- Capillary viscometer.
- Rotational viscometer
- Oscillating viscometer
- Brookfield viscometer.

Kazemi et al. [82] used Brookfield DV2EXTRA-Pro rotational viscometer to measure the viscosity of SiO_2 nanofluid. Akilu et al. [83] used a rotational rheometer (Physica MCR 302, Anton Paar) to measure the EG-based TiO_2–CuO nanofluid. Different researchers developed their theoretical correlations to calculate the viscosity of nanofluids (as shown in Table 3.2). Tiwari et al. [93] used LVDV-II Brookfield digital viscometer to calculate the viscosity of CeO_2 + MWCNT hybrid nanofluid. Anish et al. [94] used the Fenske bath apparatus for viscosity measurement of ZnO/Theminol nanofluid. Yiamsawas et al. [95] used a capillary tube viscometer for the viscosity

TABLE 3.2 Viscosity correlations of nanofluid.

Nanofluid	Viscosity correlations	Ref.
Ag-MgO/water	$\frac{\mu_{nf}}{\mu_f} = 1 + 32.795\varnothing - 7214\varnothing^2 + 71400\varnothing^3 - 0.1941 \times 10^8\varnothing^4$	[84]
Al_2O_3-CuO-TiO_2/water	$\mu_{nf} = \mu_{bf}\left(0.955 - 0.00271T + 1.858\frac{\varnothing}{100} + \left(705\frac{\varnothing}{100}\right)^{1.223}\right)$	[85]
MgO-MWCNT/EG	$\frac{\mu_{nf}}{\mu_{bf}} = \left[0.91\varnothing + 0.240\left(T^{-0.342}\varnothing^{-0.473}\right)\right]\exp\left(1.45T^{0.12}\varnothing^{0.158}\right)$	[86]
MWCNT-ZnO/engine oil	$\mu_{nf} = 796.8 + 76.26\varnothing + 12.88T + 0.7695\varnothing T + \frac{-196.9T - 16.54\varnothing T}{\sqrt{T}}$	[87]
SiC-TiO_2/diathermic oil	$\frac{\mu_{nf}}{\mu_f} = 0.312\varnothing + 1.3194$	[88]
SiO_2-oleic acid/liquid paraffin	$\frac{\mu_{nf}}{\mu_{bf}} = 6.8376 + 15.2522w + 0.038779 - 2.63029w^2$, where w is the mass fraction (wt.%)	[89]
SiO_2-TiO_2/water	$\frac{\mu_{nf}}{\mu_f} = 1.42(1+R)^{-0.1063}\left(\frac{T}{80}\right)^{0.2321}$	[90]
TiO_2-Cu/water	$\mu_{nf} = \frac{\mu_f}{(1-\varnothing_1)^{2.5} + (1-\varnothing_2)^{2.5}}$	[91]
TiO_2-CuO/EG	$\mu_{nf} = 0.9653 + 77.4567\left[\frac{\varnothing}{100}\right]^{1.1558}\left[\frac{T}{333}\right]^{0.6881}$	[83]
TiO_2-SiO_2/water	$\frac{\mu_{nf}}{\mu_f} = 37\left(0.1 + \frac{\varnothing}{100}\right)^{1.59}\left(0.1 + \frac{T}{80}\right)^{0.31}$	[92]

measurement of TiO_2/EG–water and Al_2O_3/EG–water nanofluid and used the following viscosity correlations:

$$\mu_{nf} = A\phi^B T^c \mu_f^D \tag{3.21}$$

where A, B, C, and D are constant and μ_f is the viscosity of the base fluid and defined as

$$\mu_f = 0.0003T^2 - 0.0461T + 2.3775 \tag{3.22}$$

Azmi et al. [96] proposed the following model to calculate the viscosity of different water-based nanofluids like Al_2O_3, CuO, SiO_2, ZnO, and TiO_2.

$$\frac{\mu_{nf}}{\mu_{bf}} = C_1\left(1+\frac{\phi}{100}\right)^a\left(1+\frac{T_{nf}}{100}\right)^{-\lambda}\left(1+\frac{d_p}{170}\right)^{-\sigma} \tag{3.23}$$

Hemmat et al. [97] proposed the following viscosity model to calculate the viscosity of Fe/water nanofluid:

$$\frac{\mu_{nf}}{\mu_{bf}} = 1 + \left(0.1008\phi^{0.6974} d_p^{0.44708}\right) \tag{3.24}$$

3.4.3 Specific heat

Specific heat of the nanofluid is determined by its thermal storage capacity. It is one of the important thermophysical properties of nanofluid to predict other important properties such as thermal conductivity and thermal diffusivity. The accurate measurement of specific heat is needed as its value is used to calculate the heat transfer from the fluids. Satti et al. [98] used C-Thermal apparatus (Fig. 3.13 [98]) for the measurement of specific heat of five

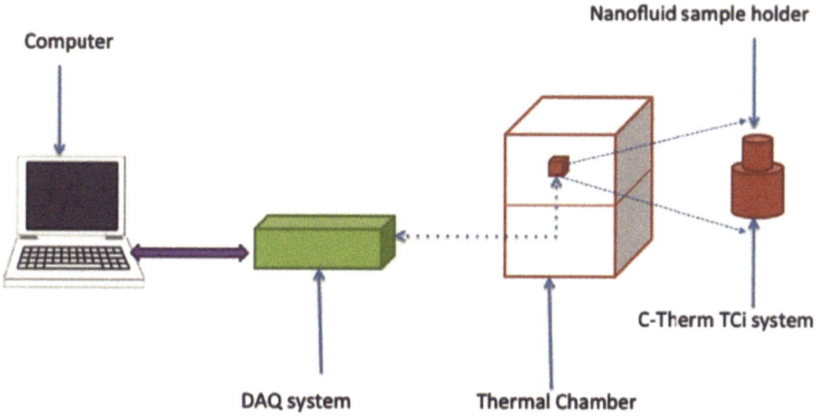

FIGURE 3.13 Experimental setup of C-thermal apparatus. Copyright Elsevier (LIC# 5100120164016) [98].

different nanofluids. The equipment followed the principle of the transient plane source method, which provides a constant heat source to the nanofluid sample. A voltage drop is sensed by the sensor on temperature increase, which inversely varies with the thermophysical properties of nanofluid. The given system effectively measure the thermal conductivity and thermal diffusivity values simultaneously. These values are further used to calculate heat capacity values using Eq. (3.25):

$$C_p = \frac{e^2}{k.\rho} \tag{3.25}$$

Further, they developed the following correlation for specific heat measurement.

$$\frac{C_{p,nf}}{C_{p,bf}} = \frac{\left[(A \times T) + B \times \left(\frac{C_{p,nf}}{C_{p,bf}}\right)\right]}{C + \phi} \tag{3.26}$$

where A, B, and C are curve fitting constants.

Qiao et al. [99] developed the following correlation for specific heat measurement of molten-salt based nanofluid:

$$C_{p,\text{sample}} = \frac{Q_{\text{sample}} - Q_{\text{Blank}}}{Q_{\text{Reference}} - Q_{\text{Blank}}} \frac{m_{\text{Reference}}}{m_{\text{sample}}} C_{p,\text{Reference}} \tag{3.27}$$

Sunder et al. [100] used the following correlation for specific heat measurement of nanodiamond–Fe_3O_4/water–EG nanofluid:

$$C_{p,(ND-Fe_3O_4)} = \frac{C_{p,ND} \times W_{ND} + C_{p,Fe_3O_4} \times W_{Fe_3O_4}}{W_{ND} + W_{Fe_3O_4}} \tag{3.28}$$

Kanti et al. [101] used differential scanning calorimeter for specific heat measurement of fly ash-based nanofluids and determined it using the following correlation:

$$C_{p,nf} = \frac{(1-\phi)(\rho C_p)_{bf} + \varphi(\rho C_p)_{np}}{(1-\phi)(\rho)_{bf} + \varphi(\rho)_{np}} \tag{3.29}$$

Barbes et al. [102] studied the specific heat model and for measurement it, they used the first law of thermodynamics to determine the specific heat using the following energy conversion correlation:

$$C_{p,nf} = \omega.C_{p,np} + (1-\omega).C_{p,bf} \tag{3.30}$$

where $C_{p,bf}$, $C_{p,nf}$, are $C_{p,np}$ are the specific heat capacities of base fluid, nanofluid, and nanoparticles, respectively, and ω is denoted as nanoparticle mass friction. Eq. (3.30) can also be rewritten as [102]:

$$C_{p,nf} = \frac{\varnothing.\rho_{np}.C_{p,np} + (1-\varphi).\rho_{bf}.C_{p,bf}}{\varphi.\rho_{np} + (1-\varphi).\rho_{bf}} \tag{3.31}$$

Ghadikolaei et al. [91] used the following correlation to measure the specific heat of TiO_2–CuO/water nanofluid:

$$(\rho C_p)_{nf} = (\rho C_p)_f (1 - \varnothing_2) \left[(1 - \varnothing_1) + \varnothing_1 \frac{(\rho C_p)_{np1}}{(\rho C_p)_f} \right] + \varnothing_2 (\rho C_p)_{np2} \quad (3.32)$$

Moldoveanu et al. [103] used DSC equipment from Mettler–Toledo for specific heat measurement of simple and hybrid oxide/water nanofluid and determine the following correlations:

$$C_{p,nf} = \left(\frac{\rho_p}{\rho_f} \right)^{0.2} \left(1 + \frac{d_p}{50} \right)^{0.4167} \left(1 - \frac{\phi}{100} \right)^{2.272} C_{p,f} \quad (3.33)$$

Osho et al. [104] used Mettler–Toledo 823 model differential scanning calorimeter for specific heat measurement

$$C_{p,nf} = \frac{\phi \rho_{np} C_{np} + \rho_{bf} C_{bf} (1 - \phi)}{\phi \rho_{np} + (1 - \phi) \rho_{bf}} \quad (3.34)$$

3.4.4 Density

Density is another important thermophysical property of nanofluid in determining the Reynolds number, flow rate, stability, pressure, and heat transfer. The density of the nanofluids varies with nanoparticle's volume concentration and temperature and it has positive trends with volume concentration while possessing negative trends with temperature. The density of nanofluid is generally determined by different density meters. Shoghl et al. [105] used DMA-35N portable density meter devices to measure the nanofluid density. Sarifpur et al. [106] used Rudolph DDM 2911 laboratory density meter to measure the density of nanofluid and compare it by experimental value. The following correlation is used to determine the density of nanofluid experimentally:

$$\rho_{nf} = \frac{m_p + m_f}{V_p + V_f + V_v} \quad (3.35)$$

where ρ is denoted as the density of nanofluid, m_p and m_f are mass of particles and base fluid, v_p and v_f is denoted as the volume of nanoparticles and base fluid, while v_v is denoted as the volume of voids between nanolayers. Kanti et al. [101] used KEM-DA 130 N density meter for density measurement of fly ash-based nanofluid and determined by the following correlation:

$$\rho_{nf} = (1 - \phi) \rho_{bf} + \phi \rho_{nf} \quad (3.36)$$

Takabi et al. [107] determined the following correlation model for measurement of the density of metal oxide-based hybrid nanofluid:

$$\rho_{nf} = (1 - \varnothing_1 - \varnothing_2)\rho_{bf} + \varnothing_1 \rho_{np1} + \varnothing_2 \rho_{np2} \tag{3.37}$$

where ρ is density, nf is hybrid nanofluid, *bf* is base fluid, and np is nanoparticle. Vijjha et al. [108] measured the density of Al_2O_3/EG, $Sb_2O_5-SnO_2$/EG−water, ZnO/EG−water, and presented the following density model for $Sb_2O_5-SnO_2$/EG−water hybrid nanofluid.

$$\rho_{nf} = \frac{(0.9848\varnothing + 0.7382)}{100}(1 - \varnothing)\rho_{bf} + \varnothing\rho_{np} \tag{3.38}$$

Vijjha et al. [109] used Anton-Paar (model DMA 45) digital density meter for measurement of density using the following correlations:

$$\rho_{nf} = \frac{(T^2 - C_2)}{C_1} \tag{3.39}$$

where T is denoted as a period of oscillation, while C_1 and C_2 are the devices constant.

3.5 Conclusion

Nanofluids are the homogeneous colloidal dispersion of nanoparticles into the base fluid. So far, distinct nanofluids have been synthesized by researchers in various heat transfer enhancement applications. Different characterization methods were used to analyze the characteristics and heat transfer behavior of nanofluids. The following conclusions have been drawn from the above study:

- The chemical methods for nanofluid synthesis are cheaper and faster in comparison to physical methods.
- As nanoparticles are commercially available on an industrial scale, the two-step process is currently the most energy-efficient and economic nanofluid synthesis method
- Characterization of nanofluid is used to analyze materials, structure, shape, size, composition, and physical and chemical properties. There are two famous characterization techniques, that is, the microscopy method and the spectroscopy method.
- Zeta potential is the potential difference between the fluid layer absorbed on the nanoparticle surface and the bulk phase of the liquid, and it estimates the stability of nanofluid.
- The THW method is an extensively used method for thermal conductivity measurement, and other methods are the 3ω method and thermal constant analyzer.

- The viscosity of nanofluids is measured by different viscometers, that is, capillary viscometer, rotational viscometer, oscillating viscometer, and Brookfield viscometer.
- Different density meter has been used sofar for density measurement of nanofluids.

Nomenclature

C_p	Specific heat
d	Distance between the sequential parallel plane of a crystal
K	Thermal conductivity (W/m.K)
m	Mass (kg)
n	Order of diffraction
p/l	The amplitude of the power per unit length
q	The magnitude of heat transmission (J)
R_0	Electric resistance
r_p	Probe radius
V	Volume
ΔT	Temperature difference (°C)

Abbreviation

DC	Direct current
DI water	Distilled water
EG	Ethylene glycol
ESCA	Electron spectroscopy
FESM	Field emission scanning microscope
FTIR	Fourier-transform infrared spectroscopy
MWCNT	Multiwall carbon nanotube
PVP	Polyvinyl pyrrolidone
SAED	Selected area electron diffraction
SPR	Surface plasmon surface
TEM	Transmission electron microscopy
THW	Transient hot-wire
TPS	Transient plane source
UV–Vis	Ultraviolet–visible spectroscopy
XRD	X-ray crystallography

Greek symbols

$\varnothing$	Volumetric concentration
μ	Viscosity
α	The temperature coefficient of the electric resistance
θ	The complementary angle of the X-ray incidence.
λ	Wavelength
ρ	Density
τ	Electrification time

subscript

bf Base fluid
n Nanofluid
r Radial
t Time

References

[1] N.K. Cakmak, Z. Said, L.S. Sundar, Z.M. Ali, A.K. Tiwari, Preparation, characterization, stability, and thermal conductivity of rGO-Fe3O4-TiO2 hybrid nanofluid: an experimental study, Powder Technol. 372 (2020) 235–245. Available from: https://doi.org/10.1016/j.powtec.2020.06.012.

[2] V. Kumar, N. Pandya, B. Pandya, A. Joshi, Synthesis of metal-based nanofluids and their thermo-hydraulic performance in compact heat exchanger with multi-louvered fins working under laminar conditions, J. Therm. Anal. Calorim. 135 (2019) 2221–2235. Available from: https://doi.org/10.1007/s10973-018-7304-x.

[3] N.S. Pandya, H. Shah, M. Molana, A.K. Tiwari, Heat transfer enhancement with nanofluids in plate heat exchangers: a comprehensive review, Eur. J. Mech. B/Fluids 81 (2020) 173–190. Available from: https://doi.org/10.1016/j.euromechflu.2020.02.004.

[4] S. Sharma, S. Tiwari, A.K. Tiwari, G. Nandan, R. Prakash, Thermal Performance Enhancement of Flat-Plate Solar Collector Using CeO2–Water Nanofluid, Springer, Singapore, 2020, pp. 109–118. Available from: https://doi.org/10.1007/978-981-15-3635-9_12.

[5] H.T. Zhu, Y.S. Lin, Y.S. Yin, A novel one-step chemical method for preparation of copper nanofluids, J. Colloid Interface Sci. 277 (2004) 100–103. Available from: https://doi.org/10.1016/j.jcis.2004.04.026.

[6] Y. Li, J. Zhou, S. Tung, E. Schneider, S. Xi, A review on development of nanofluid preparation and characterization, Powder Technol. 196 (2009) 89–101. Available from: https://doi.org/10.1016/j.powtec.2009.07.025.

[7] J. Tavares, S. Coulombe, Dual plasma synthesis and characterization of a stable copper-ethylene glycol nanofluid, Powder Technol. 210 (2011) 132–142. Available from: https://doi.org/10.1016/j.powtec.2011.03.006.

[8] C.H. Lo, T.T. Tsung, L.C. Chen, C.H. Su, H.M. Lin, Fabrication of copper oxide nanofluid using submerged arc nanoparticle synthesis system (SANSS), J. Nanopart. Res. 7 (2005) 313–320. Available from: https://doi.org/10.1007/s11051-004-7770-x.

[9] D.J. Lee, F.L. Jin, S.J. Park, Preparation of Cu nanoparticles by a pulsed wire evaporation process for conductive ink applications, Bull. Mater. Sci. 42 (2019) 1–8. Available from: https://doi.org/10.1007/s12034-018-1725-9.

[10] T.X. Phuoc, Y. Soong, M.K. Chyu, Synthesis of Ag-deionized water nanofluids using multi-beam laser ablation in liquids, Opt. Lasers Eng. 45 (2007) 1099–1106. Available from: https://doi.org/10.1016/j.optlaseng.2007.06.005.

[11] A. Khan, A. Rashid, R. Younas, R. Chong, A chemical reduction approach to the synthesis of copper nanoparticles, Int. Nano Lett. 6 (2016) 21–26. Available from: https://doi.org/10.1007/s40089-015-0163-6.

[12] S. Zeroual, P. Estellé, D. Cabaleiro, B. Vigolo, M. Emo, W. Halim, et al., Ethylene glycol based silver nanoparticles synthesized by polyol process: characterization and thermophysical profile, J. Mol. Liq. 310 (2020) 113229. Available from: https://doi.org/10.1016/j.molliq.2020.113229.

[13] A.K. Singh, V.S. Raykar, Microwave synthesis of silver nanofluids with polyvinylpyrrolidone (PVP) and their transport properties, Colloid Polym. Sci. 286 (2008) 1667–1673. Available from: https://doi.org/10.1007/s00396-008-1932-9.

[14] H.A. Mohammed, A.A. Al-Aswadi, N.H. Shuaib, R. Saidur, Convective heat transfer and fluid flow study over a step using nanofluids: a review, Renew. Sustain. Energy Rev. 15 (2011) 2921–2939. Available from: https://doi.org/10.1016/j.rser.2011.02.019.

[15] Choi. Application data (60) provisional application No. 60/066. vol. 487, 1997.

[16] L. Yang, Y. Hu, Toward TiO2 nanofluids—part 1: preparation and properties, Nanoscale Res. Lett. (2017) 12. Available from: https://doi.org/10.1186/s11671-017-2184-8.

[17] G.J. Lee, C.K. Kim, M.K. Lee, C.K. Rhee, Characterization of ethylene glycol based TiO2 nanofluid prepared by pulsed wire evaporation (PWE) method, Rev. Adv. Mater. Sci. 28 (2011) 126–129.

[18] C.H. Lo, T.T. Tsung, L.C. Chen, Shape-controlled synthesis of Cu-based nanofluid using submerged arc nanoparticle synthesis system (SANSS), J. Cryst. Growth 277 (2005) 636–642. Available from: https://doi.org/10.1016/j.jcrysgro.2005.01.067.

[19] C.-H. Lo, T.-T. Tsung, H.-M. Lin, Preparation of silver nanofluid by the submerged arc nanoparticle synthesis system (SANSS), J. Alloy. Compd. 434–435 (2007) 659–662. Available from: https://doi.org/10.1016/j.jallcom.2006.08.217.

[20] M.J. Kao, C.H. Lo, T.T. Tsung, Y.Y. Wu, C.S. Jwo, H.M. Lin, Copper-oxide brake nanofluid manufactured using arc-submerged nanoparticle synthesis system, J. Alloy. Compd. 434–435 (2007) 672–674. Available from: https://doi.org/10.1016/j.jallcom.2006.08.305.

[21] H.J. Kim, I.C. Bang, J. Onoe, Characteristic stability of bare Au-water nanofluids fabricated by pulsed laser ablation in liquids, Opt. Lasers Eng. 47 (2009) 532–538. Available from: https://doi.org/10.1016/j.optlaseng.2008.10.011.

[22] T. Khamliche, S. Khamlich, M.K. Moodley, B.M. Mothudi, M. Henini, M. Maaza, Laser fabrication of Cu nanoparticles based nanofluid with enhanced thermal conductivity: experimental and molecular dynamics studies, J. Mol. Liq. 323 (2021) 114975. Available from: https://doi.org/10.1016/j.molliq.2020.114975.

[23] A. Riahi, S. Khamlich, M. Balghouthi, T. Khamliche, T.B. Doyle, W. Dimassi, et al., Study of thermal conductivity of synthesized Al2O3-water nanofluid by pulsed laser ablation in liquid, J. Mol. Liq. (2020) 304. Available from: https://doi.org/10.1016/j.molliq.2020.112694.

[24] J. Garg, B. Poudel, M. Chiesa, J.B. Gordon, J.J. Ma, J.B. Wang, et al., Enhanced thermal conductivity and viscosity of copper nanoparticles in ethylene glycol nanofluid, J. Appl. Phys. (2008) 103. Available from: https://doi.org/10.1063/1.2902483.

[25] W. Yu, H. Xie, L. Chen, Y. Li, C. Zhang, Synthesis and characterization of monodispersed copper colloids in polar solvents, Nanoscale Res. Lett. 4 (2009) 465–470. Available from: https://doi.org/10.1007/s11671-009-9264-3.

[26] X. Wei, H. Zhu, T. Kong, L. Wang, Synthesis and thermal conductivity of Cu2O nanofluids, Int. J. Heat. Mass. Transf. 52 (2009) 4371–4374. Available from: https://doi.org/10.1016/j.ijheatmasstransfer.2009.03.073.

[27] S.A. Kumar, K.S. Meenakshi, B.R.V. Narashimhan, S. Srikanth, G. Arthanareeswaran, Synthesis and characterization of copper nanofluid by a novel one-step method, Mater. Chem. Phys. 113 (2009) 57–62. Available from: https://doi.org/10.1016/j.matchemphys.2008.07.027.

[28] H. Zhu, D. Han, Z. Meng, D. Wu, C. Zhang, Preparation and thermal conductivity of cuo nanofluid via a wet chemical method, Nanoscale Res. Lett. 6 (2011) 2–7. Available from: https://doi.org/10.1186/1556-276X-6-181.

[29] G. Paul, J. Philip, B. Raj, P.K. Das, I. Manna, Synthesis, characterization, and thermal property measurement of nano-Al95Zn05 dispersed nanofluid prepared by a two-step process, Int. J. Heat. Mass. Transf. 54 (2011) 3783−3788. Available from: https://doi.org/10.1016/j.ijheatmasstransfer.2011.02.044.

[30] W. Yu, H. Xie, A review on nanofluids: preparation, stability mechanisms, and applications, J. Nanomater. (2012) 2012. Available from: https://doi.org/10.1155/2012/435873.

[31] A. Nentwich, Production of nanoparticles and nanomaterials, Planet-Aust. At. 6 (2011) 1−4.

[32] J. Gou, J. Zhuge, F. Liang, Processing of Polymer Nanocomposites, Woodhead Publishing Limited, 2012. Available from: https://doi.org/10.1533/9780857096258.1.95.

[33] A. Rajaeiyan, M.M. Bagheri-Mohagheghi, Comparison of sol-gel and co-precipitation methods on the structural properties and phase transformation of γ and α-Al2O3 nanoparticles, Adv. Manuf. 1 (2013) 176−182. Available from: https://doi.org/10.1007/s40436-013-0018-1.

[34] M.A. Behnajady, H. Eskandarloo, N. Modirshahla, M. Shokri, Investigation of the effect of sol-gel synthesis variables on structural and photocatalytic properties of TiO2 nanoparticles, Desalination 278 (2011) 10−17. Available from: https://doi.org/10.1016/j.desal.2011.04.019.

[35] M. Leena, S. Srinivasan, Synthesis and ultrasonic investigations of titanium oxide nanofluids, J. Mol. Liq. 206 (2015) 103−109. Available from: https://doi.org/10.1016/j.molliq.2015.02.001.

[36] A.K. Tiwari, S. Javed, H.F. Oztop, Z. Said, N.S. Pandya, Experimental and numerical investigation on the thermal performance of triple tube heat exchanger equipped with different inserts with WO3/water nanofluid under turbulent condition, Int. J. Therm. Sci. 164 (2021) 106861. Available from: https://doi.org/10.1016/j.ijthermalsci.2021.106861.

[37] E. Begum Elcioglu, S.M.S. Murshed, Ultrasonically tuned surface tension and nano-film formation of aqueous ZnO nanofluids, Ultrason. Sonochem. 72 (2021) 105424. Available from: https://doi.org/10.1016/j.ultsonch.2020.105424.

[38] S.S. Chawhan, D.P. Barai, B.A. Bhanvase, Investigation on thermophysical properties, convective heat transfer and performance evaluation of ultrasonically synthesized Ag-doped TiO2 hybrid nanoparticles based highly stable nanofluid in a minichannel, Therm. Sci. Eng. Prog. (2021) 100928. Available from: https://doi.org/10.1016/j.tsep.2021.100928.

[39] S. Chakraborty, I. Sarkar, D.K. Behera, S.K. Pal, S. Chakraborty, Experimental investigation on the effect of dispersant addition on thermal and rheological characteristics of TiO2 nanofluid, Powder Technol. 307 (2017) 10−24. Available from: https://doi.org/10.1016/j.powtec.2016.11.016.

[40] G.J. Lee, C.K. Kim, M.K. Lee, C.K. Rhee, S. Kim, C. Kim, Thermal conductivity enhancement of ZnO nanofluid using a one-step physical method, Thermochim. Acta 542 (2012) 24−27. Available from: https://doi.org/10.1016/j.tca.2012.01.010.

[41] J.M. Salehi, M.M. Heyhat, A. Rajabpour, Enhancement of thermal conductivity of silver nanofluid synthesized by a one-step method with the effect of polyvinylpyrrolidone on thermal behavior, Appl. Phys. Lett. 102 (2013) 2013−2016. Available from: https://doi.org/10.1063/1.4809998.

[42] S. Aberoumand, A. Jafarimoghaddam, Experimental study on synthesis, stability, thermal conductivity and viscosity of Cu−engine oil nanofluid, J. Taiwan. Inst. Chem. Eng. 71 (2017) 315−322. Available from: https://doi.org/10.1016/j.jtice.2016.12.035.

[43] X.xiu Huang, W.gong Zhang, Study on successively preparation of nano-TiO 2 ethanol colloids by pulsed laser ablation and fluorescence property, Appl. Surf. Sci. 254 (2008) 3403−3407. Available from: https://doi.org/10.1016/j.apsusc.2007.11.027.

[44] P.S. Liu, W.P. Cai, L.X. Wan, M.Da Shi, X.D. Luo, W.P. Jing, Fabrication and characteristics of rutile TiO2 nanoparticles induced by laser ablation, Trans. Nonferrous Met. Soc. China (Engl. Ed.) 19 (2009) s743–s747. Available from: https://doi.org/10.1016/S1003-6326(10)60143-4.
[45] C.H. Liang, Y. Shimizu, T. Sasaki, N. Koshizaki, Preparation of ultrafine TiO2 nanocrystals via pulsed-laser ablation of titanium metal in surfactant solution, Appl. Phys. A Mater. Sci. Process. 80 (2005) 819–822. Available from: https://doi.org/10.1007/s00339-003-2489-6.
[46] M. Rahmati, Effects of ZnO/Water Nanofluid on the Thermal Performance of Wet Cooling Towers, Elsevier Ltd., 2021. Available from: https://doi.org/10.1016/j.ijrefrig.2021.03.017.
[47] R. Agarwal, K. Verma, N.K. Agrawal, R.K. Duchaniya, R. Singh, Synthesis, characterization, thermal conductivity and sensitivity of CuO nanofluids, Appl. Therm. Eng. 102 (2016) 1024–1036. Available from: https://doi.org/10.1016/j.applthermaleng.2016.04.051.
[48] M. Eltaweel, A.A. Abdel-Rehim, A.A.A. Attia, Energetic and exergetic analysis of a heat pipe evacuated tube solar collector using MWCNT/water nanofluid, Case Stud. Therm. Eng. 22 (2020) 100743. Available from: https://doi.org/10.1016/j.csite.2020.100743.
[49] M.A. Sharafeldin, G. Gróf, Evacuated tube solar collector performance using CeO2/water nanofluid, J. Clean. Prod. 185 (2018) 347–356. Available from: https://doi.org/10.1016/j.jclepro.2018.03.054.
[50] W. Ahmed, Z.Z. Chowdhury, S.N. Kazi, M.R. Johan, N. Akram, C.S. Oon, Effect of ZnO-water based nanofluids from sonochemical synthesis method on heat transfer in a circular flow passage, Int. Commun. Heat. Mass. Transf. (2020) 114. Available from: https://doi.org/10.1016/j.icheatmasstransfer.2020.104591.
[51] Synthesis, characterization and heat transfer study of reduced graphene oxide-Al2O3 nanocomposite based nanofluids: investigation on thermal conductivity and rheology Science Direct (n.d.) https://www.sciencedirect.com/science/article/abs/pii/S2352492820329974 (accessed 22.5.21).
[52] B. Wei, C. Zou, X. Li, Experimental investigation on stability and thermal conductivity of diathermic oil based TiO2 nanofluids, Int. J. Heat. Mass. Transf. 104 (2017) 537–543. Available from: https://doi.org/10.1016/j.ijheatmasstransfer.2016.08.078.
[53] D. Banerjee, Nanofluids and Applications to Energy Systems, vol. 4, Elsevier, 2017. Available from: https://doi.org/10.1016/B978-0-12-409548-9.10144-7.
[54] R. Sadeghi, S.G. Etemad, E. Keshavarzi, M. Haghshenasfard, Investigation of alumina nanofluid stability by UV–vis spectrum, Microfluid. Nanofluid. 18 (2015) 1023–1030. Available from: https://doi.org/10.1007/s10404-014-1491-y.
[55] J. Campbell, S. Burkitt, N. Dong, C. Zavaleta, Nanoparticle Characterization Techniques, Elsevier Inc., 2019. Available from: https://doi.org/10.1016/B978-0-12-816662-8.00009-6.
[56] A.K. Tiwari, N.S. Pandya, Z. Said, H.F. Öztop, N. Abu-Hamdeh, 4S consideration (synthesis, sonication, surfactant, stability) for the thermal conductivity of CeO2 with MWCNT and water based hybrid nanofluid: an experimental assessment, Colloids Surf. A Physicochem. Eng. Asp. (2021) 610. Available from: https://doi.org/10.1016/j.colsurfa.2020.125918.
[57] P. Mohamed Shameer, P. Mohamed Nishath, Exploration and Enhancement on Fuel Stability of Biodiesel: A Step Forward in the Track of Global Commercialization, Elsevier Ltd., 2019. Available from: https://doi.org/10.1016/B978-0-08-102791-2.00008-8.
[58] R. Sindhu, P. Binod, A. Pandey, Microbial Poly-3-Hydroxybutyrate and Related Copolymers, Elsevier B.V., 2015. Available from: https://doi.org/10.1016/B978-0-444-63453-5.00019-7.

[59] D. Titus, E. James Jebaseelan Samuel, S.M. Roopan, Nanoparticle Characterization Techniques, Elsevier Inc., 2019. Available from: https://doi.org/10.1016/b978-0-08-102579-6.00012-5.

[60] L.R. de Oliveira, S.R.F.L. Ribeiro, M.H.M. Reis, V.L. Cardoso, E.P. Bandarra Filho, Experimental study on the thermal conductivity and viscosity of ethylene glycol-based nanofluid containing diamond-silver hybrid material, Diam. Relat. Mater. 96 (2019) 216–230. Available from: https://doi.org/10.1016/j.diamond.2019.05.004.

[61] Z. Said, M. Ghodbane, L.S. Sundar, A.K. Tiwari, M. Sheikholeslami, B. Boumeddane, Heat transfer, entropy generation, economic and environmental analyses of linear fresnel reflector using novel rGO-Co3O4 hybrid nanofluids, Renew. Energy 165 (2021) 420–437. Available from: https://doi.org/10.1016/j.renene.2020.11.054.

[62] B.J. Inkson, Scanning Electron Microscopy (SEM) and Transmission Electron Microscopy (TEM) for Materials Characterization, Elsevier Ltd., 2016. Available from: https://doi.org/10.1016/B978-0-08-100040-3.00002-X.

[63] R. Desai, V. Mankad, S. Gupta, P. Jha, Size distribution of silver nanoparticles: UV-visible spectroscopic assessment, Nanosci. Nanotechnol. Lett. 4 (2012) 30–34. Available from: https://doi.org/10.1166/nnl.2012.1278.

[64] D. Sutherland, Chapter 6 characterisation, Pergamon Mater. Ser. 5 (2001) 213–249. Available from: https://doi.org/10.1016/S1470-1804(01)80009-2.

[65] H. Kimpton, E. Stulz, X. Zhang, Silver nanofluids based broadband solar absorber through tuning nanosilver geometries, Sol. Energy 208 (2020) 515–526. Available from: https://doi.org/10.1016/j.solener.2020.08.018.

[66] M. Gupta, V. Singh, R. Kumar, Z. Said, A review on thermophysical properties of nanofluids and heat transfer applications, Renew. Sustain. Energy Rev. 74 (2017) 638–670. Available from: https://doi.org/10.1016/j.rser.2017.02.073.

[67] A.K. Tiwar, V. Kumar, Z. Said, H.K. Paliwal, A review on the application of hybrid nanofluids for parabolic trough collector: recent progress and outlook, J. Clean. Prod. 292 (2021) 126031. Available from: https://doi.org/10.1016/j.jclepro.2021.126031.

[68] V. Kumar, A.K. Tiwari, S.K. Ghosh, Application of nanofluids in plate heat exchanger: a review, Energy Convers. Manag. 105 (2015) 1017–1036. Available from: https://doi.org/10.1016/j.enconman.2015.08.053.

[69] A. Kumar, A.K. Tiwari, Z. Said, A comprehensive review analysis on advances of evacuated tube solar collector using nanofluids and PCM, Sustain. Energy Technol. Assess. 47 (2021) 101417. Available from: https://doi.org/10.1016/j.seta.2021.101417.

[70] A.K. Sharma, A.K. Tiwari, A.R. Dixit, Rheological behaviour of nanofluids: a review, Renew. Sustain. Energy Rev. 53 (2016) 779–791. Available from: https://doi.org/10.1016/J.RSER.2015.09.033.

[71] N.K. Gupta, S. Mishra, A.K. Tiwari, S.K. Ghosh, A review of thermo physical properties of nanofluids, Mater. Today Proc. 18 (2019) 968–978. Available from: https://doi.org/10.1016/J.MATPR.2019.06.534.

[72] D. Ratna, Thermal properties of thermosets, Thermosets (2012) 62–91. Available from: https://doi.org/10.1533/9780857097637.1.62.

[73] W. Guo, G. Li, Y. Zheng, C. Dong, Measurement of the thermal conductivity of SiO2 nanofluids with an optimized transient hot wire method, Thermochim. Acta 661 (2018) 84–97. Available from: https://doi.org/10.1016/j.tca.2018.01.008.

[74] D.W. Oh, A. Jain, J.K. Eaton, K.E. Goodson, J.S. Lee, Thermal conductivity measurement and sedimentation detection of aluminum oxide nanofluids by using the 3ω method, Int. J.

Heat. Fluid Flow. 29 (2008) 1456–1461. Available from: https://doi.org/10.1016/j.ijheatfluidflow.2008.04.007.

[75] G. Paul, M. Chopkar, I. Manna, P.K. Das, Techniques for measuring the thermal conductivity of nanofluids: a review, Renew. Sustain. Energy Rev. 14 (2010) 1913–1924. Available from: https://doi.org/10.1016/j.rser.2010.03.017.

[76] W. Jiang, G. Ding, H. Peng, Measurement and model on thermal conductivities of carbon nanotube nanorefrigerants, Int. J. Therm. Sci. 48 (2009) 1108–1115. Available from: https://doi.org/10.1016/j.ijthermalsci.2008.11.012.

[77] D. Zhu, X. Li, N. Wang, X. Wang, J. Gao, H. Li, Dispersion behavior and thermal conductivity characteristics of Al2O3-H2O nanofluids, Curr. Appl. Phys. 9 (2009) 131–139. Available from: https://doi.org/10.1016/j.cap.2007.12.008.

[78] X.F. Li, D.S. Zhu, X.J. Wang, N. Wang, J.W. Gao, H. Li, Thermal conductivity enhancement dependent pH and chemical surfactant for Cu-H2O nanofluids, Thermochim. Acta 469 (2008) 98–103. Available from: https://doi.org/10.1016/j.tca.2008.01.008.

[79] D.K. Prajapati, M. Tiwari, Evolution Surf. Topography Dur. Wear Process. (2020). Available from: https://doi.org/10.1007/978-3-030-47451-5_9.

[80] Z. Said, M.A. Abdelkareem, H. Rezk, A.M. Nassef, H.Z. Atwany, Stability, thermophysical and electrical properties of synthesized carbon nanofiber and reduced-graphene oxide-based nanofluids and their hybrid along with fuzzy modeling approach, Powder Technol. 364 (2020) 795–809. Available from: https://doi.org/10.1016/j.powtec.2020.02.026.

[81] S. Sharma, A.K. Tiwari, S. Tiwari, R. Prakash, Viscosity of hybrid nanofluids: measurement and comparison, J. Mech. Eng. Sci. 12 (2018) 3614–3623. Available from: https://doi.org/10.15282/jmes.12.2.2018.8.0320.

[82] I. Kazemi, M. Sefid, M. Afrand, A novel comparative experimental study on rheological behavior of mono & hybrid nanofluids concerned graphene and silica nano-powders: characterization, stability and viscosity measurements, Powder Technol. 366 (2020) 216–229. Available from: https://doi.org/10.1016/j.powtec.2020.02.010.

[83] S. Akilu, A.T. Baheta, K.V. Sharma, Experimental measurements of thermal conductivity and viscosity of ethylene glycol-based hybrid nanofluid with TiO2-CuO/C inclusions, J. Mol. Liq. 246 (2017) 396–405. Available from: https://doi.org/10.1016/j.molliq.2017.09.017.

[84] M. Benzema, Y.K. Benkahla, N. Labsi, S.E. Ouyahia, M. El Ganaoui, Second law analysis of MHD mixed convection heat transfer in a vented irregular cavity filled with Ag–MgO/water hybrid nanofluid, J. Therm. Anal. Calorim. 137 (2019) 1113–1132. Available from: https://doi.org/10.1007/s10973-019-08017-x.

[85] R.R. Sahoo, V. Kumar, Development of a new correlation to determine the viscosity of ternary hybrid nanofluid, Int. Commun. Heat. Mass. Transf. 111 (2020) 104451. Available from: https://doi.org/10.1016/j.icheatmasstransfer.2019.104451.

[86] O. Soltani, M. Akbari, Effects of temperature and particles concentration on the dynamic viscosity of MgO-MWCNT/ethylene glycol hybrid nanofluid: experimental study, Phys. E Low-Dimens. Syst. Nanostruct. 84 (2016) 564–570. Available from: https://doi.org/10.1016/j.physe.2016.06.015.

[87] M. Asadi, A. Asadi, Dynamic viscosity of MWCNT/ZnO-engine oil hybrid nanofluid: an experimental investigation and new correlation in different temperatures and solid concentrations, Int. Commun. Heat. Mass. Transf. 76 (2016) 41–45. Available from: https://doi.org/10.1016/j.icheatmasstransfer.2016.05.019.

[88] B. Wei, C. Zou, X. Yuan, X. Li, Thermo-physical property evaluation of diathermic oil based hybrid nanofluids for heat transfer applications, Int. J. Heat. Mass. Transf. 107

(2017) 281–287. Available from: https://doi.org/10.1016/j.ijheatmasstransfer.2016.11.044.

[89] Z. Li, S. Asadi, A. Karimipour, A. Abdollahi, I. Tlili, Experimental study of temperature and mass fraction effects on thermal conductivity and dynamic viscosity of SiO2-oleic acid/liquid paraffin nanofluid, Int. Commun. Heat. Mass. Transf. (2020) 110. Available from: https://doi.org/10.1016/j.icheatmasstransfer.2019.104436.

[90] K.A. Hamid, W.H. Azmi, M.F. Nabil, R. Mamat, K.V. Sharma, Experimental investigation of thermal conductivity and dynamic viscosity on nanoparticle mixture ratios of TiO2-SiO2 nanofluids, Int. J. Heat. Mass. Transf. 116 (2018) 1143–1152. Available from: https://doi.org/10.1016/j.ijheatmasstransfer.2017.09.087.

[91] S.S. Ghadikolaei, M. Yassari, H. Sadeghi, K. Hosseinzadeh, D.D. Ganji, Investigation on thermophysical properties of Tio2–Cu/H2O hybrid nanofluid transport dependent on shape factor in MHD stagnation point flow, Powder Technol. 322 (2017) 428–438. Available from: https://doi.org/10.1016/j.powtec.2017.09.006.

[92] M.F. Nabil, W.H. Azmi, K. Abdul Hamid, R. Mamat, F.Y. Hagos, An experimental study on the thermal conductivity and dynamic viscosity of TiO2-SiO2 nanofluids in water: ethylene glycol mixture, Int. Commun. Heat. Mass. Transf. 86 (2017) 181–189. Available from: https://doi.org/10.1016/j.icheatmasstransfer.2017.05.024.

[93] A.K. Tiwari, N.S. Pandya, Z. Said, S.H. Chhatbar, Y.A. Al-Turki, A.R. Patel, 3S (Sonication, surfactant, stability) impact on the viscosity of hybrid nanofluid with different base fluids: an experimental study, J. Mol. Liq. 329 (2021) 115455. Available from: https://doi.org/10.1016/j.molliq.2021.115455.

[94] M. Anish, J. Jayaprabakar, N. Joy, A. Prabhu, B. Krishna Kanth Reddy, E. Govinda Rajulu, Measurement of temperature-dependent thermal conductivity, viscosity and density of ZnO-Theminol 55 Nanofluids, Mater. Today Proc. 21 (2020) 148–151. Available from: https://doi.org/10.1016/j.matpr.2019.04.206.

[95] T. Yiamsawas, O. Mahian, A.S. Dalkilic, S. Kaewnai, S. Wongwises, Experimental studies on the viscosity of TiO2 and Al2O3 nanoparticles suspended in a mixture of ethylene glycol and water for high temperature applications, Appl. Energy 111 (2013) 40–45. Available from: https://doi.org/10.1016/j.apenergy.2013.04.068.

[96] W.H. Azmi, K.V. Sharma, R. Mamat, A.B.S. Alias, I. Izwan Misnon, Correlations for thermal conductivity and viscosity of water based nanofluids, IOP Conf. Ser. Mater. Sci. Eng. (2012) 36. Available from: https://doi.org/10.1088/1757-899X/36/1/012029.

[97] M. Hemmat Esfe, S. Saedodin, S. Wongwises, D. Toghraie, An experimental study on the effect of diameter on thermal conductivity and dynamic viscosity of Fe/water nanofluids, J. Therm. Anal. Calorim. 119 (2015) 1817–1824. Available from: https://doi.org/10.1007/s10973-014-4328-8.

[98] J.R. Satti, D.K. Das, D. Ray, Specific heat measurements of five different propylene glycol based nanofluids and development of a new correlation, Int. J. Heat. Mass. Transf. 94 (2016) 343–353. Available from: https://doi.org/10.1016/j.ijheatmasstransfer.2015.11.065.

[99] G. Qiao, M. Lasfargues, A. Alexiadis, Y. Ding, Simulation and experimental study of the specific heat capacity of molten salt based nanofluids, Appl. Therm. Eng. 111 (2017) 1517–1522. Available from: https://doi.org/10.1016/j.applthermaleng.2016.07.159.

[100] L. Syam Sundar, S. Mesfin, E. Venkata Ramana, Z. Said, A.C.M. Sousa, Experimental investigation of thermo-physical properties, heat transfer, pumping power, entropy generation, and exergy efficiency of nanodiamond + Fe3O4/60:40% water-ethylene glycol hybrid nanofluid flow in a tube, Therm. Sci. Eng. Prog. (2021) 21. Available from: https://doi.org/10.1016/j.tsep.2020.100799.

[101] P. Kanti, K.V. Sharma, C.G. Ramachandra, W.H. Azmi, Experimental determination of thermophysical properties of Indonesian fly-ash nanofluid for heat transfer applications, Part. Sci. Technol. 39 (2021) 597–606. Available from: https://doi.org/10.1080/02726351.2020.1806971.

[102] B. Barbés, R. Páramo, E. Blanco, M.J. Pastoriza-Gallego, M.M. Piñeiro, J.L. Legido, et al., Thermal conductivity and specific heat capacity measurements of Al 2O3 nanofluids, J. Therm. Anal. Calorim. 111 (2013) 1615–1625. Available from: https://doi.org/10.1007/s10973-012-2534-9.

[103] G.M. Moldoveanu, A.A. Minea, Specific heat experimental tests of simple and hybrid oxide-water nanofluids: proposing new correlation, J. Mol. Liq. 279 (2019) 299–305. Available from: https://doi.org/10.1016/j.molliq.2019.01.137.

[104] I. Wole-Osho, E.C. Okonkwo, D. Kavaz, S. Abbasoglu, An experimental investigation into the effect of particle mixture ratio on specific heat capacity and dynamic viscosity of Al2O3-ZnO hybrid nanofluids, Powder Technol. 363 (2020) 699–716. Available from: https://doi.org/10.1016/j.powtec.2020.01.015.

[105] S. Nabati Shoghl, J. Jamali, M. Keshavarz Moraveji, Electrical conductivity, viscosity, and density of different nanofluids: an experimental study, Exp. Therm. Fluid Sci. 74 (2016) 339–346. Available from: https://doi.org/10.1016/j.expthermflusci.2016.01.004.

[106] M. Sharifpur, S. Yousefi, J.P. Meyer, A new model for density of nanofluids including nanolayer, Int. Commun. Heat. Mass. Transf. 78 (2016) 168–174. Available from: https://doi.org/10.1016/j.icheatmasstransfer.2016.09.010.

[107] B. Takabi, S. Salehi, Augmentation of the heat transfer performance of a sinusoidal corrugated enclosure by employing hybrid nanofluid, Adv. Mech. Eng. (2014) 2014. Available from: https://doi.org/10.1155/2014/147059.

[108] R.S. Vajjha, D.K. Das, B.M. Mahagaonkar, Density measurement of different nanofluids and their comparison with theory, Pet. Sci. Technol. 27 (2009) 612–624. Available from: https://doi.org/10.1080/10916460701857714.

[109] R.S. Vajjha, D.K. Das, Measurements of specific heat and density of Al2O3 nanofluid, AIP Conf. Proc. 1063 (2008) 361–370. Available from: https://doi.org/10.1063/1.3027181.

Chapter 4

Thermophysical and rheological properties of unitary and hybrid nanofluids

Hatice Mercan[1], Ali Celen[2] and Tolga Taner[3]

[1]*Department of Mechatronics Engineering, Yildiz Technical University, Istanbul, Turkey,* [2]*Department of Mechanical Engineering, Faculty of Engineering, Erzincan Binali Yıldırım University, Erzincan, Turkey,* [3]*Department of Motor Vehicles and Transportation Technology, Aksaray University, Aksaray, Turkey*

4.1 Introduction

When a thermal system operates, the heat is exchanged by using a working fluid. The working fluid of such systems can be a single fluid or a mixture or a nanosuspension. Ethylene glycol, ethanol, water, engine oil, ammonia, and halogenated hydrocarbons are typical examples of working fluids frequently used in industrial applications. Thermophysical properties of the working fluid determine the efficiency limits of a thermal system and the desire for a more efficient system leads researchers to engineer a better performing working fluid. The addition of nanosize particles with a considerably low concentration enhanced the thermophysical properties, such as thermal conductivity, specific heat capacity, and convective heat transfer coefficient significantly. The common nanoparticles are a range of metals, metal oxides, metal carbides, nanotubes, and carbon. If a single type of nanoparticle is added to the base fluid the blend is called a unitary nanofluid, if more than one type of nanoparticle is added; the blend is called a hybrid nanofluid. Although nanofluids are sophisticated suspensions, spreading, stability, chemical and thermal suitableness of the nanoparticles should be guaranteed during the preparation procedures. To maintain this preferred elongated stability, the selection of a suitable concentration accompanied with surfactants addition and/or sonication procedures are commonly used remedies in the literature.

The properties like thermal conductivity, viscosity, density, heat capacity, and surface tension are the central parameters that affect the performance of the working fluid and the overall efficiency due to the better thermal

Advances in Nanofluid Heat Transfer. DOI: https://doi.org/10.1016/B978-0-323-88656-7.00005-2

conductivity of the nanoparticle and the increased heat transfer surface because of the small size of the particles. The relative motion between the nanoparticles and microturbulence within the flow results not only in stronger spreading but also enhanced mass and heat transfer. Although the thermal capacity and performance of the working fluid improve with nanoparticle inclusion, the viscosity of the colloidal dispersion increases considerably. The viscosity increase results in a higher pressure drop and as a result an increased pumping power, which is not desired during the operation of the thermal systems. The density difference between the nanoparticles and the base fluid affects the stability, pressure drop, and thermal and rheological properties of the nanoblend. In heat transfer applications, estimating the density of the fluid is directly related to the pressure drop. It is evident that the addition of nanoparticles increases the density of the working fluid. The thermal conductivity alteration of working fluids with the addition of nanoparticles has been investigated by many researchers in the literature. It is reported that there is no consensus on the global effect of the aforementioned operating conditions on the thermal conductivity of generalized nanoblend behavior. Thus every blend for every operating condition needs to be investigated separately in detail.

The impact of concentration, temperature, and geometrical properties of nanoparticles and stability improvement techniques in terms of viscosity, thermal conductivity, density, and specific heat capacity are compared for different base fluids and nanoparticle combinations by using recent studies presented in the literature. In this chapter, the effects of nanofluid types and operation conditions on these thermophysical properties are reported in detail and the model equations that are proposed in the literature are tabulated. In addition, this chapter highlights the significance of nanofluids with an overview of industrial applications such as evaporators and cooling systems. Thus this study will pave the way for more efficient use of nanofluids in thermal applications, especially for the industrial processes.

4.2 Thermophysical properties

4.2.1 Thermal conductivity

In heat transfer applications, the thermal conductivity of the fluid is one of the most important parameter for the estimation of heat transfer capacity for both quiescent and inflow conditions. Improving the thermal capacity is desired for the equipments requiring higher heat load demand. The size and efficiency directly related to thermal capacity, and lower sizes with higher efficiencies are preferred in many industrial applications. Nanoparticle inclusion to a base fluid generally improves the thermal conductivity. One common observation is that the enhancement in thermal conductivity is more intense for higher particle fractions at elevated temperature values.

The relative thermal conductivity is the ratio of the conductivity of nanofluid to the base fluid, which is a measure of enhancement compared to the base fluid at the mean time. The escalation of the relative thermal conductivity is reported for increasing concentration. The models that are proposed for the determination of the thermal conductivity in the literature for various colloidal dispersions are listed in Table 4.1. In addition, the enhancement of thermal conductivity of Al_2O_3 and SiO_2 nanofluids at various particle sizes are given in Tables 4.2 and 4.3, respectively. The recent studies on the determination of the thermal conductivity of nanofluids are summarized as follows:

In an experimental study, Hussein et al. [28] compared the thermal conductivity of SiO_2/water, Al_2O_3/water, and TiO_2/water nanofluids for increasing temperature values. The unitary nanofluids displayed higher thermal conductivity values compared to base fluid and the thermal conductivity enhanced with increasing temperature values. They reported Al_2O_3/water showed the highest thermal conductivity and TiO_2 showed the lowest thermal conductivity.

A similar observation regarding the effect of concentration and temperature on thermal conductivity is valid for hybrid nanofluids [2,4,29]. Moldoveanu investigated $Al_2O_3-SiO_2$/water [18] and $Al_2O_3-TiO_2$/water [4] hybrid nanofluids for increasing total particle concentration and temperature. Esfe investigated Al_2O_3−CNTs/water hybrid nanofluid thermal conductivity changes under thermal and concentration effects. They all reported that increasing concentration and temperature augmented the thermal conductivity at different rates. It is observed that the $Al_2O_3-SiO_2$/water hybrid nanofluid is more sensitive to a temperature rise with increasing concentration. The slope of the thermal conductivity curve becomes steeper at higher constant concentration. For the $Al_2O_3-TiO_2$/water and Al_2O_3−CNTs/water hybrid nanofluids on the other hand, the constant concentration curves are almost parallel to each other. A recent study [30] investigated the thermophysical properties of ZnO−MgO/W−EG hybrid nanofluid for 0.025%−0.1% concentration and 30°C−50°C temperature intervals. A simultaneous rise is reported for increasing temperature and concentration. They reported a temperature rise not only reduced the clustering at higher concentration but also the Brownian motion enhancement augmented the thermal conductivity up to 17.5%. Ilyas et al. [31] investigated the f-ZnO/oil nanofluid for increasing concentration and temperature [31] (see Fig. 4.1). The temperature rise improved the thermal conductivity at all concentrations. The concentration rise showed a nonlinear behavior due to the nonhomogeneous distribution of the nanoparticles.

Thermal conductivity models have evolved through the years. The initial models were modifications of the Einstein model which only includes the concentration of the nanofluid. The thermal conductivity enhancement with concentration is almost always underestimated by these common conventional models. In Fig. 4.2 the experimental data and the Maxwell model prediction of thermal conductivity enhancement of f-ZnO/oil nanofluid for

TABLE 4.1 Thermal conductivity models of the unitary and hybrid nanofluids.

Researchers and operation conditions	Thermal conductivity
Maxwel model for nanofluids [1]	$\frac{k_{nf}}{k_f} = \frac{k_{np}+2k_{bf}+2\varphi_v(k_{np}-k_{bf})}{k_{np}+2k_{bf}-\varphi_v(k_{np}-k_{bf})}$
Esfe et al. [2], Al_2O_3/H_2O, $26 \le T(°C) \le 55$, $0.25\% \le \phi \le 5\%$	$\frac{k_{nf}}{k_f} = 0.911 \times T^{\left(-0.0415-\frac{0.404}{\ln(\varphi)}\right)}$
Chon et al. [3], Al_2O_3/H_2O, $20 \le T(°C) \le 70$, $1\% \le \phi \le 4\%$, $d_p = 11, 47, 150$ nm	$\frac{k_{nf}}{k_f} = 1 + 64.7 \times \varphi^{0.7460}\left(\frac{d_{BF}}{d_p}\right)\left(\frac{k_p}{k_{BF}}\right)^{0.7476} \times Pr^{0.9955}Re^{1.2321}$, $Pr = \frac{\mu}{\rho_{BF}\alpha}$, $Re = \frac{\rho_{BF}k_bT}{3\pi\mu^2 l_{BF}}$, $\mu = A \times 10^{B/(T-C)}$, where φ is the volume concentration, d_p is the nanoparticle diameter, d_{BF} is the base fluid molecular diameter, k_p is the thermal conductivity of nanoparticle, k_{BF} is the thermal conductivity of base fluid, Pr is the Prandtl number, Re is the Reynolds number, ρ_{Bf} is the density of the base fluid, μ is the base fluid viscosity, $k_b = 1.3807 \times 10^{-23}$ J/K is the Boltzmann constant, and $l_{BF} = 0.17$nm is the mean free path of the nanoparticles in water.
Moldoveanu et al. [4], Al_2O_3/H_2O, Room temperature, $1\% \le \phi \le 4\%$	$\frac{k_{nf}}{k_f} = 4.54\varphi + 1.01$
Moldoveanu et al. [4], TiO_2/H_2O, Room temperature, $1\% \le \phi \le 4\%$	$\frac{k_{nf}}{k_f} = 1.57\varphi + 1.06$
Esfe et al. [2], Al_2O_3–CNTs/H_2O, $1\% \le \phi \le 3\%$	$\frac{k_{nf}}{k_f} = \frac{-214.83+T}{346.58-106.98\varphi} + \frac{227.69}{T}$
Moldoveanu et al. [4], Al_2O_3–TiO2/H_2O, $293.15 \le T(K) \le 323.15$, $0.5\% \le \varphi_{tot} \le 2.5\%$	$k_{hnf} = -0.347 + 0.003T + 6.639\varphi_{tot} - 109.024\varphi_{tot}^2$, where $\varphi_{tot} = \varphi_{alumina} + \varphi_{titania}$

Buongiorno [5], Al_2O_3/H_2O	$\frac{k_{hnf}}{k_{bf}} = 1 + 7.47\varphi$
Buongiorno [5], TiO_2/H_2O	$\frac{k_{hnf}}{k_{bf}} = 1 + 2.92\varphi - 11.99\varphi^2$
Timofeeva et al. [6], Al_2O_3/H_2O	$\frac{k_{hnf}}{k_{bf}} = 1 + 3\varphi$
Maiga et al. [7], All nanofluids	$\frac{k_{hnf}}{k_{bf}} = 1 + 2.72\varphi + 4.97\varphi^2$
Duangthongsuk et al. [8], TiO_2/H_2O at T = 25°C	$\frac{k_{hnf}}{k_{bf}} = 1.0204 + 0.0249\varphi$
Vajjha and Das [9,10], Al_2O_3/H_2O–EG, Zn/H_2O–EG, CuO/H_2O–EG nanofluids	$k_{nf} = \left(2.8217 \times 10^{-2}\varphi + 3.917 \times 10^{-3}\right)\frac{T}{T_0} + \left(-3.0669 \times 10^{-2}\varphi - 3.91123 \times 10^{-3}\right)$
Feng et al. [11], All nanofluids	$\frac{k_{nf}}{k_{bf}} = \left(1 - \varphi_{p,e}\right)\frac{k_{p,eq} + 2k_{bf} + 2\left(k_{p,e} - k_{bf}\right)(1+\beta)^3\varphi_p}{k_{p,eq} + 2k_{bf} - \left(k_{p,e} - k_{bf}\right)(1+\beta)^3\varphi_p} + \varphi_{p,eq}\left[\left(1 - \frac{3}{2}\varphi_{p,e}\right) + \frac{3\varphi_{p,eq}}{\left(1 - \frac{k_{bf}}{k_{p,e}}\right)}\left[\frac{1}{\left(1 - \frac{k_{bf}}{k_{p,e}}\right)} ln\frac{\frac{d_p}{2} + d_{is}}{\left(\frac{r_p}{2} + d_{is}\right)\left(\frac{k_{bf}}{k_{p,e}}\right)} - 1\right]\right]$, where $k_{p,eq}$ is the equivalent thermal conductivity, d_p particle diameter and d_{is} is the interfacial shell thickness.
Hamid et al. [12], SiO_2–TiO_2 (20:80, 40:60, 50:50, 60:40, and 80:20) /water, 1%, 30°C–80°C	$\frac{k_{hnf}}{k_f} = 1.17(1+\varphi)^{-0.1151}\left(\frac{T}{80}\right)^{0.0437}$, where φ is the volume fraction of TiO_2 in the mixture
Asadi et al. [13], Al_2O_3-MWCNT/oil, 0.125%-1.5% - 25°C–50°C	$k_{hnf} = 0.1534 + 0.00026T + 1.1193\varphi$

(Continued)

TABLE 4.1 (Continued)

Researchers and operation conditions	Thermal conductivity
Esfe et al. [14], MWCNTS–Al_2O_3/EG 0.02%-0.8%, 25°C–50°C	$\frac{k_{hnf}}{k_f} = 0.00281 + 5.28956 \times 10^{-4} T + 0.17384\varphi + 5.31721 \times 10^{-3} T\varphi - 0.23762\varphi^2 - 5.06549 \times 10^{-3} T\varphi^2 + 0.19937\varphi^3$
Li et al. [15], SiO_2–oleic acid/liquid paraffin, 0.005–5wt.%, 25°C–75°C	$\frac{k_{hnf}}{k_{bf}} = 0.756951 + 0.00735396w - 8.26715 \times 10^{-5} T^2 + 0.0123973T$, where w is mass concentration (wt.%)
Sundar et al. [16] ND + Fe_3O_4–60:40 H_2O/EG0%–0.2%, $T_{min} = 20°C$, $T_{max} = 60°C$	$k_{hnf} = 1.041 \times k_{bf} \left[(1+\varphi)^{0.39} \times \left[\frac{T_{min}}{T_{max}} \right]^{0.383 \times 10^{-1}} \right]$

TABLE 4.2 The enhancement of thermal conductivity of Al_2O_3/water nanofluids at various particle sizes.

Researchers	Particle size (nm)	Volume concentration (φ)	Enhancement
Esfe et al. [2]	5	5%	32%
Timofeeva et al. [6]	11	5%	8%
Zhang et al. [17]	20	5%	15%
Timofeeva et al. [6]	20	5%	7%
Wang et al. [18]	28	5.5%	16%
Li and Peterson [19]	36	6%	28%
Wong and Bhaskar [20]	36	8.47%	22%
Das et al. [21]	38	4%	24.3%
Timofeeva et al. [6]	40	5%	10%
Iqbal et al. [22]	40	1%	10.13%
Chon et al. [3]	47	4%	28%
Li and Peterson [19]	47	6%	26%
Yoo et al. [23]	50	1%	4%
Putra et al. [24]	131	4%	24%

TABLE 4.3 The enhancement of thermal conductivity of SiO_2/water nanofluids at various particle sizes.

Researchers	Volume concentration (φ)	Particle size (nm)	Enhancement
Yan et al. [25]	1%	2–10	28%
	3%	2–10	33%
	5%	2–10	44%
Yan et al. [26]	5%	8–166	6.8%
Igbal et al. [22]	1%	15	6.5%
Niu et al. [27]	0.5%	30	41.5%
	0.5%	50	40%

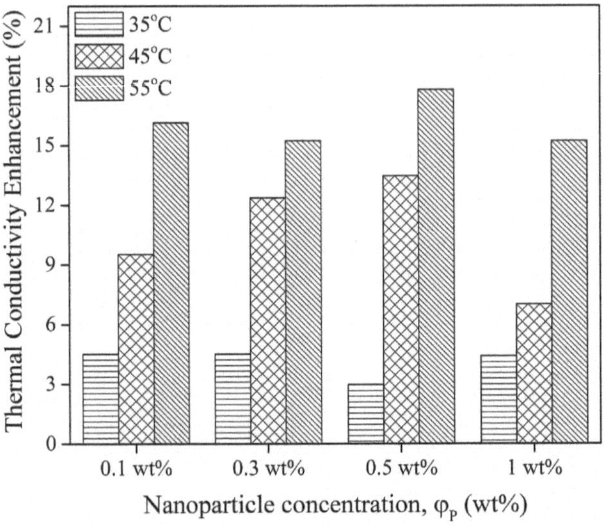

FIGURE 4.1 Thermal conductivity enhancement for increasing temperature of f-ZnO/oil nanofluids [31].

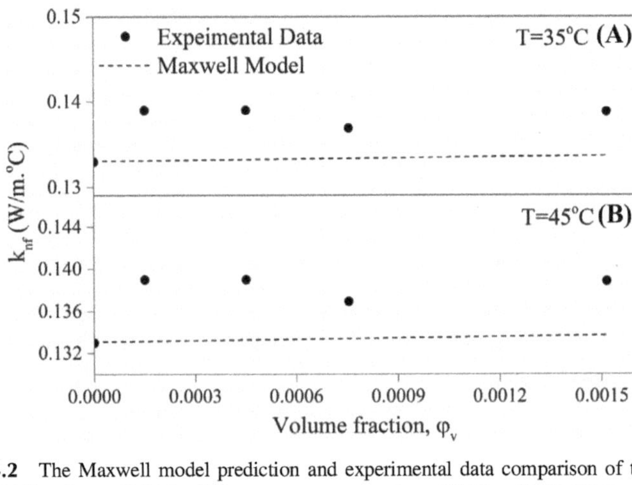

FIGURE 4.2 The Maxwell model prediction and experimental data comparison of thermal conductivity of f-Zn-oil nanofluid for different temperature values, (A) T = 35°C, (B) T = 45°C [31].

increasing concentration at two different temperature values are shown [31]. It is observed that the enhancement in the thermal conductivity is predicted to be smaller by the Maxwell model [31]. It is observed that the thermal conductivity of the nanofluids alters with the Brownian motion of the nanoparticles in the base fluid, nanoparticle shape, size and surface functionalization, as well as agglomeration and temperature. Unfortunately there is no common model to predict the thermal conductivity of the unitary or hybrid nanofluids. Some typical thermal conductivity models for unitary and hybrid nanofluids present in the literature are listed in Table 4.1. Thermal conductivity models contain the concentration and temperature as well as nanoparticle diameter and Brownian effects, see Table 4.1.

The particle size is almost always a nonuniform distribution in a nanofluid blend and typically is a Gaussian distribution. In a study [32] the particle distribution in a SiO_2-methanol nanofluid was reported to have approximately 190 nm particle diameter on average with a Gauss distribution. In Tables 4.2 and 4.3 the enhancement of thermal conductivity of Al_2O_3 and SiO_2 nanofluids with water as a base fluid is listed for various nanoparticle sizes, respectively. In the literature, some researchers reported a thermal conductivity rise for decreasing nanoparticle size—Al_2O_3/water by [33] and Al_2O_3/DI water by [34]. It was concluded that the thermal conductivity rise originates from enhanced Brownian motion due to decreased nanoparticle size. Esfe et al. [35] reported a 32% enhancement in the thermal conductivity of Al_2O_3/water nanofluid compared to base fluid with 5 nm particle size. On the other hand, Timofeeva et al. [6] reported 7%, 8%, and 10% enhancement of the thermal conductivity for particle sizes of 20, 11, and 40 nm, respectively. Chon et al. [3] and Li and Peterson [19] observed similar results for particle size of 47 nm, where the enhancement is around 28%–26%. For higher particle sizes the enhancement becomes notable and reaches up to 30% for particle size as high as 120 nm. Some studies reported that the thermal conductivity increases with increasing particle size (a-SiC/water nanofluids [36] and SiO_2/water nanofluids [37]), and remains unchanged for diameters larger than 50 nm (Al_2O_3/water and Al_2O_3/EG nanofluids [38]). As the particle size decreases the heat transfer surface becomes larger which makes it harder for heat flow and results in a drop of the heat transfer coefficient. Increasing particle size shows different enhancement rates for thermal conductivity in different base fluids. Timofeeva et al. [39] compared the thermal conductivity enhancement of SiC nanofluids with water and EG/water base fluids for increasing average particle sizes at identical volumetric concentration. Thermal conductivity enhancement is reported to be more intense in EG/water base fluid for increasing particle size. This enhancement is clarified by the lower interfacial thermal resistance of EG/water [39].

The shape of nanoparticles influences the thermal conductivity of the nanofluids considerably. Nanoparticles can have a variety of shapes like cylinders, rectangles, blades, bricks, spheres, and arbitrary shapes like a banana. Hamilton and Crosser [40] introduced a shape factor, n, which is a measure of the sphericity of the nanoparticle. The shape factor is $\frac{3}{\varnothing}$, where $\varnothing$ is the ratio of the area of the sphere and the area of the particle with same volumes. The decrease in sphericity leads to an increase in the particle's surface and thus affects the thermal conductivity. Xie et al. [41] conducted an experimental study for spherical and cylindrical shaped nanoparticles for identical concentrations of SiC/water nanofluids and reported a significantly higher thermal conductivity for the cylindrical shaped colloidal dispersion.

Rod-shaped TiO_2/DI water showed higher conductivity compared to spherical TiO_2/DI water [42], and [43] reported an order for the particle shapes from high to low thermal conductivity of Al_2O_3 nanofluids as cylinders, bricks, platelets, and blades, where the platelets and blades showed similar results.

4.2.2 Specific heat capacity

Specific heat capacity is defined as the heat which can increase one unit temperature of a mass unit of a substance. Determination of heat capacity is critical because it is an important parameter for the prediction of heat transfer capacity. In nanofluid applications, specific heat capacity can decrease or increase according to the selected particle and the type of the base fluid [44].

Specific heat capacity of the nanofluids is generally measured by the differential scanning calorimeter (DSC). In this method, the reference material (generally water) and the sample material temperature rise is measured due to heating and the sample heat capacity is determined. The majority of the measurements reported in the literature employ DSC technique due to its accurate and practical character. Additionally the specific heat capacity of a nanofluid can be predicted by the equations given in the literature. One of the pioneering studies to model the specific heat capacity of the nanofluids was proposed by Pak et al. [45] and it is given in Eq. (4.1):

$$\mathrm{c_{p,nf}} = \varphi \mathrm{c_{p,np}} + (1 - \varphi)\mathrm{c_{p,bf}} \tag{4.1}$$

where $\mathrm{c_{nf}}$, φ, $\mathrm{c_{p,np}}$, and $\mathrm{c_{p,bf}}$ are specific heat capacity of nanofluid, particle volume fraction, specific heat capacity of particle, and specific heat capacity of base fluid, respectively.

Xuan et al. [46] improved the specific heat capacity model by including the nanoparticle and base fluid densities which showed a better prediction performance compared to Pak et al.'s [45] model, and it is given as follows:

$$c_{p,nf} = \frac{\varphi c_{p,np}\rho_{np} + (1 - \varphi)\rho_{bf}C_{p,bf}}{\varphi \rho_{np} + (1 - \varphi)\rho_{bf}} \tag{4.2}$$

where c_{nf}, φ, $c_{p,np}$, ρ_{np}, ρ_{bf}, and $c_{p,bf}$ are specific heat capacity of nanofluid, particle volume fraction, specific heat capacity of particle, density of nanoparticle, density of base fluid, and specific heat capacity of base fluid, respectively.

Both Eq. (4.1) and Eq. (4.2) can be used for the estimation of specific heat capacity of unitary nanofluids, however a modification of these models is necessary for hybrid nanofluids where the blend contains more than one nanoparticle. Sundar et al. [47] proposed a new strategy for evaluating the hybrid nanofluids specific heat capacity by determination of averaged nanoparticle specific heat capacity as follows [47]:

$$c_{p,hnp} = \frac{\rho_{np,1}w_1 + \rho_{np,2}w_2}{w_1 + w_2} \tag{4.3}$$

where $c_{p,np}$, $\rho_{np,1}$, w_1, $\rho_{np,2}$, and w_2 are averaged specific heat capacity of hybrid nanoparticles, density of nanoparticle 1, weight percentage of nanoparticle 1, density of nanoparticle 2, and weight percentage of nanoparticle 2, respectively. The averaged $c_{p,hnp}$ value should be used in the models of Pak and Cho [45] (Eq. 4.1) or Xuan and Roetzel [46] (Eq. 4.2). Some of the recent studies related to specific heat of unitary and hybrid nanofluids are listed in Table 4.4 and summarized as follows:

Wole-Osho et al. [51] experimentally investigated specific heat capacity of Al_2O_3-ZnO/H_2O hybrid nanofluids with various mixture ratios using a DSC. The experiments were conducted for a range of various temperature and volume concentration values in the range of 25°C–65°C and 0.33%–1.67%, respectively. It is shown that the specific heat capacity of hybrid nanofluids rises with an increase of temperature and the highest value of it is observed for an equal mixture ratio of Al_2O_3 and ZnO nanoparticles. In addition, they also proposed a polynomial model for the determination of specific heat capacity, see Table 4.5.

Moldoveanu, and Minea [52] conducted an experimental study to develop a single equation for the estimation of the specific heat of three different unitary and two different hybrid nanofluids, see Table 4.5. They performed their experiments for Al_2O_3, SiO_2, and TiO_2 nanoparticles and their hybrid blends. In the study, the relative specific heat of mono and hybrid nanofluids is compared for different volume concentrations, between 0.01% and 0.03%, under room temperature conditions. The experiments revealed that the relative specific heat of Al_2O_3–SiO_2/H_2O hybrid nanofluid is higher than Al_2O_3/H_2O and SiO_2/H_2O nanofluids. In addition, it is reported that Al_2O_3/H_2O nanofluid has higher relative specific heat compared to TiO_2/H_2O and Al_2O_3–TiO_2/H_2O hybrid nanofluids.

Sundar et al. [16] presented a study on thermophysical properties, heat transfer characteristics, and exergy analysis of nanodiamond-Fe_3O_4/H_2O–EG hybrid nanofluid flowing in a tube. They first determined the thermophysical properties for different temperatures (in the range of 20°C–60°C) and volume concentrations (in the range 0.05%–0.2%). They stated that the Nusselt number can be improved approximately up to 11% with the use of hybrid nanofluid with a volume concentration of 0.2% for the tested operation conditions. In other results of the study, they developed an equation for the determination of thermophysical properties having highest deviations of 6%.

Kannaiyan et al. [53] produced Al_2O_3-CuO/H_2O–EG hybrid nanofluids and tested them in order to study the variation of thermophysical properties with temperature (in the range of 20°C–60°C) and volume concentrations (in the range of 0.05%–0.2%). The experimental results of specific heat are compared with the equation given in the literature and it was observed that there is not a satisfactory agreement between experimental and theoretical results for the tested hybrid nanofluid. They stated that this discrepancy can

TABLE 4.4 Some of the recent studies related to specific heat capacity of unitary and hybrid nanofluids.

Researcher	Particle type	Particle size (nm)	Base fluid	Volume concentration (φ)	Surfactant	Temperature range (°C)
Popa et al. [48]	Al_2O_3	10	H_2O, EG	0.24%–5%	-	10–50
	CuO	29				
Sekhar and Sharma [49]	Al_2O_3	47	H_2O	0.01%–1%	-	25–45
Vajjha and Das [9,10]	Al_2O_3	44	H_2O, EG/H_2O	2%–10%	-	42–90
	ZnO	77				
	SiO_2	20				
Satti et al. [50]	Al_2O_3	15–45	PG	0.5%–6%	-	-30–90
	ZnO	36–76				
	CuO	30				
	TiO_2	15				
	SiO_2	30				
Wole-Osho et al. [51]	Al_2O_3	29	H_2O	0.33%–1.67%	-	25–65
	ZnO	70				
Moldoveanu and Minea [52]	Al_2O_3	45	H_2O	1%–3%		Room temperature
	SiO_2	20				
	TiO_2	30				
	Al_2O_3/SiO_2	45–20				
	Al_2O_3/TiO_2	45–30				

Sundar et al. [16]	Nanodiamond/Fe_3O_4	62.41	H_2O/EG	0.05%–0.2%	-	20–60
Kannaiyan et al. [53]	Al_2O_3–CuO	17–20	H_2O/EG	0.05%–0.2%	-	20–60
Çolak et al. [54]	Al_2O_3	46	H_2O	0.0125%–0.2%	Arabic Gum	25–65
	Cu	65				
	Al_2O_3/Cu	46–65				

TABLE 4.5 Some correlations for determination of specific heat capacity of unitary and hybrid nanofluids.

Researchers	Specific heat capacity correlations
Wole-Osho et al. [51]	$c_{p,nf} = \left(-0.29993 + 0.00227 \times T_f + 1.52321 \times 10^{-6} \times T_f^2 + 2.82505 \times R - 3.05882 \times R^2 - 2.95905 \times \varphi - 27.55833 \times \varphi^2\right) \times c_{p,f}$
Moldoveanu and Minea [52]	$c_{p,nf} = \left(\frac{\rho_p}{\rho_f}\right)^{0.2} \times \left(1+\frac{d_p}{50}\right)^{0.4167} \times \left(1-\frac{\varphi}{100}\right)^{2.272} \times c_{p,f}$
Sundar et al. [16]	$c_{p,hnf} = c_{p,bf} \times \left[(1+\varphi)^{-0.148*10^{-1}} \times \left[\frac{T_{min}}{T_{max}}\right]^{-0.537 \times 10^{-4}}\right]$
Satti et al. [50]	$\frac{c_p}{c_{p,0}} = 0.6294 + 0.3707 \times \left(\frac{T}{T_0}\right)$ $c_{p,0} = 3250 \frac{J}{kgK}$
Vajjha and Das [9,10]	Al_2O_3/H_2O $\frac{c_{p,nf}}{c_{p,bf}} = \frac{\left((0.0008911 \times T) + 0.5179 \times \left(\frac{c_{p,np}}{c_{p,bf}}\right)\right)}{(0.4250 + \varphi)}$ SiO_2/H_2O $\frac{c_{p,nf}}{c_{p,bf}} = \frac{\left((0.001769 \times T) + 1.1937 \times \left(\frac{c_{p,np}}{c_{p,bf}}\right)\right)}{(0.8021 + \varphi)}$ ZnO/H_2O $\frac{c_{p,nf}}{c_{p,bf}} = \frac{\left((0.0004604 \times T) + 0.9855 \times \left(\frac{c_{p,np}}{c_{p,bf}}\right)\right)}{(0.299 + \varphi)}$

Sekhar and Sharma [49]	$c_p = 0.8429 \times \left(1 + \frac{T_{nf}}{50}\right)^{-0.3037} \times \left(1 + \frac{d_p}{50}\right)^{0.4167} \times \left(1 + \frac{\varphi}{100}\right)^{2.272}$	
Popa et al. [48]	$Al_2O_3/H_2O{-}EG \; \frac{c_{p,nf}}{c_{p,bf}} = \frac{\left((-0.00015564609 \times T) + 3.86103789525 \times \left(\frac{c_{p,np}}{c_{p,bf}}\right)\right)}{(1.15981407441 + \varphi)}$ $Al_2O_3/H_2O \quad \frac{c_{p,nf}}{c_{p,bf}} = \frac{\left((-0.00354913844 \times T) + 7.97557769531 \times \left(\frac{c_{p,np}}{c_{p,bf}}\right)\right)}{(0.40507654022 + \varphi)}$ $CuO/H_2O \quad \frac{c_{p,nf}}{c_{p,bf}} = \frac{\left((-0.00036213307 \times T) + 2.22570776506 \times \left(\frac{c_{p,np}}{c_{p,bf}}\right)\right)}{(0.17019394822 + \varphi)}$	

be overcome by considering nanolayer formation between bulk fluid and the nanoparticle.

Colak et al. [54] presented an experimental study for the determination of the specific heat capacity of water-based unitary and hybrid nanofluids with Al_2O_3 and Cu nanoparticles. In experiments, the differential thermal analysis method is employed and the specific heat capacities of Al_2O_3/H_2O, Cu/H_2O, and Al_2O_3-Cu/H_2O are specified. The experimental results revealed that the specific heat capacity increases with increasing temperature and reduces with an increase in volume concentration. In addition, it is seen that Cu/H_2O nanofluid has the lowest specific heat capacity value compared to other tested nanofluids.

Satti et al. [50] investigated the variation of specific heat capacity of PG and H_2O-based nanofluids with Al_2O_3, ZnO, CuO, TiO_2, and SiO_2 nanoparticles in a large range of temperature and volume concentrations, between 30°C and 90°C and 0.5%–6%, respectively. It was observed that the nanofluid with the higher volume concentration had the lowest capacity for all tested nanofluids. In addition, the experimental results revealed that the specific heat of nanofluids is not affected by nanoparticle size and it is necessary to develop a new correlation for accurate prediction. They proposed a new correlation by using a database with 610 data points and stated that this correlation can predict the experimental data in the range of ±5%, see Table 4.5.

Vajjha and Das [9,10] established an experimental setup for the development of new correlations for c_p by using nanofluids with Al_2O_3, ZnO, and SiO_2 nanoparticles. In the study, H_2O and H_2O/EG mixture was selected as the base fluid and the effects of both temperature and volume concentration on specific heat capacity were studied for different operation conditions. The developed correlations (see Table 4.5) have maximum error values of 5%, 4.4%, and 3.1% for nanofluids having Al_2O_3, ZnO, and SiO_2 nanoparticles, respectively.

Sekhar et al. [49] investigated viscosity and specific heat capacity properties of Al_2O_3/H_2O nanofluids with volume concentrations up to 1% and temperature between 25°C and 45°C. They observed that specific heat capacity of the nanofluid reduces with the increase of volume concentration because of an improvement in thermal diffusivity. They also proposed a generalized correlation (see Table 4.5) for the estimation of specific heat capacity by using the available experimental data for different nanofluids with nanoparticles such as Al_2O_3, TiO_2, CuO, and SiO_2. They included the particle size in their correlation which showed good agreement with the experimental data with deviations in the ranges between +10% and −8%.

Popa et al. [48] measured the specific heat capacity of CuO/H_2O, Al_2O_3/H_2O, and Al_2O_3/EG and investigated the variation of it with temperature and volume concentrations. They observed that the specific heat capacity of nanofluids having an EG-based fluid is lower than nanofluid with H_2O as a base fluid. In addition, the coefficients in the equation of Vajjha and Das [11] were updated in order to obtain a new equation with a better prediction performance.

Fig. 4.3A represents the effect of nanofluid type and temperature on specific heat capacity for volume concentration of 0.33%. It can be seen from the figure that the specific heat capacity of nanofluids increased with elevated temperatures where the specific heat capacity of base fluid (water) has no significant change with increasing temperature. Also, it should be noted that the specific heat capacity values of hybrid nanofluids are significantly lower than unitary nanofluids and more sensitive to temperature alteration. Fig. 4.3B shows the variation of specific heat capacity of CuO/PG with temperature and volume concentration. It is observed that the volume concentration increment in nanofluid decreases the specific heat capacity for all tested temperatures and a temperature rise increases the c_p as expected. Fig. 4.3C depicts the comparison of experimental and predicted specific heat capacity results of hybrid nanofluids and it can be seen that the specific heat capacity of $Al_2O_3-Si_2O_3/H_2O$ hybrid nanofluid can be predicted with a lower

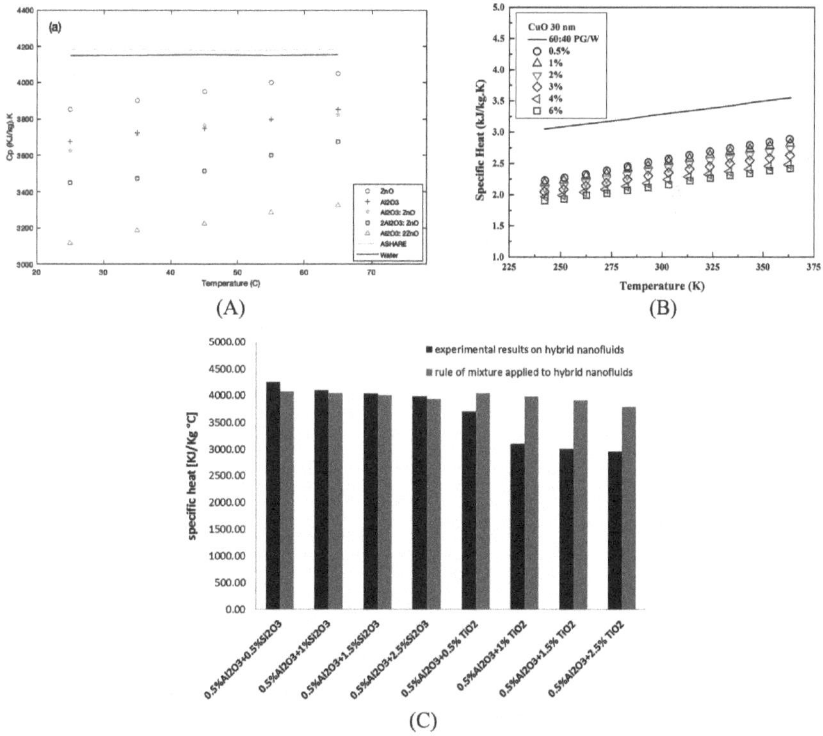

FIGURE 4.3 The variation of density with temperature for (A) different nanofluids [51]; (B) different volume concentrations [50]; (C) the comparison of experimental and predicted specific heat capacity results of hybrid nanofluids [52].

deviation (3%) with the application of mixture rule, where the specific heat capacity of $Al_2O_3-TiO_2/H_2O$ hybrid nanofluid has a deviation in the range between 4% and 14%.

Some correlations for determination of specific heat capacity of unitary and hybrid nanofluids are given in the Table 4.5.

4.2.3 Density

Density is defined as the ratio of mass to volume and it can be considered as one of the important thermophysical properties of incompressible substances. In nanofluid applications, it is known that the density of nanoparticles, which can be selected during preparation of the nanofluid, is higher than the base fluid density and it has serious effects on stability, thermal performance, and pressure drop. In the design stages of thermal equipment that operate with nanofluids, it is important to determine density accurately due to the aforementioned effects.

The density of a nanofluid can be determined by using a density meter, calculation of ratio between mass and volume, or using the developed correlations readily available in the literature. Since the correlations show rather good agreement with experiments in the prediction of the density of nanofluids, the researchers generally prefer the models of experimental density measurements. In the literature, the density of unitary nanofluids can be predicted by using the equation proposed by Pak et al. [45] and it is given as follows:

$$\rho_{nf} = \varphi\rho_{np} + (1-\varphi)\rho_{bf} \tag{4.4}$$

where ρ_{nf}, φ, ρ_{np}, and ρ_{bf} are the density of nanofluid, particle volume fraction, density of particle, and density of base fluid, respectively.

In addition, the density of hybrid nanofluids can be estimated by using equation given below [55]:

$$\rho_{hnf} = \varphi_1\rho_{np,1} + \varphi_2\rho_{np,2} + (1-(\varphi_1+\varphi_2))\rho_{bf} \tag{4.5}$$

where ρ_{hnf}, φ_1, $\rho_{np,1}$, φ_2, $\rho_{np,2}$, and ρ_{bf} are density of hybrid nanofluid, particle volume fraction of first nanofluid having nanoparticle 1, density of particle 1, particle volume fraction of second nanofluid having nanoparticle 2, density of nanoparticle 2, and density of base fluid, respectively.

Some of the recent studies related to the density of unitary and hybrid nanofluids are given in Table 4.6 and summarized as follows:

Vidhya [30] determined a ZnO–MgO hybrid nanofluid by using an H_2O and EG mixture at 40:60% as a base fluid. In the study, experiments were conducted for temperatures of 30–80°C and concentrations of 0.0125%–0.1% and the effect of temperature and concentration on density of ZnO–MgO hybrid nanofluids were investigated. It was concluded that the density of a nanofluid has the highest value for the lowest temperature and highest volume concentration.

TABLE 4.6 Some of the recent studies related to density of unitary and hybrid nanofluids.

Researcher	Particle type	Particle size (nm)	Base fluid	Volume concentration φ	Surfactant	Temperature range (°C)
Vidhya et al. [30]	ZnO/MgO	225	H_2O–EG	0.0125%–0.1%	Cetyl trimethyl ammonium bromide	30–80
Yarmand et al. [56]	GNP/Ag	2000	H_2O	0.02%–0.1%	-	20–40
Wanatasanappan et al. [57]	Al_2O_3/CuO	8/24	H_2O–EG	0.1%	Linear alkyl benzene sulfonate (LAS)	30
Hozien et al. [58]	TiO_2	14	H_2O	0.25%	-	30–60
	ZnO	20				
	Ag	16				
Kumar et al. [59]	Al_2O_3–MWCNT	-	H_2O	0.25%–2.0%	Cetyl trimethyl ammonium bromide	25–50
	TiO_2–WCNT	-				
	ZnO–MWCNT	-				
	CeO_2–MWCNT	-				
Said et al. [60]	CNF	501	H_2O	0.04%	-	20–65
	F–CNF	333				
	rGO	570				
	F–CNF/rGO	370				

(Continued)

TABLE 4.6 (Continued)

Researcher	Particle type	Particle size (nm)	Base fluid	Volume concentration φ	Surfactant	Temperature range (°C)
Kishore et al. [61]	Cu–GnP		H_2O	0.01%–0.012%	Copper coated trimethylammonium bromide	43–100
Asokan et al. [62]	Al_2O_3		H_2O–EG	0.02%–0.06%	Polyvinyl pyrrolidone (PVP)	30
	CuO					
	Al_2O_3–CuO					

Yarmand et al. [56] conducted experimental studies to determine thermophysical properties and heat transfer characteristics of graphene nanoplatelets (GNP)−Ag/H_2O hybrid nanofluid flowing in a tube having diameter of 10 mm. The density of hybrid nanofluid having volume concentrations of 0.02%−0.1% and temperature of 20°C−40°C was tested and the results showed that the density reduces with increasing temperature. In addition, they stated that the density of nanofluid can be increased with increased concentration of the nanoparticle.

Wanatasanappen et al. [57] studied the effect of nanoparticle mixture ratio in a hybrid nanofluid on thermophysical properties. They investigated Al_2O_3−CuO/H_2O−EG hybrid nanofluid for volume concentration of 1.0% and at 30°C. The hybrid nanofluid with a nanoparticle mixture ratio (Al_2O_3: CuO) of 0.6:0.4 showed the highest density augmentation compared to the base fluid as high as 4.8%. They also stated that it is not possible to exhibit a general trend related to the variation of density with nanoparticle mixture ratio.

Hozien et al. [58] investigated the heat transfer of water-based nanofluids having TiO_2, ZnO, and Ag nanoparticles flowing in helically coiled tube for volume concentration of 0.25%. In the study, heat transfer as well as density enhancement was evaluated for temperature in the range of 30−60 °C and the highest increase was evaluated to be up to 24% for nanofluid having Ag nanoparticles compared to water. The density of Ag/H_2O exhibits the highest value at the temperature range and the trend is approximately linear.

Kumar et al. [59] researched the thermophysical properties of Al_2O_3-MWCNT/H_2O, TiO_2−WCNT/H_2O, ZnO−MWCNT/H_2O, and CeO_2−MWCNT/H_2O hybrid nanofluids and used an artificial neural network approach for prediction of the properties. In the study, the effect of volume concentration (in the range of 0.25%−2.0%) and temperature (25°C and 50°C) on densities of hybrid nanofluids was determined and the results showed that a reduction in hybrid nanofluids is approximately up to 2% for the tested temperatures.

Said et al. [60] prepared three different nanofluids [carbon nanofiber (CNF)/H_2O, functionalized CNF (F−CNF)/H_2O, and reduced graphene oxide (rGO)/H_2O] and one hybrid nanofluid [rGo coated over F-CNF (F−CNF-rGO)/H_2O] with volume concentrations of 0.04%. They experimentally determined the thermophysical properties of the nanofluids and performed a comparison between the predicted values, which were evaluated by using fuzzy modeling approach. The variations of densities of nanofluid with temperature were estimated and it was seen that the highest density increase was obtained for CNF/H_2O nanofluid at 0.19% at a temperature of 20°C.

Kishore et al. [61] produced a Cu−GnP/H_2O hybrid nanofluid with volume concentrations of 0.01% and 0.02% in order to investigate the thermophysical properties and stability, and compared their results with H_2O. They also studied the effect of nanoparticle mixture ratio (30:70 Cu−GnP and 70:30 Cu−GnP) on the density of the hybrid nanofluid and it was observed that the Cu−GnP/H_2O hybrid with a mixture ratio of 70:30 Cu−GnP exhibited the highest density.

Asokan et al. [62] investigated the thermal properties of DI H_2O and EG mixture based two unitary nanofluids (with Al_2O_3 and CuO nanoparticles) and one hybrid nanofluid (with Al_2O_3-CuO nanoparticles). They determined the densities of nanofluids with volume concentrations in the range between 0.02% and 0.06% for operation temperature of 30 °C. It was shown that the density of a unitary nanofluid with Al_2O_3 nanoparticles is slightly higher than a hybrid nanofluid with Al_2O_3-CuO nanoparticles.

As can be seen in Fig. 4.4A, the density of a nanofluid decreases with an increase in temperature because of the increase in volume of the base fluid as well as the increased molecular motion within the nanofluid. In addition, the density of a nanofluid increases with the increase of volume concentration since the density of solid nanoparticles used for the production of a nanofluid is higher than the base fluid. Fig. 4.4B represents the effect of temperature and nanoparticle type on density of nanofluids. It can be observed that density slightly increases with the usage of nanofluid instead of base fluid and it reduces with increasing temperature as expected. The highest rise in the density is observed for the largest nanoparticle size where the particle size of CNF = 510 nm. Fig. 4.4C compares the density of unitary and hybrid nanofluid for different volume concentrations and it is revealed that the density of a unitary nanofluid is slightly higher than hybrid nanofluid. It can be concluded that in the view of the density, material/material combination selection is very important for nanofluids.

4.2.4 Viscosity

One common example to show the effect of solid particles in a fluid flow is the flow around a circular object. Even in laminar regimes, the wake of the spherical object shows a noteworthy alteration in the flow field for larger diameters, with almost no viscosity change in the fluid. However for smaller diameter particle inclusion with a certain concentration, where the particle diameter is smaller than 1 μm, due to higher colloidal forces between the particles, the viscosity change becomes more significant. Nanofluids with both low and high viscosity base fluids, like water and EG, respectively, display Newtonian behavior at lower nanoparticle concentrations [63−65]. For a higher volumetric concentration of nanoparticles water-based nanofluids start to display non-Newtonian behavior. Tseng et al. [66] investigated the rheological behavior of Al_2O_3/H_2O nanofluid with an average particle size of 37 nm, volumetric concentration of 0.01−0.16, and pH 2 to 11. They reported a pseudoplastic behavior and very low yield stress at shear rates between 1 and 100 s^{-1} at lower concentrations. At higher concentrations shear thinning behavior is observed and a power law model can be used as a constitutive relation.

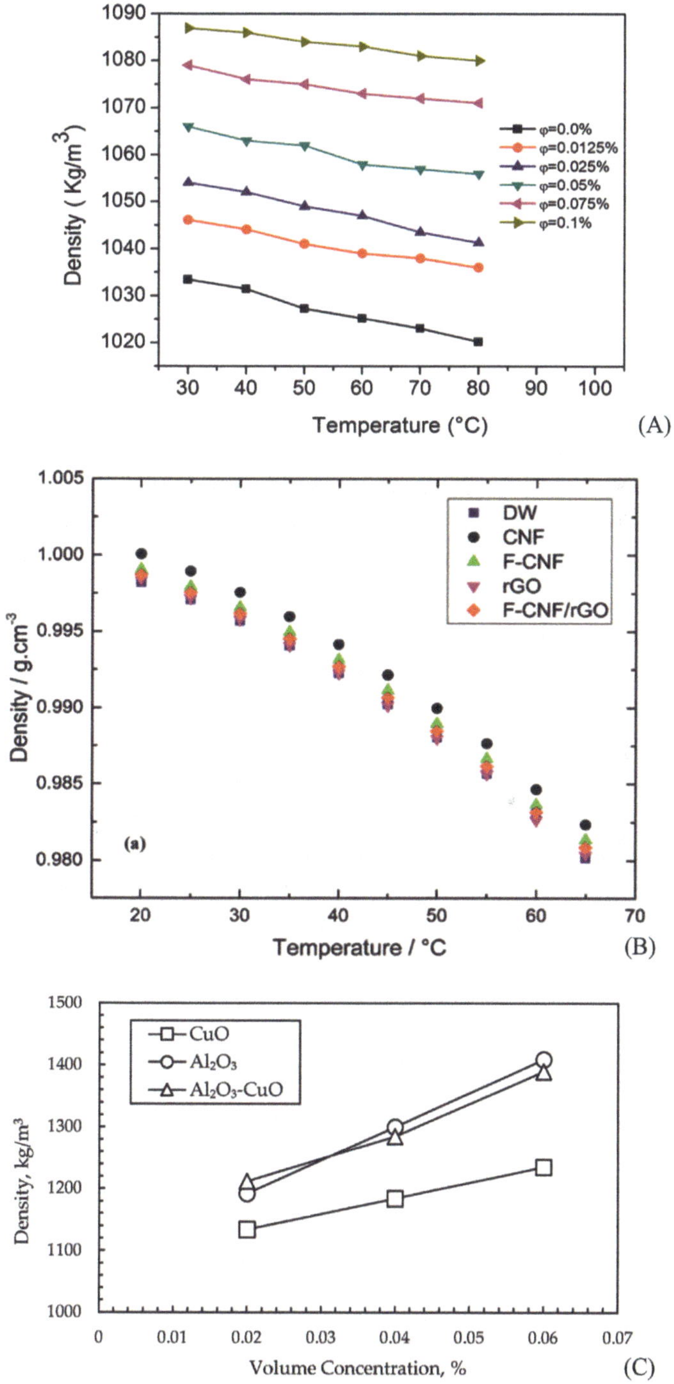

FIGURE 4.4 The variation of density with temperature (A) for different volume concentrations [30], (B) for different nanofluids [60], and (C) the variation of density with nanofluid and hybrid nanofluid [62].

In heat transfer applications, it is important to predict the viscosity of the fluid since it is directly related to pressure drop. It is also known that the addition of the nanoparticles increases the viscosity of the working fluid. Common observations can be summarized as the nanofluid viscosity increases with increasing particle concentration and decreases with increasing temperature [27,28,67]. Some selected viscosity models proposed in the literature for various unitary and hybrid nanofluids are listed in Table 4.7 and it is observed that these models are generally a function of base fluid viscosity and the volume concentration of the nanofluid.

In a series of experimental studies, Mahbubul et al. [74,75] reported a considerable viscosity increase in nanorefrigerants with increasing concentration and a decrease with increasing temperature. Mahbubul et al. [76] investigated the nanorefrigerant 0.5−2.0 vol.% Al_2O_3/R141b at 5°C−30°C and described a 179-fold increase in viscosity at 2%vol. and 5°C. Mahbubul et al. [67] studied experimentally the thermal performance of the nanorefrigerant, 5 vol.% Al_2O_3/R−134a, for the temperature range of 283−308K. The viscosity of the nanofluid was modeled using the Brinkman model which is a generalized Newtonian model (see Table 4.7). It was observed that the viscosity augmented 13.6% with a volume fraction increase at the same temperature. A series of studies [77−80] investigated the effect of four different nanoparticles on the viscosity of R134-a at 1%−5%vol. concentration and 300K−325K temperature. The nanoparticles were CuO, Al_2O_3 [78], SWCNT [79], and TiO_2 [80], where the universal remark is the viscosity increase with increasing volume concentration and viscosity decrease at higher temperatures. Stability also affects the viscosity of the nanofluid in time. The duration and temperature of sonication decreases the viscosity. At higher temperatures Brownian motion is enhanced which leads to lower viscosity. The intermolecular adhesion weakens with increased temperature and increased sonication time, which leads to lower viscosity [81].

In an experimental study [27], the viscosity was reported for SiO_2/H_2O nanofluid at 0.5% concentration for three different particle sizes, 15, 30, and 50 nm. The viscosity showed a drop with increasing particle size. In another study (2015) Al_2O_3/H_2O nanofluid showed a viscosity rise with increasing particle mean diameter. They reported that nanoparticle size of 200 nm had a lower viscosity than a 250 nm nanofluid. The results in the literature about the effect of the nanoparticle size on the viscosity are not consistent so one cannot draw a comprehensive statement about viscosity alteration with particle size.

The water-based nanofluids exhibit Newtonian behavior which can be modeled with the well-known Einstein equation for viscosity [28]. An experimental study [28] compared the rheological effect of Al_2O_3, TiO_2, and SiO_2 nanoparticles on water as a base fluid. The viscosity of all nanofluids reduced with increasing temperature and the highest viscosity value is observed for SiO_2 and the lowest is observed for Al_2O_3 nanofluid.

TABLE 4.7 Viscosity models of the unitary and hybrid nanofluids.

Researchers and operation conditions	Viscosity
Al_2O_3/H_2O, TiO_2/H_2O, SiO_2/H_2O for $\varphi < 5\%$, Hussein et al. [28]	$\mu_{nf} = \mu_{bf}(1 + 2.5\varphi)$
$Al_2O_3/R-134a, 283 < T(°C) < 308$, $\varphi = 5\%$, [67]	$\mu_{nf} = \mu_{bf} \frac{1}{(1-\varphi)^{2.5}}$
Al_2O_3/water [18]	$\frac{\mu_{nf}}{\mu_f} = 1 + 7.3\varphi + 123\varphi^2$
$SiO_2/EG-H_2O(60:40)$, $2\% < \varphi < 10\%$, $-35 < T(°C) < 50$ [68]	$Log(\mu_{nf}) = Ae^{-BT}$, where $A = 0.1193(\varphi)^3 - 1.9289(\varphi)^2 - 2.245\varphi + 167.17 B = -7 \times 10^{-6}(\varphi)^2 - 0.0004(\varphi) + 0.0192$
TiO_2/water [8]	$\frac{\mu_{nf}}{\mu_f} = 1.0226 + 0.0477\varphi - 0.0112\varphi^2$ for $T = 15^oC$ $\frac{\mu_{nf}}{\mu_f} = 1.0130 + 0.0920\varphi - 0.0150\varphi^2$ for $T = 25^oC$ $\frac{\mu_{nf}}{\mu_f} = 1.0180 + 0.01120\varphi - 0.0177\varphi^2$ for $T = 35^oC$
TiO_2/water, [5]	$\frac{\mu_{nf}}{\mu_f} = 1 + 5.45\varphi + 108.2\varphi^2$
Al_2O_3/water, [5]	$\frac{\mu_{nf}}{\mu_f} = 1 + 39.11\varphi + 533.9\varphi^2$
TiO_2/water, [29]	$\frac{\mu_{nf}}{\mu_f} = 1.5056 - 0.3202\varphi + 0.2302\varphi^2$
Al_2O_3/water, [29]	$\frac{\mu_{nf}}{\mu_f} = 2.3792 - 1.5449\varphi + 0.6152\varphi^2$
$Al_2O_3-TiO_2$/water, [29]	$\frac{\mu_{nf}}{\mu_f} = 1.2921 - 0.1011 + 0.3371\varphi^2$
Al_2O_3/water, [69]	$\frac{\mu_{nf}}{\mu_f} = 1 + 1631\left(\frac{\varphi}{1-\varphi}\right)^{2.8}$
$AlN-Al_2O_3$/water, 1%–4% [70]	$\frac{\mu_{hnf}}{\mu_{bf}} = \left(1 + \frac{\varphi}{100}\right)^{11.3}\left(1 + \frac{T_{hnf}}{70}\right)^{-0.038}\left(1 + \frac{d_p}{170}\right)^{-0.061}$

(Continued)

TABLE 4.7 (Continued)

Researchers and operation conditions	Viscosity
MWCNT–SiO_2(50:50)/water-EG(50:50), 0.0625%–2%, 27.5°C–50°C [71]	$\mu = m\dot{\gamma}^{n-1}$ where $m = 0.01125 + \left(\frac{38.19 - 0.3T}{7.655 + 0.6953T}\right)(0.01138\varphi + 0.5529\varphi^2 - 0.3613\varphi^3 + 0.07\varphi^4)$, consistency index $n = 0.8543 + \left(\frac{-3.03 + 1.418T}{15.8 + 0.391T}\right)(-0.7366\varphi + 0.8519\varphi^2 - 0.4552\varphi^3 + 0.08871\varphi^4)$, power law index
MWCNT–ZnO/Engine oil, 0.125%–1%, 5°C–55°C Asadi et al. [72]	$\mu_{hnf} = 796.8 + 76.26\varphi + 12.88T + 0.7695\varphi T + \frac{-196.9T - 16.54\phi T}{\sqrt{T}}$
TiO_2–Cu/Water (0.02%–0.02%) [73]	$\mu_{hnf} = \frac{\mu_f}{(1-\varphi_1)^{2.5}(1-\varphi_2)^{2.5}}$

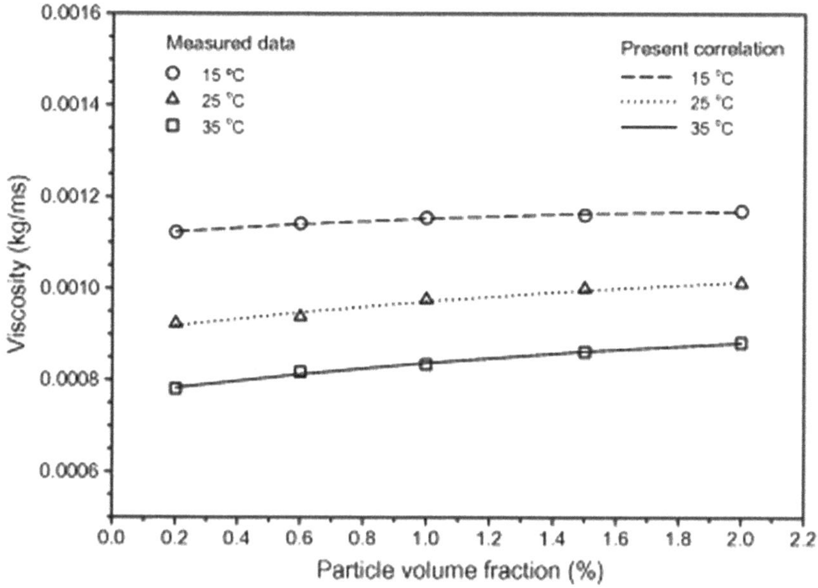

(A)

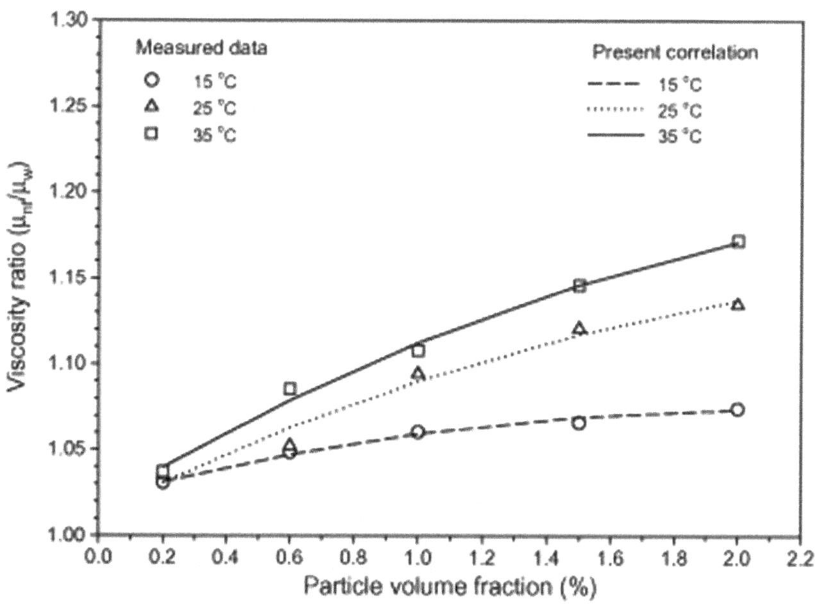

(B)

FIGURE 4.5 (A) Visocity alteration; (B) visocity ratio alteration—of TiO_2/water nanofluid with concentration and temperature—comparison of the experimental and the proposed model results (see Table 4.7) [8].

The relative viscosity is the ratio of the nanofluid viscosity to base fluid viscosity which is used as an enhancement measure for a viscosity rise or drop. Moldoveanu et al. [29] investigated experimentally the effect of concentration on relative viscosity of water-based alumina and titania nanofluids and compared the results with the hybrid nanofluid. It was reported that the concentration augmentation improved the relative viscosity of all unitary and hybrid nanoblends. The models proposed in the literature demonstrate a similar trend. The discrepancy between the enhancement curves may be caused by the size and shape difference of the nanoparticles used in particular studies.

The temperature effect for increasing concentration on relative viscosity of TiO_2/water nanofluid is shown in Fig. 4.5. In this figure the model and experimental result comparison is shown [8]. Although it is reported that the viscosity of the TiO_2/water nanofluid decreased with increasing temperature the relative viscosity showed an adverse effect on viscosity which may be caused by the nanofluid preparation procedure and particle morphology.

4.3 Conclusion

The main scope of this chapter is to highlight the alteration in thermophysical properties of working fluids with one or more types of nanoparticle inclusion and to present the most recent models for the prediction of thermophysical properties of the nanoblends available in the literature. The three main properties that affect the thermophysical properties of the unitary and hybrid nanofluids are the volumetric concentration, the operating temperature, and the particle type. The particle size on the other hand also influences those properties. However, due to dissimilar production procedures of the nanoparticles it needs further and detailed investigation during experimental studies. The important conclusions can be presented as follows:

- Nanofluids inherently have low concentration and generally it is lower than 2%, however up to 5% is acceptable, to avoid agglomeration and sedimentation, which are the main reasons for an unstable suspension. The volumetric concentration of the nanoparticles increase enhances the thermal conductivity, and augments the density, viscosity, and thus the pressure drop, and reduces the specific heat capacity for the nanofluids.
- In application, the unitary and hybrid nanofluids are generally employed in single−phase flows. H_2O, EG, and H_2O/EG are the most commonly used base fluids and thus the operation temperature does not exceed 80°C in real life applications. In the operating temperature interval, $5°C < T < 80°C$, both the thermal conductivity and the specific heat capacity increase with increasing temperature, and density and viscosity reduce with temperature rise.
- Particle type has a significant effect on the thermophysical properties. The different nanoparticles enhance the thermophysical properties of the

nanoblends to different levels. Corrosion performance, clogging, cost-effectiveness, availability, and compactness (the surface volume ratio of the nanoparticle) of the nanoparticle are the main criteria for the selection of the nanoparticles. It is generally reported that the hybrid nanofluids perform better in terms of thermal conductivity compared to unitary nanofluids and display a higher viscosity rise, density rise, and lower stability. However, the selection of a compatible pair of nanoparticles for the production of the hybrid nanofluids also plays an important role. For example, the high thermal conductivity gap between two nanoparticles results in a lower overall thermal conductivity of the hybrid nanofluid compared to unitary nanoblends.

- Particle size is seldom uniform in a nanofluid and usually the average diameter value is used in the reports. In an ideal case the diameter distribution should display low variance and the nanoparticle should be pure, however due to different suppliers globally, these conditions might not be satisfied properly. For this reason it is difficult to draw a rule of thumb between particle size and thermophysical properties.
- The nanoparticle inclusion to the base fluid alters the thermal conductivity, density, specific heat capacity, and viscosity rather rapidly, thus the classical models such as Maxwell and Einstein models proposed for suspensions fail to predict the result accurately. Additionally, so far, there is no global model which can predict the thermophysical properties with a good prediction performance. For that reason it is necessary to select and use the models that have been developed for a specific operation condition interval and particular particle type.

Environment-friendly refrigerants are becoming more popular because of their low ozone depleting and low global warming potentials and it is necessary to lower their prices in order to be competitive with traditional refrigerant alternatives. Besides a reduction in prices, the performance of these refrigerants can be enhanced with the addition of nanoparticles notably. Thus the heat transfer characteristics of nanofluids during phase change such as condensation and boiling characteristics are also exceedingly important. Especially for hybrid nanorefrigerants, more experimental studies should be encouraged. It is expected that this study will be useful for researchers working on this subject.

Nomenclature

c_p	specific heat capacity at constant pressure
d	diameter
DSC	differential scanning calorimeter
EG	ethylene glycol
l_{BF}	the mean free path of the nanoparticles in water
k	thermal conductivity

k_b Boltzmann constant
m consistency index
n power law index
Pr Prandtl number
Re Reynolds number
T temperature
w mass concentration

Subscripts

1,2 different nanoparticles in a hybrid nanofluids
bf base fluid
eq equivalent
f fluid
is interfacial shell
max maximum
min minimum
nf nanofluid
np nanoparticle
p particle
tot total
v volume

Greek Letters

μ viscosity
ρ density
ϕ volume fraction
$\emptyset$ ratio of the area of the sphere and the area of the particle with same volume

References

[1] J.C. Maxwell, A Treatise on Electricity and Magnetism: Pt. III. Magnetism. Pt. IV. Electromagnetism, Vol. 2, Clarendon Press, 1881.

[2] M.H. Esfe, S. Saedodin, W.M. Yan, M. Afrand, N. Sina, Study on thermal conductivity of water-based nanofluids with hybrid suspensions of CNTs/Al_2O_3 nanoparticles, J. Therm. Anal. Calorim. 124 (1) (2016) 455–460.

[3] C.H. Chon, K.D. Kihm, S.P. Lee, S.U. Choi, Empirical correlation finding the role of temperature and particle size for nanofluid (Al_2O_3) thermal conductivity enhancement, Appl. Phys. Lett. 87 (15) (2005) 153107.

[4] G.M. Moldoveanu, A.A. Minea, G. Huminic, A. Huminic, Al_2O_3/TiO_2 hybrid nanofluids thermal conductivity, J. Therm. Anal. Calorim. 137 (2) (2019) 583–592.

[5] J. Buongiorno, Convective transport in nanofluids, J. Heat. Transf. 128 (2006) 240–250.

[6] E.V. Timofeeva, A.N. Gavrilov, J.M. McCloskey, Y.V. Tolmachev, S. Sprunt, L.M. Lopatina, et al., Thermal conductivity and particle agglomeration in alumina nanofluids: experiment and theory, Phys. Rev. E 76 (6) (2007) 061203.

[7] S.E.B. Maiga, S.J. Palm, C.T. Nguyen, G. Roy, N. Galanis, Heat transfer enhancement by using nanofluids in forced convection flows, Int. J. Heat. Fluid Flow. 26 (4) (2005) 530–546.

[8] W. Duangthongsuk, S. Wongwises, Measurement of temperature-dependent thermal conductivity and viscosity of TiO_2-water nanofluids, Exp. Therm. Fluid Sci. 33 (4) (2009) 706–714.

[9] R.S. Vajjha, D.K. Das, Experimental determination of thermal conductivity of three nanofluids and development of new correlations, Int. J. Heat. Mass. Transf. 52 (2009) 4675–4682. Available from: https://doi.org/10.1016/j.ijheatmasstransfer.2009.06.027.

[10] R.S. Vajjha, D.K. Das, Specific heat measurement of three nanofluids and development of new correlations, J. Heat. Transf. 131 (7) (2009).

[11] Y. Feng, B. Yu, P. Xu, M. Zou, The effective thermal conductivity of nanofluids based on the nanolayer and the aggregation of nanoparticles, J. Phys. D: Appl. Phys. 40 (2007) 3164–3171. Available from: https://doi.org/10.1088/0022-3727/40/10/020.

[12] K.A. Hamid, W.H. Azmi, M.F. Nabil, R. Mamat, K.V. Sharma, Experimental investigation of thermal conductivity and dynamic viscosity on nanoparticle mixture ratios of TiO_2-SiO_2 nanofluids, Int. J. Heat. Mass. Transf. 116 (2018) 1143–1152.

[13] A. Asadi, M. Asadi, A. Rezaniakolaei, L.A. Rosendahl, M. Afrand, S. Wongwises, Heat transfer efficiency of Al_2O_3-MWCNT/thermal oil hybrid nanofluid as a cooling fluid in thermal and energy management applications: an experimental and theoretical investigation, Int. J. Heat. Mass. Transf. 117 (2018) 474–486.

[14] M.H. Esfe, S. Esfandeh, M. Rejvani, Modeling of thermal conductivity of MWCNT-SiO_2 (30:70%)/EG hybrid nanofluid, sensitivity analyzing and cost performance for industrial applications, J. Therm. Anal. Calorim. 131 (2) (2018) 1437–1447.

[15] Z. Li, S. Asadi, A. Karimipour, A. Abdollahi, I. Tlili, Experimental study of temperature and mass fraction effects on thermal conductivity and dynamic viscosity of SiO_2-oleic acid/liquid paraffin nanofluid, Int. Commun. Heat. Mass. Transf. 110 (2020) 104436.

[16] L.S. Sundar, S. Mesfin, E.V. Ramana, Z. Said, A.C. Sousa, Experimental investigation of thermo-physical properties, heat transfer, pumping power, entropy generation, and exergy efficiency of nanodiamond + Fe_3O_4/60: 40% water-ethylene glycol hybrid nanofluid flow in a tube, Therm. Sci. Eng. Prog. 21 (2021) 100799.

[17] X. Zhang, H. Gu, M. Fujii, Effective thermal conductivity and thermal diffusivity of nanofluids containing spherical and cylindrical nanoparticles, J. Appl. Phys. 100 (4) (2006) 044325.

[18] X. Wang, X. Xu, S.U. Choi, Thermal conductivity of nanoparticle-fluid mixture, J. Thermophys. Heat. Transf. 13 (4) (1999) 474–480.

[19] C.H. Li, G.P. Peterson, The effect of particle size on the effective thermal conductivity of Al_2O_3-water nanofluids, J. Appl. Phys. 101 (4) (2007) 044312.

[20] K.F. Wong, T. Bhaskar, Transport properties of alumina nanofluids, ASME Int. Mech. Eng. Congr. Exposition 47640 (2006) 251–260.

[21] S.K. Das, N. Putra, P. Thiesen, W. Roetzel, Temperature dependence of thermal conductivity enhancement for nanofluids, J. Heat. Transf. 125 (4) (2003) 567–574.

[22] S.M. Iqbal, C.S. Raj, J.J. Michael, A.M. Irfan, A comparative investigation of Al_2O_3/H_2O, SiO_2/H_2O and ZrO_2/H_2O nanofluid for heat transfer applications, Dig. J. Nanomater. Biostruct. (DJNB) 12 (2) (2017).

[23] D.H. Yoo, K.S. Hong, T.E. Hong, J.A. Eastman, H.S. Yang, Thermal conductivity of Al_2O_3/water nanofluids, J. Korean Phys. Soc. 51 (1) (2007).

[24] N. Putra, W. Roetzel, S.K. Das, Natural convection of nano-fluids, Heat. Mass. Transf. 39 (8–9) (2003) 775–784.

[25] S. Yan, F. Wang, Z. Shi, R. Tian, Heat transfer property of SiO_2/water nanofluid flow inside solar collector vacuum tubes, Appl. Therm. Eng. 118 (2017) 385–391.

[26] S. Yan, H. Zhang, F. Wang, R. Ma, Y. Wu, R. Tian, Analysis of thermophysical characteristic of SiO_2/water nanofluid and heat transfer enhancement with field synergy principle, J. Renew. Sustain. Energy 10 (6) (2018) 063704.

[27] Y.F. Niu, W.L. Zhao, Y.Y. Gong, Preparation and thermophysical properties of SiO2 nanofluids, Advanced Materials Research, Vol. 771, Trans Tech Publications Ltd, 2013, pp. 125–128.

[28] A.M. Hussein, R.A. Bakar, K. Kadirgama, K.V. Sharma, Experimental measurement of nanofluids thermal properties, Int. J. Automot. Mech. Eng. 7 (2013) 850.

[29] G.M. Moldoveanu, C. Ibanescu, M. Danu, A.A. Minea, Viscosity estimation of Al_2O_3, SiO_2 nanofluids and their hybrid: an experimental study, J. Mol. Liq. 253 (2018) 188–196.

[30] R. Vidhya, T. Balakrishnan, B. Suresh Kumar, Investigation on thermophysical properties and heat transfer performance of heat pipe charged with binary mixture based ZnO-MgO hybrid nanofluids, Mater. Today: Proc. 37 (Part 2) (2021) 3423–3433. ISSN 2214–7853.

[31] S.U. Ilyas, M. Narahari, J.T.Y. Theng, R. Pendyala, Experimental evaluation of dispersion behavior, rheology and thermal analysis of functionalized zinc oxide-paraffin oil nanofluids, J. Mol. Liq. 294 (2019) 111613.

[32] R.M. Mostafizur, et al., Investigation on stability and viscosity of SiO_2–CH_3OH(methanol) nanofluids, Int. Commun. Heat. Mass. Transf. (2016). Available from: https://doi.org/10.1016/j.icheatmasstransfer.2016.01.001.

[33] S. Dong, X. Chen, An improved model for thermal conductivity of nanofluids with effects of particle size and Brownian motion, J. Therm. Anal. Calorim. 129 (2017) 1255–1263. Available from: https://doi.org/10.1007/s10973-017-6256-x.

[34] J. Xu, B. Yu, M. Zou, P. Xu, A new model for heat conduction of nanofluids based on fractal distributions of nanoparticles, J. Phys. D: Appl. Phys. 39 (2006) 4486–4490. Available from: https://doi.org/10.1088/0022-3727/41/13/139801.

[35] M. Hemmat Esfe, S. Saedodin, O. Mahian, S. Wongwises, Thermal conductivity of AlO/water nanofluids, J. Therm. Anal. & Calorim. 117 (2) (2014).

[36] E.V. Timofeeva, D.S. Smith, W. Yu, D.M. France, D. Singh, J.L. Routbort, Particle size and interfacial effects on thermo-physical and heat transfer characteristics of water-based α-SiC nanofluids, Nanotechnology 21 (2010) 215703. Available from: https://doi.org/10.1088/0957-4484/21/21/215703.

[37] G. Chen, W. Yu, D. Singh, D. Cookson, J. Routbort, Application of SAXS to the study of particle-size-dependent thermal conductivity in silica nanofluids, J. Nanopart. Res. 10 (2008) 1109–1114. Available from: https://doi.org/10.1007/s11051-007-9347-y.

[38] M.P. Beck, Y. Yuan, P. Warrier, A.S. Teja, The effect of particle size on the thermal conductivity of alumina nanofluids, J. Nanopart. Res. 11 (2009) 1129–1136. Available from: https://doi.org/10.1007/s11051-008-9500-2.

[39] E.V. Timofeeva, W. Yu, D.M. France, D. Singh, J.L. Routbort, Base fluid and temperature effects on the heat transfer characteristics of SiC in ethylene glycol/H_2O and H_2O nanofluids, J. Appl. Phys. 109 (1) (2011) 014914.

[40] R.L. Hamilton, O.K. Crosser, Thermal conductivity of heterogeneous two-component systems, Ind. Eng. Chem. Fundamentals 1 (3) (1962) 187–191.

[41] H. Xie, J. Wang, T. Xi, Y. Liu, Thermal conductivity of suspensions containing nanosized SiC particles, Int. J. Thermophys. 23 (2002) 571–580. Available from: https://doi.org/10.1023/A:1015121805842.

[42] S.M.S. Murshed, K.C. Leong, C. Yang, Enhanced thermal conductivity of TiO_2-water based nanofluids, Int. J. Therm. Sci. 44 (2005) 367–373. Available from: https://doi.org/10.1016/j.ijthermalsci.2004.12.005.

[43] E.V. Timofeeva, J.L. Routbort, D. Singh, Particle shape effects on thermophysical properties of alumina nanofluids, J. Appl. Phys. 106 (2009) 014304. Available from: https://doi.org/10.1063/1.3155999.

[44] I.M. Shahrul, I.M. Mahbubul, S.S. Khaleduzzaman, R. Saidur, M.F.M. Sabri, A comparative review on the specific heat of nanofluids for energy perspective, Renew. Sustain. Energy Rev. 38 (2014) 88–98.

[45] B.C. Pak, Y.I. Cho, Hydrodynamic and heat transfer study of dispersed fluids with submicron metallic oxide particles, Exp. Heat. Transf. Int. J. 11 (2) (1998) 151–170.

[46] Y. Xuan, W. Roetzel, Conceptions for heat transfer correlation of nanofluids, Int. J. Heat. Mass. Transf. 43 (19) (2000) 3701–3707.

[47] L.S. Sundar, M.K. Singh, A.C. Sousa, Turbulent heat transfer and friction factor of nanodiamond-nickel hybrid nanofluids flow in a tube: an experimental study, Int. J. Heat. Mass. Transf. 117 (2018) 223–234.

[48] C.V. Popa, C.T. Nguyen, I. Gherasim, New specific heat data for Al_2O_3 and CuO nanoparticles in suspension in water and ethylene glycol, Int. J. Therm. Sci. 111 (2017) 108–115.

[49] Y.R. Sekhar, K.V. Sharma, Study of viscosity and specific heat capacity characteristics of water-based Al_2O_3 nanofluids at low particle concentrations, J. Exp. Nanosci. 10 (2) (2015) 86–102.

[50] J.R. Satti, D.K. Das, D. Ray, Specific heat measurements of five different propylene glycol based nanofluids and development of a new correlation, Int. J. Heat. Mass. Transf. 94 (2016) 343–353.

[51] I. Wole-Osho, E.C. Okonkwo, D. Kavaz, S. Abbasoglu, An experimental investigation into the effect of particle mixture ratio on specific heat capacity and dynamic viscosity of Al_2O_3-ZnO hybrid nanofluids, Powder Technol. 363 (2020) 699–716.

[52] G.M. Moldoveanu, A.A. Minea, Specific heat experimental tests of simple and hybrid oxide-water nanofluids: proposing new correlation, J. Mol. Liq. 279 (2019) 299–305.

[53] S. Kannaiyan, C. Boobalan, A. Umasankaran, A. Ravirajan, S. Sathyan, T. Thomas, Comparison of experimental and calculated thermophysical properties of alumina/cupric oxide hybrid nanofluids, J. Mol. Liq. 244 (2017) 469–477.

[54] A.B. Çolak, O. Yildiz, M. Bayrak, A. Celen, S. Wongwises, Experimental study on the specific heat capacity measurement of water-based Al_2O_3-cu hybrid nanofluid by using differential thermal analysis method, Curr. Nanosci. 16 (6) (2020) 912–928.

[55] L.S. Sundar, E.V. Ramana, M.P.F. Graça, M.K. Singh, A.C. Sousa, Nanodiamond-Fe_3O_4 nanofluids: preparation and measurement of viscosity, electrical and thermal conductivities, Int. Commun. Heat. Mass. Transf. 73 (2016) 62–74.

[56] H. Yarmand, S. Gharehkhani, G. Ahmadi, S.F.S. Shirazi, S. Baradaran, E. Montazer, et al., Graphene nanoplatelets–silver hybrid nanofluids for enhanced heat transfer, Energy Convers. Manag. 100 (2015) 419–428.

[57] V.V. Wanatasanappan, M.Z. Abdullah, P. Gunnasegaran, Thermophysical properties of Al_2O_3-CuO hybrid nanofluid at different nanoparticle mixture ratio: an experimental approach, J. Mol. Liq. 313 (2020) 113458.

[58] O. Hozien, W.M. El-Maghlany, M.M. Sorour, Y.S. Mohamed, Experimental study on thermophysical properties of TiO_2, ZnO and Ag water base nanofluids, J. Mol. Liq. 334 (2021) 116128.

[59] V. Kumar, A. Pare, A.K. Tiwari, S.K. Ghosh, Efficacy evaluation of oxide-MWCNT water hybrid nanofluids: an experimental and artificial neural network approach, Colloids Surf. A: Physicochem. Eng. Asp. 620 (2021) 126562.

[60] Z. Said, M.A. Abdelkareem, H. Rezk, A.M. Nassef, H.Z. Atwany, Stability, thermophysical and electrical properties of synthesized carbon nanofiber and reduced-graphene oxide-based nanofluids and their hybrid along with fuzzy modeling approach, Powder Technol. 364 (2020) 795–809.

[61] P.S. Kishore, V. Sireesha, V.S. Harsha, V.D. Rao, A.B. Solomon, Preparation, characterization and thermo-physical properties of Cu-graphene nanoplatelets hybrid nanofluids, Mater. Today: Proc. 27 (2020) 610–614.

[62] N. Asokan, P. Gunnasegaran, V.V. Wanatasanappan, Experimental investigation on the thermal performance of compact heat exchanger and the rheological properties of low concentration mono and hybrid nanofluids containing Al_2O_3 and CuO nanoparticles, Therm. Sci. Eng. Prog. 20 (2020) 100727.

[63] L. Chen, H. Xie, W. Yu, Y. Li, Rheological behaviors of nanofluids containing multi-walled carbon nanotube, J. Dispers. Sci. Technol. 32 (4) (2011) 550–554.

[64] Y. Ding, H. Chen, Y. He, A. Lapkin, M. Yeganeh, L. Šiller, et al., Forced convective heat transfer of nanofluids, Adv. Powder Technol. 18 (6) (2007) 813–824.

[65] S. Kim, C. Kim, W.H. Lee, S.R. Park, Rheological properties of alumina nanofluids and their implication to the heat transfer enhancement mechanism, J. Appl. Phys. 110 (3) (2011) 034316.

[66] W.J. Tseng, C.H. Wu, Aggregation, rheology and electrophoretic packing structure of aqueous $A1_2O_3$ nanoparticle suspensions, Acta Mater. 50 (15) (2002) 3757–3766.

[67] I.M. Mahbubul, A. Saadah, R. Saidur, M.A. Khairul, A. Kamyar, Thermal performance analysis of Al_2O_3/R-134a nanorefrigerant, Int. J. Heat. Mass. Transf. 85 (2015) 1034–1040.

[68] P.K. Namburu, D.P. Kulkarni, A. Dandekar, D.K. Das, Experimental investigation of viscosity and specific heat of silicon dioxide nanofluids, Micro Nano Lett. 2 (3) (2007) 67–71.

[69] M. Chandrasekar, S. Suresh, R. Srinivasan, A.C. Bose, New analytical models to investigate thermal conductivity of nanofluids, J. Nanosci. Nanotechnol. 9 (1) (2009) 533–538.

[70] S. Akilu, A.T. Baheta, K.V. Sharma, Experimental measurements of thermal conductivity and viscosity of ethylene glycol-based hybrid nanofluid with TiO_2-CuO/C inclusions, J. Mol. Liq. 246 (2017) 396–405.

[71] H. Eshgarf, N. Sina, M.H. Esfe, F. Izadi, M. Afrand, Prediction of rheological behavior of MWCNTs–SiO_2/EG–water non-Newtonian hybrid nanofluid by designing new correlations and optimal artificial neural networks, J. Therm. Anal. Calorim. 132 (2) (2018) 1029–1038.

[72] M. Asadi, A. Asadi, Dynamic viscosity of MWCNT/ZnO–engine oil hybrid nanofluid: an experimental investigation and new correlation in different temperatures and solid concentrations. Inte. Commun. Heat Mass Transf. 76 (2016) 41–45.

[73] S.S. Ghadikolaei, M. Yassari, H. Sadeghi, K. Hosseinzadeh, D.D. Ganji, Investigation on thermophysical properties of TiO_2–Cu/H_2O hybrid nanofluid transport dependent on shape factor in MHD stagnation point flow, Powder Technol. 322 (2017) 428–438.

[74] I.M. Mahbubul, S.A. Fadhilah, R. Saidur, K.Y. Leong, M.A. Amalina, Thermophysical properties and heat transfer performance of Al_2O_3/R-134a nanorefrigerants, Int. J. Heat. Mass. Transf. 57 (1) (2013) 100–108.

[75] I.M. Mahbubul, R. Saidur, M.A. Amalina. Investigation of viscosity of R123-TiO_2 nanorefrigerant, in: Proceedings of Regional Tribology Conference 2011: RTC, 2011, p. 1.

[76] I.M. Mahbubul, R. Saidur, M.A. Amalina, Thermal conductivity, viscosity and density of R141b refrigerant based nanofluid, Procedia Eng. 56 (2013) 310–315.

[77] O.A. Alawi, N.A.C. Sidik, Influence of particle concentration and temperature on the thermophysical properties of CuO/R134a nanorefrigerant, Int. Commun. Heat. Mass. Transf. 58 (2014) 79–84.

[78] O.A. Alawi, N.A.C. Sidik, Mathematical correlations on factors affecting the thermal conductivity and dynamic viscosity of nanorefrigerants, Int. Commun. Heat. Mass. Transf. 58 (2014) 125–131.

[79] O.A. Alawi, N.A.C. Sidik, The effect of temperature and particles concentration on the determination of thermo and physical properties of SWCNT-nanorefrigerant, Int. Commun. Heat. Mass. Transf. 67 (2015) 8–13.

[80] O.A. Alawi, N.A.C. Sidik, A.S. Kherbeet, Measurements and correlations of frictional pressure drop of TiO_2/R123 flow boiling inside a horizontal smooth tube, Int. Commun. Heat. Mass. Transf. 61 (2015) 42–48.

[81] A. Bhattad, J. Sarkar, P. Ghosh, Improving the performance of refrigeration systems by using nanofluids: a comprehensive review, Renew. Sustain. Energy Rev. 82 (2018) 3656–3669.

Chapter 5

Comparison of physical properties enhancement in various heat transfer nanofluids by MXene

Lingenthiran Samylingam[1], Kumaran Kadirgama[2,3], Navid Aslfattahi[4], M. Samykano[2] and R. Saidur[1,5]

[1]Research Center for Nano-Materials and Energy Technology (RCNMET), School of Engineering and Technology, Sunway University, Selangor Darul Ehsan, Malaysia, [2]Faculty of Mechanical Engineering, Universiti Malaysia Pahang, Pekan, Malaysia, [3]Automotive Engineering Centre, Universiti Malaysia Pahang, Pekan, Malaysia, [4]Department of Fluid Mechanics and Thermodynamics, Faculty of Mechanical Engineering, Czech Technical University in Prague, Prague, Czech Republic, [5]Department of Engineering, Lancaster University, Lancaster, United Kingdom

Nomenclature

k_{nf}	thermal conductivity of nanofluid (W/m.K)
T	temperature (K)
Wt.%	nanoparticle weight fraction
°C	Celsius
vol.%	volume percentage

Subscripts

bf	base fluid
th	thermal
nf	nanofluid

Abbreviations

2D	two-dimensional
C	carbon
DEG	diethylene glycol
DI	deionized
F	fluorine

Advances in Nanofluid Heat Transfer. DOI: https://doi.org/10.1016/B978-0-323-88656-7.00007-6

FESEM field emission scanning electron microscopy
HCL hydrochloric acid
HF hydrofluoric acid
N nitrogen
NF nanofluid
O oxygen
OH hydroxide
PEG poly(ethylene) glycol
POME palm oil methyl ester
SO soybean oil
TGA thermogravimetric analysis
UV-Vis ultraviolet visible spectroscopy

5.1 Introduction

Formulation of nanofluids is based on the suspension of various types of nanoparticles into traditional heat transfer fluids [1]. The emergence of excellent physical properties within formulated nanofluids is related to the dispersion condition of nanoparticles as additives [2,3]. Murshed et al. [4] reported on the promising performance of developed nanofluids in the open literature in terms of physical properties. Murshed et al. [4] also elucidated on the better performance of developed nanofluids over the base fluids. Determination of physical properties related to developed heat transfer nanofluids is extremely dependent on the stability of the dispersed nanoparticles [5]. Furthermore, loading concentration and stability of the developed nanofluids play significant roles toward the improved physical properties.

Nanoparticles show significant impact in the advancement of heat transfer nanofluid performances. Carbon nanotubes and two-dimensional (2D) graphene nanomaterials are the widely utilized nanoparticles to advance the effective physical characteristics of nanofluid suspensions. In recent studies, 2D nanomaterials have attained enormous attention for their unique structure and extraordinary thermal, optical, and electrical properties. An emerging 2D nanomaterial, MXene was discovered at Drexel University in 2011 by Naguib et al. [6]. The superior properties of this promising nanomaterial over other materials has led the researchers to further explore the efficiency of this nanomaterial experimentally. MXenes family materials have exhibited better physical properties with respect to potential applications. $M_{n+1}X_nT_x$ ($n = 1 - 3$) represents the general formula of MXene family materials, where M stands for early transition metals, X stands for C and/or N atoms, and T_x represents the surface termination groups such as hydroxide (−OH), oxygen (−O), and fluorine (−F). Specific surface area, adsorption ability, and hydrophilicity are some of the advantageous aspects of MXene family materials which play significant roles in energy-related applications [7]. Carbides, nitrides, and carbonitrides as transition metals are considered as the main precursors in order to derive MXene family materials [6]. According to the ongoing research activities and established techniques, only a few types of MXene family materials have been explored, while science is expecting more emerging nanomaterials under this category. Layers of SP elements from three-dimensional material (MAX phases) should be etched in order to produce emerging nanomaterial, MXene. High capacitance, energy storage properties, and promising mechanical properties are considered as some important factors which give immense priority for MXene family materials over other types of nanoparticles [8–11].

Recently, nanofluids have attracted great interest in thermal related applications based on the exhibition of significant advancements in physical properties. The most challenging part related to the preparation of stable heat transfer nanofluids by MXene is confirmed to be associated with the proper dispersion of MXene nanomaterial as an additive. Homogenous nanofluid with the incorporation of MXene nanomaterial should be attained through the proper dispersion of MXene. Aslfattahi et al. [12] developed silicone oil/MXene heat transfer nanofluids with extraordinary physical properties. According to the experimental results, the highest enhancement of 64% was achieved in thermal conductivity for the highest loading concentration of MXene over the base fluid at 150°C. In another study, Samylingam et al. [13] formulated olein palm oil-based MXene nanofluid, and evaluated the physical properties of this new class of heat transfer nanofluid. Rubbi et al. [14] revealed an enhancement of almost 61% in thermal conductivity for the developed soybean oil (SO)-based nanofluid due to the presence of MXene nanomaterial as additive. Aqueous polyethylene glycol-based MXene nanofluid was developed by Aslfattahi et al. [15] with enhanced physical properties as a novel heat transfer nanofluid. Das et al. [16] proved enhanced physical properties for the formulated aqueous ionic liquid-based MXene heat transfer nanofluid. Based on the conducted studies, the 0.2 wt.% loading concentration of MXene as an additive advanced the thermal conductivity value up to a maximum of 47%. In another study, palm oil methyl ester (POME)-based MXene nanofluid was developed by Rahmadiawan et al. [17]. Based on experiments, an extraordinary enhancement of ~176% was reported at 65°C for the loading concentration of 0.1 wt.%. Abdelrazik et al. [18] investigated water-based MXene nanofluids with their respective application in solar thermal system and enhanced physical properties. Bao et al. [19] proved an enhancement of almost 65% in thermal conductivity for the developed nanofluid with the inclusion of 5 vol.% of single layer MXene nanomaterial. In the research study performed by Bakthavatchalam et al. [20], a novel heat transfer nanofluid with the mixture of diethylene glycol (DEG) and ionic liquid incorporated with MXene nanomaterial was developed which exhibited enhanced physical properties due to the presence of MXene nanomaterial. This chapter attempts to cover all investigations related to the developed nanofluids with the incorporation of MXene nanomaterial, including the preparation methods and adopted techniques related to the physical properties' characterizations. Afterwards, a comprehensive comparative study with consideration of physical properties of developed nanofluids with the inclusion of MXene is given followed by the main focus on the impact of MXene nanomaterials in all formulated advanced nanofluids.

5.2 Methodology

5.2.1 Preparation of MXene nanomaterial as an additive to heat transfer nanofluids

The synthesis process of MXene nanomaterial as an additive to heat transfer nanofluids has been conducted through four different techniques, which are represented in detail in Table 5.1. FESEM images of the MXene (Ti_3C_2) flakes which are used as additives in the developed heat transfer nanofluids by MXene are demonstrated in Fig. 5.1. As it can be observed from Fig. 5.1,

TABLE 5.1 Detailed explanation on the synthesis/type of utilized MXene as additive for development of heat transfer nanofluids by MXene.

Heat transfer nanofluids by MXene	Type of MXene nanomaterial	Synthesis technique
Silicone oil/MXene [12]	Ti_3C_2	Selective etching technique using hydrofluoric acid and Ti_3AlC_2
Olein palm oil/MXene [13]	Ti_3C_2	Selective etching technique using HCL mixture with lithium fluoride
Soybean oil/MXene [14]	Ti_3C_2	Selective etching technique using 2 M solution of NH_4HF_2 as etchant
Aqueous PEG/MXene [15]	Ti_3C_2	Selective etching technique using HCL mixture with lithium fluoride
Aqueous ionic liquid/MXene [16]	Ti_3C_2	Selective etching technique using 2 M solution of NH_4HF_2 as etchant
Palm oil methyl ester/MXene [17]	Ti_3C_2	Selective etching technique using HCL mixture with lithium fluoride
Water-based MXene [18]	Ti_3C_2	Selective etching technique using HCL mixture with lithium fluoride
Ionic liquid + DEG/MXene [20]	Ti_3C_2	Selective etching technique using HCL mixture with sodium fluoride
Ethylene glycol/MXene [19]	Ti_3C_2	Selective etching technique using hydrofluoric acid and Ti_3AlC_2

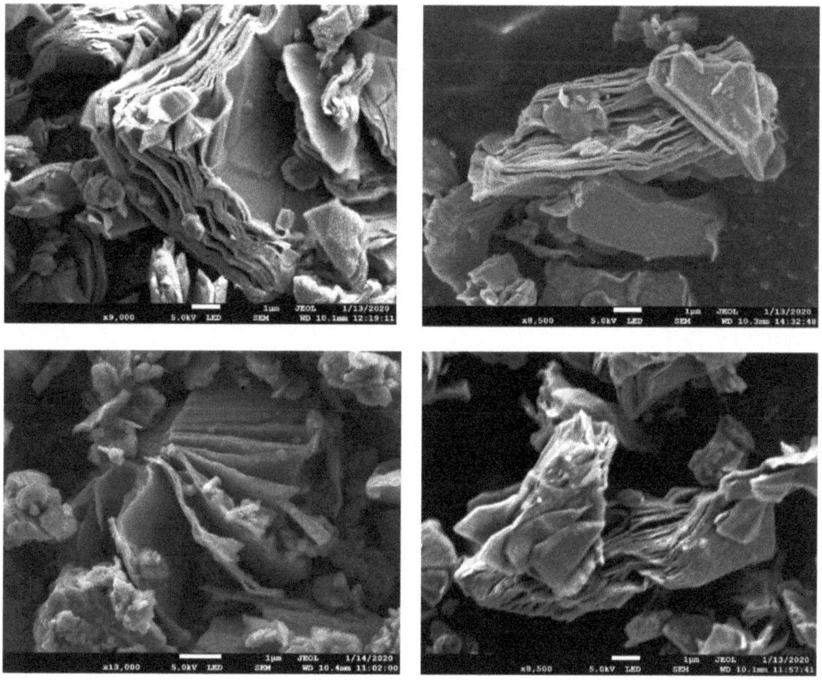

FIGURE 5.1 FESEM images of the as-synthesized Ti_3C_2 nanomaterial.

the successful exfoliation of MXene nanomaterial from the precursor MAX phase material is proven.

5.2.2 Preparation of various heat transfer nanofluids by MXene

5.2.2.1 MXene-based new class of silicone oil heat transfer nanofluid

Aslfattahi et al. [12] formulated three different concentrations of silicone oil/ MXene heat transfer nanofluid following an accurate protocol for sample preparation. The authors have developed a new method for the addition of MXene in order to boost the stability of the advanced heat transfer nanofluid. The dilution process of silicone oil was conducted by using various types of solvents with a ratio of 1:1 for all types of utilized solvents. The authors concluded there was a better dilution condition by using toluene after comprehensive evaluation of the dilution process through all solvents. Afterwards, the selected solvent (toluene) was adopted in order to dilute the silicone oil, followed by the addition of MXene nanomaterial in three different concentrations. Probe sonicator and magnet stirring were used to homogenize the added nanoparticles with the diluted silicone oil. Since then, the evaporation process was conducted with the consideration of the boiling point of toluene in order to remove the utilized solvent from the formulated heat transfer nanofluids. The authors assessed the stability of the formulated silicone oil/MXene heat transfer nanofluids through visual inspection and UV–Vis analysis for consecutive 2 weeks and revealed a good condition in terms of stability. The authors stated that the reason behind the good stability of prepared heat transfer nanofluids by MXene is based on an enhanced interface interaction between additive and base fluid due to an extraordinary specific surface area of MXene nanomaterial.

5.2.2.2 Olein palm oil with MXene as new class of heat transfer nanofluid

Samylingam et al. [13] developed olein palm oil/MXene with six different concentrations. The suspension of MXene nanomaterial into the olein palm oil was processed through a probe sonicator and magnet stirring. The hot plate magnet stirrer was utilized to conduct the stirring process at 700 rpm for 30 minutes. Afterwards, the probe sonication process was continued for 60 minutes using a power of 70% in order to homogenize the added MXene nanomaterial into olein palm oil. An identical protocol was adopted for all six developed olein palm oil/MXene heat transfer nanofluids with different concentrations.

5.2.2.3 Soybean oil/MXene heat transfer nanofluid

Rubbi et al. [14] synthesized SO/MXene heat transfer nanofluid using a two-step method. The formulated heat transfer nanofluids by MXene were prepared in three different concentrations. In order to develop stable heat transfer nanofluids, two stabilization techniques including magnet stirrer and probe sonication were considered precisely. The process of stirring was conducted using a hot plate magnet stirrer at a fixed temperature of 70°C for 30 minutes, followed by continuous probe sonication for 30 minutes to agitate the intermolecular forces between MXene nanomaterials and disperse uniformly. Ultrasonic probe sonication boosts the homogeneity and stability of the suspensions by breaking clusters.

5.2.2.4 Aqueous poly (ethylene) glycol (PEG)-based MXene nanofluid

Aslfattahi et al. [15] formulated a novel aqueous PEG-based MXene heat transfer nanofluid with the ratio of 70% PEG and 30% DI water. Two different loading concentrations of heat transfer nanofluids by MXene were developed precisely. The procedure of stirring was conducted at 1000 rpm for 10 minutes using a hot plated magnet stirrer in order to obtain a homogenous aqueous solution followed by addition of MXene nanomaterial to the prepared solution. Afterwards, the stirring process was continued for an additional 30 minutes to disperse the added MXene nanomaterial uniformly.

Finally, the probe sonication process was adopted for 60 minutes using 70% power to ensure the high dispersion of MXene nanomaterial as an additive. An identical process was considered for both developed heat transfer nanofluids by MXene nanomaterial.

5.2.2.5 *Aqueous ionic liquid/MXene nanofluid*

Das et al. [16] formulated an aqueous ionic liquid/MXene heat transfer nanofluid by mixing DI water and ionic liquid (80 vol.% of DI water), followed by magnetic stirring process at 45°C and 700 rpm. The purpose of adding water to the developed solution is for the sake of balancing the high viscosity of the utilized ionic liquid. In addition, the hydrophilic nature of MXene nanomaterial will play a significant role in obtaining higher stability due to the presence of water in the prepared solution. The authors developed three loading concentrations of heat transfer nanofluids by MXene using two stabilization techniques. The process of stirring was conducted for 60 minutes at a fixed temperature of 50°C using a hot plate magnet stirrer, followed by probe sonication for 30 minutes in order to form stable advanced heat transfer nanofluids.

5.2.2.6 *Palm oil methyl ester-based MXene heat transfer nanofluid*

Rahmadiawan et al. [17] developed POME-based MXene heat transfer nanofluids in five different loading concentrations. The two-step technique was adopted to formulate the abovementioned heat transfer nanofluids. The magnet stirrer process was continued for 120 minutes at a fixed temperature of 60°C using a hot plate magnet stirrer, followed by a probe sonication process for additional 120 minutes to ensure stabilization of the prepared heat transfer nanofluids with consideration of fixed temperature maintenance during sonication process.

5.2.2.7 *Water-based MXene nanofluid*

Abdelrazik et al. [18] prepared water/MXene heat transfer nanofluids at five different concentrations by using two different surfactants in order to evaluate the impact of utilized surfactants on the stability of formulated heat transfer nanofluids. As a standard procedure, the authors adopted a two-step technique for the preparation of heat transfer nanofluids by MXene. The calculated amounts of surfactant and MXene nanomaterial were added to the DI water solution, followed by stirring at 600 rpm for continuous 30 minutes. For the sake of better stabilization and achievement of homogenous heat transfer nanofluid, a probe sonication process was performed continuously for 60 minutes using 70% power. The same procedure was conducted for all five different developed water-based MXene heat transfer nanofluids.

5.2.2.8 Diethylene glycol and ionic liquid mixture-based MXene nanofluid

Bakthavatchalam et al. [20] prepared the mixture of ionic liquid and DEG as a solution followed by adding MXene nanomaterial in order to develop heat transfer nanofluids by MXene. The homogenous solution is obtained through a two-step technique including a magnet stirrer and probe sonication. The magnet stirring process was conducted for 30 minutes followed by probe sonication for an additional 90 minutes using 60% power. The formulated nanofluids were prepared at four different loading concentrations and the ratio of ionic liquid and DEG mixture was 1:9.

5.2.2.9 Ethylene glycol (EG)-based MXene heat transfer nanofluid

Bao et al. [19] formulated ethylene glycol/MXene heat transfer nanofluids through a two-step technique. The authors developed the heat transfer nanofluids using single-layer and multilayer MXene nanomaterials at various loading concentrations. All prepared samples were probe sonicated continuously for 180 minutes in order to obtain highly stabilized heat transfer nanofluids.

5.2.3 Physical properties

5.2.3.1 Thermal conductivity measurement

Thermal conductivity analysis for the formulated silicone oil-based MXene heat transfer nanofluids has been processed using THB 500 equipment from Linseis. Stabilization of the samples prior to the measurement process is a crucial point which has been taken under consideration seriously. The developed heat transfer nanofluids are measured using Linseis Hot Pont sensors as a better choice to investigate liquids and prevention of convection. The authors have conducted the thermal conductivity measurements in the temperature range of 25°C–150°C. For the olein palm oil/MXene heat transfer nanofluids, the thermal conductivity measurements are processed using TEMPOS (Meter Group). KS-3 sensor with thermal conductivity range of 0.02–2 W/mK in the accuracy range of ± 10% was selected to measure the thermal conductivity of developed olein palm oil/MXene heat transfer nanofluids at the temperatures of 25°C, 40°C, 55°C and 70°C. The thermal conductivity measurement of SO/MXene heat transfer nanofluids is conducted in similar way to the developed olein palm oil/MXene heat transfer nanofluids. The thermal conductivity analysis of the formulated aqueous PEG-based MXene heat transfer nanofluids was performed using TEMPOS thermal analyzer and KS-3 sensor at three different temperatures, 25°C, 35°C, and 45°C. In order to gain the temperature uniformity, the sensor and prepared samples were given 30 minutes to achieve temperature equilibrium before any measurement. The same procedure of formulated poly(ethylene) glycol-based MXene heat transfer nanofluids was adopted in order to measure the thermal conductivity of aqueous ionic liquid/MXene heat transfer nanofluids. Thermal conductivity measurements of the developed POME-based MXene heat transfer nanofluids were investigated using TEMPOS and KS-3 sensor within temperature ranging from 25°C to 85°C. Thermal conductivity analysis of the mixture of ionic liquid and DEG/MXene heat transfer nanofluid was processed in a similar technique with POME-based MXene heat transfer nanofluid in the temperature range of 20°C–80°C. C-THERM TCi thermal conductivity analyzer was considered for the thermal conductivity measurement of ethylene glycol/MXene heat transfer nanofluids within the temperature range of 10°C–60°C. The measurements repeated three times ensured the accuracy of the obtained experimental results.

5.2.3.2 Viscosity measurement of various heat transfer nanofluids by MXene

Aslfattahi et al. [12] adopted a rheometer to measure the viscosity of formulated silicone oil/MXene heat transfer nanofluids. The authors evaluated the shear rate as a function of temperature at a broad range of temperatures. All prepared heat transfer nanofluids were taken into consideration for the T-Ramp measurements as well. The same procedures were conducted for the formulated olein palm oil/MXene and developed SO/MXene heat transfer nanofluids at a broad range of temperatures. The viscosity evaluation of aqueous PEG-based MXene heat transfer nanofluids was performed at temperatures of 25°C, 50°C, and 85°C. The temperature range from 20°C to 50°C was selected by Das et al. [16] to measure the viscosity of the developed aqueous ionic liquid/MXene heat transfer nanofluids. The viscosity analysis of the developed POME-based MXene heat transfer nanofluids in five different loading concentrations was conducted according to ISO 3219 and DIN 53019 standards. Different temperatures and shear rates were considered to investigate the flow behavior. Bakthavatchalam et al. [20] performed the viscosity and rheology analysis of the mixture of ionic liquid and DEG/MXene heat transfer nanofluids in a similar way to that adopted by Das et al. [16]. Bao et al. [19] used LV DV-II pro Brookfield viscometer with the precision of $\pm 1\%$ and the reproducibility of $\pm 0.2\%$ to measure the rheological performance of formulated ethylene glycol/MXene heat transfer nanofluids. The uniformity of the temperature was maintained during the process using water bath heating.

5.2.3.3 Thermal stability test using thermal stability analysis

Thermal stability analysis (TGA) of the formulated silicone oil/MXene heat transfer nanofluids was performed using Perkin Elmer TGA 4000. The decomposition temperature measurement of the prepared samples has been processed at a temperature range of 30°C−800°C. The identical amount of all loading concentrations was utilized in order to maintain comparable investigation. The same procedure was adopted by Samylingam et al. [13] to measure thermal stability of formulated olein palm oil/MXene heat transfer nanofluids. Rubbi et al. [14] investigated the thermal degradation temperature of developed SO/MXene heat transfer nanofluids using TGA 4000 (Perkin Elmer). The analysis is performed from 30 to 500°C for each sample to examine the complete degradation profile of the developed heat transfer nanofluids. This analysis is advantageous as it requires a small amount to evaluate the thermal stability of the bulk sample and provides highly accurate data up to 1000°C. About 15 mg of each nanofluid sample is heated at the rate of 10°C/min up to 500°C in a compact ceramic furnace to measure the decomposition temperature of the nanofluids. Das et al. [16] adopted a similar procedure to Rubbi et al. [14] to measure the thermal degradation temperature of formulated aqueous ionic liquid/MXene heat transfer nanofluids. TGA for the developed POME-based MXene heat transfer nanofluids in five different loading concentrations was conducted in the temperature range of 30°C−550°C. Bakthavatchalam et al. [20] performed the thermal degradation analysis for the mixture of ionic liquid and DEG/MXene heat transfer nanofluids in the temperature range of 30°C−650°C.

5.3 Results and discussion

5.3.1 Thermal conductivity of various heat transfer nanofluids by MXene

Aslfattahi et al. [12] assessed the impact of temperature variation on the thermal conductivity of formulated silicone oil/MXene heat transfer nanofluids as illustrated in Fig. 5.2. The standard deviations for the conducted measurements were found to be ± 0.002. The authors proved the dependency of thermal conductivity for developed silicone oil/MXene heat transfer nanofluids on both factors including temperature and loading concentrations of the additive. According to the experimental results, a maximum thermal conductivity value of $\sim$0.203 W/m K was achieved at 25°C followed by additional enhancement up to $\sim$0.263 W/m K at 150°C. This outstanding advancement is related to the advantageous aspects of MXene nanomaterial, such as its extraordinary basal plane as an additive [21,22]. At 150°C, a maximum enhancement of $\sim$64% in thermal conductivity is achieved for the loading concentration of 0.1 wt.%. The authors proved that the impact of MXene loading concentrations is more effective than the temperature increase on the thermal conductivity enhancement of formulated silicone oil/MXene heat transfer nanofluids. Based on the experimental results, the dominant impact of MXene nanomaterial over temperature was proved in order to advance the physical properties of the developed heat transfer nanofluids.

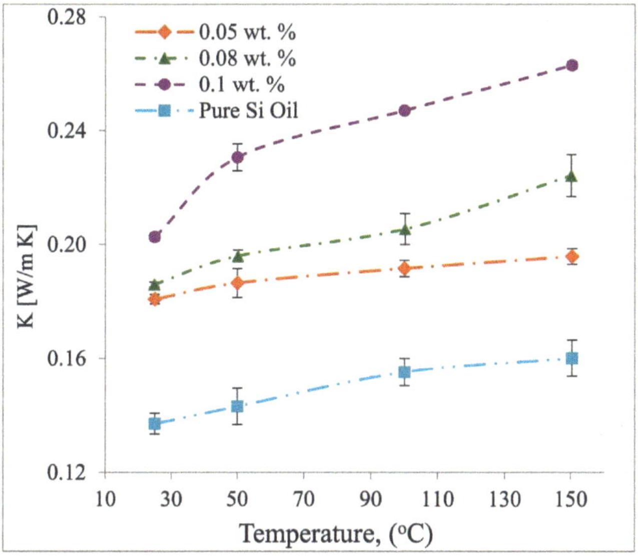

FIGURE 5.2 Experimental results for thermal conductivity of silicone oil-based MXene heat transfer nanofluids as a function of temperature for various loading concentrations of MXene nanomaterial [12].

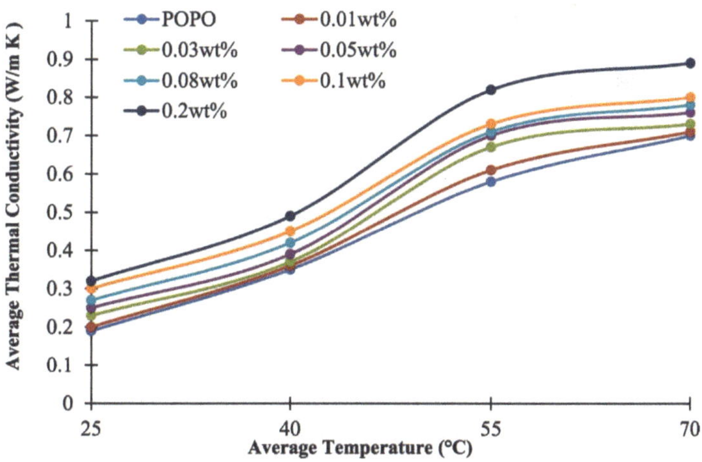

FIGURE 5.3 Thermal conductivity of formulated olein palm oil/MXene heat transfer nanofluids at different concentrations as a function of temperature [13].

Samylingam et al. [13] assessed the thermal conductivity of formulated olein palm oil/MXene heat transfer nanofluids at average temperatures of 25°C, 40°C, 55°C, and 70°C, as shown in Fig. 5.3. Due to the influential factors, such as motion of free particles and their molecular vibrations, it was expected to achieve enhancement in the physical properties of developed heat transfer nanofluid by MXene [23]. The expectation on enhancement is based on the suspension of nanoparticles' Brownian motion, which improves with a further increase in temperature [24]. The enhancement in Brownian motion with temperature is due to the gain in kinetic energy [25]. In addition, MXene nanomaterial acts as nanofillers which makes the possibility to obtain better heat transfer nanofluids due to their extraordinary surface area properties [26,27]. Furthermore, the liquid layering with base fluid can be attained perfectly due to the proficient hydrophilic nature of MXene nanomaterial, which is an additional advantage of this novel nanomaterial. Authors reported the highest enhancement of 68.5% in thermal conductivity at 25°C for developed olein palm oil/MXene heat transfer nanofluid with the inclusion of 0.2 wt.% MXene nanomaterial. According to the findings of the authors, the improvement percentage contributed by the concentration of MXene nanomaterial is more significant compared to the temperature increment. The authors reported the importance of additive nanomaterial over the temperature variation in order to advance the physical properties.

Rubbi et al. [14] assessed the thermal conductivity properties of formulated SO/MXene heat transfer nanofluids with different loading concentrations of additives. Based on the acquired experimental results, the highest enhancement of 60.82% at 55°C was achieved for the highest loading concentration of MXene. A linear correlation between the increase of thermal conductivity and addition of MXene nanomaterial was observed. The obtained results are attributed to the high thermal conductivity of MXene (Ti_3C_2) nanosheets at greater weight fraction in addition to a greater conductive surface area with a negligible thickness of Ti_3C_2 flakes. In addition, intermolecular interactions between Ti_3C_2 flakes with SO significantly impacted the thermal conductivity enhancement. Besides, larger surface area and higher aspect ratio of Ti_3C_2 nanosheets are influential factors that enhance the physical properties of the developed heat transfer nanofluids.

Aslfattahi et al. [15] examined the thermal conductivity of the developed PEG/MXene heat transfer nanofluids in order to evaluate the impact of MXene nanomaterial over physical properties of the formulated heat transfer nanofluids. The measured thermal conductivity values revealed the significance of loading concentration of MXene nanomaterial to reach promising enhancement for the respective physical properties. The highest thermal conductivity value of ~0.60 W/mK was achieved at 25°C (0.1 wt.%), followed by reaching ~1.17 at 45°C. Better heat conduction through MXene basal plane might be the reason behind this promising enhancement [22]. According to the obtained results, the formulated aqueous PEG/MXene heat transfer nanofluids exhibited a promising enhancement of ~29% in thermal conductivity at 45°C. Based on the obtained results, dominancy of MXene nanomaterial loading concentration over temperature was proved in terms of efficiency in thermal conductivity enhancement, which proves the influential factor related to this promising nanomaterial to develop advanced heat transfer nanofluids

Das et al. [16] assessed the thermal conductivity property of aqueous ionic liquid [MMIM][DMP] + MXene heat transfer nanofluids at a temperature range of 25°C–60°C. Fig. 5.4, demonstrates the acquired thermal conductivity enhancements as well as the original experimental results. Based on the experimental results, the thermal conductivity property is enhanced effectively due to the inclusion of MXene nanomaterial in the developed heat transfer nanofluids. Promising specifications related to MXene nanomaterial, such as extremely high surface area and efficient interaction with base

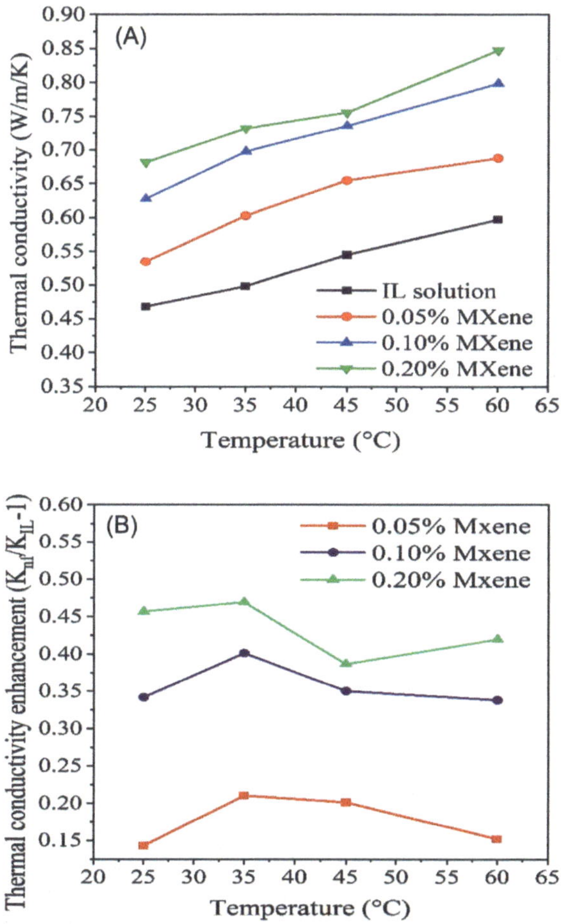

FIGURE 5.4 (A) Acquired experimental results for thermal conductivity with consideration of variation in loading concentration and temperature. (B) Achieved enhancement ratio in thermal conductivity with concentration of variation in temperature over base solution [16].

fluid, are the prominent reasons that enhance this physical property. Based on the experimental results, the addition of 0.2 wt.% MXene nanomaterial as an additive within the advanced heat transfer nanofluid caused a maximum enhancement of 47% at 35°C. The authors concluded that the inclusion of a higher amount of MXene nanomaterial might contribute to better enhancements in thermal conductivity.

Rahmadiawan et al. [17] obtained very promising thermal conductivity enhancement for the developed POME/MXene heat transfer nanofluids due to advantages factors related to MXene nanomaterial. The highest thermal conductivity enhancement of ~176% was acquired for the formulated POME/MXene heat transfer nanofluid with the inclusion of 0.1 wt.% MXene nanomaterial over based fluid at 65°C. According to the acquired experimental results, thermal conductivity at 25°C for base fluid was 0.166 W/mK, followed by promising enhancement up to 0.367 W/mK for developed heat transfer nanofluid by 0.1 wt.% MXene nanomaterial. Bakthavatchalam et al. [20] assessed the thermal conductivity enhancement of developed mixture of DEG and ionic liquid/MXene heat transfer nanofluids at the temperature ranges from 25°C to 80°C. Authors reported a valuable enhancement of 57.45% in thermal conductivity for the developed heat transfer nanofluids by MXene using the highest loading concentration of additive. Bao et al. [19] reported the thermal conductivity enhancement of 64.9% for the formulated EG/MXene heat transfer nanofluids due to the inclusion of single layer MXene nanomaterial.

5.3.2 Viscosity analysis of various heat transfer nanofluids by MXene

The rheological performance of silicone oil/MXene heat transfer nanofluids is investigated as a function of shear rate from 25°C to 125°C. According to the obtained results, developed silicone oil/MXene heat transfer nanofluids act as Newtonian fluids. This Newtonian rheological performance might be attributed to the alignment of fluid molecules due to spindle rotation [28]. In addition, the independency of the viscosity with the rate of shear force was proved for the formulated silicone oil/MXene heat transfer nanofluid. Samylingam et al. [13] evaluated the rheological properties of the developed olein palm oil/MXene heat transfer nanofluids with the shear rates ranging from 0 to 100 1/s at varying temperatures of 25°C, 50°C, and 75°C respectively. According to the acquired results, at a fixed temperature, the viscosity of olein palm oil/MXene heat transfer nanofluids remains almost the same with the increasing shear rate from 0 to 100 1/s which proved its performance as a Newtonian fluid. The Newtonian shear in these nanofluids: when the spindle rotation increases, the alignment of the fluid molecules is not deflected [28]. Followed by this phenomenon when the shear rate increased further, this rotation does not create any change to the viscosity. The independency of viscosity with shear rate force was realized for developed heat transfer nanofluids. Rubbi et al. [14] measured the viscosity of SO/MXene heat transfer nanofluids at the temperature range of 25°C–85°C. Fig. 5.5A, demonstrates the experimental results for the viscosity of formulated SO/MXene heat transfer nanofluids. The heat transfer nanofluids' viscosity increased with inclusion of MXene nanomaterial but linearly decreased as with ascended temperature. This is due to weakening intermolecular forces between the fluid and particles. The maximum enhancement of 13.28% was achieved for heat transfer nanofluid with loading concentration of 0.125 wt.% at 25°C. Based on Fig. 5.5B, the Newtonian rheological performance is revealed for the formulated SO/MXene heat transfer nanofluid. The results indicate the Newtonian rheological behavior of the NF as constant shear viscosity resulted within the operating range of shear rate.

The viscosity analysis for the formulated aqueous PEG/MXene heat transfer nanofluids proved the Newtonian rheological performance for all loading concentrations. The acquired experimental results exhibited a decrease of ~85% in viscosity for the developed heat transfer nanofluids by MXene at 85°C. A maximum enhancement of 32% in viscosity property was observed for the developed heat transfer nanofluid at 25°C which might be attributed to the inclusion of MXene nanomaterial. Das et al. [16] assessed the viscosity of aqueous ionic liquid heat transfer nanofluids from 25°C to 60°C. Based on the experimental results, the inclusion of 0.2 wt.% of MXene nanomaterial into the heat transfer nanofluid enhanced the viscosity value from 2.694 to 3.066 mPa.s which is reasonable due to the addition of nanoparticles to base fluid. This enhancement is related to the increment in shear stress due to the inclusion of MXene nanomaterial. The viscosity analysis for developed POME/MXene heat transfer nanofluids revealed the

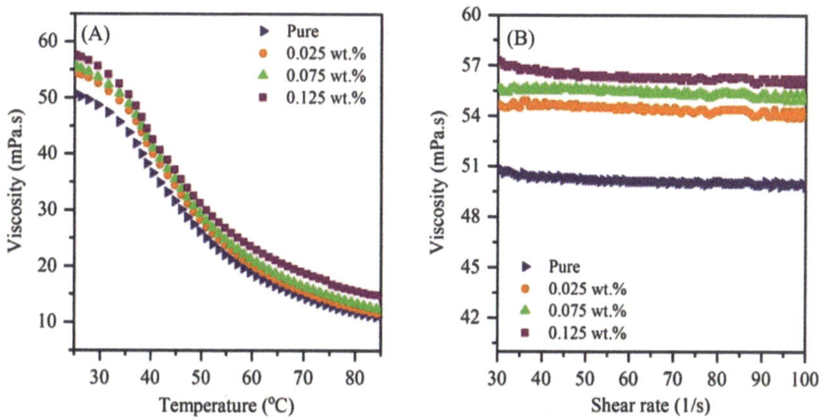

FIGURE 5.5 Viscosity measurements of Soybean oil/MXene heat transfer nanofluids (A) against increasing temperature (B) against increasing shear rate [14].

Newtonian rheological performance. The measured viscosities for the formulated heat transfer nanofluids by MXene exhibited steady performance over the various loading concentrations. Based on the acquired results, the increased loading concentration of additives caused reduction in the viscosity for the evaluated samples. Bakthavatchalam et al. [20] assessed the viscosity of developed aqueous DEG + ionic liquid/MXene heat transfer nanofluids as a function of temperature from 25°C to 75°C. According to the experimental results, for the formulated heat transfer nanofluids' viscosity decreased from 38 to 6 mPa.s which is related to the measurement condition in elevated temperatures. From the flow behavior of developed heat transfer nanofluids by MXene, the viscosity is extremely dependent on the variation of shear stress. Therefore, it can be concluded that the formulated Ionanofluids prove to be a non-Newtonian pseudoplastic nature. It can also be noted that the low shear rate caused higher viscosity, which might be attributed to the irregular arrangement of nanoparticle offering high flow resistance at the beginning. Bao et al. [19] revealed a Newtonian rheological performance for the advanced EG/MXene hear transfer nanofluids with the inclusion of both M-$Ti_3C_2T_x$ and S-$Ti_3C_2T_x$ nanomaterials as additives. The authors proved the similar Newtonian rheological performance for both single and multi-layer MXene nanomaterial as additive.

5.3.3 Thermal Stability of various heat transfer nanofluids by MXene

Thermal durability analysis of formulated silicone oil/MXene heat transfer nanofluids indicated enhancement in thermal degradation due to the presence of MXene nanomaterial which might be attributed to the creation of better bonding between MXene nanomaterial as additive and base fluid. A maximum enhancement of 9.8% in thermal stability was achieved for the highest loading concentration of MXene nanomaterial as an additive. According to the experimental results, the degradation step becomes smoothed compared to base fluids due to the presence of MXene nanomaterial. The enhancement in the activation energy due to the high specific surface area of MXene nanomaterial might lead to an enhancement in degradation temperature for the developed heat transfer nanofluids [29]. The TGA of developed olein palm oil/MXene heat transfer nanofluids with six concentrations was investigated using 10°C/min heating rate and an identical procedure for sample measurements. The thermal degradation for the temperature range analyzed does not show any kind of decomposition for the developed nanofluids. Rubbi et al. [14] evaluated the thermal durability of developed SO/MXene heat transfer nanofluids with various weight fractions for the temperature range of 30°C–80°C. No significant changes in the temperature profiles are noted for each heat transfer nanofluid sample in the working temperature range (as shown in Fig. 5.6). In addition, it was observed that thermal degradation of

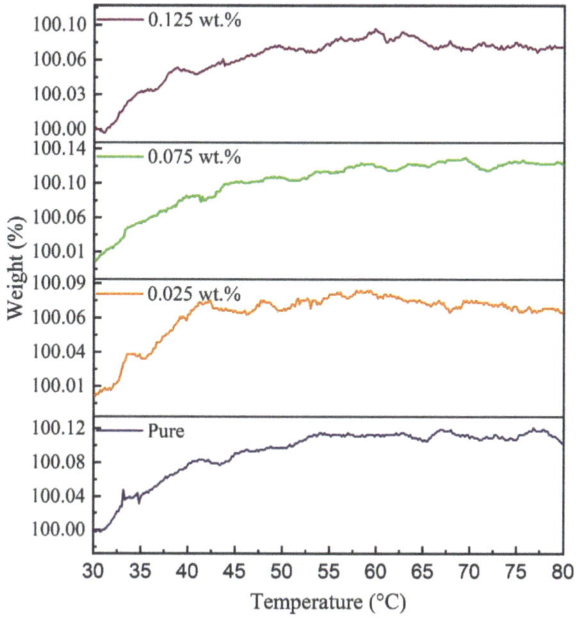

FIGURE 5.6 Thermogravimetric analysis of formulated soybean oil/MXene heat transfer nanofluids [14].

the heat transfer nanofluids starts approximately at 320°C and the degradation rate increases noticeably from 400°C to higher temperature. The sudden degradation of the heat transfer nanofluids ends at 490°C, which represents the completion of decomposition. The results convey the indication of a high temperature stand of formulated SO/MXene heat transfer nanofluids. The addition of Ti_3C_2 particles has a minor effect on the decomposition of formulated heat transfer nanofluids, as no notable shift is found up to 320°C for any of the samples. The steady outcomes can be attributed to low weight concentration (0.025–0.125 wt.%) of the suspensions. TGA analysis demonstrations proved that the heat transfer nanofluids begin to decompose around 300°C. It indicates that the heat transfer nanofluids can be used for a medium temperature range (up to 200°C) energy conversion system as well.

The thermal stability evaluation for the developed aqueous ionic liquid/MXene heat transfer nanofluids proved the stability for all samples up to 45°C. According to the obtained experimental results, the presence of MXene nanomaterial boosts up the thermal durability of formulated heat transfer nanofluids. This promising performance is related to the presence of hydroxyl groups on the surface area of MXene nanomaterial which might cause the creation of better bonding between the base fluid and additive in order to improve the thermal stability of developed heat transfer nanofluids.

Rahmadiawan et al. [17] proved a better performance for the formulated POME/MXene heat transfer nanofluids in terms of thermal resistance. The addition of MXene nanomaterial (0.1 wt.%) into POME resulted in the highest decomposition temperature value (~280°C) which was about a 1.4% enhancement. Higher thermal resistance of MXene nanomaterial as an additive is the main effective factor in order to achieve this promising enhancement. The thermal stability results for DEG + ionic liquid/MXene heat transfer nanofluids exhibited an enhancement in the initial decomposition temperature. The results showed that the addition of MXene nanomaterial enhanced the thermal stability of the as-prepared heat transfer nanofluids by MXene, followed by complete degradation after 200°C.

5.4 Conclusion

This chapter attempts to provide a comprehensive comparison between the highly efficient formulated nanofluids with the incorporation of MXene as a promising additive. Various types of developed nanofluids with improved physical properties due to the inclusion of MXene nanomaterial are considered as the main focus of this chapter in order to provide insightful perceptions related to the impact of MXene within the formulated heat transfer nanofluids. Formulated silicone oil/MXene heat transfer nanofluids exhibited the highest improvement of ~64% in thermal conductivity due to the inclusion of 0.1 wt.% MXene nanomaterial. For the developed olein palm oil/MXene heat transfer nanofluids, the authors reported the highest enhancement of 68.5% in thermal conductivity at 25°C with the inclusion of 0.2 wt.% MXene nanomaterial. Soybean oil/MXene heat transfer nanofluids showed the highest enhancement of 60.82% at 55°C for the highest loading concentration of MXene. The formulated aqueous PEG/MXene heat transfer nanofluids with the inclusion of 0.1 wt.% MXene nanomaterial showed an improvement of ~29% at 45°C. A maximum of 47% enhancement in thermal conductivity occurs at 35°C for 0.20% loading concentration of Ti_3C_2 nanomaterial for the developed aqueous ionic liquid/MXene heat transfer nanofluids. The highest thermal conductivity enhancement of ~176% was acquired for the formulated POME/MXene heat transfer nanofluid with the inclusion of 0.1 wt.% MXene nanomaterial over base fluid at 65°C. The formulated DEG + ionic liquid/MXene heat transfer nanofluids proved the enhancement of 57.45% in thermal conductivity with the inclusion of 0.4 wt.% MXene nanomaterial. Thermal conductivity enhancement of 64.9% was reported for the formulated EG/MXene heat transfer nanofluids due to the inclusion of single-layer MXene nanomaterial. The conducted comparative study in terms of viscosity analysis proved the Newtonian rheological performance for most of the developed heat transfer nanofluids by MXene. For the developed silicone oil/MXene heat transfer nanofluids, a maximum enhancement of 9.8% in thermal stability was achieved for the highest loading concentration of MXene nanomaterial. For the formulated olein palm oil/MXene and SO/MXene heat transfer nanofluids, the experimental results proved no significant change in thermal degradation temperature. Developed ionic liquid/MXene heat transfer nanofluids proved better thermal durability performance due to the presence of MXene nanomaterial. The addition of MXene nanomaterial into POME with a maximum loading of 0.1 wt.% using ultrasonication resulted in the highest decomposition temperature value (~280°C) and it was about 1.4% enhancement over the base fluid temperature. The thermal stability results for DEG + ionic liquid/MXene heat transfer nanofluids exhibited enhancement in the initial decomposition temperature with constant weight loss.

Acknowledgment

The authors would like to thank the Minister of Higher Education under Fundamental Research Grant Scheme (FRGS) No. FRGS/1/2019/TK07/UMP/02/3 and Universiti Malaysia Pahang under grant number RDU192209.

References

[1] S. Choi, et al., Developments and applications of non-Newtonian flows, ASME FED. 66 (1995) 99–105.

[2] M. Mehrali, et al., Full-spectrum volumetric solar thermal conversion via graphene/silver hybrid plasmonic nanofluids, Appl. Energy 224 (2018) 103–115.

[3] Aslfattahi, N., et al., Experimental investigation of thermal stability and enthalpy of eutectic alkali metal solar salt dispersed with MgO nanoparticles. 2019 10(6) p. 1112–1119.

[4] S. Murshed, et al., Investigations of thermal conductivity and viscosity of nanofluids, Int. J. Therm. Sci. 47 (5) (2008) 560–568.

[5] F. Wang, et al., Surfactant-free ionic liquid-based nanofluids with remarkable thermal conductivity enhancement at very low loading of graphene, Nanoscale Res. Lett. 7 (1) (2012) 314.

[6] M. Naguib, et al., Two-dimensional nanocrystals produced by exfoliation of Ti3AlC2, Adv. Mater. 23 (37) (2011) 4248–4253.

[7] M. Li, et al., Effect of different dimensional carbon materials on the properties and application of phase change materials: a review, Appl. Energy 242 (2019) 695–715.

[8] D. Sun, et al., Two-dimensional Ti3C2 as anode material for Li-ion batteries, Electrochem. Commun. 47 (2014) 80–83.

[9] Q. Hu, et al., MXene: a new family of promising hydrogen storage medium, J. Phys. Chem. A 117 (51) (2013) 14253–14260.

[10] M. Khazaei, et al., Novel electronic and magnetic properties of two-dimensional transition metal carbides and nitrides, Adv. Funct. Mater. 23 (17) (2013) 2185–2192.

[11] Aslfattahi, N., et al., Experimental investigation of energy storage properties and thermal conductivity of a novel organic phase change material/MXene as a new class of nanocomposites. 2020 27 p. 101115.

[12] N. Aslfattahi, et al., MXene based new class of silicone oil nanofluids for the performance improvement of concentrated photovoltaic thermal collector, Sol. Energy Mater. Sol. Cell 211 (2020) 110526.

[13] L. Samylingam, et al., Thermal and energy performance improvement of hybrid PV/T system by using olein palm oil with MXene as a new class of heat transfer fluid, Sol. Energy Mater. Sol. Cell 218 (2020) 110754.

[14] F. Rubbi, et al., Performance optimization of a hybrid PV/T solar system using Soybean oil/MXene nanofluids as A new class of heat transfer fluids, Sol. Energy 208 (2020) 124–138.

[15] N. Aslfattahi, et al. Thermal conductivity and rheological investigation of aqueous poly (ethylene) glycol/MXene as a novel heat transfer fluid, in: AIP Conference Proceedings, AIP Publishing LLC, 2021.

[16] L. Das, et al., Improved thermophysical properties and energy efficiency of aqueous ionic liquid/MXene nanofluid in a hybrid PV/T solar system, Nanomaterials 10 (7) (2020) 1372.

[17] D. Rahmadiawan, et al., MXene based palm oil methyl ester as an effective heat transfer fluid, J. Nano Res. (2021).

[18] A. Abdelrazik, et al., Optical, stability and energy performance of water-based MXene nanofluids in hybrid PV/thermal solar systems, Sol. Energy 204 (2020) 32–47.

[19] Z. Bao, et al., Ti3C2Tx MXene contained nanofluids with high thermal conductivity, super colloidal stability and low viscosity, Chem. Eng. J. 406 (2021) 126390.

[20] B. Bakthavatchalam, et al., Optimization of thermophysical and rheological properties of mxene ionanofluids for hybrid solar photovoltaic/thermal systems, Nanomaterials 11 (2) (2021) 320.

[21] T.T. Baby, et al., Investigation of thermal and electrical conductivity of graphene based nanofluids, J. Appl. Phys. 108 (12) (2010) 124308.

[22] M. Mehrali, et al., Investigation of thermal conductivity and rheological properties of nanofluids containing graphene nanoplatelets, Nanoscale Res. Lett. 9 (1) (2014) 15.

[23] J. Akhter, et al., Characterization and stability analysis of oil-based copper oxide nanofluids for medium temperature solar collectors, Materialwiss. Werkst. 50 (3) (2019) 311–319.

[24] L. Samylingam, et al., Thermal analysis of cellulose nanocrystal-ethylene glycol nanofluid coolant, Int. J. Heat. Mass. Transf. 127 (2018) 173–181.

[25] K. Kadirgama, et al., Thermal analysis of SUS 304 stainless steel using ethylene glycol/ nanocellulose-based nanofluid coolant, Int. J. Adv. Manuf. Technol. 97 (5–8) (2018) 2061–2076.

[26] J. Taha-Tijerina, et al., Theoretical prediction of physical properties (viscosity) on 2D-based nanofluids, Ing. Invest. Tecnol. 18 (1) (2017) 101–109.

[27] J. Taha-Tijerina, et al., 2D-based nanofluids: materials evaluation and performance, in two-dimensional materials-synthesis, Characterization and Potential Applications, InTech, 2016, pp. 153–198.

[28] M.N. Rashin, et al., Synthesis and viscosity studies of novel ecofriendly ZnO–coconut oil nanofluid, Exp. Therm. Fluid Sci. 51 (2013) 312–318.

[29] Yao, F., et al., Thermal decomposition kinetics of natural fibers: activation energy with dynamic thermogravimetric analysis. 2008 93(1) p. 90–98.

Chapter 6

Numerical modeling of nanofluids' flow and heat transfer

Tehmina Ambreen[1], Arslan Saleem[2] and Cheol Woo Park[1]
[1]*School of Mechanical Engineering, Kyungpook National University, Bukgu Daegu, South Korea,* [2]*School of Engineering, Cardiff University, Cardiff, Wales, United Kingdom*

Nomenclature

Abbreviations

2D	two-dimensional
3D	three-dimensional
CFD	computational fluid dynamics
CVFEM	control volume finite element method
EDL	electrical double layer
FDM	finite difference method
FEM	finite element method
FVM	finite volume method
GNP	graphene nanoplatelets
LDM	lattice Boltzmann method
MWCNT	multiwalled carbon nanotubes
VOF	volume of fluid

Symbols

A	area, m^2
a	acceleration, m/s^2
A′	Hamaker constant, J
Au	gold
Ag	silver
$\mathbf{Al_2O_3}$	aluminum oxide
B	magnetic field, T
C	constant
$\mathbf{C_{Br}}$	Brownian coefficient, kg/m/s
$\mathbf{C_c}$	Cunningham correction factor
$\mathbf{C_{drag}}$	coefficient of drag

Advances in Nanofluid Heat Transfer. DOI: https://doi.org/10.1016/B978-0-323-88656-7.00015-5

C_f	inertial coefficient
C_p	specific heat, J/kg$^{/}$K
C_{th}	thermophoresis coefficient, kg/m/s/K
Cu	copper
CuO	Copper oxide
d	diameter, m
d_{ij}	deformation tensor, s
E	electrical field, V/m
E-E	Eulerian model
F	force, N
f	friction factor
Fe	iron
Fe_2O_3	iron oxide
g	gravitation acceleration, m/s^2
G_f	fluid shear force, N/m
h	convective heat transfer coefficient, W/m^2/K
Hom	homogenous model
k	thermal conductivity, W/m/K
K_B	Boltzmann constant, J/K
k_d	thermal conductivity enhancement, W/m$^{/}$K
K_n	Knudsen number
k_{tr}	turbulent thermal conductivity
L-E	Eulerian–Lagrangian model
m	mass, kg
M	molar mass, kg/mol
Mix	mixture model
MgO	magnesium oxide
N	Avogadro's number
Nu	Nusselt number
P	pressure, N/m^2
Pe	Peclet number
P_r	Prandtl number
Q	heat flux, W/m^2
q	electrical conductivity, S/m
r	radius, m
R	position, m
Re	Reynolds number
SiO_2	silicon dioxide
T	temperature, K
t	time, s
TiO_2	titanium dioxide
z	minimum distance between two spheres at the contact point, m
ZnO	zinc oxide
ZrO_2	zirconium oxide

Greek symbols

ξ	particle surface potential, V
$K^{(-1)}$	Debye length, m V

μ	viscosity, kg/m/s
β	nanoparticle aspect ratio
$\bar{\tau}$	stress−strain tensor
ρ	density, kg/m^3
ψ	nanoparticle mass fraction
φ	nanoparticle volume fraction
$'\Omega$	particle rotational velocity, m/s
η_r	relative viscosity of nanofluid
$\mathbf{V}$	volume, m^3
v	kinematic viscosity, m^2/s
z_p	zeta potential, m V
ζ_i	independent Gaussian random variables

Subscripts

EDL	electrical double layer
f	liquid phase
h	hydraulic diameter
i	*i*th phase
in	inlet
local	Local
mass	Mass
M, lift	Magnus lift
n	nanoparticles
nf	nanofluid
p	nanoparticle phase
r	relative
S, lift	Saffman lift
slip	Slip
T	reference temperature
t	time
vdw	van der Waals force
w	wall

6.1 Introduction

With the advancement of power and availability of computational techniques, the numerical methods offer speedy and affordable solutions to fluid flow, heat transfer, and other physical problems. These numerical techniques are in continuous advancement and have led to the growing use of realistic mathematical models in most design and analysis applications. In many engineering applications, numerical techniques are replacing analytical and experimental methods. In addition to rapid solutions and reduced cost compared to experimental procedures, the computational methods provide a detailed set of information throughout the domain of interest. Moreover, the computational simulations also offer the advantages of flow solutions at the actual scale of the system under the real operating conditions, while most experimental procedures require scaling up or down the system.

Most fluid dynamics and other physical phenomena are modeled by the respective governing equations derived by implementing the laws of conservation of mass, momentum, and energy. These transport equations describe the physics of the problem and are generally in the form of partial differential equations. A complete problem description includes governing equations, physical model, boundary conditions, and material properties. The mathematical models for the specific applications are established by considering certain approximations and assumptions. Therefore a proper understanding of the applicability criterion and limitations of these computational methods is essential to realize their full potential.

The performance evaluation of the nanofluids through numerical methods can be much more affordable and swift compared to experimentation involving complex and expensive processes and materials. In the nanofluid-based experimental analysis, the stability of the nanofluids is a critical factor influencing the accuracy of results. Nanoparticle destabilization in the hosting fluid causes anomalies in the nanofluid thermophysical properties, which consequently affect the calibrated data [1]. The influence of the experimental limitations (such as the high cost of nanofluid preparation and characterization, and nanoparticles' aggregation and sedimentation) on the hydrothermal results of the nanofluids can be eliminated by adopting proper numerical approaches to simulate nanofluid flows.

This chapter summarizes the commonly adopted numerical methods to compute the nanofluids' flow physics and thermal characteristics. Accurate modeling of nanofluids' flow and thermal activity requires a proper realization of the underlying physical mechanisms. Therefore the potential physical mechanisms defining the heat transfer attributes of the nanofluids and the associated nanofluid thermophysical properties are briefly explained. The description of the single-phase and multiphase (Eulerian–Eulerian and Eulerian–Lagrangian models) mathematical models to simulate nanofluids' flow and heat transfer attributes is presented in detail. Additionally, a brief overview of the macro-, micro-, and mesoscale computational fluid dynamic techniques to solve the associated governing equations is presented in this chapter.

6.2 Heat transfer enhancement mechanism of nanofluid

Accurate modeling of nanofluids' flow and heat transfer is essentially dependent on the proper realization of the underlying physical mechanisms determining the heat transfer attributes of the nanofluids. For the case of particulate flows, a wide range of mechanisms is involved in determining their heat transfer and fluid flow characteristics. However, only a few of these nanoscale mechanisms are significant enough to be considered for modeling nanofluids due to the small concentration of the nanosized particles. The possible mechanisms reported in the literature can be broadly classified as:

1. Particle–particle interactions
2. Particle–liquid interactions
3. External forces on particles

An overview of these interactions is given in this section.

6.2.1 Particle–particle interactions

The particle–particle interactions are significant for the dense and wall-mounted nanofluid flows in which nanoparticle motions are governed by the collisions and continuous contact with other nanoparticles. However, for the dilute solutions, the hosting fluid mainly governs the nanoparticle motions. A particle–liquid solution can be classified as dense or dilute based on its volume fraction. The nanoparticle volume fraction is the volume of nanoparticles divided by the total volume of the mixture's constituents prior to the mixing.

$$\varphi = \frac{\mathcal{V}_p}{\mathcal{V}_p + \mathcal{V}_f} = \frac{\left(\frac{m_p}{\rho_p}\right)}{\left(\frac{m_p}{\rho_p}\right) + \left(\frac{m_f}{\rho_f}\right)} \tag{6.1}$$

The volume fraction and mass fraction of the nanoparticles are interconvertible as:

$$\varphi = \frac{\psi\rho_f}{\rho_p + \psi\rho_f - \psi\rho_p} \tag{6.2}$$

A particle–liquid solution can be regarded as dilute if the nanoparticle volume fraction is $\varphi \leq 0.001$. The dense flow ($\varphi > 0.001$) is classified as collision-dominated and contact-dominated based on the collisions of the particles [2]. In the collision-dominated flows ($0.001 < \varphi < 0.1$), particles rebound with different trajectories after colliding with each other. While, in the contact-dominated flow ($\varphi > 0.1$), the particles are in constant contact with each and particle motion is governed by the contact forces (Fig. 6.1). It should be noted that for the case of actual nanofluid samples, both dilute and dense phases may coexist in the same flow, as nanoparticles are highly prone to forming clusters and sedimentation.

The particle–particle interactions significant for the nanofluid ($\varphi > 0.001$) flow include van der Waals force, electrical double-layer (EDL) force, and particle agglomeration (Fig. 6.2).

The attractive forces between molecules or tiny particles of sizes ranging from 0.2 to 10 nm are called van der Waals forces [2]. Generally, van der

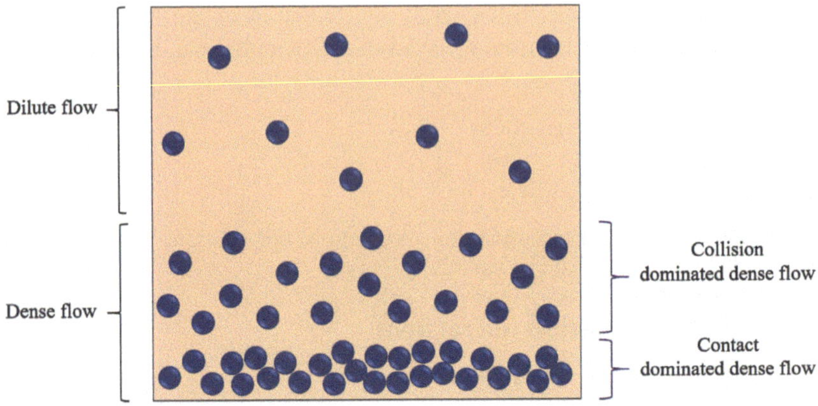

FIGURE 6.1 Dilute and dense flows.

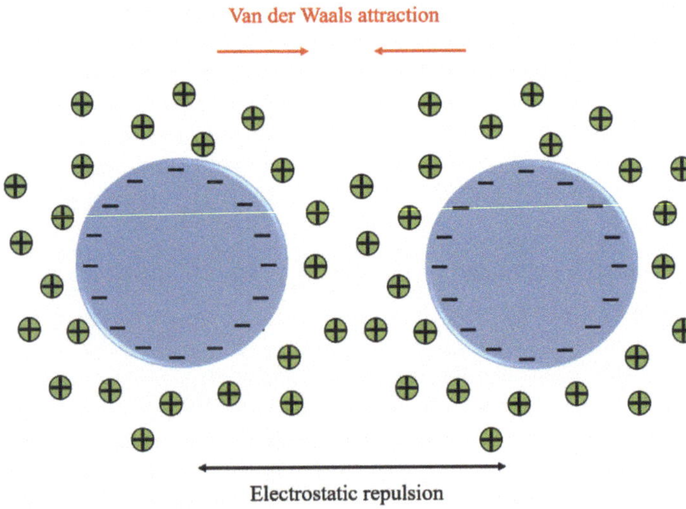

FIGURE 6.2 Particle–particle interactions.

Waals forces between surfaces of different materials are estimated by the Hamaker theory [3]. The van der Waals force per unit area between two spheres of the diameter d_1 and d_2 (Fig. 6.3) is estimated as:

$$F_{\text{vdw}} = \frac{A}{12z}\left(\frac{D_1.D_2}{D_1 + D_2}\right) \tag{6.3}$$

where z denotes the minimum distance between two spheres at the contact point. A' is the Hamaker constant and a few of its values for different materials are enlisted in Table 6.1.

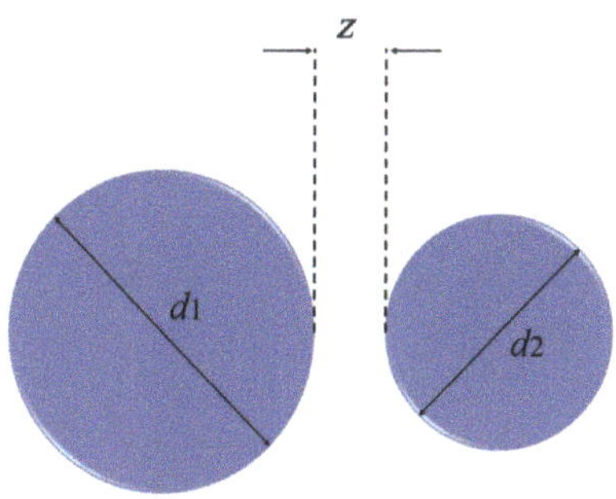

FIGURE 6.3 Two spheres of the diameter d_1 and d_2.

TABLE 6.1 Hamaker constant values for different materials [2,4].

Material	Hamaker constant [J/10^{-20}]	Material	Hamaker constant [J/10^{-20}]
Al_2O_3	15.5	Silica (fused)	6.5
Au, Ag. Pt	~40	TiO_2	19.5
Cu	28.4	Ethanol	4.2
Polystyrene	6.15–6.6	Decane	4.8
Water	4.38	hexadecane	5.1

In nanofluids, the van der Waals forces are primarily responsible for the nanoparticle cluster formation into the solution because of their enlarged particle surface area [5,6].

When the charged particles are immersed in an electrolyte solution such as salt water, the fluid ions form an EDL encapsulating the charged particles [7–10], which induces a small range repulsive force between particles called the EDL force. The EDL consists of surface charge (charged ions embedded on the particle surface), stern layer (counterions attached to particle surface), and diffusion layer (a film of the surfactant in the solution) as shown in Fig. 6.4. The Hogg–Healy–Fuerstenau correlation [11] for estimating EDL force is defined as:

$$F_{EDL} = \frac{dE_{Edl}}{dz_P} \tag{6.4}$$

The electrostatic interaction potential energy E_{Edl} between two particles is defined as:

$$E_{Edl} = \left[\frac{\varepsilon r_1 r_2 (\xi_1^2 + \xi_2^2)}{4(r_1 + r_2)}\left[\frac{2\xi_1\xi_2}{(\xi_1^2 + \xi_2^2)}\ln\left(\frac{1 + \exp(-\mathcal{K} z_p)}{1 - \exp(-\mathcal{K} z_p)}\right) + \ln 1 - \exp(-\mathcal{K} z_p)\right]\right. \tag{6.5}$$

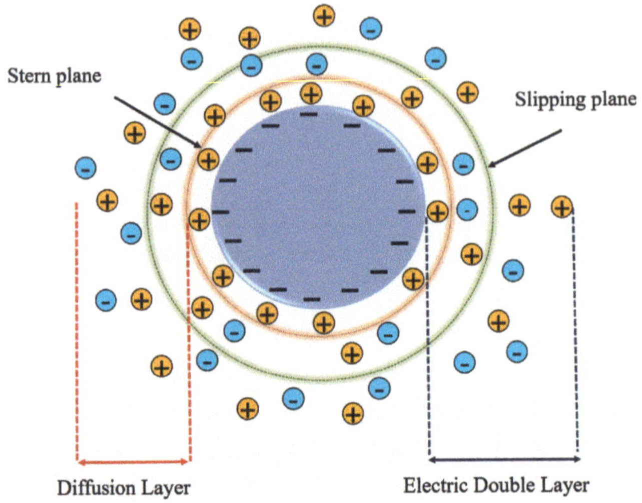

FIGURE 6.4 Electrical double layer around a particle.

where r_1 and r_2 are the radius of two particles, ξ is the particle surface potential, z_p is the zeta potential, and $\mathscr{K}^{-1}$ is the double layer thickness around the particle, called the Debye length. The repulsive nature of the EDL forces prevents the agglomeration of the nanoparticles and thereby improves the stability of the nanofluid solution [10].

The EDL protect the suspended particles against the rapid aggregations in the aqueous fluid. Such particles with identical surface charges are kinetically stable in the aqueous fluid. However, in the absence of repulsive forces or when the repulsive electrostatic (EDL) forces between the particles are overweighed by the attractive van der Waals forces, it results in the formation of particle agglomerations and aggregates [12]. If energy dissipated during the collisions of the particles is larger than the impact kinetic energy, the particles are unable to rebound and thereby particle clusters are formed (Fig. 6.5). Such a phenomenon is more perceptible at the larger nanoparticle concentration and small nanoparticle sizes because of the enlarged particle effective surface area. The particle clusters can be of various shapes ranging from spherical to chain-like structures. Due to the irregular shapes of these clusters, it is practically impossible to determine the particle–particle and particle–liquid interactions. In most of the literature studies, particle agglomeration is regarded as an undesirable phenomenon [13–15] and tends to be reduced by introducing additives or surfactants as well as particle surface treatment. Thereby most of the studies based on the modeling of the nanofluids disregard particle agglomerations by considering a uniform dispersion of the nanofluids. However, depending on the particular applications, the particle agglomerations can either be desirable in a controlled manner or undesirable [12].

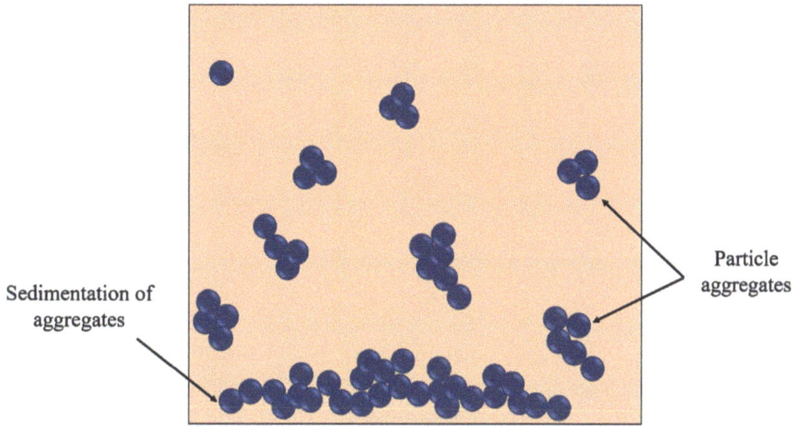

FIGURE 6.5 The uncharged suspended particles are prone to form agglomerates in the hosting fluid.

6.2.2 Particle–liquid interactions

Particle–liquid interactions refer to the energy and momentum exchanges among particles and the hosting fluid and are accountable for the interphase coupling. The expected significant particle–liquid interactions for the nanofluids include thermophoresis, Brownian force, drag force, pressure gradient force, lift force, and virtual mass force.

The erratic random movement of particles in the hosting fluid is called Brownian motion. In nanofluids, Brownian motion optimizes the heat dissipation capacity directly by the nanoparticle diffusions and indirectly by actuating the nanoconvections of the neighboring base fluids [16–19]. Its intensity is perceptible in the regions of high fluid temperature and varies inversely with the particle size and fluid viscosity [20]. Its value can be calculated by the following correlation [21]:

$$F_{\text{Brownian}} = m_p \zeta_i \sqrt{\frac{\pi}{\Delta t} \frac{216 \nu_f K_B T}{\pi^2 \rho_f d_p^5 \left(\frac{\rho_p}{\rho_f}\right)^2 C_c}} \tag{6.6}$$

where ζ_i is the independent Gaussian random variables with zero mean and a variance of unity, Δt is the integration time step size for integrating the equations of particle motion, K_B is the lattice Boltzmann constant, ν_f is the base fluid kinematic viscosity, T is the temperature, and C_c is the Cunningham correction factor.

Thermophoresis or the Soret effect is a temperature gradient phenomenon in which particles in the hotter region move towards the colder region due to the temperature difference (Fig. 6.6). Thereby, the heat transfer process is

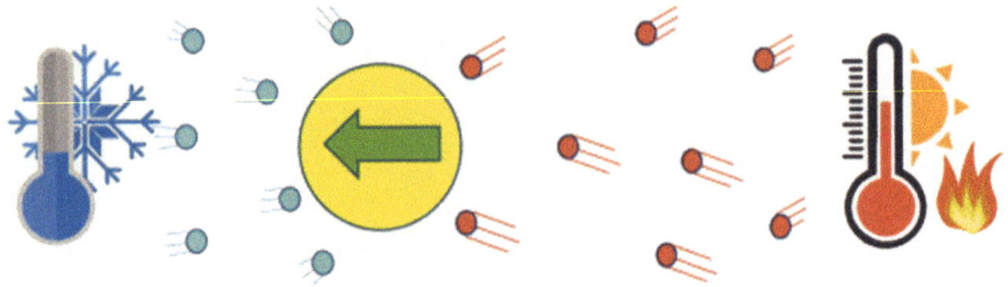

FIGURE 6.6 Thermophoresis mechanism schematic [22].

augmented by these moving particles. This effect is significant for very low fluid velocity, such as natural convection [22]. The thermophoresis force can be calculated by the following expression [23].

$$F_{\text{thermophoresis}} = \frac{7.02\pi d_p \mu_f^2 (K + 2.18K_n)}{\rho_f (1 + 3.42K_n)(1 + 2K + 4.36K_n)} \times \frac{\nabla T}{T_f} \tag{6.7}$$

where $K_n = \frac{2\lambda}{d_s}$ is the Knudsen number.

When a particle is moving with respect to the hosting fluid, the drag force (fluid resistance or fluid friction) is the force acting opposite to the relative motion of the particle. For the case of nanofluids, F_{drag} value can be determined based on the Stokes law [10] as:

$$F_{\text{drag}} = -3\pi d_p \mu_f f(\vec{V}_p - \vec{V}_f) \tag{6.8}$$

where d_p, μ_f, $\vec{V}_p$, and $\vec{V}_f$ are the diameter of the particle, viscosity of the hosting fluid, and velocities of the particle, and the hosting fluid, respectively. The f is the friction factor dependent that is on the fluid inertia, slip flow, and crowding of the fluid particles. Eqs. (6.6)–(6.8) can be rewritten in the form of drag coefficient (C_{drag}) as:

$$C_{\text{drag}} = \frac{F_{\text{drag}}}{\frac{1}{2}\rho_f \vec{V}_{slip}^2 A_p} = \frac{24f}{Re_p} \tag{6.9}$$

where ρ_f, $\vec{V}_{\text{slip}}$, and A_p are the base fluid density, slip velocity magnitude, and nanoparticle surface area, respectively. The $Re_p = \rho_f V_{\text{slip}} d_p / \mu_f$ is the Reynolds number of the nanoparticles.

The lift (F_{lift}) force acing on a moving particle due to the surrounding liquid comprises two types, called Saffman lift force and the Magnus lift force. The shear-induced force acting on the particle when the particle and the hosting liquid are moving in the same direction is called the Saffman lift force. Whereas a sidewise force acting on a spinning particle suspended in fluid when it is spinning with respect to the surrounding fluid is called the

Magnus lift force. The Saffman and Magnus lift forces can be estimated by Eqs. (6.6)–(6.10) and (6.6)–(6.11), respectively [10]:

$$F_{S,\text{lift}} = 6.46\mu_f \left(\frac{d_p}{2}\right)^2 \vec{V}_{\text{slip}} \left(\frac{\rho_f G_f}{\mu_f}\right)^{1/2} \tag{6.10}$$

$$F_{M,\text{lift}} = \pi\rho_f \left(\frac{d_p}{2}\right)^3 \vec{V}_{\text{slip}} \left('\Omega - \frac{11}{8\pi} G_f\right) \tag{6.11}$$

where G_f and $'\Omega$ are the fluid shear force and particle rotational velocity, respectively.

The pressure-gradient force (F_{pressure}) on the particle is the force resulting from the difference in pressure across the particle surface. Its value can be estimated as [24]:

$$F_{\text{pressure}} = \rho_f \frac{D\vec{V}_f}{D_t} \left(\frac{\pi}{6} d_p^3\right) \tag{6.12}$$

where $\frac{D}{D_t}$ is the material derivative.

The virtual mass or the added mass force is the inertia produced due to the fluid (surrounding the particle) disturbance caused by the acceleration or deceleration of moving particles. It is estimated as:

$$F_{\text{virtual}} = -C_{\text{mass}}\rho_f \left(\frac{d\vec{V}_p}{dt} - \frac{D\vec{V}_f}{D_t}\right) \left(\frac{\pi}{6} d_p^3\right) \tag{6.13}$$

6.2.3 External forces on particles

The important external forces acting on the nanoparticles that can influence the heat transfer rate of the nanofluid are reduced gravity and Lorentz forces.

The force acting on the particles due to the gravitational and buoyancy forces is called reduced gravity force. It can be estimated by the following correlation [24].

$$F_{\text{gravity}} = g\left(\rho_p - \rho_f\right) \left(\frac{\pi}{6} d_p^3\right) \tag{6.14}$$

A few research studies disregard the influence of the reduced gravity force while modeling nanofluid flows considering the ultrafine size of the nanoparticles [10].

When an external magnetic and electric field is applied to the system, the force acting on the charged particles in the electrically conductive fluid is called Lorentz force. This force can be estimated as:

$$F_{\text{Lorentz}} = q\left(\vec{E} + \vec{V} x \vec{B}\right) x \vec{B} \tag{6.15}$$

where q is the fluid electrical conductivity, $\vec{E}$ represents the electrical field, $\vec{B}$ denotes the magnetic field, and $\vec{V}$ is the fluid velocity vector.

6.3 Thermophysical properties of nanofluids

The thermophysical properties alteration of the solution with nanoparticles addition contributes to the hydrothermal characteristics of the nanofluid. Therefore it is essential to consider the influence of these properties for accurate modeling of the nanofluids' flow and heat transfer characteristics. The values of these thermophysical properties are determined either by the experimental data or through the theoretical correlations. A brief overview of these thermophysical properties of the nanofluids and their respective property correlations are presented in this section.

6.3.1 Thermal conductivity

Thermal conductivity is the most important thermophysical property of the nanofluids that determines their thermal activity. It is sensitive to factors such as particle material, morphology, and concentration, base fluid properties, operational temperature, and solution stability (Fig. 6.7) [26].

The research trends demonstrate a direct dependence of nanofluid thermal conductivity on the nanoparticle concentration despite a few controversies (Fig. 6.8). However, this thermal conductivity enhancement with the nanoparticle addition attains a maximum value and the further addition of nanoparticles deteriorates instead of enhances the nanofluid thermal conductivity [27]. The high operating temperature also augments nanofluid thermal conductivity and such a characteristic is particularly favorable for nanofluid usage in high-temperature applications, i.e., solar collectors. Such a trend of nanofluid thermal conductivity is generally associated with the intensifying

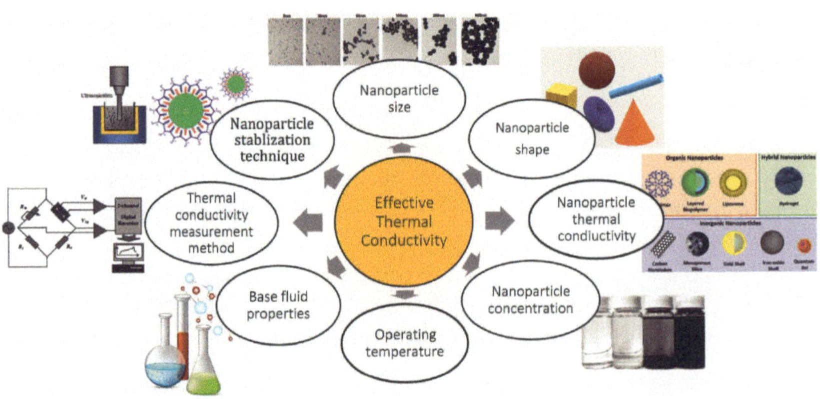

FIGURE 6.7 Factors determining the thermal conductivity of nanofluids [25].

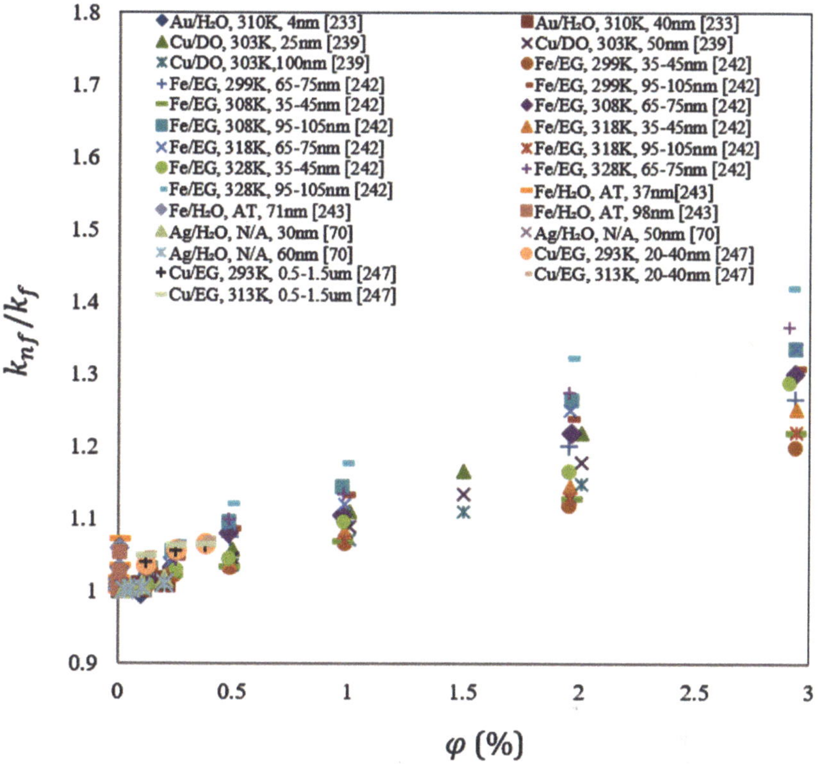

FIGURE 6.8 Nanofluid thermal conductivity variation with particle volume fraction [25].

nanoparticle Brownian motion at high temperatures [28–40]. Such a thermal conductivity trend has been observed also for the nanotube-based nanofluids [34,41–44]. However, some reports state that the variation of thermal conductivity with operational temperature is relatively less significant compared to nanoparticle concentration [45].

The thermal conductivity of the nanofluids is also affected by the particle morphology (size and shape). The research trends show that the reduction in nanoparticle size optimizes the thermal conductivity of the nanofluids for the particle size range less than 100 nm, and this tendency is more perceptible at the larger nanoparticle concentration and operational temperature. However, when the nanoparticle size range is larger than 100 nm, the variation in thermal conductivity with nanoparticle size becomes insignificant, as shown in Fig. 6.9 [25].

In terms of the nanoparticle shape-dependent thermal conductivity of the nanofluids, the majority of the literature studies show an increasing trend of the nanofluid thermal conductivity with the aspect ratio of the nanoparticles [29,46,47]. Moreover, the greater thermal conductivity of the nanotube-based

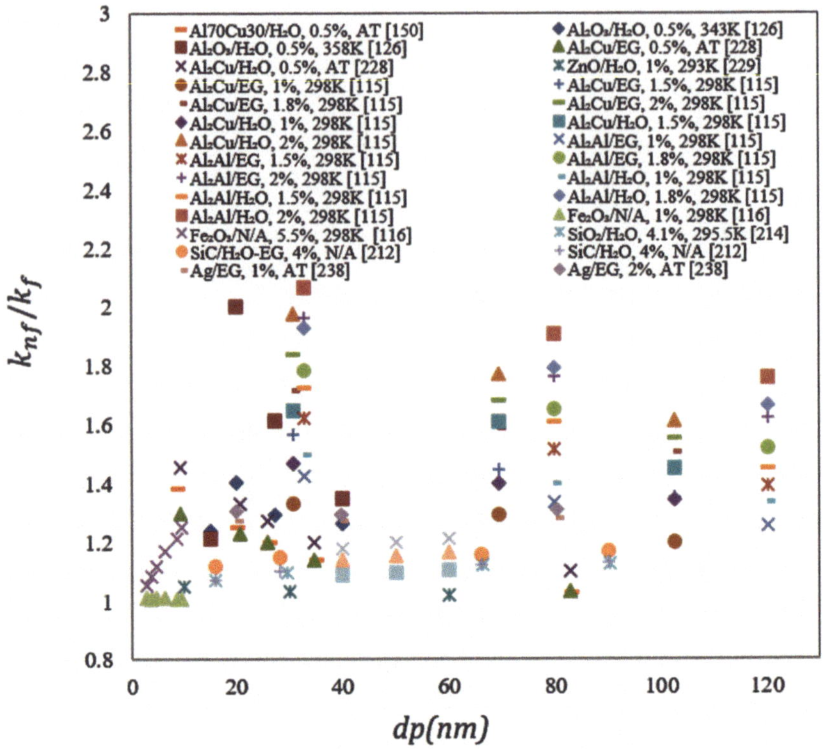

FIGURE 6.9 Nanofluid thermal conductivity variation with particle size [25].

nanofluid also demonstrates such particle shape-dependent thermal conductivity trends of nanofluids [25]. Fig. 6.10 shows the different shapes of the platinum nanoparticles [48].

The thermal conductivity of the nanofluids is sensitive towards the nanoparticle material and thermophysical properties of the hosting fluid. The dispersion of the highly thermal conductive nanoparticles results in the larger augmentation of the nanofluid thermal conductivity [37,39,49,50]. Moreover, the thermal conductivity of the hosting fluid contributes to the thermal conductivity augmentation of the solution [1].

The stability of the nanoparticles into the base fluid determines the nanoparticle agglomeration and sedimentation in the nanofluids. The particle agglomeration and sedimentation influence the thermal conductivity of the nanofluids, as discussed in Section 6.2. In general, the nanofluid stability is ensured by introducing additives or surfactants along with the sonication treatment to adjust the pH value of the solution [51,52]. Therefore the selection of appropriate additives is a critical step to suppress the particle aggregations and sedimentation [53]. The stability of the nanofluids can also be enhanced by extending sonication time up to certain limits. A brief sonication

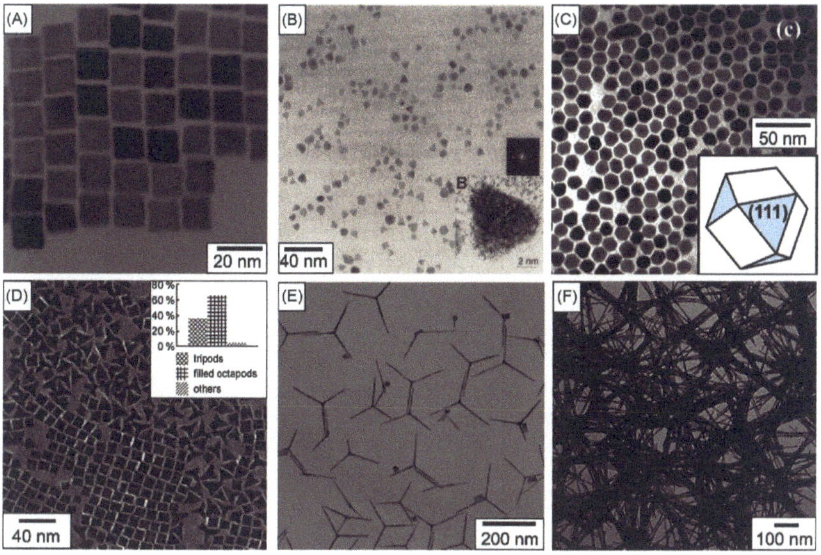

FIGURE 6.10 Platinum nanoparticles of different shapes [48].

results in the poor dispersion of the nanoparticles and cluster formation, while a rigorous sonication may alter the size and shape of the nanoparticles. Therefore sonication for a moderate period is usually recommended [36, 54–56]. The research shows that the difference in thermal conductivity measurement techniques, such as transient, steady-state, and thermal comparator methods, also influences the thermal conductivity results [29,57–59].

A few of the thermal conductivity empirical correlations based on the theoretical and analytical reports are listed in Tables 6.2 and 6.3, respectively. Most of these correlations include empirical constants of proportionality and the values of these constants are determined from the experimental data [26].

6.3.2 Viscosity

The viscosity of the nanofluids is important in determining their flow nature and the pressure drop across the system. The addition of the nanoparticles results in the augmentation of the viscosity. Studies relevant to the rheology of the nanofluids show that nanofluids behave as Newtonian or non-Newtonian fluids depending on factors such as nanoparticle concentration, material and morphology, temperature, base fluid viscosity, shear rate, shear stress, and nanofluid stability. Most of the literature studies demonstrate that the viscosity of the nanofluids observes a direct correlation with the concentration of the nanoparticles [37,80–82] and base fluid viscosity [83]. Such a

TABLE 6.2 Empirical correlations for estimation of nanofluid thermal conductivity based on experimental data.

Reference	Nanofluid	Nanoparticle size (nm), fluid temperature T (K)	Nanofluid thermal conductivity model
[60]	$\frac{Al_2O_3}{H_2O}, 1 \le \varphi \le 4\%$	$d_p = 11, 47, 150, 294 \le T \le 344 -$	$k_{nf} = \left[1 + \text{const.}\left(\frac{1}{d_p}\right)^{0.369}\left(\frac{T^{1.2321}}{10^{\frac{2.46428}{T-140}}}\right)\right]k_f$
[35]	Al_2O_3/H_2O, Al_2O_3/EG, $2 \le \varphi \le 4\%$	$8 \le d_p \le 282$, $T = 298$	$k_{nf} = [4.4134\varphi(1 - e^{-0.025d_p})]k_f + k_p$
[61]	$\frac{Fe_2O_3}{\frac{N}{A}}$, $\varphi = 1\%, 5.5\%$	$2.8 \le d_p \le 9.5$, $T = 298$	$k_{nf} = \left[\left(1 + \text{є}(1/)^{2.5}Pr^{0.3333}\varphi\right)\left(\frac{1+2\theta\varphi}{1-\theta\varphi}\right)\right]k_f$
[57]	$ZnO/H_2O - EG$, $\varphi = 2\%–4\%$	$d_p = 29, 77$, $298 \le T \le 363$	$k_{nf} = \frac{k_p + 2k_f - 2(k_f - k_p)\varphi}{k_p + 2k_f + 2(k_f - k_p)\varphi}k_f + 5{*}10^4\beta\varphi\rho_f(C_p)_f\sqrt{\left(\frac{k_BT}{\rho_p d_p}\right)}f(T,\varphi)$
[62]	Al_2O_3/H_2O, Al_2O_3/EG, $1 \le \varphi \le 3$	$11 \le d_p \le 150$, $T = 293, 323$	$k_{nf} = \left[\left(1 + 0.135\varphi^{0.467}\left(\frac{k_p}{k_f}\right)^{0.273}\left(\frac{T}{20}\right)^{0.547}\left(\frac{100}{d_p}\right)^{0.234}\right)k_f\right.$
[63]	Ag/EG, $\varphi = 1\%, 2\%$	$d_p = 20–80$, AT	$\frac{k_{nf}(d_p,T,\varphi)}{k_f(T)} = \left[\frac{k_p(d_p,T)}{k_f(T)}\right]^{\varphi}$
[64]	SiO_2/H_2O, $1.96 \le \varphi \le 12.85\%$	$d_p = 10–60$, $T = 293$	$\frac{k_{nf} - k_O}{k_\infty - k_O} = 1 - exp\left[\frac{-6.3\left(\frac{r-r_O}{r_O}\right)^{1.4}}{\varphi^{-0.3}\left(\frac{d_p}{G_p}\right)^{-0.2}}\right]$
[65]	Fe/H_2O, $0.000313 \le \varphi \le 0.01\%$	$d_p = 37, 71, 98$, AT	$k_{nf} = \left[1 + \left(0.26876\varphi^{0.99288}d_p^{-0.35106}\right)\right]k_f$

[66]	MgO/H_2O, $1 \leq \varphi \leq 3\%$	$d_p = 40, 50, 60,\ T = 293$	$k_{nf} = [1.1507 + 0.05\varphi - 0.013d_p + 0.0038T - 0.0049\varphi d_p + 0.0006\varphi T - 0.0019d_pT - 0.0057\varphi^2 - 0.0012T^2 - 0.0013\varphi d_pT + 0.0013\varphi^2 d_p]k_f$
[67]	SiO_2/H_2O, Al_2O_3/H_2O, TiO_2/H_2O, ZrO_2/H_2O, $\varphi = 2\%$	$d_p = 10-100\ (SiO_2)$, $d_p = 50-150\ (Al_2O_3)$, $d_p = 71-150\ (TiO_2)$, $d_p = 44, 105(ZrO_2)$, $T = 298K$	$k_{nf} = \left[1 + \left(0.0193 + 0.00383\frac{\rho_p}{\rho_f}\sqrt{\frac{\varphi d_p}{d_f}}\right)\right]k_f$

TABLE 6.3 Empirical correlations for estimation of nanofluid thermal conductivity based on theoretical studies.

Reference	Effective thermal conductivity correlation
[68]	$k_{nf}=\left(\frac{k_p+2k_f-2\varphi(k_f-k_p)}{k_p+2k_f+2\varphi(k_f-k_p)}+\frac{\rho_p\varphi c_p}{2k_f}\sqrt{\frac{2k_BT}{3\pi d_p\mu_p}}\right)k_f$
[69]	$k_{nf}=\left(\frac{k_p+2k_f+2\varphi(k_f-k_p)}{k_p+2k_f-\varphi(k_f-k_p)}\right)k_f+5^{*}10^4\beta_f\varphi\rho_f(C_p)_p\ldots{}^{*}\sqrt{\frac{k_BT}{\rho_pd_p}}f(T,\varphi)$ $f(T,\varphi)=(-6.04\varphi+0.4705)T+(1722.3\varphi-134.63)$
[19]	$k_{nf}=\left(1+\frac{2k_BT^{*}}{\pi\mu d_p^2}C\frac{\varphi d_f}{k_f(1-\varphi)d_p}\right)k_f,\ C=\psi\lambda_p nk_B$
[70]	$k_{nf}=\left[1+3\Theta\varphi\left(1+\frac{2\delta}{d_p}\right)^3+\frac{3\Theta^2\left(\varphi\left(1+\frac{2\delta}{d_p}\right)^3\right)^2}{1-\Theta\left(\varphi\left(1+\frac{2\delta}{d_p}\right)^3\right)}\right]k_f,\ \Theta=\frac{\left(\frac{k_l-k_f}{k_l+2k_f}\right)^2\left[\left(1+\frac{2\delta}{d_p}\right)^3\left(\frac{k_l-k_f}{k_l+2k_f}\right)-\left(\frac{k_p-k_l}{k_p+2k_l}\right)\right]}{\left(1+\frac{2\delta}{d_p}\right)^3+2\left(\frac{k_l-k_f}{k_l+2k_f}\right)\left(\frac{k_p-k_l}{k_p+2k_l}\right)}$
[71]	$k_{nf}=\left[1+F(Pe)+3\Theta\varphi\left(1+\frac{2\delta}{d_p}\right)^3+\frac{3\Theta^2\left(\varphi\left(1+\frac{2\delta}{d_p}\right)^3\right)^2}{1-\Theta\left(\varphi\left(1+\frac{2\delta}{d_p}\right)^3\right)}\right]k_f,\ \Theta=\frac{\left(\frac{k_l-k_f}{k_l+2k_f}\right)^2\left[\left(1+\frac{2\delta}{d_p}\right)^3\left(\frac{k_l-k_f}{k_l+2k_f}\right)-\left(\frac{k_p-k_l}{k_p+2k_l}\right)\right]}{\left(1+\frac{2\delta}{d_p}\right)^3+2\left(\frac{k_l-k_f}{k_l+2k_f}\right)\left(\frac{k_p-k_l}{k_p+2k_l}\right)}$ $F(Pe)=0.0556Pe+0.1649Pe^2-0.0391Pe^3+0.0034Pe^4$
[72]	$k_{nf}=(1-\varphi)k_f+\beta_kk_p\varphi+18^{*}10^6\frac{d_f}{d_p}k_f\left(\frac{k_BT}{3\pi\mu_p\mu_fl_f}\right)^2\mathrm{Pr}_{\varphi_p}$
[73]	$k_{nf}=\left(\frac{k_p+2k_f-2\varphi(k_f-k_p)}{k_p+2k_f+\varphi(k_f-k_p)}+cNu\frac{d_f}{Pr}\frac{(2-L_f)L_f}{(1-L_f)^2}\frac{\left[\left(\frac{d_{p,max}}{d_{p,min}}\right)^{1-L_f}-1\right]^2}{\left[\left(\frac{d_{p,max}}{d_{p,min}}\right)^{1-L_f}-1\right]^2}\frac{1}{d_{p,avg}}\right)k_f,\ L_f=D-\ln\varphi/\ln\left(\frac{d_{p,max}}{d_{p,min}}\right)$

[7]	$k_{nf} = k_f \left[\frac{(k_p + 2k_f + 2(k_p - k_f)\varphi}{(k_p + 2k_f - (k_p - k_f)\varphi} \right] + \frac{\varsigma \lambda_p C_p}{3} \frac{k_B T}{3\pi\mu d_p \lambda_f} \left[1 + \frac{3\pi\mu d_p \lambda_f \sqrt{(Aq^2 \lambda_p)}}{k_B T \sqrt{\varepsilon_f m_p \left(\frac{d_p}{2} \varphi^{-\frac{1}{3}} \right)^2}} \right]$
[74]	$Re = \frac{\rho d_p}{\mu} \left[\sqrt{\lvert v^2 \rvert} + \frac{1}{2} (d_p \sqrt{\lvert \omega^2 \rvert}) \right]$
[75]	$k_{nf} = \frac{(k_p - k_{lr})\varphi k_{lr} [2б_1^3 - б^3 + 1] + (k_p + 2k_{lr}) б_1^3 (\varphi б^3 (k_{lr} - k_f) + k_f]}{б_1^3 (k_p + 2k_{lr}) - (k_p - k_{lr})\varphi [б_1^3 + б^3 - 1]}, б = 1 + \frac{2\delta}{d_p}, б_1 = 1 + \frac{\delta}{d_p}$
[76]	$k_{nf} = [1 + \mathcal{Q} Re^e Pr^{0.333} \varphi] \left[\frac{(k_p(1+2b) + 2k_z) + 2\varphi([k_p(1-b) - k_z])}{(k_p(1+2b) + 2k_z) - \varphi(k_p(1-b) - k_z)} \right] k_f, b = \frac{2R_b k_z}{d_p}$
[77]	$k_{nf} = \frac{k_p + 2k_{lr} + 2(k_p - k_{lr}) \left(\frac{d_p}{d_p + 2\delta} \right)}{k_p + 2k_{lr} - (k_p - k_{lr}) \left(\frac{d_p}{d_p + 2\delta} \right)} \left(\frac{2k_{lr}}{d_p + 2\delta} \right)$
[78]	$k_{nf} = \left[\frac{k_f \varphi k_{lr} (k_p - k_{lr})(2б_1^3 - б^3 + 1) + (k_p + 2k_{lr}) б_1^3 (\varphi б^3 (k_{lr} - k_f) + k_f)}{б_1^3 (k_p + 2k_{lr}) - (k_p - k_{lr})\varphi (б_1^3 + б^3 - 1)} \right] + [\varphi^2 б^6 k_f \left(3\Lambda^2 + \frac{3\Lambda^2}{4} + \frac{9\Lambda^3}{16} \frac{(k_{cp} + 2k_f)}{(2k_{cp} + 3k_f)} + \frac{3\Lambda^4}{2^6} + \ldots \right)$ $+ \left[\frac{1}{2} \rho_{cp} (C_p)_{cp} d_s \sqrt{\frac{12 k_B T (1 - 1.5б^3 \varphi)}{\pi \rho_{cp} б^3 d_p^3}} + \frac{G_T}{3\pi\mu б d_p d_s} \right] б = 1 + \frac{2\delta}{d_p},$ $б_1 = 1 + \frac{\delta}{d_p}$
[79]	$k_{nf} = \left[1 + J \left(\frac{\varphi_p^a}{d_p^b} \right) \left(\frac{T}{T_o} \right)^c \right] k_f$

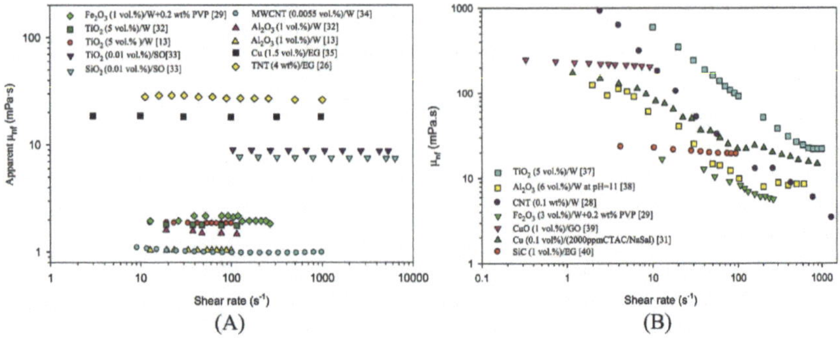

FIGURE 6.11 Literature data on shear rate dependent viscosity of the (A) Newtonian nanofluids and (B) non-Newtonian nanofluids at room temperature [84].

trend is more noticeable for the less stable nanofluids where nanoparticles agglomerate and eventually sedimentation is caused. Some studies show that at the low particle concentration, nanofluids show Newtonian behavior, however, at the moderate to larger particle loading, these fluids demonstrate non-Newtonian shear thinning nature, as depicted in Fig. 6.11 [84].

For the case of nanoparticle size-dependent viscosity of the nanofluids, a group of scholars state that decreasing the nanoparticle size improves the viscosity of the nanofluids, whereas on the contrary other reports claim the opposite trend. Therefore nanofluid viscosity dependence on the nanoparticle size remains inconclusive due to controversies in the literature [85]. Similarly, a mixed trend of the temperature-dependent viscosity of nanofluids is observed in the literature studies. Some studies demonstrated a direct correlation between the temperature and nanofluid viscosity, while others reported an inverse dependence of the nanofluid viscosity on the temperature [84].

The stability of the nanofluids also affects the viscosity of the nanofluid. The unstable and nonuniform dispersion of the nanofluids results in the formation of the nanoparticle clusters which contribute to viscosity augmentation. This phenomenon becomes more characteristic at the higher nanoparticle concentration and lower nanoparticle size. The stable and unstable nanoparticle clusters adjoin and result in the formation of thick agglomerates over time. Eventually, these agglomerates settle down in the form of sediments and further contribute to the viscosity augmentation of the nanofluids [86].

A few empirical correlations reported by different scholars for the estimation of the nanofluid viscosity based on the experimental and theoretical investigations are listed in Tables 6.4 and 6.5, respectively. Most of these correlations are a function of nanoparticle concentration and temperature, while a few also consider the influence of nanoparticle size on nanofluid viscosity. However, none of these correlations or theoretical models is applicable for all the combinations of nanofluids and operational conditions.

TABLE 6.4 Experimental correlations for estimation of nanofluid viscosity.

Reference	Nanofluid type	Effective viscosity model
[87]	TiO_2/H_2O	$\mu_{nf} = (a + b\varphi + c\varphi^2)\mu_f$
[88]	Cu/H_2O	$\ln\left(\frac{\mu_{nf}}{\mu_{bf}}\right) = A\left(\frac{1}{T}\right) - B$
[89]	Al_2O_3/H_2O or EG	$\mu_{nf} = \left(1 + 7.3\varphi + 123\varphi^2\right)\mu_f$
[90]	$\frac{CuO}{H_2O}$:EG(60:40)	$\log\frac{\mu_{nf}}{\mu_f} = \left(165.56 - 29.643\varphi + 1.8375\varphi^2\right) \times \exp\left[\left(0.0186 - 0.001\varphi + 4 \times 10^{-6}\varphi^2\right)T\right]\log\frac{\mu_{nf}}{\mu_f} = Ae^{-BT}$
[91]	Al_2O_3/H_2O	$\mu_{nf} = \mu_f 0.904e^{14.83\varphi}$ $\mu_{nf} = \mu_f\left(1 + 2.5\varphi + 150\varphi^2\right)$ $\mu_{nf} = \mu_f(1.125 - 0.0007T)$ $\mu_{nf} = \mu_f\left(2.1275 - 0.0215T + 0.002T^2\right)$
[92]	Al_2O_3/H_2O:EG	$\mu_{nf} = \mu_f\left(Ae^{\left(\frac{B}{T} + C\varphi\right)}\right)$

TABLE 6.5 Theoretical correlations for estimation of nanofluid viscosity.

Reference	Effective viscosity model
[93]	$\mu_{nf} = \left(1 + \pi\varphi_p + k_H\varphi_p^2\right) = \left(1 + 2.5\varphi_p + 6.2\varphi_p^2\right)$
[94]	$\mu_{nf} = \frac{1}{(1-\varphi_p)^{2/5}} = \left(1 + 2.5\varphi_p + 4.375\varphi_p^2 + \ldots\right)$
[95]	$\mu_{nf} = \mu_f\left[\frac{1 + 1.25\varphi_p}{1 - \frac{\varphi_p}{0.78}}\right] = \left(1 + 2.5\varphi_p + 4.75\varphi_p^2 + \ldots\right)$
[96]	$\mu_{nf} = \left(1 + 2.5\varphi_p\right)$
[97]	$\mu_{nf} = \left(\frac{9}{8}\frac{\left(\frac{\varphi_p}{\varphi_{p,\max}}\right)^{\frac{1}{3}}}{1 - \left(\frac{\varphi_p}{\varphi_{p,\max}}\right)^{\frac{1}{3}}}\right)$
[98]	$\mu_{nf} = \left(1 + 2.5\varphi_p\right) + \left[\frac{4.5}{\frac{h}{r_p}\cdot\left(2 + \frac{h}{r_p}\right)^2}\right]$
[99]	$\mu_{nf} = \frac{1}{1 - 2.5\varphi_p} = \left(1 + 2.5\varphi_p + 6.25\varphi_p^2 + \ldots\right)$
[100]	$\mu_{nf} = \mu_f\left[1 + \frac{1.25\varphi_p}{1 - \varphi_p}\varphi_p\right] = \left(1 + 2.5\varphi_p + 2.5\varphi_p^2 + \ldots\right)$
[101]	$\mu_{nf} = \exp\left(\frac{2.5\varphi_p}{1 - k\varphi_p}\right) = \left\{1 + \frac{2}{5\varphi_p} + \left[\frac{3}{125} + \left(\frac{2}{5k}\right)\right]\varphi_p^2 + \ldots\right\}$

6.3.3 Specific heat

The specific heat of the nanofluids is important in modeling the heat transfer as it estimates the heat carried by the nanofluid. Compared to studies relevant to the nanofluid thermal conductivity and viscosity, only a few reports are available in the literature regarding the specific heat of the nanofluids. The majority of these studies demonstrate an inverse dependence of the nanofluid density on the nanoparticle concentration and temperature. It increases with the nanoparticle size and vice versa. The addition of the surfactants and additives augments the density of the nanofluids. The specific heat of the nanoparticles and base fluid directly influences the specific heat of the nanofluid [102]. Most of the numerical and experimental studies estimated nanofluid specific heat through the classical mixture model [103], which establishes a linear relationship between the specific heats of the base fluid and nanoparticles based on their volume fractions in the mixture.

$$(C_p)_{nf} = (1 - \varphi)(C_p)_f + \varphi(C_p)_p \tag{6.16}$$

TABLE 6.6 Empirical correlations for estimation of nanofluid specific heat.

Reference	Nanofluid specific heat model
[104]	$(\rho C_p)_{nf} = (1-\varphi)(\rho C_p)_f + \varphi(\rho C_p)_p$, $\rho_{nf} = (1-\varphi)\rho_f + \varphi\rho_p$
[105]	$(C_p)_{nf} = Q/m\left(\frac{\Delta t}{\Delta T}\right)$
[106]	$(C_p)_{nf} = \frac{\left[(1-\varphi)\rho_f (C_p)_f + \varphi\rho_p (C_p)_p\right]}{[(1-\varphi)\rho_f + (1-\varphi)\rho_p]}$
[42]	$\frac{(C_p)_f - (C_p)_{nf}}{(C_p)_f} = (0.0128T + 1.8382)\varphi^{0.4779}$
[107]	$(C_p)_{nf} = 0.8429\left(1+\frac{T}{50}\right)^{-0.3037}\left(1+\frac{d_p}{50}\right)^{0.4167}\left(1+\frac{\varphi}{100}\right)^{2.272}$
[108]	$(C_p)_{nf} = 2.62 - 6\times 10^{-3}T + 2\times 10^{-5}T^2$

A few empirical correlations for the determination of the specific heat of the nanofluids reported in the literature are listed in Table 6.6.

6.3.4 Density

The suspension of the nanoparticles into the base fluid results in the increase of the density of the solution. Generally, the density of nanofluids is directly proportional to the nanoparticle concentration. This characteristic is important in modeling the nanofluid's flow and heat transfer characteristics because it directly affects the Reynolds number, flow rate, pressure drop, pumping power, and Darcy friction factor. Very few empirical correlations for the determination of nanofluid density are available in the literature. Most of the numerical and experimental studies estimated nanofluid density through the classical mixture model [103] as:

$$\rho_{nf} = (1-\varphi)\rho_f + \varphi\rho_p \tag{6.17}$$

A few other empirical correlations for nanofluid density estimation are listed in Table 6.7.

It should be noted that there are huge controversies in the literature studies concerning the thermophysical properties of nanofluids. Even studies reporting similar trends of thermophysical properties depending on their influencing parameters show different rates of variation [25]. The nanoparticle instability is one of the prime reasons behind such anomalies.

6.4 Mathematical models to simulate nanofluids

6.4.1 Single-phase model

Despite their two-phase nature, nanofluids can be modeled as single-phase fluid after consideration of appropriate assumptions. The single-phase model

TABLE 6.7 Empirical correlations for estimation of nanofluid density.

Reference	Nanofluid thermal density
[109]	$\rho_{nf(\mathrm{MWCNT-COOH})} = -0.00443T^2 - 0.096\varphi T - 0.02468T + 10.4\varphi + 1000.15$ $\rho_{nf(\mathrm{F-GNP})} = -0.00518T^2 + 25.6\varphi T - 0.0184T + 0.01741\varphi + 2.456\varphi + 999.88$
[110]	$\rho_{nf} = \frac{m_p + m_f}{\left(\mathcal{V}_f + \mathcal{V}_p \frac{(r_p + t_{nlyr})^3}{r_p^3}\right)}$

is subsequently divided into homogeneous, thermal dispersion, and Buongiorno models [10].

6.4.1.1 *Homogenous model*

The homogenous approximation of the nanofluids is the simplest approach to model nanofluid forced convection. This approach is based on the assumptions of a stable and homogenous dispersion of the ultrafine nanoparticles into the hosting fluid. The mixture of nanoparticles and hosting fluid is regarded as a single-phase fluid of enhanced thermophysical characteristics. Both the phases are presumed to be at thermal and hydrodynamic equilibrium by discounting all the interphase slip mechanisms. The enhanced thermophysical characteristics of the nanofluids are calculated using theoretical thermophysical property correlations or experimental values [10]. The governing equations for the single-phase model can be expressed as:

Continuity equation:

$$\nabla \cdot \left(\rho_{nf}\vec{V}\right) = 0 \tag{6.18}$$

Momentum equation:

$$\rho_{nf}(\vec{V}.\nabla)\vec{V} = -\nabla P + \nabla.\left[\mu_{nf}\left(\nabla\vec{V} + \nabla\vec{V}^T\right)\right] \tag{6.19}$$

Energy equation:

$$\nabla.\left((\rho C_p)_{nf}\vec{V}T\right) = \nabla.\left(k_{nf}\nabla T\right) \tag{6.20}$$

Eqs. (6.6)–(6.19) is the momentum conservation equation for the nanofluid with the temperature-dependent viscosity. For the case of nanofluids with constant viscosity, Eqs. (6.6)–(6.19) are solved without $\nabla\vec{V}^T$ term. The accuracy of the homogenous approach is primarily determined by the accurate estimation of the thermophysical properties (ρ_{nf}, C_p, μ_{nf}, k_{nf}) of the nanofluids. The thermophysical traits can be assumed to be either constant or temperature-dependent based on the problem.

6.4.1.2 Thermal dispersion model

Firstly proposed by Xuan and Roetzel [104], the thermal dispersion model is an improved version of the homogenous model. It accounts for the effect of slip mechanisms between the nanoparticles and hosting fluid phases in terms of thermal dispersion. The random motion of the nanoparticles into the base fluid is regarded as the thermal dispersion, which ameliorates the heat transfer rate of the nanofluids. In this model, the fluid velocity ($\vec{V}$) and temperature (T) are given as:

$$\vec{V} = \vec{V}^{v} + \vec{V}' \tag{6.21}$$

$$T = T^{v} + T' \tag{6.22}$$

where $\vec{V}'$ and T' signify the small variabilities in the fluid velocity and temperature caused by the nanoparticle movements. The mean $\vec{V}^{v}$ and T^{v} values of the working fluid are given as:

$$\vec{V}^{v} = \frac{1}{\mathcal{V}_f} \int_{\mathcal{V}_f} \vec{V} d\mathcal{V} \tag{6.23}$$

$$T^{v} = \frac{1}{\mathcal{V}_f} \int_{\mathcal{V}_f} T d\mathcal{V} \tag{6.24}$$

In the above equations, $\mathcal{V}$ is the working fluid volume. By assuming a negligible boundary layer in-between the fluid and nanoparticles, the energy equation can be rewritten as:

$$\nabla.\left((\rho C_p)_{nf} \vec{V}^{v} T^{v}\right) = -\nabla.\left(k_{nf} \nabla T^{v}\right) - \nabla.(\rho C_p)_{nf} \vec{V}' T' \tag{6.25}$$

The heat flux generated due to the thermal dispersion is estimated as:

$$Q_{\mathbf{d}} = \nabla.(\rho C_p)_{nf} \vec{V}' T' = -k_{\mathbf{d}} \nabla T^{v} \tag{6.26}$$

Here $k_{\mathbf{d}}$ represents the thermal conductivity enhancement caused by the dispersion. After inserting $k_{\mathbf{d}}$ into Eqs. (6.6)–(6.25), the final energy equation can be written as:

$$\nabla.\left((\rho C_p)_{nf} \vec{V}^{v} T^{v}\right) = \nabla.\left(k_{nf} + k_{\mathbf{d}}\right) \nabla T^{v} \tag{6.27}$$

For the estimation of $k_{\mathbf{d}}$, Xuan and Roetzel [104] proposed the following correlation based on the thermal dispersion for the case of forced convection in the porous media.

$$k_{\mathbf{d}} = C^{*} (\rho C_p)_{nf} V r = C_l (\rho C_p)_{nf} V R d_p r \tag{6.28}$$

where r denotes the radius of the tube containing nanofluid. While C_l is the constant which is estimated by the experimental data. d_p and φ are the nanoparticle diameter and volume fraction, respectively.

Later on, Khaled et al. [111] suggested following $k_{\mathbf{d}}$ correlation in their study based on the heat transfer enhancement due to dispersion effects.

$$k_{\mathbf{d}} = C_l\left(\rho C_p\right)_{nf}\varphi d_p V_m \tag{6.29}$$

where C_l is unknown constant and V_m denotes the average bulk fluid velocity.

Another expression for the determination of the radial $k_{\mathbf{d}}$ was proposed by Mojarrad et al. [112] on the basis of their research regarding dispersion heat transfer enhancement of aqueous-based alumina nanofluids in a circular tube.

$$k_{\mathbf{d}} = C_l\left(\rho C_p\right)_{nf}\frac{r\varphi}{d_p}\left(\frac{\partial T}{\partial r}\right) \tag{6.30}$$

In the above equation, the effect of the velocity variation was incorporated indirectly by inducing the temperature gradient into the equation.

Mehdi and Seyed [113] also presented a model for the estimation of the radial $k_{\mathbf{d}}$ based on their study on the thermal performance of aqueous based alumina nanofluids under forced convection inside a circular tube.

$$k_{\mathbf{d}} = C_l\left(\rho C_p\right)_{nf}\left(\frac{\partial V_x}{\partial r}\right)\varphi(r)rd_p \tag{6.31}$$

In the above equation $\varphi(r)$ represents the distribution of the nanoparticle volume fraction in the radial direction.

6.4.1.3 Buongiorno model

Buongiorno [114] attempted to improve the homogenous and thermal dispersion models by taking into account seven slip phenomena comprising thermophoresis, gravity, Brownian diffusion, Magnus effect, inertia, diffusiophoresis, and fluid drainage devoid of turbulent effects. It was concluded that only thermophoresis and particle Brownian motion were significant enough to be considered for modeling the nanofluids. Therefore incorporating these two slip mechanisms, he proposed a two-component four-equation nonhomogeneous equilibrium model for mass, momentum, and heat transport of nanofluids. The nanofluid was assumed to be the dilute mixture of nanoparticles and the base fluid. Both nanoparticle and fluid phases were assumed to be at thermal equilibrium with each other. The transport equations of this model are:

Continuity equation:

$$\nabla\cdot\left(\rho_{nf}\vec{V}\right) = 0 \tag{6.32}$$

Momentum equation:

$$\rho_{nf}(\vec{V}.\nabla)\vec{V} = -\nabla P + \nabla.\left[\mu_{nf}\left(\nabla\vec{V} + \nabla\vec{V}^{T}\right)\right] \tag{6.33}$$

Energy equation:

$$\nabla.\left((\rho C_p)_{nf}\vec{V}T\right) = \nabla.\left(k_{nf}\nabla T\right) + (\rho C_p)_{nf}\left[C_{Br}\nabla\varphi.\nabla T + C_{th}\frac{(\nabla T + \nabla T)}{T}\right] \tag{6.34}$$

Nanoparticles conservation equation:

$$\vec{V}.\nabla\varphi = \nabla.\left(C_{Br}\nabla\varphi + C_{th}\frac{\nabla T}{T}\right) \tag{6.35}$$

In Eqs. (6.32)–(6.35), C_{Br} and C_{th} are the nanoparticle Brownian diffusion and thermophoresis coefficients respectively, and are defined as:

$$C_{Br} = \frac{K_B T}{3\pi\mu_{nf} d_p} \tag{6.36}$$

$$C_{th} = 0.26\left(\frac{k_{nf}}{2k_{nf} + k_p}\right)\left(\frac{\mu_{nf}}{\rho_{nf}}\right)\varphi \tag{6.37}$$

In Eq. (6.36), K_B is the Boltzmann constant.

In literature studies, the single-phase homogenous model is the most commonly adopted approach to simulate nanofluid due to its simplicity and low computational cost. Most of the earlier numerical studies on the nanofluid flows were conducted using this approach [115–126]. A few studies based on the comparative analysis of the homogenous and multiphase models to simulate nanofluid thermophysical attributes also emphasized the accuracy of this model. Moraveji and Esmaelli [127] evaluated homogenous and multiphase models (constant and temperature-dependent properties) to simulate Al_2O_3/water($\varphi = 1\%, 4\%$) nanofluid in a uniformly heated tube. Their results showed that single and Eulerian–Lagrangian multiphase models estimated quite similar convective heat transfer results with the maximum of 11% difference at $\varphi = 1\%$. Akbari et al. [128] analyzed homogenous and Eulerian–Eulerian multiphase models [volume of fluid (VOF), mixture, Eulerian] to simulate the turbulent forced convection of Al_2O_3/water ($\varphi = 0.02 - 0.5\%$) and Cu/water($\varphi = 0.1\%, 2\%$) in the horizontal tubes. The results were compared with the experimental data of two literature studies. Their analysis showed that that the homogenous model estimated the most accurate results at all the particle concentrations and flow rates compared to other tested models. The Eulerian–Eulerian models predicted higher Nusselt number values compared to the experimental data. Moreover, the study showed that the computed hydrothermal results were less sensitive towards the Reynolds number variation unlike the experimental results. In a computational study based on the laminar flow of Al_2O_3/water($\varphi = 0.2\%, 0.5\%$) nanofluid in a shell and tube system by using homogeneous model, Akhtari et al. [120] reported a maximum deviation of 15.6% between the numerical and experimental results. Ahmad and Vafai [129] studied the forced convection of the Al_2O_3/water($\varphi = 0.1\%, 0.25\%$) nanofluid in the developing region of the tube by utilizing homogenous, mixture, and Eulerian–Lagrangian models. Their results showed that the mixture and Eulerian–Lagrangian models overestimated the heat transfer coefficient results while the homogenous model estimated accurate results. On the other hand, multiple studies claim that multiple particle–particle, particle–liquid, and external forces also contribute to determining the thermal performance of the nanofluids in addition to their enhanced thermal conductivity [130–132]. Therefore at the large nanoparticle concentration when the particle–particle and particle–liquid interactions become significant [130–132], this model fails to interpret the flow and heat transfer characteristics of the nanofluid. While at low nanoparticle volume fractions, this model estimates results with reasonable accuracy [133,134]. Sinan et al. [135] compared the dispersion model with the homogenous model for predicting the Al_2O_3/water($\varphi = 0.6 - 1.6\%$) nanofluid heat transfer characteristics in the entrance region of the uniformly heated and reported the superiority of the dispersion model in terms of the accuracy. Mehdi et al. [113] evaluated the homogenous, thermal dispersion model with the Eulerian–Lagrangian model to simulate the Al_2O_3/water($\varphi = 1 - 3\%$) nanofluid forced convection in tube. In the thermal dispersion mode, a nonuniform nanoparticle distribution across the tube was assumed. Their results showed that results predicted by the thermal dispersion model were in more agreement with the experimental data compared to the Euleria–Lagrangian and homogenous models.

6.4.2 Multiphase Eulerian–Eulerian model

The multiphase Eulerian–Eulerian models are categorized as mixture, VOF, and Eulerian models, based on the interactions between different phases and interphase coupling.

6.4.2.1 Mixture model

In the multiphase mixture model, the governing equations are solved for the nanoparticle and base fluid mixture by considering the weak interaction between the phases. Each phase is assumed to demonstrate its peculiar velocity field vector and volume fraction. The base fluid impacts the nanoparticles via drag force while the nanoparticles distress the liquid, decreasing its mean momentum and enhancing its turbulence dissipation. The thermophysical characteristics of the mixture are explicated as the arithmetic mean of all the phases weighing their corresponding volume fractions. Both liquid and nanoparticles are presumed to share the same pressure. A negligible interaction between nanoparticles is presumed. The shape of the nanoparticles is regarded as spherical and of uniform size. The concentration of the nanoparticle phase is computed using a scalar formulation through the phase slip correction [136]. Of all multiphase models, the computational cost of this model is the least of all the multiphase models. The transport equations are:

Continuity equation:

$$\nabla\cdot\left(\rho_m \vec{V}_m\right) = 0 \tag{6.38}$$

Momentum equation:

$$\rho_m\left(\vec{V}_m.\nabla\right)\vec{V}_m = -\nabla P + \nabla.\left[\mu_m\left(\nabla\vec{V}_m + \nabla\vec{V}_m{}^T\right)\right] + \nabla.\left(\sum\nolimits_{i=1}^{n}\varphi_i\rho_i\vec{V}_{dr,i}\vec{V}_{dr,i}\right) \tag{6.39}$$

Energy equation:

$$\nabla.\sum\nolimits_{i=1}^{n}\left(\varphi_i\rho_i C_{p,i}\vec{V}_i T_i\right) = \nabla.\left(\sum\nolimits_{i=1}^{n} k_{nf}\nabla T_i\right) \tag{6.40}$$

Equation of volume fraction for nanoparticle phase (p):

$$\nabla.\varphi_p\rho_p\vec{V}_m = -\nabla.\varphi_p\rho_p\vec{V}_{dr,p} \tag{6.41}$$

The sum of volume fraction of all the phases is unity, which can be written as $\sum_{i=1}^{n}\varphi_i = 1$. Where n represents the total number of phases.

The mixture velocity is estimated by $\vec{V}_m = \frac{\varphi_f\vec{V}_f\rho_f + \varphi_p\vec{V}_p\rho_p}{\rho_m}$ and the drift velocity of the particle (secondary) phase is calculated using the expression shown below.

$$\vec{V}_{dr,p} = \vec{V}_{pf} - \vec{V}_m \tag{6.42}$$

where $\vec{V}_{pf} = \vec{V}_p - \vec{V}_f$ is the slip velocity between the primary (base fluid) and secondary (particle) phases. When the flow is laminar, the slip velocity can be modeled by using the correlation presented by Manninen et al. [137].

$$\vec{V}_{pf} = \frac{\rho_p d_p^2}{18\mu_f F_{drag}}\left(\frac{\rho_p - \rho_m}{\rho_p}\right)\vec{a} \tag{6.43}$$

where acceleration factor ($\vec{a}$) is defined as,

$$\vec{a} = \vec{g} - \left(\vec{V}_m.\nabla\right)\vec{V}_m \tag{6.44}$$

Schiller and Naumann [138] correlation is incorporated to predict the drag force based on Reynolds number;

$$F_{\text{drag}} = \begin{cases} 1 + 0.15Re_p^{0.687} & Re_p \leq 1000 \\ 0.0183Re_p & Re_p > 1000 \end{cases} \tag{6.45}$$

where

$$Re_p = \frac{\vec{V}_m d_p}{v_m} \tag{6.46}$$

In Eqs. (6.6–6.46), d_p signifies the diameter of a spherical nanoparticle. The mixture characteristics such as density (ρ_p), viscosity, and thermal conductivity are determined as:

$$\rho_p = \sum_{i=1}^{n} \varphi_i \rho_i \tag{6.47}$$

$$\mu_p = \sum_{i=1}^{n} \varphi_i \mu_i \tag{6.48}$$

$$k_p = \sum_{i=1}^{n} \varphi_i (k_i + k_{tr}) \tag{6.49}$$

In the above equation, k_{tr} is the turbulent thermal conductivity.

6.4.3 Eulerian model

Separate continuity, momentum, and energy conservation equations are solved for all primary and secondary phases in the Eulerian model instead of equations for a mixture. The interphase coupling is considered strong which allows the modeling of nanofluids for a wide range of nanoparticle concentrations. All the phases share the pressure and interphase exchange coefficients that determine the coupling between all the phases. The momentum transfer among different phases is modeled using the drag and Saffman lift force, while particle–particle contact is determined in terms of collision forces. The volume fraction is integrated throughout the computational domain to estimate the volume of individual phases. The sum of volume fraction of all the phases results in unity, that is, $\sum_{i=1}^{n} \varphi_i = 1$. This model is the most complicated of all the Eulerian–Eulerian models and requires the highest computational cost. The governing equations of i_{th} (nanoparticle/base fluid) phase are [10]:

Continuity equation:

$$\nabla \cdot \left(\varphi_i \rho_i \vec{V}_i\right) = 0 \tag{6.50}$$

Momentum equation:

$$\rho_i \varphi_i \left(\vec{V}_i . \nabla\right) \vec{V}_i = -\varphi_i \nabla P + \nabla . \left[\mu_i \varphi_i \left(\nabla \vec{V}_i + \nabla \vec{V}_i^{\,T}\right)\right] + \sum_{i=1}^{n} S_{pf} + F_{\text{other},i} \tag{6.51}$$

Energy equation:

$$\nabla . \left(\varphi_i \rho_i C_{p,i} \vec{V}_i T_i\right) = \nabla . (k_i \nabla T_i) - \left[\mu_i \varphi_i \left(\nabla \vec{V}_i + \nabla \vec{V}_i^{\,T}\right)\right] : \nabla \vec{V}_i + \sum_{i=1}^{n} Q_{pf} \tag{6.52}$$

In Eqs. (6.6–6.51), F_{other} are the other forces acting on the ith phase, such as body, wall lubrication, lift, and virtual mass forces. The S_{pf} defines the interphase interactive forces between the nanoparticle phase (p) and base fluid (f) phase. It is calculated as:

$$S_{pf} = \sum_{p=1}^{n} C_{pf}(\vec{V}_p - \vec{V}_f) \tag{6.53}$$

where C_{pf} represents interphase momentum exchange coefficient. The C_{pf} expression for the case of dilute nanofluids is proposed by Wen and Yu [139] as:

$$C_{pf} = \frac{3}{4} C_{\text{drag}} \frac{\left(\varphi_p \varphi_p \rho_f \left|\vec{V}_p - \vec{V}_f\right|\right)}{d_p} \varphi_f^{-265} \tag{6.54}$$

In the above equation, the drag coefficient C_{drag} is determined as:

$$C_{\text{drag}} = \frac{24\left(1 + 0.15(\varphi_f Re_p)^{0.678}\right)}{\varphi_f Re_p} \tag{6.55}$$

The Reynolds number of nanoparticles Re_p is calculated as:

$$Re_p = \frac{d_p \rho_p \left|\vec{V}_p - \vec{V}_f\right|}{\mu_f} \tag{6.56}$$

In Eqs. (6.6)–(6.52), Q_{pf} defines the interphase heat exchange coefficient.

$$Q_{pf} = h_{pf}(T_p - T_f) \tag{6.57}$$

The h_{pf} is the interphase heat exchange coefficient calculated as:

$$h_{pf} = 6k_f \varphi_f \varphi_p Nu_p / d_p^2 \tag{6.58}$$

where Nu_p (Nusselt number) is determined by Ranz and Marshall correlation [140];

$$Nu_p = 2 + 0.6 Re_p^{0.5} Pr_f^{0.333} \tag{6.59}$$

where $Pr_f = \frac{C_{pf}\mu_f}{k_f}$ defines the Prandtl number of the liquid phase.

6.4.4 Volume of fluid model

The continuity equation for the volume fraction of the fluid phase is resolved to track the interface between the phases. The velocity components and temperature is shared by all phases and a single momentum and energy equation is solved for that purpose. This model is used for simulating immiscible fluids where the interphase between the phases is required to be traced.

Considering the uniform dispersion of the nanosized nanoparticles into the base fluid, this model is not an impropriate choice to simulate nanofluid flow. However, this model is briefly discussed in this section due to some relevant literature studies considering this model for nanofluid modeling. The governing equations are [10];

Continuity equation of the *i*th (nanoparticle/base fluid) phase:

$$\nabla\cdot\left(\varphi_i\rho_i\vec{V}_i\right)=0 \tag{6.60}$$

The sum of volume fraction of all the phases can be represented using $\sum_{i=1}^{n}\varphi_i=1$, where n represents the total number of phases.

Momentum equation (temperature-dependent viscosity):

$$\rho_{nf}(\vec{V}.\nabla)\vec{V}=-\nabla P+\nabla.\left[\mu_{nf}\left(\nabla\vec{V}+\nabla\vec{V}^{T}\right)\right] \tag{6.61}$$

Energy equation:

$$\nabla.\left((\rho C_p)_{nf}\vec{V}T\right)=\nabla.\left(k_{nf}\nabla T\right) \tag{6.62}$$

Analogous to the mixture model, the thermophysical characteristics of the solution are explicated as the arithmetic mean of all the phases weighing their corresponding volume fractions.

Hejazian and Moraveji [141] analyzed the homogenous and multiphase mixture model results of the TiO_2/water ($\varphi=0.1-0.25\%$) nanofluid for the turbulent laminar flow in a tube. The results indicated that the mixture model estimated more accurate results compared to the homogenous model. Akbari et al. [142] simulated laminar forced convection of the Al_2O_3/water ($\varphi=0.5-2\%$) nanofluid in a tube subjected to constant heat flux by utilizing homogenous model and multiphase Eulerian–Eulerian models (mixture, VOF, and Eulerian). The study showed all three Eulerian–Eulerian models estimated indistinguishable and accurate convective heat transfer coefficient results compared to the single-phase model. However, all the models predicted accurate hydrodynamic fields. Majid et al. [143] also endorsed Eulerian and mixture models to accurately simulate turbulent forced convection of Al_2O_3/water($\varphi=0.1-2\%$) nanofluid in a horizontal tube. Among the studied models, the mixture model was considered to be better in predicting the average Nusselt number of the nanofluids. Sinan et al. [135] regarded the Eulerian model as the more accurate model for predicting the Al_2O_3/water($\varphi=0.6\%-1.6\%$) nanofluid heat transfer characteristics in the flow developing section of the uniformly heated tube, compared to the mixture model. Lofti et al. [144] reported the superiority of the mixture model for accurately simulating the forced convection of Al_2O_3/water nanofluid in a horizontal tube subjected to uniform surface heat flux. This conclusion was made based on the comparative analysis of the homogenous, Eulerian, and mixture models. A similar conclusion was made by Zhong et al. [145] based on the magnetohydrodynamic flow of the Fe_3O_4/water($\varphi=0\%-2\%$)

nanofluid in a porous ribbed microchannel under external magnetic field. They compared the homogenous and mixture models to simulate the thermophysical performance of nanofluid. The comparative studies conducted by Fard et al. [146], Behzadmehr et al. [147], and Akbaridoust et al. [148] also confirmed better nanofluid heat transfer results of multiphase Eulerian–Eulerian models compared to single-phase models. On the contrary, Ehsan et al. [149] simulated TiO_2/water ($\varphi = 2.3\%$) nanofluid using Eulerian–Eulerian mixture, Eulerian, and homogenous models and compared the computed results with their experimental data for the nanofluid flow in a solar heat exchanger. The study showed that all the studied models predicted inaccurate results.

6.4.5 Multiphase Eulerian–Lagrangian model

In the Eulerian–Lagrangian or discrete phase model (DPM), paths of the individual nanoparticles are determined using the Lagrangian approach while the transport equations of the primary phase (base fluid) are solved by employing the Eulerian approximation. The computational cost of this model is the highest of all of the two-phase models. The momentum and energy exchange between the primary and particulate phases is simulated by solving the source terms in the momentum (S_m) and energy (S_e) equations of the primary phase. The transport equations of the primary phase can be stated as [10]:

Continuity equation:

$$\nabla\cdot\left(\rho_f \vec{V}_f\right) = 0 \tag{6.63}$$

Momentum equation:

$$\rho_f\left(\vec{V}_f.\nabla\right)\vec{V}_f = -\nabla P + \nabla.\left[\mu_f\left(\nabla\vec{V}_f + \nabla\vec{V}_f^{\,T}\right)\right] + \vec{S}_M \tag{6.64}$$

Energy equation:

$$\nabla.\left((\rho C_p)_f \vec{V}_f T_f\right) = \nabla.\left(k_f \nabla T_f\right) + S_E \tag{6.65}$$

where S_M represents the momentum source term defining the interactions between the primary and particulate phases. The S_E denotes the energy source term solving the interphase energy exchange. These source terms are computed as:

$$\vec{S}_M = \sum n_p \frac{m_p}{\delta \mathcal{V}} \frac{dV_p}{dt} \tag{6.66}$$

$$S_E = \sum n_p \frac{m_p}{\delta \mathcal{V}} \left(C_p\right)_p \frac{dT_p}{dt} \tag{6.67}$$

where n_p defines the number of nanoparticles, m_p is the mass of the nanoparticle and $\delta\mathcal{V}$ is the volume of a grid cell. For the case of nanofluids, the momentum interactions generally comprise particle Brownian motion

($F_{Brownian}$), drag (F_{drag}) and Saffman lift ($F_{S,lift}$), Magnus lift ($F_{S,lift}$), thermophoresis ($F_{thermophoresis}$), virtual mass ($F_{virtual}$), reduced gravity ($F_{gravity}$), and pressure gradient ($F_{pressure}$) forces:

$$\frac{dV_p}{dt} = F_{drag} + F_{gravity} + F_{Brownian} + F_{thermophoresis} + F_{S,lift} + F_{M,lift} + F_{virtual} + F_{pressure} + \ldots \tag{6.68}$$

In energy source term S_E, the interphase heat exchange is calculated as:

$$m_p (C_p)_p \frac{dT_p}{dt} = h_p A_p (T_f - T_p) \tag{6.69}$$

$$N_{u_p} = \frac{h_p d_p}{k_f} = 2 + 0.6 Re_p^{0.5} Pr^{0.33} \tag{6.70}$$

where Nu is determined using the Ranz and Marshall model [140].

Behroyan et al. [150] compared the single-phase model, Eulerian–Eulerian (mixture, Eulerian), and Eulerian–Lagrangian models to compute turbulent forced convection of the Cu/water($\varphi = 0.1\%, 2\%$) nanofluid in a tube under constant heat flux. Their study demonstrated that the Eulerian–Lagrangian model estimated the most accurate Nusselt number results. The non-Newtonian single-phase model underpredicted the results, while the Eulerian–Eulerian model estimated accurate results only for the particle volume fraction of 0.5%. Tehmina and Kim [133] evaluated homogenous, Eulerian–Eulerian (VOF, mixture, Eulerian), and Eulerian–Lagrangian models to determine the thermophysical characteristics of Al_2O_3/water ($\varphi = 0.25\% - 2\%$) and TiO_2/water($0.24\% - 1.18\%$) nanofluid for laminar flow in the micro- and minichannels, respectively. The study showed that the Eulerian–Lagrangian approximation of the nanofluids determined the most accurate convective heat transfer over all the considered nanoparticle concentrations and flow rates. The single-phase model assessed the accurate results of the dilute nanofluids at low Reynolds number, however, at the larger nanoparticle concentration and Reynolds number, it underestimated the convective heat transfer results. The Eulerian–Eulerian models also assessed accurate convective heat transfer results for the dilute nanofluids; however these models overvalued the results at the larger nanoparticle concentration and Reynolds number (Fig. 6.12). All the studied models predicted reasonable pressure drop results with a slightly high deviation at the larger nanoparticles concentration.

Kumar and Puranik [151] also endorsed the Eulerian–Lagrangian model for accurately simulating the turbulent forced convection of the Al_2O_3/water($\varphi = 1\%, 2\%$), TiO_2/water($\varphi = 0.2\% - 2\%$) and Cu/water ($\varphi = 1\%$) nanofluid within horizontal tube subjected to uniform heat flux, in comparison with the homogenous model. Onyiriuka et al. [152] evaluated

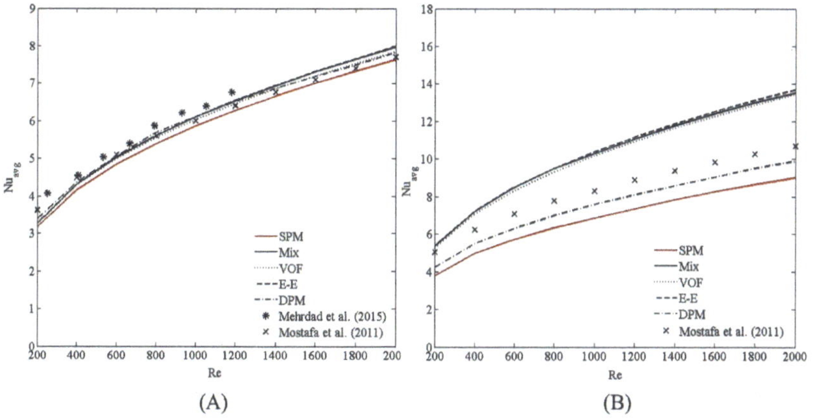

FIGURE 6.12 Deviation of average Nusselt number values of Al_2O_3/water nanofluid estimated by homogenous (SPM), mixture (Mix), VOF, Eulerian (E-E), and Eulerian–Lagrangian (DPM) models for particle volume fraction (A) $\varphi = 0.25\%$ and (B) $\varphi = 2\%$ [133].

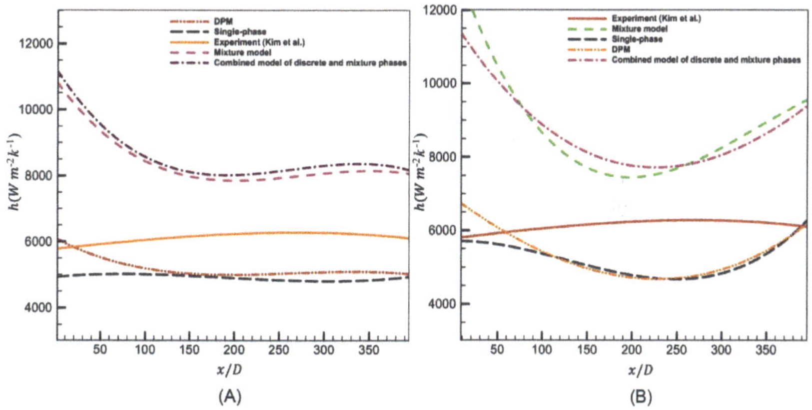

FIGURE 6.13 Comparison of two-phase models with (A) constant and (B) temperature-dependent thermophysical characteristics [152].

the accuracy of the homogenous, mixture, discrete (Eulerian–Lagrangian), and combined mixture–DPMs to numerically simulate forced convection of the Al_2O_3/water($\varphi = 1\% - 3\%$) nanofluids in the laminar and turbulent ranges by considering constant and temperature-dependent thermophysical properties. Their analysis showed that discrete phase modeling of the nanofluids provided the most accurate heat transfer results. The mixture and combined mixture–dpm models estimated unrealistic and overestimated results (Fig. 6.13). The homogenous model predicted better results for the nanofluids with the constant thermophysical properties compared to the temperature-dependent thermophysical properties of nanofluids.

Ying et al. [153] performed a comprehensive comparison of the single-phase homogenous, thermal dispersion, Eulerian–Lagrangian, and Eulerian–Eulerian (Mixture, Eulerian) models to simulate forced convection of the Al_2O_3/water nanofluids. Their study showed that all the studied models estimated accurate heat transfer results when the nanoparticle concentration was low ($\varphi < 0.5\%$). However, at the larger nanoparticle concentrations ($\varphi > 0.5\%$), the Eulerian–Eulerian models overpredicted the results while the single-phase models underestimated the thermal fields of the nanofluid. The Eulerian–Lagrangian approximation of the nanofluids provided the results with minimum deviation from the experimental data at the larger particle concentrations. Similar results were reported by the study conducted by Ambreen et al. [134] who compared the homogenous, Eulerian–Eulerian (VOF, mixture, and Eulerian), and Eulerian–Lagrangian models to model the forced convection of MWCNT/water and MWCNT − GNP/water-based nanofluid in a straight tube (particle mass fractions ranging from 0.075%–0.25%). Their analysis showed that the homogenous and all the multiphase (Eulerian–Eulerian and Eulerian–Lagrangian) models predicted accurate hydrothermal results of the nanofluids. However, at the larger nanoparticle concentration, the Eulerian–Lagrangian model obtained the most accurate results compared to the other tested models.

6.5 Numerical techniques to simulate nanofluid

The numerical techniques to simulate nanofluid can be classified as macro-, micro-, and mesoscale techniques (Fig. 6.14) [154]. This section presents a basic overview of these techniques.

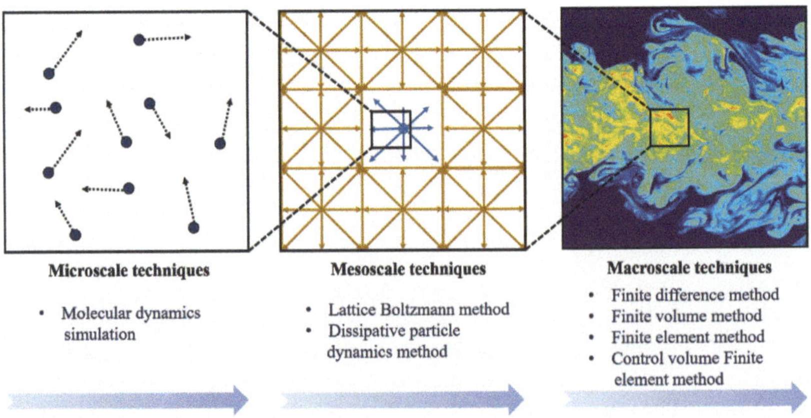

FIGURE 6.14 Computational fluid dynamic techniques to simulate nanofluid flow.

6.5.1 Macroscale techniques

At the macroscopic level, fluid is approximated as a continuum despite its molecular contents and interactions. The fluid is assumed to be distributed across the computational domain and comprises properties such as velocity, temperature, and density. The fluid molecular properties are assessed through its transport coefficients such as thermal conductivity and viscosity [155]. The macroscale techniques are based on transforming partial differential equations into algebraic equations through discretization. The popular macroscale techniques employed for the nanofluid flow problems are finite difference method (FDM), finite volume method (FVM), finite element method (FEM), and control volume finite element method (CVFEM).

6.5.1.1 Finite difference method

The FDM is the oldest approach to obtain the solution of partial differential equations. It is based upon the application of the local Taylor expansion to approximate the differential equations. Compared to other methods, it is a relatively straightforward method that involves the solution of partial differential equations over a group of interlinked points within the computational domain, called nodes. The imposed boundary conditions are enforced over different edges or zones of the computational domain depending on the problem. The entire framework of the interlinked points is called mesh or grid. The derivatives of the partial differential equations are estimated by employing a difference approximation which is derived utilizing the Taylor series expansion [156]. In the schematic representation of the 2D computational domain and mesh shown in Fig. 6.15, the nodes are the blue and red solid

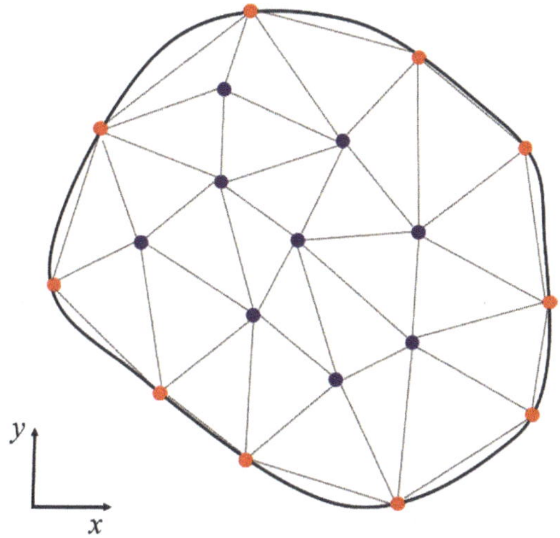

FIGURE 6.15 A 2D computational domain and mesh (Finite difference method).

circles. Red circles at the boundary of the domain are the boundary nodes, while the blue nodes are the interior nodes.

In FDM, the partial differential equations are solved without converting the transport equations into the alternative form. The actual partial differential equations are solved at different locations of the computational domain. Therefore the obtained solution is referred to as a strong form of solution [156]. This is an easy and effective method for the structured mesh, however, it is not suitable for the solution of complex flows and structures when a body-fitted grid is required to discretize the computational domain [154].

6.5.1.2 Finite volume method

In the FVM, the governing partial differential equations are satisfied over the finite control volumes instead of nodes. In this method, the first step is splitting the computational domain into several control volumes called cells (Fig. 6.16). Generally, the shapes of these cells are polygons and polyhedrons in the 2D and 3D computational domains, respectively. These cells are enclosed by straight edges and plane surfaces in 2D and 3D domains, respectively. Therefore the curve bounding surfaces are approximated with the straight edges or plane faces which are called cell faces or faces. The cell vortices are referred to as nodes which are similar to nodes of the FDM. The value of the desired parameter is calculated at the centers of the cell called the cell centers [156]. In Fig. 6.16, the cell centers are highlighted as the green stars, while the blue and red circles are the nodes that are similar to the FDM method.

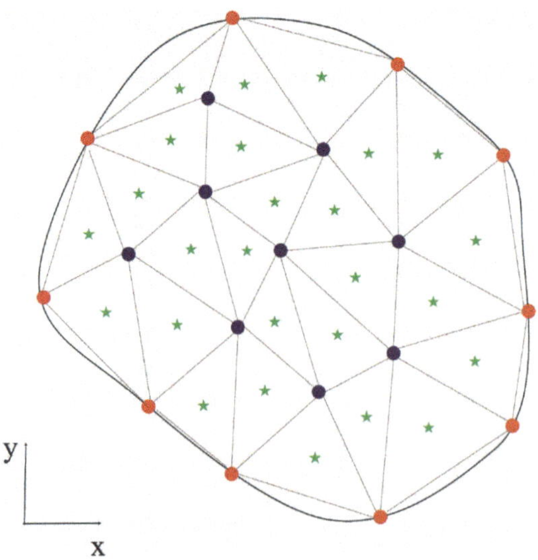

FIGURE 6.16 A 2D computational domain and mesh (Finite difference method).

The finite-volume equations derivation starts with integrating the partial differential equations over the control cells of the computational domain. The solution of partial differential equations obtained through the FVM is considered weak because in this method the partial differential equations are firstly integrated over the control volumes and then solved, instead of directly solving them. Moreover, the boundary conditions can also not be satisfied directly due to the absence of cell centers at the boundary. In nanofluid-related numerical studies, it is the most commonly used technique for nanofluid flow problems.

6.5.1.3 Finite element method

The FEM involves two basic steps: initially, the strong form of partial differential equations is converted into their weak form of the weighted integrals, following by their solution over the computational domain discretized by finite-sized elements. The derivation of the weak partial differential equations is achieved by multiplying them with the weight functions and afterward integrating the equation over the whole computational domain. One of the advantages of the FEM technique is its flexibility for the choice of basic functions and dealing with complex geometries. However, the main disadvantage of this approach is undesirable round-off errors and error accumulation. However, solution accuracy can be improved by reducing the element sizes or by increasing the order of the basic function.

6.5.1.4 Control volume finite element method

The control volume FEM (CVFEM) is the hybrid between the standard FVM and the FEM. It combines the interesting features of the standard FEM and the FVM. The flexibility of the FEM to discretize complex geometry is combined with the conservative formulation of the FVM, in which the variables are interpreted in terms of sources, fluxes, and forces [157]. To enforce the local conservation, the partial differential equations are integrated over control volumes which are created by joining the midpoints of the edges or faces of the elements of an unstructured finite element mesh (Fig. 6.17).

In nanofluid-related simulations, CVFEM is introduced and mostly utilized by Sheikholeslami and his colleagues for natural and forced convection of nanofluids by utilizing single-phase and two-phase models [157–161].

6.5.2 Microscale techniques

The microscale technique (molecular dynamic simulations) describes how the velocity and position of the system atoms and molecules change over time. Considering molecules and atoms as fluid particles, their dynamic and interactions are simulated over a period [154]. The solution of differential equations of motion provides the positions of the atoms. The atomic

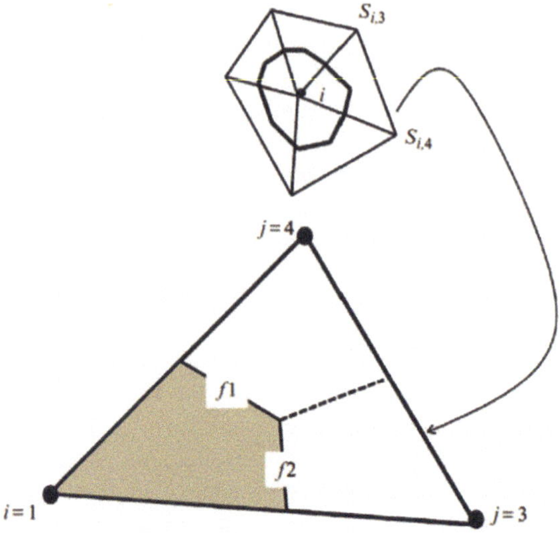

FIGURE 6.17 A triangular element and the corresponding control volume [157].

positions of the atoms are coupled in time that provides data about the dynamics of the atoms in a motion picture. The operating mechanism of the molecular dynamic model entirely hinges on classical mechanics without any assumptions. In this method, the force $\boldsymbol{F_l}$ acting on a single ith particle calculated from the interatomic potential U between the atoms is:

$$\boldsymbol{F_i} = \nabla_i U \tag{6.71}$$

The acceleration of atom of ith particle of mass (m) is determined through Newton's second law of motion as:

$$\boldsymbol{F_i} = m\boldsymbol{a_i} \tag{6.72}$$

The velocity $\boldsymbol{V_{i,f}}$ and $\boldsymbol{R_{i,f}}$ position of the ith particle at time t is calculated as:

$$\boldsymbol{V_{i,f}} = \boldsymbol{V_{i,i}} + \boldsymbol{a_i}t^2 \tag{6.73}$$

$$\boldsymbol{R_{i,f}} = \boldsymbol{R_{i,i}} + \boldsymbol{V_{i,f}}t + \frac{1}{2}\boldsymbol{a_i}t^2 \tag{6.74}$$

where $\boldsymbol{V_{i,i}}$ and $\boldsymbol{R_{i,i}}$ are the initial velocity and position of the ith particle, respectively. After updating the particle velocity and position, the Eqs. (6.71)–(6.74) are solved again for time t + Δt for the total number of system particles (Fig. 6.18). The mean values of the characteristics such as thermal transport characteristics are estimated by statistical mechanics due to a large number of system particles [162].

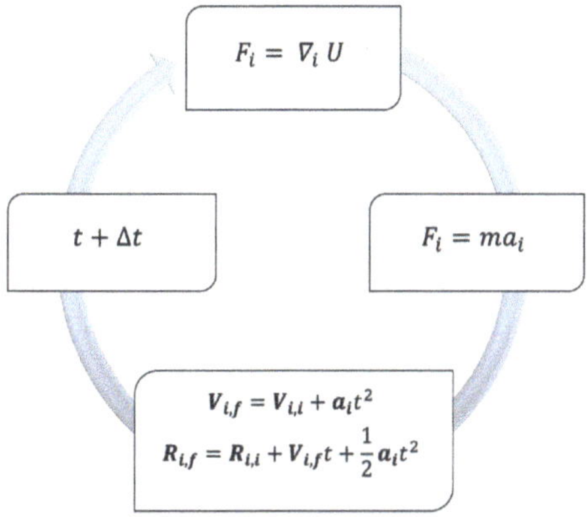

FIGURE 6.18 Working principle of molecular dynamics.

6.5.3 Mesoscale techniques

The mesoscale computational techniques bridge the gap between the macro- and microscale approaches by simulating the system at a scale in-between the two approaches. The mesoscale techniques incorporate the advantages of macro- and microscale techniques. The two mesoscale techniques which have been used to simulate nanofluid flows are the lattice Boltzmann method and dissipative particle dynamic (DPD) techniques.

6.5.3.1 *Lattice Boltzmann method*

The lattice Boltzmann method considers the behavior of a packet of particles as a distribution function instead of treating each particle separately. The property of the packet of particles is denoted by the distribution function which acts as the representative of the group of particles. In this method, a special form of the Boltzmann equation, referred to as the lattice Boltzmann equation, is discretized both in space and time. This equation is written as [163]:

$$f_a = (\boldsymbol{x} + \boldsymbol{e}_a \Delta \mathrm{t}, \mathrm{t} + \Delta \mathrm{t}) - f_a(\boldsymbol{x}, t) = \Omega_a(\boldsymbol{V}, \rho, \tau) \tag{6.75}$$

where f_a denotes the discrete distribution function which defines the packet of particles having statistically the same momentum in a direction a. Such momentum is signified by a set of basic lattice velocities e. In Eqs. (6.6)–(6.75), the $\boldsymbol{x}$ and t denote the positional vector and time, respectively. Ω denotes the collision operation, which is dependent on the fluid local velocity $\boldsymbol{V}$ and density ρ. τ is relaxation time which is associated with the phase physical properties such as molecular diffusivity and kinematic viscosity.

The lattice Boltzmann method is easily applicable on the complex domains, multiphase problems, and multicomponent flows without the tracing the interface between the different phases. It can be adaptive to the paralleled processing because of the explicit nature and locality of this method. However, this method requires more computational cost compared to the macroscopic Navier–Stokes solvers.

6.5.3.2 Dissipative particle dynamics

DPD is an off-lattice mesoscopic technique involving a packet of particles moving across the continuous space in discrete time. These particles signify the whole fluid regions of molecules instead of individual atoms and the atomistic details are not addressed. The particles' degree of freedom is replaced by the simplified pairwise dissipative and random forces to conserve local momentum. The nonbonded total force which acts on a DPD *i*th particle is calculated by summation of all n number of particles located in a fixed cut-off distance of three pairwise-additive forces.

$$f_i = \sum_{N \neq i} f_{in}^{C} + f_{in}^{D} + f_{in}^{R} \tag{6.76}$$

In the above equation, the first, second, and third terms are the conservative, dissipative, and random force, respectively. The conservative force estimates the chemical identity of the beads while the dissipative and random forces form the thermostat to keep the system average temperature at a constant value. The key feature of the nonbonded force is that it conserves the momentum locally, therefore the hydrodynamic modes of fluid are taken into account even for a small number of particles. The primary advantage of this method is that it provides access to a longer time and length scale compared to the conventional molecular dynamic simulation [164].

6.6 Conclusion

In this chapter mathematical models, classified as single-phase, Eulerian–Eulerian, and Eulerian–Lagrangian models, to model nanofluids' flow and thermal attributes are presented in detail. The single-phase models included the homogenous model, thermal dispersion, and Buongiorno models while the Eulerian–Eulerian models comprised the mixture, VOF, and Eulerian models. The physical mechanisms involved in determining the thermal activity of the nanofluid, such as the particle–particle, particle–liquid, and external forces on the particles, are discussed to establish an understanding of the underlying mechanism of nanofluids. Moreover, a brief overview of the thermophysical property trends dependent on their influencing factors and the correlations to estimates these properties are presented. A brief overview of the macro-, micro-, and mesoscale computational fluid dynamic techniques to solve the respective governing equations is also presented. The key findings of this chapter can be summarized as:

The particle–particle interactions can be neglected while modeling the dilute nanofluids when the nanoparticle concentrations are less than 10%.

The homogenous model is the simplest model to simulate nanofluid flows and requires the least computational cost. It predicts the flow and thermal fields of the nanofluids with reasonable accuracy for the dilute nanofluids when the particle–particle, particle–liquid, and other forces are not significant. However, at the larger nanoparticle concentrations, it fails to incorporate the mechanisms responsible for the thermal activity of the nanofluids and thereby underestimates their heat transfer characteristics.

Among the three Eulerian–Eulerian models, the mixture model is the simplest approach, while the Eulerian model is the most complex one and requires high computational resources. The Eulerian–Eulerian models also accurately predict the thermal and hydrodynamic performance of the nanofluids at low particle concentrations. However, at larger particle concentrations these models overpredict the thermal fields due to the high estimation of volume-averaged thermal conductivity of nanofluids.

The Eulerian–Lagrangian approach provides the most realistic modeling of the nanofluids by incorporating the thermal enhancement mechanism of the nanofluids. Studies show that heat transfer and pressure drop results estimated by using this approach are close to the experimental values at both the small and larger nanoparticle concentrations. However, this model requires high computational resources and time.

In most of the nanofluid-based numerical studies, nanofluids are assumed as uniformly dispersed into the base fluid. The accuracy of the numerical results is estimated relative to the experimental results. However, there are huge controversies in the reported heat transfer and thermophysical property results of the nanofluids. Even studies reporting similar trends of thermophysical property and hydrothermal characteristics of the nanofluids as a function of their influencing parameters show different rates of variation. Such discrepancies are primarily caused by the instability and agglomeration of the suspended nanoparticles. The particle clusters can be of various shapes ranging from spherical to chain-like structures. Due to the irregular shapes of these clusters, it is practically impossible to determine their particle–particle and particle–liquid interactions.

References

[1] D. Dhinesh Kumar, A. Valan Arasu, A comprehensive review of preparation, characterization, properties and stability of hybrid nanofluids, Renew. Sustain. Energy Rev. 81 (May 2017) (2018) 1669–1689.

[2] C.T. Crowe, J.D. Schwarzkopf, M. Sommerfeld, Y. Tsuji, Multiphase Flows with Droplets and Particles, CRC Press, 2011.

[3] H.C. Hamaker, The London-van der Waals attraction between spherical particles, Physica 4 (10) (1937) 1058–1072.

[4] Cosgrove T. Colloid Science: Principles, Methods and Applications. 2009. 1–288.

[5] J. Huang, X. Wang, Q. Long, X. Wen, Y. Zhou, L. Li, Influence of pH on the stability characteristics of nanofluids, Symp. Photonics Optoelectron SOPO 2009 (2009) 2–4.

[6] B.H. Mahmoud, L.F. Mortimer, M. Fairweather, J. Peakall, D. Harbottle, H.P. Rice, Multiscale modelling of ceramic nanoparticle interactions and their influence on the thermal conductivity of nanofluids, J. Phys. D Appl. Phys. 53 (1) (2020).
[7] J.Y. Jung, J.Y. Yoo, Thermal conductivity enhancement of nanofluids in conjunction with electrical double layer (EDL), Int. J. Heat Mass Transf. 52 (1–2) (2009) 525–528.
[8] A.P. Gavrilyuk, I.L. Isaev, V.S. Gerasimov, S.V. Karpov, Physical principles of the formation of a nanoparticle electric double layer in metal hydrosols, Colloid Polym. Sci. 298 (1) (2020) 1–7.
[9] Q. Zhao, H. Xu, L. Tao, Two-layer nanofluid flow and heat transfer in a horizontal microchannel with electric double layer effects and magnetic field, Int. J. Numer. Methods Heat Fluid Flow (2020).
[10] O. Mahian, L. Kolsi, M. Amani, P. Estellé, G. Ahmadi, C. Kleinstreuer, et al., Recent advances in modeling and simulation of nanofluid flows-Part I: fundamentals and theory, Phys. Rep. 790 (2019) 1–48.
[11] R. Hogg, T.W. Healy, D.W. Fuerstenau, Mutual coagulation of colloidal dispersions, Trans. Faraday Soc. 62 (615) (1966) 1638–1651.
[12] Dahman Y. Nanotechnology and Functional Materials for Engineers. 2017. 1–268.
[13] P. Keblinski, R. Prasher, J. Eapen, Thermal conductance of nanofluids: is the controversy over? J. Nanopart. Res. 10 (7) (2008) 1089–1097.
[14] W. Evans, R. Prasher, J. Fish, P. Meakin, P. Phelan, P. Keblinski, Effect of aggregation and interfacial thermal resistance on thermal conductivity of nanocomposites and colloidal nanofluids, Int. J. Heat Mass Transf. 51 (5–6) (2008) 1431–1438.
[15] Keblinski P, Phillpot R, Choi S, Eastman A. Mechanisms of heat flow in suspensions of nano-sized particles (nanofluids). 2002; 45: 855–863.
[16] K.N. Shukla, T.M. Koller, M.H. Rausch, A.P. Fröba, Effective thermal conductivity of nanofluids—a new model taking into consideration Brownian motion, Int. J. Heat Mass. Transf. 99 (2016) 532–540.
[17] B. Xiao, Y. Yang, L. Chen, Developing a novel form of thermal conductivity of nanofluids with Brownian motion effect by means of fractal geometry, Powder Technol. 239 (2013) 409–414.
[18] R. Prasher, P. Bhattacharya, P.E. Phelan, Thermal conductivity of nanoscale colloidal solutions (nanofluids), Phys. Rev. Lett. 94 (2) (2005) 3–6.
[19] D.H. Kumar, H.E. Patel, V.R.R. Kumar, T. Sundararajan, T. Pradeep, S.K. Das, Model for heat conduction in nanofluids, Phys. Rev. Lett. 93 (14) (2004) 144301.
[20] G. Puliti, S. Paolucci, M. Sen, Nanofluids and their properties, Appl. Mech. Rev. 64 (3) (2012) 030803.
[21] A. Li, G. Ahmadi, Dispersion and deposition of spherical particles from point sources in a turbulent channel flow, Aerosol Sci. Technol. 16 (1992) 209–226.
[22] M. Hemmat Esfe, S. Esfandeh, M.H. Kamyab, History and introduction, Hybrid Nanofluids for Convection Heat Transfer. INC (2020) 1–48.
[23] L. Talbot, R.K. Cheng, R.W. Schefer, D.R. Willis, Thermophoresis of particles in a heated boundary layer, J. Fluid. Mech. 101 (04) (1980) 737.
[24] J.S. Marshall, S. Li, Adhesive Particle Flow: A Discrete-Element Approach Winter Conditions in Six European Shallow Lakes: A Comparative Synopsis, Cambridge University Press, 2012, pp. 1–342.
[25] T. Ambreen, M.H. Kim, Influence of particle size on the effective thermal conductivity of nanofluids: a critical review, Appl. Energy 264 (September 2019) (2020).

[26] S.M. Vanaki, P. Ganesan, H.A. Mohammed, Numerical study of convective heat transfer of nanofluids: a review, Renew. Sustain. Energy Rev. 54 (2016) 1212–1239.

[27] A. Ceylan, K. Jastrzembski, S.I. Shah, Enhanced solubility Ag-Cu nanoparticles and their thermal transport properties, Metall. Mater. Trans. A 37 (7) (2006) 2033–2038.

[28] S.H. Kim, S.R. Choi, D. Kim, Thermal conductivity of metal-oxide nanofluids: particle size dependence and effect of laser irradiation, J. Heat Transfer. 129 (3) (2007) 298.

[29] J. Buongiorno, D.C. Venerus, N. Prabhat, T. McKrell, J. Townsend, R. Christianson, et al., A benchmark study on the thermal conductivity of nanofluids, J. Appl. Phys. 106 (9) (2009) 094312.

[30] H.T. Zhu, C.Y. Zhang, Y.M. Tang, Novel synthesis and thermal conductivity of CuO nanofluid, J. Appl. Phys. Chem. C 111 (4) (2007) 1646–1650.

[31] S.M.S. Murshed, K.C. Leong, C. Yang, Enhanced thermal conductivity of TiO2—Water based nanofluids, Int. J. Therm. Sci. 44 (4) (2005) 367–373.

[32] T.P. Teng, Y.-H. Hung, T.C. Teng, H.-E. Mo, H.-G. Hsu, The effect of alumina/water nanofluid particle size on thermal conductivity, Appl. Therm. Eng. 30 (14–15) (2010) 2213–2218.

[33] H. Xie, J. Wang, T. Xi, Y. Liu, Thermal conductivity of suspensions containing nanosized SiC particles, Int. J. Thermophys. 23 (2) (2002) 571–580.

[34] Y. Ding, H. Alias, D. Wen, R.A. Williams, Heat transfer of aqueous suspensions of carbon nanotubes (CNT nanofluids), Int. J. Heat Mass Transf. 49 (1–2) (2006) 240–250.

[35] M.P. Beck, Y. Yuan, P. Warrier, A.S. Teja, The effect of particle size on the thermal conductivity of alumina nanofluids, J. Nanopart. Res. 11 (5) (2009) 1129–1136.

[36] M.J. Assael, I.N. Metaxa, J. Arvanitidis, D. Christofilos, C. Lioutas, Thermal conductivity enhancement in aqueous suspensions of carbon multi-walled and double-walled nanotubes in the presence of two different dispersants, Int. J. Thermophys. 26 (3) (2005) 647–664.

[37] S.M.S. Murshed, K.C. Leong, C. Yang, Investigations of thermal conductivity and viscosity of nanofluids, Int. J. Therm. Sci. 47 (5) (2008) 560–568.

[38] E.V. Timofeeva, A.N. Gavrilov, J.M. McCloskey, Y.V. Tolmachev, S. Sprunt, L.M. Lopatina, et al., Thermal conductivity and particle agglomeration in alumina nanofluids: experiment and theory, Phys. Rev. E 76 (6) (2007) 061203.

[39] X. Zhang, H. Gu, M. Fujii, Effective thermal conductivity and thermal diffusivity of nanofluids containing spherical and cylindrical nanoparticles, J. Appl. Phys. 100 (4) (2006) 593–599.

[40] M.-S. Liu, M. Ching-Cheng Lin, I.-T. Huang, C.-C. Wang, Enhancement of thermal conductivity with carbon nanotube for nanofluids, Int. Commun. Heat Mass Transf. 32 (9) (2005) 1202–1210.

[41] D. Wen, Y. Ding, Effective thermal conductivity of aqueous suspensions of carbon nanotubes (carbon nanotube nanofluids), J. Thermophys. Heat Transf. 18 (4) (2004) 481–485.

[42] M. Fakoor Pakdaman, M.A.A. Akhavan-Behabadi, P. Razi, An experimental investigation on thermo-physical properties and overall performance of MWCNT/heat transfer oil nanofluid flow inside vertical helically coiled tubes, Exp. Therm. Fluid. Sci. 40 (2012) 103–111.

[43] M. Xing, J. Yu, R. Wang, Experimental investigation and modelling on the thermal conductivity of CNTs based nanofluids, Int. J. Therm. Sci. 104 (2016) 404–411.

[44] A. Shahsavar, M.R. Salimpour, M. Saghafian, M.B. Shafii, An experimental study on the effect of ultrasonication on thermal conductivity of ferrofluid loaded with carbon nanotubes, Thermochim. Acta 617 (2015) 102–110.

[45] B. Buonomo, O. Manca, L. Marinelli, S. Nardini, Effect of temperature and sonication time on nanofluid thermal conductivity measurements by nano-flash method, Appl. Therm. Eng. 91 (2015) 181–190.

[46] Y. Yang, E.A. Grulke, Z.G. Zhang, G. Wu, Thermal and rheological properties of carbon nanotube-in-oil dispersions, J. Appl. Phys. 99 (11) (2006) 114307.

[47] K.J. Lee, S.H. Yoon, J. Jang, Carbon nanofibers: a novel nanofiller for nanofluid applications 3, Small (2007) 1209–1213.

[48] Z. Peng, H. Yang, Designer platinum nanoparticles: control of shape, composition in alloy, nanostructure and electrocatalytic property, Nano Today 4 (2) (2009) 143–164.

[49] S.K. Das, N. Putra, P. Thiesen, W. Roetzel, Temperature dependence of thermal conductivity enhancement for nanofluids, J. Heat Transfer 125 (4) (2003) 567.

[50] Y. Hwang, J.K. Lee, C.H. Lee, Y.M. Jung, S.I. Cheong, C.H. Lee, et al., Stability and thermal conductivity characteristics of nanofluids, Thermochim. Acta 455 (1–2) (2007) 70–74.

[51] H. Xie, W. Yu, Y. Li, L. Chen, Discussion on the thermal conductivity enhancement of nanofluids, Nanoscale Res. Lett. 6 (1) (2011) 1–12.

[52] H. Xie, J. Wang, T. Xi, Y. Liu, F. Ai, Q. Wu, Thermal conductivity enhancement of suspensions containing nanosized alumina particles, J. Appl. Phys. 91 (7) (2002) 4568–4572.

[53] T. Paramethanuwat, N. Bhuwakietkumjohn, S. Rittidech, Y. Ding, Experimental investigation on thermal properties of silver nanofluids, Int. J. Heat Fluid Flow 56 (2015) 80–90.

[54] M.J. Assael, C.-F. Chen, I. Metaxa, W.A. Wakeham, Thermal conductivity of suspensions of carbon nanotubes in water, Int. J. Thermophys. 25 (4) (2004) 971–985.

[55] T.-K. Hong, H.-S. Yang, C.J. Choi, Study of the enhanced thermal conductivity of Fe nanofluids, J. Appl. Phys. 97 (6) (2005) 064311.

[56] J. Philip, P.D. Shima, B. Raj, Evidence for enhanced thermal conduction through percolating structures in nanofluids, Nanotechnology 19 (30) (2008) 305706.

[57] R.S. Vajjha, D.K. Das, Experimental determination of thermal conductivity of three nanofluids and development of new correlations, Int. J. Heat Mass Transf. 52 (21–22) (2009) 4675–4682.

[58] K.S. Suganthi, K.S. Rajan, Metal oxide nanofluids: review of formulation, thermophysical properties, mechanisms, and heat transfer performance, Renew. Sustain. Energy Rev. 76 (March) (2017) 226–255.

[59] K. Han, W.-H. Lee, C. Kleinstreuer, J. Koo, Critical invalidation of temperature dependence of nanofluid thermal conductivity enhancement, J. Heat Transfer 135 (5) (2013) 051601.

[60] C.H. Chon, K.D. Kihm, S.P. Lee, S.U.S.S. Choi, Empirical correlation finding the role of temperature and particle size for nanofluid (Al 2O 3) thermal conductivity enhancement, Appl. Phys. Lett. 87 (15) (2005) 1–3.

[61] P.D. Shima, J. Philip, B. Raj, Role of microconvection induced by Brownian motion of nanoparticles in the enhanced thermal conductivity of stable nanofluids, Appl. Phys. Lett. 94 (22) (2009) 1–4.

[62] H.E. Patel, T. Sundararajan, S.K. Das, An experimental investigation into the thermal conductivity enhancement in oxide and metallic nanofluids, J. Nanopart. Res. 12 (3) (2010) 1015–1031.

[63] P. Warrier, A. Teja, Effect of particle size on the thermal conductivity of nanofluids containing metallic nanoparticles, Nanoscale Res. Lett. 6 (1) (2011) 247.

[64] B.B. Chengzhen Sun, Lu. J.L. Wen-Qiang, Shear-rate dependent effective thermal conductivity of H 2 O + SiO 2 nanofluids, Phys. Fluids 25 (5) (2013) 052002.

[65] Hemmat M, Seyfolah E. An experimental study on the effect of diameter on thermal conductivity and dynamic viscosity of Fe/water nanofluids. 2015;1817–1824.

[66] M. Hemmat, H. Rostamian, A. Shabani-samghabadi, M. Hemmat Esfe, H. Rostamian, A. Shabani-samghabadi, et al., Application of three-level general factorial design approach for thermal conductivity of MgO/water nanofluids, Appl. Therm. Eng. 127 (2017) 1194–1199.

[67] M.I. Pryazhnikov, A.V. Minakov, V.Y. Rudyak, D.V. Guzei, Thermal conductivity measurements of nanofluids, Int. J. Heat Mass Transf. 104 (2017) 1275–1282.

[68] Y. Xuan, Q. Li, W. Hu, Q. Li, Q. Li, W. Hu, et al., Aggregation structure and thermal conductivity of nanofluids, Thermodynamics 49 (4) (2003) 1038–1043. <file:///Users/catarina/Catarina/Biblioteca/pdf/Nanoparticles gelation%5CXuan.pdf>.

[69] J. Koo, C. Kleinstreuer, A new thermal conductivity model for nanofluids, J. Nanopart. Res. 6 (6) (2004) 577–588.

[70] H. Xie, M. Fujii, X. Zhang, Effect of interfacial nanolayer on the effective thermal conductivity of nanoparticle-fluid mixture, Int. J. Heat. Mass. Transf. 48 (14) (2005) 2926–2932.

[71] Y. Ren, H. Xie, A. Cai, Effective thermal conductivity of nanofluids containing spherical nanoparticles, J. Phys. D Appl. Phys. 38 (21) (2005) 3958–3961.

[72] S. Pil Jang, S.U.S. Choi, Effects of various parameters on nanofluid thermal conductivity, J. Heat Transf. 129 (5) (2007) 617.

[73] J. Xu, B. Yu, A new model for heat conduction of nanofluids based on fractal distributions of nanoparticles, J. Phys. D Appl. Phys. 39 (2006) 4486–4490.

[74] D. Song, Y. Yang, D. Jing, International Journal of Heat and Mass Transfer Insight into the contribution of rotating Brownian motion of nonspherical particle to the thermal conductivity enhancement of nanofluid, Int. J. Heat Mass Transf. 112 (2017) 61–71.

[75] K.C.C. Leong, C. Yang, S.M. Murshed, A model for the thermal conductivity of nanofluids—the effect of interfacial layer, J. Nanopart. Res. 8 (2) (2006) 245–254.

[76] R. Prasher, P. Bhattacharya, P.E. Phelan, Brownian-motion-based convective-conductive model for the effective thermal conductivity of nanofluids, J. Heat Transf. 128 (6) (2006) 588.

[77] X.F. Zhou, L. Gao, Thermal conductivity of nanofluids: effects of graded nanolayers and mutual interaction, J. Appl. Phys. 103 (8) (2008).

[78] S.M.S. Murshed, K.C. Leong, C. Yang, A combined model for the effective thermal conductivity of nanofluids, Appl. Therm. Eng. 29 (11–12) (2009) 2477–2483.

[79] S.M. Hosseini, M. Reza, M. Goodarzi, A.A.A.A. Alrashed, New temperature, interfacial shell dependent dimensionless model for thermal conductivity of nanofluids, Int. J. Heat Mass Transf. 114 (2017) 207–210.

[80] M. Saeedinia, M.A. Akhavan-Behabadi, P. Razi, Thermal and rheological characteristics of CuO-Base oil nanofluid flow inside a circular tube, Int. Commun. Heat Mass Transf. 39 (1) (2012) 152–159.

[81] M. Abareshi, S.H. Sajjadi, S.M. Zebarjad, E.K. Goharshadi, Fabrication, characterization, and measurement of viscosity of α-Fe2O3-glycerol nanofluids, J. Mol. Liq. 163 (1) (2011) 27–32.

[82] B. Bakthavatchalam, K. Habib, R. Saidur, B. Baran, K. Irshad, Comprehensive study on nano fl uid and ionano fluid for heat transfer enhancement: a review on current and future perspective distilled water, J. Mol. Liq. 305 (2020) 112787.

[83] L. Syam Sundar, E. Venkata Ramana, M.K. Singh, A.C.M. Sousa, Thermal conductivity and viscosity of stabilized ethylene glycol and water mixture Al2O3 nanofluids for heat transfer applications: an experimental study, Int. Commun. Heat Mass Transf. 56 (2014) 86–95.

[84] S.M.S. Murshed, P. Estellé, A state of the art review on viscosity of nanofluids, Renew. Sustain. Energy Rev. 76 (August 2016) (2017) 1134–1152.

[85] H.D. Koca, S. Doganay, A. Turgut, I.H. Tavman, R. Saidur, I.M. Mahbubul, Effect of particle size on the viscosity of nanofluids: a review, Renew. Sustain. Energy Rev. 82 (April 2016) (2018) 1664–1674.

[86] A. Arshad, M. Jabbal, Y. Yan, D. Reay, A review on graphene based nanofluids: preparation, characterization and applications, J. Mol. Liq (2019) 444–484.

[87] W. Duangthongsuk, S. Wongwises, Measurement of temperature-dependent thermal conductivity and viscosity of TiO2-water nanofluids, Exp. Therm. Fluid. Sci. 33 (4) (2009) 706–714.

[88] D.P. Kulkarni, D.K. Das, G.A. Chukwu, Temperature dependent rheological property of copper oxide nanoparticles suspension (nanofluid), J. Nanosci. Nanotechnol. 6 (4) (2006) 1150–1154.

[89] Maïga SEB, Nguyen CT, Galanis N, Roy G, BS El, Tam C, et al. Heat transfer behaviours of nanofluids in a uniformly heated tube, in: Superlattices and Microstructures Academic Press; Mar 1, 2004, pp. 543–557.

[90] P.K. Namburu, D.P. Kulkarni, D. Misra, D.K. Das, Viscosity of copper oxide nanoparticles dispersed in ethylene glycol and water mixture, Exp. Therm. Fluid Sci. 32 (2) (2007) 397–402.

[91] C.T. Nguyen, F. Desgranges, G. Roy, N. Galanis, T. Maré, S. Boucher, et al., Temperature and particle-size dependent viscosity data for water-based nanofluids—hysteresis phenomenon, Int. J. Heat Fluid Flow 28 (6) (2007) 1492–1506.

[92] B.C. Sahoo, R.S. Vajjha, R. Ganguli, G.A. Chukwu, D.K. Das, Determination of rheological behavior of aluminum oxide nanofluid and development of new viscosity correlations, Pet. Sci. Technol. 27 (15) (2009) 1757–1770.

[93] G.K.K. Batchelor, The effect of Brownian motion on the bulk stress in a suspension of spherical particles, J. Fluid. Mech. 83 (1) (1977) 97–117.

[94] H.C. Brinkman, The viscosity of concentrated suspensions and solutions, J. Chem. Phys. 20 (4) (1952) 571.

[95] H. Eilers, Die Viskosität von Emulsionen hochviskoser Stoffe als Funktion der Konzentration, Kolloid-Zeitschrift 97 (3) (1941) 313–321.

[96] A. Einstein, Eine neue Bestimmung der Moleküldimensionen, Ann. Phys. 324 (2) (1906) 289–306.

[97] N.A. Frankel, A. Acrivos, On the viscosity of a concentrated suspension of solid spheres, Chem. Eng. Sci. 22 (6) (1967) 847–853.

[98] A.L. Graham, On the viscosity of suspensions of solid spheres, Appl. Sci. Res. 37 (3–4) (1981) 275–286.

[99] T.S. Lundgren, Slow flow through stationary random beds and suspensions of spheres, J. Fluid. Mech. 51 (2) (1972) 273–299.

[100] N. Saitô, Concentration dependence of the viscosity of high polymer solutions, I. J. Phys. Soc. Japan 5 (1) (1950) 4–8.

[101] M. Mooney, A shearing disk plastometer for unvulcanized rubber N IDEAL plastometer for unvulcanized rubber, Ind. Eng. Chem. Anal. Ed. 2 (1934) 147–151.

[102] I.M. Shahrul, I.M. Mahbubul, S.S. Khaleduzzaman, R. Saidur, M.F.M. Sabri, A comparative review on the specific heat of nanofluids for energy perspective, Renew. Sustain. Energy Rev. 38 (2014) 88–98.

[103] B.C.B. Pak, Y.I.Y. Cho, Hydrodynamic and heat transfer study of dispersed fluids with submicron metallic oxide particles, Exp. Heat Transf. 11 (2) (1998) 151–170.

[104] Y. Xuan, W. Roetzel, Conceptions for heat transfer correlation of nanofluids, Int. J. Heat Mass Transf. 43 (19) (2000) 3701–3707.

[105] D.P. Kulkarni, R.S. Vajjha, D.K. Das, D. Oliva, Application of aluminum oxide nanofluids in diesel electric generator as jacket water coolant, Appl. Therm. Eng. 28 (14–15) (2008) 1774–1781.

[106] B.X. Wang, L.P. Zhou, X.F. Peng, X.Z. Du, Y.P. Yang, On the specific heat capacity of CuO nanofluid, Adv. Mech. Eng. (2010).

[107] Y.R. Sekhar, K.V. Sharma, Study of viscosity and specific heat capacity characteristics of water-based Al2O3nanofluids at low particle concentrations, J. Exp. Nanosci. 10 (2) (2015) 86–102.

[108] M. Ghazvini, M.A. Akhavan-Behabadi, E. Rasouli, M. Raisee, Heat transfer properties of nanodiamond-engine oil nanofluid in laminar flow, Heat Transf. Eng. 33 (6) (2012) 525–532.

[109] E. Montazer, E. Salami, H. Yarmand, Z.Z. Chowdhury, M. Dahari, S.N. Kazi, et al., Development of a new density correlation for carbon-based nanofluids using response surface methodology, J. Therm. Anal. Calorim. 132 (2) (2018) 1399–1407.

[110] M. Sharifpur, S. Yousefi, J.P. Meyer, A new model for density of nanofluids including nanolayer, Int. Commun. Heat Mass Transf. 78 (2016) 168–174.

[111] A.R.A. Khaled, K. Vafai, Heat transfer enhancement through control of thermal dispersion effects, Int. J. Heat Mass Transf. 48 (11) (2005) 2172–2185.

[112] M. Salemi, M. Ali, M.S. Mojarrad, A. Keshavarz, A. Shokouhi, Nanofluids thermal behavior analysis using a new dispersion model along with single-phase, Heat Mass Transf. 49 (9) (2013) 1333–1343.

[113] M. Bahiraei, S.M. Hosseinalipour, Thermal dispersion model compared with euler-lagrange approach in simulation of convective heat transfer for nanoparticle suspensions, J. Dispers. Sci. Technol. 34 (12) (2013) 1778–1789.

[114] J. Buongiorno, Convective transport in nanofluids, J. Heat Transfer. 128 (3) (2006) 240.

[115] Akbarinia A, Behzadmehr A. Numerical study of laminar mixed convection of a nanofluid in horizontal curved tubes. 2007;27:1327–1337.

[116] Rashmi W, Ismail AF, Khalid M, Faridah Y. CFD studies on natural convection heat transfer of Al_2O_3 -water nanofluids. 2011;1301–1310.

[117] M. Keshavarz, M. Darabi, S. Mohammad, H. Haddad, R. Davarnejad, Modeling of convective heat transfer of a nano fl uid in the developing region of tube fl ow with computational fl uid dynamics, Int. Commun. Heat Mass Transf. 38 (9) (2011) 1291–1295.

[118] S. Özerinç, A.G. Yazcoglu, S. Kakaç, Numerical analysis of laminar forced convection with temperature-dependent thermal conductivity of nanofluids and thermal dispersion, Int. J. Therm. Sci. 62 (2012) 138–148.

[119] M.K. Moraveji, S. Mohammad, H. Haddad, M. Darabi, Modeling of forced convective heat transfer of a non-Newtonian nano fl uid in the horizontal tube under constant heat fl ux with computational fl uid dynamics, Int. Commun. Heat Mass Transf. 39 (7) (2012) 995–999.

[120] M. Akhtari, M. Haghshenasfard, M.R. Talaie, Numerical and experimental investigation of heat transfer of α-Al_2O_3/water nanofluid in double pipe and shell and tube heat exchangers, Numer. Heat Transf. A: Appl. 63 (12) (2013) 941–958.

[121] B.H. Salman, H.A. Mohammed, A.S. Kherbeet, Numerical and experimental investigation of heat transfer enhancement in a microtube using nanofluids, Int. Commun. Heat Mass Transf. 59 (2014) 88–100.

[122] W.I.A. Aly, Numerical study on turbulent heat transfer and pressure drop of nanofluid in coiled tube-in-tube heat exchangers, Energy Convers. Manag. 79 (2014) 304–316.

[123] A. Kumar, P. Ghosh, J. Sarkar, H. Dahiya, Numerical investigation of heat transfer and fluid flow in plate heat exchanger using nano fluids, Int. J. Therm. Sci. 85 (2014) 93–103.

[124] Salman BH, Mohammed HA, Munisamy KM, Kherbeet AS, Ikram-uniten J. Three-dimensional numerical investigation of nanofluids flow in microtube with different values of heat flux. 2015; 44(7): 599–619.

[125] Al-rashed MH, Dzido G, Korpy M, Smo J, Wójcik J. Microelectronics reliability investigation on the CPU nano fluid cooling. 2016;63:159–165.

[126] W.C. Chen, W.T. Cheng, Numerical simulation on forced convective heat transfer of titanium dioxide/water nanofluid in the cooling stave of blast furnace, Int. Commun. Heat Mass. Transf. 71 (2016) 208–215.

[127] M. Keshavarz Moraveji, E. Esmaeili, Comparison between single-phase and two-phases CFD modeling of laminar forced convection flow of nanofluids in a circular tube under constant heat flux, Int. Commun. Heat Mass Transf. 39 (8) (2012) 1297–1302.

[128] M. Akbari, N. Galanis, A. Behzadmehr, Comparative assessment of single and two-phase models for numerical studies of nanofluid turbulent forced convection, Int. J. Heat Fluid Flow 37 (2012) 136–146.

[129] A. Albojamal, K. Vafai, Analysis of single phase, discrete and mixture models, in predicting nanofluid transport, Int. J. Heat Mass Transf. 114 (2017) 225–237.

[130] D. Wen, Y. Ding, Effect of particle migration on heat transfer in suspensions of nanoparticles flowing through minichannels, Microfluid Nanofluidics 1 (2) (2005) 183–189.

[131] W. Williams, J. Buongiorno, L.-W. Hu, Experimental investigation of turbulent convective heat transfer and pressure loss of alumina/water and zirconia/water nanoparticle colloids (nanofluids) in horizontal tubes, J. Heat Transfer 130 (4) (2008) 042412.

[132] A. Khalili, M. Yildiz, A. Kos, Numerical modeling of convective heat transfer of thermally developing nano fluid flows in a horizontal microtube, Int. J. Thermal Sci. 109 (2016) 54–69.

[133] T. Ambreen, M.-H. Kim, Comparative assessment of numerical models for nanofluids' laminar forced convection in micro and mini channels, Int. J. Heat Mass Transf. 115 (2017) 513–523.

[134] T. Ambreen, A. Saleem, C.W. Park, Homogeneous and multiphase analysis of nanofluids containing nonspherical mwcnt and GNP nanoparticles considering the influence of interfacial layering, Nanomaterials 11 (2) (2021) 1–22.

[135] S. Göktepe, K. Atalik, H. Ertürk, Comparison of single and two-phase models for nanofluid convection at the entrance of a uniformly heated tube, Int. J. Therm. Sci. 80 (1) (2014) 83–92.

[136] M.R. Safaei, A. Jahanbin, A. Kianifar, S. Gharehkhani, A.S. Kherbeet, M. Goodarzi, et al., Mathematical modeling for nanofluids simulation: a review of the latest works, Modeling and Simulation in Engineering Sciences, InTech, 2016.

[137] Manninen M, Taivassalo V. On the Mixture Model for Multiphase Flow, 1996.

[138] L.V. Schiller, A drag coefficient correlation, Z Vereines Ingenieure 77 (1933) 318–320.

[139] C.Y. Wen, Y.H. Yu, Mechanics of fluidization, Chem. Eng. Progress. Symp. Ser. 62 (1966) 100–111.

[140] W. Ranz, Evaporation from drops 1, Chem. Eng. Prog. (1952).
[141] M. Hejazian, M.K. Moraveji, A comparative analysis of single and two-phase models of turbulent convective heat transfer in a tube for TiO 2 nanofluid with CFD, Numer. Heat Transf. Part A Appl. 63 (10) (2013) 795–806.
[142] M. Akbari, N. Galanis, A. Behzadmehr, Comparative analysis of single and two-phase models for CFD studies of nanofluid heat transfer, Int. J. Therm. Sci. 50 (8) (2011) 1343–1354.
[143] M. Hejazian, M.K. Moraveji, A. Beheshti, Comparative study of Euler and mixture models for turbulent flow of Al2O3nanofluid inside a horizontal tube, Int. Commun. Heat Mass Transf. 52 (2014) 152–158.
[144] R. Lotfi, Y. Saboohi, A.M. Rashidi, Numerical study of forced convective heat transfer of nanofluids: comparison of different approaches, Int. Commun. Heat Mass Transf. 37 (1) (2010) 74–78.
[145] J.F. Zhong, S.N. Sedeh, Y.P. Lv, B. Arzani, D. Toghraie, Investigation of Ferro-nanofluid flow within a porous ribbed microchannel heat sink using single-phase and two-phase approaches in the presence of constant magnetic field, Powder Technol. 387 (2021) 251–260.
[146] M. Haghshenas Fard, M.N. Esfahany, M.R. Talaie, Numerical study of convective heat transfer of nanofluids in a circular tube two-phase model vs single-phase model, Int. Commun. Heat Mass Transf. 37 (1) (2010) 91–97.
[147] A. Behzadmehr, M. Saffar-Avval, N. Galanis, Prediction of turbulent forced convection of a nanofluid in a tube with uniform heat flux using a two phase approach, Int. J. Heat Fluid Flow 28 (2) (2007) 211–219.
[148] F. Akbaridoust, M. Rakhsha, A. Abbassi, M. Saffar-Avval, Experimental and numerical investigation of nanofluid heat transfer in helically coiled tubes at constant wall temperature using dispersion model, Int. J. Heat Mass Transf. 58 (1–2) (2013) 480–491.
[149] E. Ebrahimnia-Bajestan, M. Charjouei Moghadam, H. Niazmand, W. Daungthongsuk, S. Wongwises, Experimental and numerical investigation of nanofluids heat transfer characteristics for application in solar heat exchangers, Int. J. Heat Mass Transf. 92 (2016) 1041–1052.
[150] I. Behroyan, P. Ganesan, S. He, S. Sivasankaran, Turbulent forced convection of Cu-water nanofluid: CFD model comparison, Int. Commun. Heat Mass Transf. 67 (2015) 163–172.
[151] N. Kumar, B.P. Puranik, Numerical study of convective heat transfer with nanofluids in turbulent flow using a Lagrangian-Eulerian approach, Appl. Therm. Eng. 111 (2017) 1674–1681.
[152] E.J. Onyiriuka, A.I. Obanor, M. Mahdavi, D.R.E. Ewim, Evaluation of single-phase, discrete, mixture and combined model of discrete and mixture phases in predicting nanofluid heat transfer characteristics for laminar and turbulent flow regimes, Adv. Powder Technol. 29 (11) (2018) 2644–2657.
[153] Z. Ying, B. He, D. He, Y. Kuang, J. Ren, B. Song, Comparisons of single-phase and two-phase models for numerical predictions of Al2O3/water nanofluids convective heat transfer, Adv. Powder Technol. 31 (7) (2020) 3050–3061.
[154] O. Mahian, L. Kolsi, M. Amani, P. Estellé, G. Ahmadi, C. Kleinstreuer, et al., Recent advances in modeling and simulation of nanofluid flows—part II: applications, Phys. Rep. 791 (2019) 1–59.
[155] Z. Guo, C. Shu, Lattice Boltzmann method and its applications in engineering, Advances in Computational Fluid Dynamics, 3, World Scientific, 2013.

[156] S. Mazumder, Numerical Methods for Partial Differential Equations: Finite Difference and Finite Volume Methods, 2015 1–461 p.

[157] M.S. Kandelousi, D.D. Ganji, Control volume finite element method (CVFEM), Hydrothermal Analysis in Engineering Using Control Volume Finite Element Method, Academic Press, 2015, pp. 1–12.

[158] D.D.D. Vo, M. Hedayat, T. Ambreen, S.A.S.A. Shehzad, M. Sheikholeslami, A. Shafee, et al., Effectiveness of various shapes of Al2O3 nanoparticles on the MHD convective heat transportation in porous medium: CVFEM modelling, J. Therm. Anal. Calorim. 139 (2) (2020) 1345–1353.

[159] S.A. Shehzad, M. Sheikholeslami, T. Ambreen, A. Shafee, Convective MHD flow of hybrid-nanofluid within an elliptic porous enclosure, Phys. Lett. Sect. A Gen. At. Solid State Phys. 384 (28) (2020) 126727.

[160] S.A. Shehzad, M. Sheikholeslami, T. Ambreen, A. Shafee, H. Babazadeh, M. Ahmad, Heat transfer management of hybrid nanofluid including radiation and magnetic source terms within a porous domain, Appl. Nanosci. 10 (12) (2020) 5351–5359.

[161] H. Babazadeh, T. Ambreen, S.A. Shehzad, A. Shafee, Ferrofluid non-Darcy heat transfer involving second law analysis: an application of CVFEM, J. Therm. Anal. Calorim. 143 (2020).

[162] S.L. Lee Effects of Nanoparticle Aggregation, Particle Size and Temperature of nanoflUids Using Molecular Dynamics Simulation, 2016.

[163] F. Strniša, T. Urbič, P. Žnidaršič-Plazl, I. Plazl, Process intensification and miniaturization of chemical and biochemical processes, Comput. Aided Chem. Eng. 46 (2019) 1801–1806.

[164] E. Moeendarbary, T.Y. Ng, M. Zangeneh, Dissipative particle dynamics: introduction, methodology and complex fluid applications—a review, Int. J. Appl. Mech. 1 (4) (2009) 737–763.

Chapter 7

Recent advances in machine learning research for nanofluid heat transfer in renewable energy

Zafar Said[1,2], Maham Sohail[1] and Arun Kumar Tiwari[3]
[1]Department of Sustainable and Renewable Energy Engineering, University of Sharjah, Sharjah, United Arab Emirates, [2]U.S.-Pakistan Center for Advanced Studies in Energy (USPCAS-E), National University of Sciences and Technology (NUST), Islamabad, Pakistan, [3]Mechanical Engineering Department, Institute of Engineering and Technology, Dr. A.P.J. Abdul Kalam Technical University, Lucknow, Uttar Pradesh, India

Nomenclature

b	the bias rate
f	the relationship between inputs and the target
K	number of nodes
MSE	mean squared error
nj	*j*th node input
R2	regression coefficient
X	model input
w	the weight vector
θj	hidden node's threshold
ω**ji**	load at the node connection
ξ	standard deviation
σ^2	the squared bandwidth

7.1 Introduction

Machine learning (ML) is a subgroup of Artificial Intelligence (AI), specified as a class of algorithms that can acquire data and trends and foresee results for advanced decision-making without understanding core physics [1,2]. This class of algorithms allows the system to develop from knowledge and learning without specific reprogramming. ML has been promising in several fields, such as

Advances in Nanofluid Heat Transfer. DOI: https://doi.org/10.1016/B978-0-323-88656-7.00011-8

medicine, robotics, and other scientific areas [3]. ML-based data analysis increases the output and precision of flow analysis, and opens up novel prospects like quantitative predictions of flow characteristics, depending on the qualitative data [4].

ML can be categorized into three key classifications:

1. Supervised learning: It consists of predicting a mapping function, say $f(x)$, that gets input, "x". The reliability of this mapping function will define the precision of the output projected. It has two types, which are classification and regression. Classification is described as where the output is categorized, whereas regression is described as where the projected output is incessant.
2. Unsupervised learning: It describes novel data patterns and distribution. It is used where not much labeled data is available. For instance, to assemble parallel data into an individual group. These can then provide an input to ML algorithms regarding such data.
3. Semisupervised learning: This is the event when labels are accessible for limited test data. This is considered as the most prevalent yet challenging case. It is difficult to discover a universal computational structure that is relevant in other methods as well.

   The ML process can be split up into three stages:
   1. Data collection and representation
   2. Model selection and validation
   3. Model optimization

7.1.1 Data collection and representation

Featurization or feature engineering refers to the process of transforming unrefined data into a suitable algorithm. ML prospers by presenting satisfactory solutions within restricted time and reserves in this concept. Feature space incorporates data that are significant for ML algorithm [5].

7.1.2 Model selection and validation

Various ML models or algorithms are presented in the research community. Such models can be as basic as fitting to a linear regression curve and as sophisticated as neural networks. Some preliminary research areas where ML algorithm have been employed are the chemical and pharmaceutical sciences. Some common algorithms and their potential use cases are presented in Fig. 7.1. Some commonly used ML algorithms are as follows:

1. Regressors
2. Neural networks
3. Transfer learning
4. ML toolkits
5. Natural language processing for materials literature

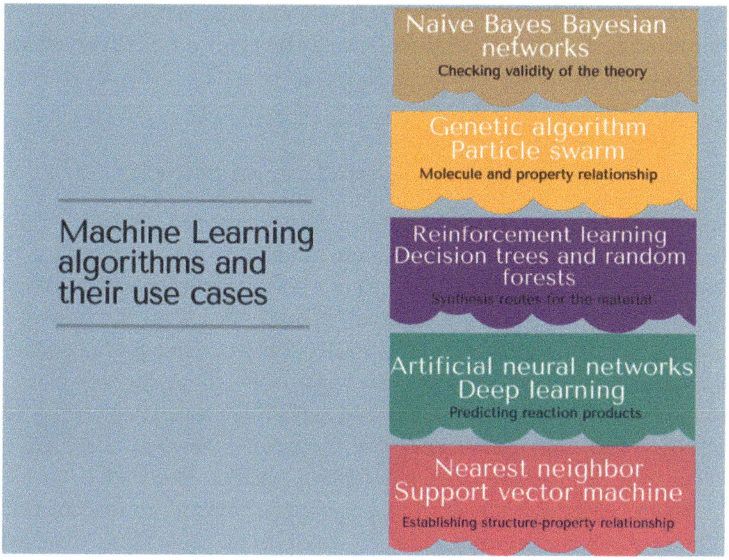

FIGURE 7.1 Examples of machine learning models [5].

7.1.3 Model optimization

Authentication or test data sets estimate the ML model performance. Errors in predicting the output from a well-known input set in supervised learning are identified as prediction errors, including bias errors and variance errors. Bias errors indicate how much the predicted output is different from the authentic output of test data. On the other hand, variance errors calculate the difference between algorithm outputs when different training data sets are considered.

The model should be chosen and optimized thoroughly by introducing a dilemma between bias and variance to prevent the over- and underfitting pitfalls. The different metrics for the model are:

1. mean absolute error and mean absolute relative error;
2. loss function;
3. learning rate;
4. coefficient of determination R^2; and
5. receiver operating characteristic curve.

7.2 Machine learning techniques

The multilayer perceptron (MLP) ANN, radial basis function (RBF), adaptive neuro fuzzy inference system (ANFIS), and the least square support vector machine (LSSVM) have been generally advanced for model analysis and are discussed in the following sections. These techniques can discover the hidden links between variables that impact the nanofluid properties and foresee the given factors based on these relationships. These techniques have shown encouraging outcomes by increasing the precision with which thermophysical properties can be predicted [6,7].

7.2.1 Multilayer perception artificial neural network

Modern-day systems can operate complex mathematical and symbolic calculations at a fast rate. But they are a long way from the human mind in operating perceptual responsibilities like language and image recognition. Artificial neural networks (ANN) have been employed across several industry and scientific fields over the past few years. The development of AI established on a biological neural network (BNN) has advanced to an ANN. Similar to the BNN, the ANN is an interconnection of nodes. ANN resolves complicated pattern-oriented complexities of both categorization and time-series forms. The connections between nodes are of two kinds: one-way connection without loop-back, and a loop-back in which the node's output can be input to prior or equivalence nodes. The fundamental node model in the ANN is presented in Fig. 7.2. Each neural network has three essential elements: node character, network topology, and learning rules. Several learning proposals have been introduced for ANNs to accomplish learning objectives. The normally investigated learning techniques include error reduction and closest neighbor techniques. MLP is the most studied kind of ANN.

MLP includes distinct nodes in three essential layers. They are regarded as the input, hidden, and output layers. A weigh vector is practical for each node to connect with other nodes at the following layer. The nodes are added in the initial layer and utilized as the next layer's input [8].

By taking into account model vector X as the input, n_j refers to jth node input:

$$n_j = \sum\nolimits_{i=1}^{n} \omega_{ji} x_i + \theta_j; \quad j = 1, 2, \ldots, K$$

In the above equation, K, θj, and ω_{ji} represent the nodes quantity, jth hidden node's maximum, and load at the node connection.

f function makes the inputs in the subsequent layer as:

$$y_j = f(n_j) = f\left(\sum\nolimits_{i=1}^{n} \omega_{ji} x_i + \theta_j\right); \quad j = 1, 2, \ldots, K$$

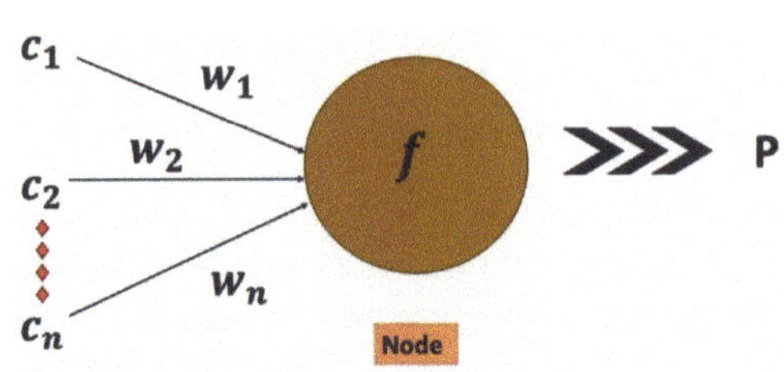

FIGURE 7.2 Basic model of a single node.

These functions have certain attributes when applied in systems. By multiplying the output by linking load, the output will be achieved. The relevant training algorithm is the backpropagation algorithm employed in the MLP-ANNs training stage to normalize bias and weigh values [9]. The MLP-ANN structure is displayed in Fig. 7.3. ANN structure needs the developer to create determinations depending on each other, such as input, training and testing data sets, learning algorithm, network design, and revolution purpose. For instance, the ANN and learning algorithm will demonstrate the input value type. It is vital to stick to an approach or definite steps when modeling ANNs such as [10]:

1. Define input variables
2. Define data to use
3. Split data into training and testing
4. Describe the network structure
5. Choose an algorithm and convert variables to inputs
6. Train – redo (ANN error < satisfactory value)
7. Test – on hold-out model to confirm ANN conclusion

ANN developers should conduct widespread and significant understanding into ANNs modeling. ANN advancements involve thorough considerate measures in developing ANNs.

7.2.2 Adaptive neuro fuzzy inference system

Jang [11] developed the ANFIS technique, a positive neural network technique to clarify approximation troubles [12]. It provides the mapping relationship between the input and output values using a hybrid learning technique to define the optimum supply of membership purposes [13]. This technique comprises five levels, with a defined action specified for each

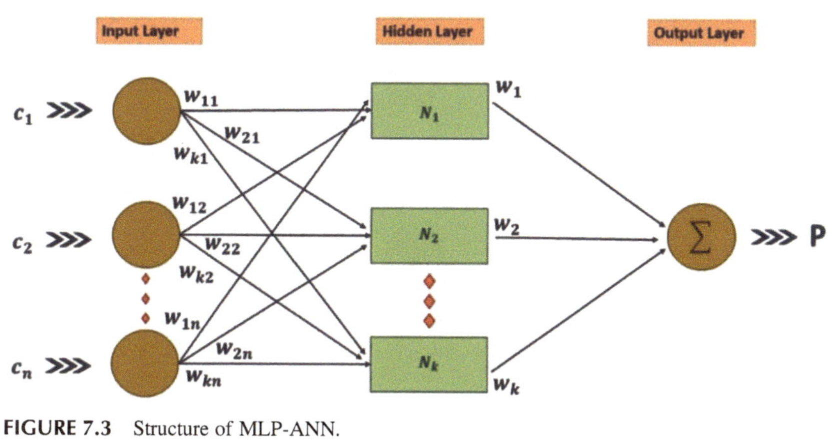

FIGURE 7.3 Structure of MLP-ANN.

level. The initial level transforms inputs into fuzzy sets and the membership in variables vary from zero to one. The second level represent input signals. In the third level, standardized firing intensity is established for every single node. In the penultimate level, the outputs are made defuzzy. The outputs are received in the last level that contains one node utilized to sum total the input from the former level. The structure is displayed in Fig. 7.4.

7.2.3 Radial basis function network

RBF has promising advantages such as uncomplicated configuration, elevated projection properties, and rapid training practice. Researchers worked on the use of RBF to solve function and pattern classification challenges. RBF is well-suited for such challenges due to their straightforward topological arrangement and their capability to demonstrate learning explicit behavior [14]. This network type is known as feed forward. There are three major layers: input, hidden, and output layers in RBF algorithm [15]. The inputs are given to nodes and transferred to the next layer. Finally, weighted links transfer values to the last layer. The important juncture is its hidden layer in which RBF generates a vector multiplied by linked bias. The structure is shown in Fig. 7.5. The input is forecast to a novel zone in the hidden layer [16]. The following expression is applied to determine the *j*th neurons output:

$$Z_j = Z(\|X - \Delta_j\|) = \exp\left(-\frac{\|X-\Delta_j\|^2}{2\xi_j^2}\right)$$

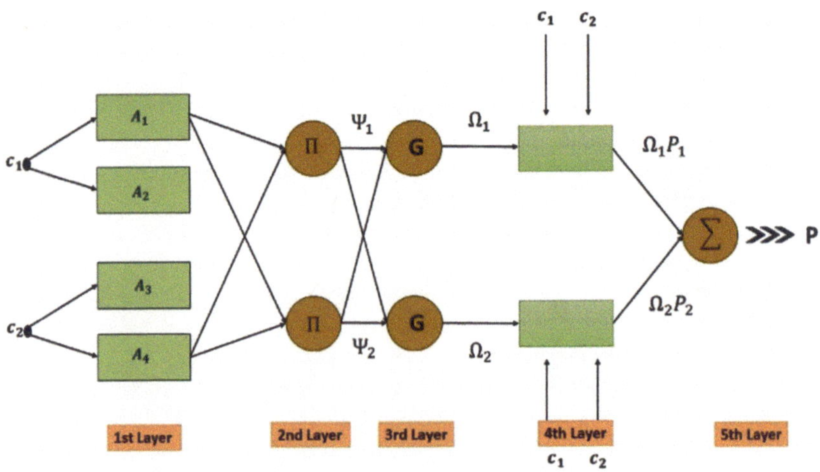

FIGURE 7.4 Structure of ANFIS model.

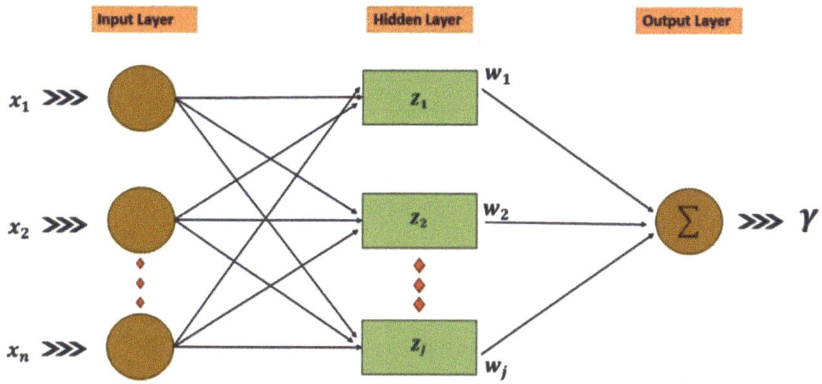

FIGURE 7.5 Structure of radial basis function model.

The standard deviation can be estimated as:

$$\xi = \frac{\theta_m}{\sqrt{\Lambda}}$$

In the last layer, the weighted signals sum is accumulated from the preceding layer.

$$\gamma = \sum_{j=1}^{A} \omega_j Z_j$$

7.2.4 Least square support vector machine

The support vector machine (SVM) technique was introduced in 1999, and it defines the regression and function. It has been applied in model recognition, regression, and distribution evaluation for discrete values. It cannot resolve the overfitting difficulties either in a conventional manner or in a feed-forward manner. The LSSVM approach was introduced to overcome this restriction, and it can be computed as follows:

$$f(x) = w^T x + b$$

where f refers to the relationship between input and the target, w refers to the weight vector, ϕ projects x into the characteristic vector, and b shows bias rate [17–19].

RBF is a standard kernel in regression errors [20]:

$$K(x, x_k) = \exp\left(-\|x_k - x\|^2/\sigma^2\right)$$

σ^2 refers to the squared bandwidth. It can be computed using the optimization technique.

The outputs' mean squared error (MSE) can be expressed as [21,22]:

$$\text{MSE} = \frac{\sum_{i=1}^{n} (\text{predicted value} - \text{actual value})^2}{ns}$$

$$R^2 = 1 - \frac{\sum_{i=1}^{i=n} \left(y_i^{\text{actual value}} - y_i^{\text{predicted value}} \right)^2}{\sum_{i=1}^{i=n} \left(y_i^{\text{actual value}} - y^{\text{actual value}} \right)^2}$$

R-squared refers to a parameter to evaluate the model accuracy.

7.3 Nanofluid heat transfer and machine learning

Nanofluids have appeared as potential substitutes for the fluids presently employed in thermal systems [23–25]. Literature studies on nanofluids cover an extensive spectrum, including hydrodynamics, thermodynamics, wind turbines, solar collectors, geothermal heat exchangers, hydroelectric rotors, and ocean power plants [26–28]. The elevated thermophysical fluid properties, such as remarkable thermal conductivity, show exceptional preference in heat transfer and thermal management systems [29,30]. Increased viscosity is unfavorable for thermal management due to the elevated pressure drop. The nanofluids' thermophysical properties depend on different factors such as volume concentration, nanoparticles' size and shape, temperature, and base fluid properties [31–33]. The most effective factors such as volume concentration and temperature been explored in research community [34,35]. Increasing temperature increases thermal conductivity and reduces dynamic viscosity, whereas an increase in volume concentration increases thermal conductivity and dynamic viscosity [36,37]. Peng et al. [36] used the ANN approach to determine the thermal conductivity of Al_2O_3 + Cu/EG nanofluids at nanoparticle concentrations of 0.125 to 2.0 vol.% and temperature ranging from 25 to 50°C. The ANN approach presented good agreement with experimental data. Improved thermophysical properties of nanofluids are employed in heat exchanger systems with nonlinear impacts on the thermal and fluid flow behavior. [38,39]. Measuring nanofluids' thermophysical properties experimentally consumes significant time and is tedious [40,41]. Numerous models have been introduced for projecting properties. Applied algorithms and input variables influence the attribute and accuracy of such models, and data points are applied for the modeling technique [40,42,43]. Therefore ML techniques can be introduced in nanofluids' study.

Said et al. [44] introduced a theoretical analysis and examined the nanofluid performance of CuO-water in heat exchangers. It was noted that the overall and the convective heat transfer coefficient showed 7% and 11.39% improvements, respectively. Wang et al. [45] predicted the heat transfer rate for heat exchangers using ANN approach. It was reported that the predicted heat transfer rates using ANN technique well fit the experimental data, as displayed in Fig. 7.6, showing that ANN approach is appropriate in predicting

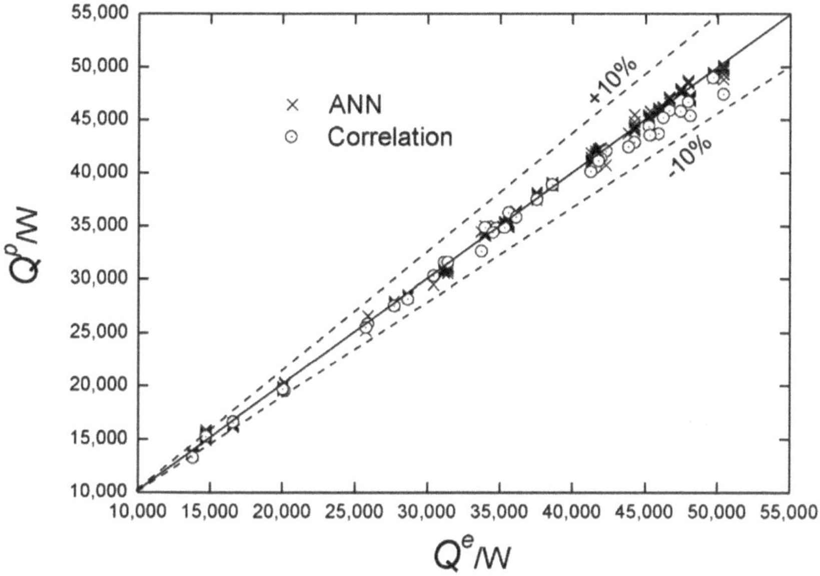

FIGURE 7.6 Comparison of artificial neural network and correlation for shell-and-tube heat exchanger. *From Wang, Q., et al., Prediction of heat transfer rates for shell-and-tube heat exchangers by artificial neural networks approach. J. Therm. Sci., 2006. 15(3): p. 257–262. Copyright Elsevier (Lic# 5098691121284).*

heat transfer rates from experimental correlations. El-said et al. [46] predicted atmosphere injection impact on heat exchangers' thermohydraulic behavior using ML algorithms such as RVFL, SVM, KNN, and SMO. The RVFL model has excellent forecasting behavior with superior precision (Fig. 7.7). Fig. 7.7 shows that it has the maximum coefficient for the three projected responses in contrast to the other models. Mohammadi et al. [47] proposed a three-dimensional numerical study of fluid flow using the computational fluid dynamics technique. From Fig. 7.8, it can be seen that the ANNs can acknowledge the wire inserts models quite well, along with the "$y = x$" correlation output line. Rico et al. [48] studied the performance of a triple concentric-tube heat exchanger numerically using ANN and from the Fig. 7.9 it can be seen that the value of regression coefficient (R^2) obtained is 0.9924 for the heat transfer rate and the absolute average relative deviation is 1.91% for the heat transfer rate. Rezaei et al. [49] investigated the heat transfer impacts of partition angle and Rayleigh number in the cavity, utilizing ANFIS of natural convection. It was noticed that the relative errors were observed at 0.005% and 1.735% for the testing and training values, respectively. Tahseen et al. [50] predicted thermal and pressure drop characteristics for in-line tube alignment in a cross-flow using ANFIS. Fig. 7.10 shows plots of temperature contour with the Reynolds number from 10 to 320 and it indicates the lowest atmosphere temperature at the entrance to a higher air temperature linked to

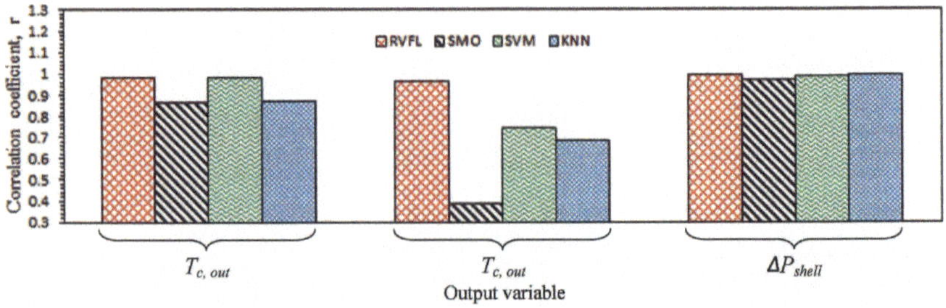

FIGURE 7.7 Correlation coefficient. *From El-Said, E.M.S., M. Abd Elaziz, and A.H. Elsheikh, Machine learning algorithms for improving the prediction of air injection effect on the thermo-hydraulic performance of shell and tube heat exchanger. Appl. Therm. Eng., 2021. 185: p. 116471. Copyright Elsevier (Lic# 5098691353386).*

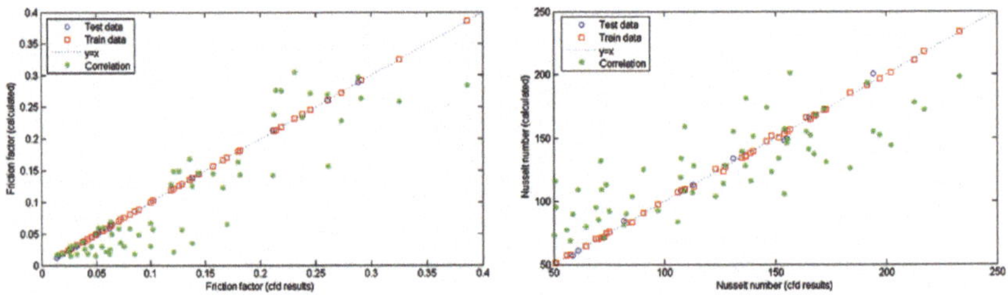

FIGURE 7.8 Comparison of artificial neural network with Nusselt number and friction factor [47].

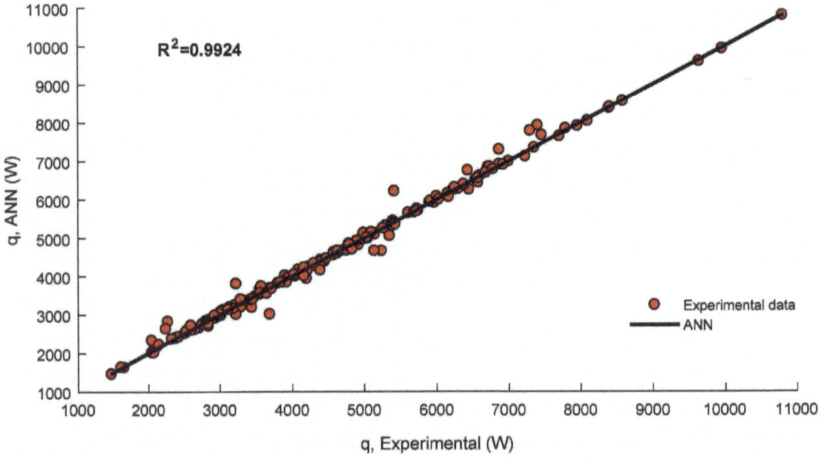

FIGURE 7.9 Experimental against predicted q. *From Moya-Rico, J.D., et al., Characterization of a triple concentric-tube heat exchanger with corrugated tubes using Artificial Neural Networks (ANN). Appl. Therm. Eng., 2019. 147: p. 1036–1046. Copyright Elsevier (Lic# 5098700030000).*

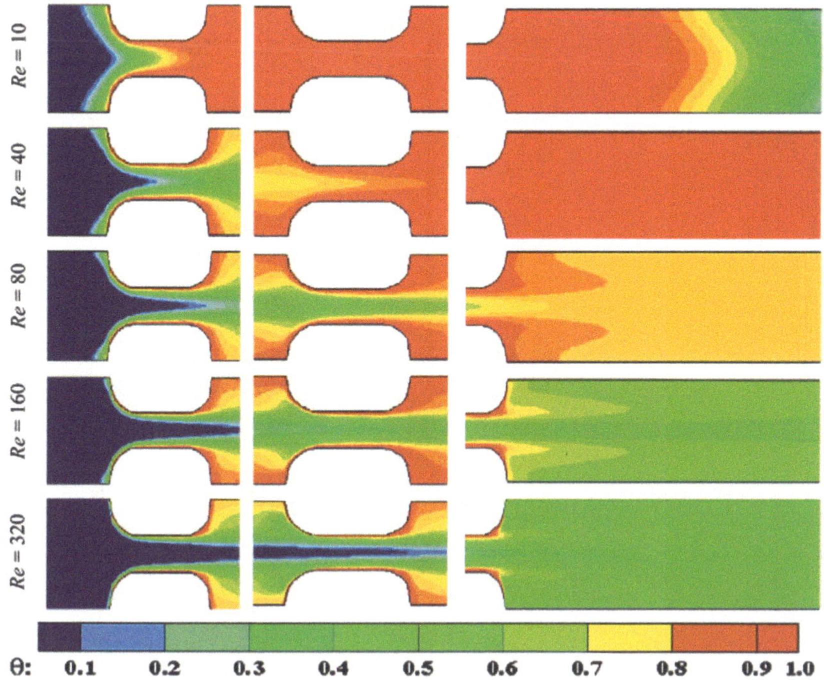

FIGURE 7.10 Isotherm plot for Reynolds number from 10 to 320. *From Tahseen, T.A., M. Ishak, and M.M. Rahman, Performance predictions of laminar heat transfer and pressure drop in an in-line flat tube bundle using an adaptive neuro-fuzzy inference system (ANFIS) model. Int. Commun. Heat. Mass. Transf., 2014. 50: p. 85–97. Copyright Elsevier (Lic# 5098700317272).*

the hot tube surface. The results between modeling and numerical values were achieved with a mean error for Nusselt number and pressure drop below 1.9% and 2.97%, respectively, as shown in Fig. 7.11. Mohammadi et al. [51] investigated nanofluids' energy and entropy generation using ANFIS for helically coiled pipe under a persistent magnetic field using flattened and microfins pipes experimentally and numerically. The results were projected with average relative errors for Nusselt number, friction factor, and entropy generation of 0.67%, 4.48%, and 2.83%, respectively. Sadeghi et al. [52] predicted the performance of constructed tubular collector by using CuO-water nanofluid using ANN and verified it using MLP and RBF models. Figs. 7.12 and 7.13 show the regression diagram of energy efficiency using MLP and RBF modeling. The MLP technique represents a superior projection of the systems' performance parameters than the RBF technique due to minimal errors; the RBF model projected has benefits in its minimalism and rapid processing as compared to MLP. Mensour et al. [53] used MLP modeling to estimate monthly solar insolation using climatic statistics. Fischer et al. [54] compared different techniques to model solar collectors.

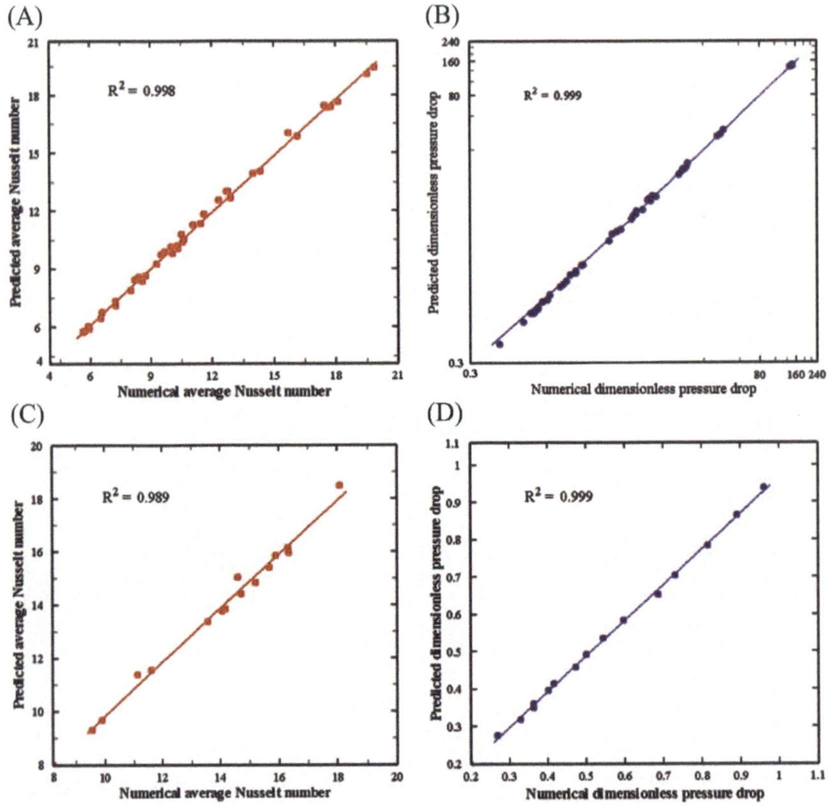

FIGURE 7.11 The comparison between numerically and estimated average Nusselt number and pressure drop using ANFIS for **(A) and (B)** training; and **(C) and (D)** testing values. *From Tahseen, T.A., M. Ishak, and M.M. Rahman, Performance predictions of laminar heat transfer and pressure drop in an in-line flat tube bundle using an adaptive neuro-fuzzy inference system (ANFIS) model. Int. Commun. Heat. Mass. Transf., 2014. 50: p. 85–97. Copyright Elsevier (Lic# 5098700317272).*

It was reported that analytical techniques help in estimating the solar efficiency more precisely than standard techniques. Boukelia et al. [55] reported an investigation to predict and optimize levelized power cost of two thermal power plants and working medium using the ANN method. Kalogirou [56] studied ANN to estimate the flat plate solar collectors performance parameters. The ANN methodology was reported to project the considerable thermal parameters of the solar system accurately than standard techniques. From these studies, it is reported that the ANN model has been widely utilized in predicting the heat transfer rate of heat exchangers and solar collectors with better precision.

Table 7.1 below presents a summary of the ML techniques implemented for the thermal performance of the nanofluids.

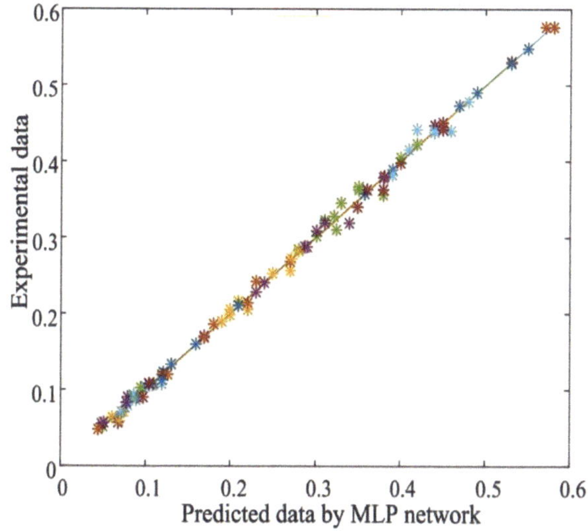

FIGURE 7.12 Prediction of energy efficiency by MLP modeling. *From Sadeghi, G., et al., Energy and exergy evaluation of the evacuated tube solar collector using Cu2O/water nanofluid utilizing ANN methods. Sustain. Energy Technol. Assess., 2020. 37: p. 100578. Copyright Elsevier (Lic# 5098700478713).*

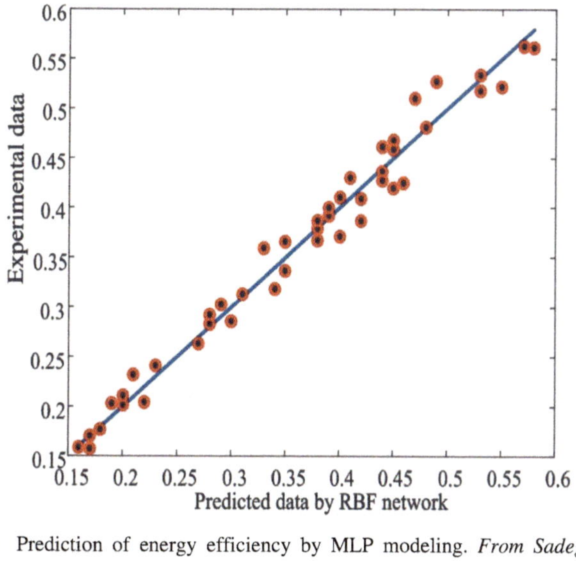

FIGURE 7.13 Prediction of energy efficiency by MLP modeling. *From Sadeghi, G., et al., Energy and exergy evaluation of the evacuated tube solar collector using Cu2O/water nanofluid utilizing ANN methods. Sustain. Energy Technol. Assess., 2020. 37: p. 100578. Copyright Elsevier (Lic# 5098700478713).*

TABLE 7.1 Summary of different studies on the nanofluids' thermal performance using machine learning techniques.

Researchers	Technique	Nanofluid	Input parameters	Output parameters	Findings
Nasirzadehroshenin et al. [57]	ANN-GA	CuO-CNT-water	Reynolds number, volume concentration, twist ratios, cavity diameter ratio	Exergy efficiency	R^2 values of 0.92.
Saeedan et al. [58]	ANN	CNT-water, Cu-water, CuO-water	Reynolds number, particle concentration, nanoparticle properties	Heat transfer coefficient, pressure drop	R^2 values of 0.9944 and 0.9991.
Maddah et al. [59]	ANN	SiO_2-water	Reynolds number, volume concentration, twist ratios, Prandtl number	Exergy efficiency	R^2 value of 0.9779
Al-Waeli et al. [60]	ANN	SiC-water/paraffin wax	Insolation, surrounding temperature	PV/T system power	R^2 value of 0.9562.
Kalani et al. [61]	ANN	ZnO-water	Surrounding and inlet temperature, incident radiation	Outlet temperature, electrical efficiency of PV/T system	R^2 values of 0.9934 and 0.9906.
Esfe et al. [62]	ANN; RBF	Ag-water	Volume concentration, Reynolds number	Nusselt number, pressure drop	R^2 for relative Nusselt number and pressure drop are 99.76% and 99.54%.
Yan et al. [63]	MLP-ANN; LSSVM; ANFIS	CNT-water	Prandtl number, volume fraction, and number of helical coiled heat exchanger	Nusselt number	LSSVM approach has the superior performance.

7.4 Machine learning of nanofluids' thermophysical properties and thermal performance

Precise measurements and prediction of nanofluids' thermophysical characteristics are essential for heat transfer processes' thermal and hydrodynamic performances [64,65]. Several parameters affect the properties, such as temperature, volume concentration, nanoparticles' size and shape, pH value, host fluid, and preparation technique [20,66]. In recent years, the major concern of the studies has been to model nanofluids' thermal conductivity with better precision [67]. Vatani et al. [68] investigated a data set experimentally with the latest findings and contrasted it with the representative standard and theoretical models such as Maxwell [69] and Bruggeman [70], Leong et al. [71], Chon et al. [72], and many others reported that they could not predict the data points precisely and they are yet to be approved as universally consistent approaches. Numerical correlations are desired for computational cost and minimalism [73]. Thermal conductivity of liquids is one of the significant elements that affect thermal processes. Several investigations have been conducted to predict nanofluids' thermal conductivity, depending on the factors such as temperature, nanoparticle concentration, nanoparticle type, and addition of surfactant. Akhgar et al. [74] proposed a correlation for thermal conductivity of a hybrid using TiO_2 + MWCNT- H_2O + EG nanofluids. Longo et al. [75] projected the thermal conductivity of oxide-H_2O-based nanofluids using ANN and reported that the model's regression coefficient (R2) was 0.1862 with 3-input, indicating an unsatisfactory outcome. The authenticity of this network and the ANN precision should be examined further. Hojjat et al. [76] introduced ANN analysis to project the thermal conductivity of three non-Newtonian nanofluids. The temperature, particle loading, and the base fluid's thermal conductivity were regarded as input variables. It was reported that the feed-forward ANN model had an improved trend with the experimental values in comparison to other models (Fig. 7.14). Boobalan et al. [77] predicted the thermal conductivity of alumina–silica hybrid nanofluids using ANN as a function of temperature and volume fraction. Fig. 7.15 compares the predicted and experimental values, predicting remarkable accuracy for modeling of the thermal conductivity. Ahmadloo et al. [78] investigated predicting the thermal conductivity ratio of nanofluids using ANN and introduced an ideal ANN model that shows a satisfactory agreement in predicting experimental values with a coefficient of 0.995 with 1.26% mean error (Fig. 7.16).

Nanofluids' dynamic viscosity has a major impact on the fluid movement and subsequently the pressure drop. Owing to its influence on Reynolds number, and the dependence of Nusselt number, it affects nanofluids' thermal performance. The nanofluids' viscosity should be determined accurately, including Reynolds number and Prandtl number. Due to such consequences, literature studies have focused on the parameters concerning viscosity. The literature has reported that dynamic viscosity reduces with a temperature increase, and

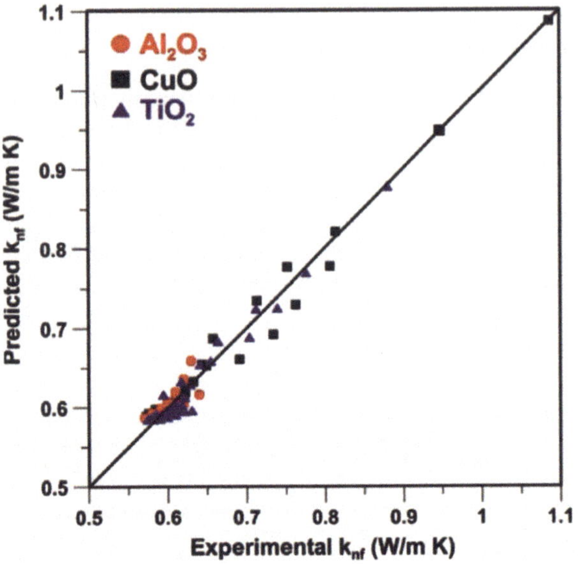

FIGURE 7.14 Experimental and predicted values of thermal conductivity. *From Hojjat, M., et al., Thermal conductivity of non-Newtonian nanofluids: experimental data and modeling using neural network. Int. J. Heat. Mass. Transf., 2011. 54(5): p. 1017–1023. Copyright Elsevier (Lic# 5098700602164).*

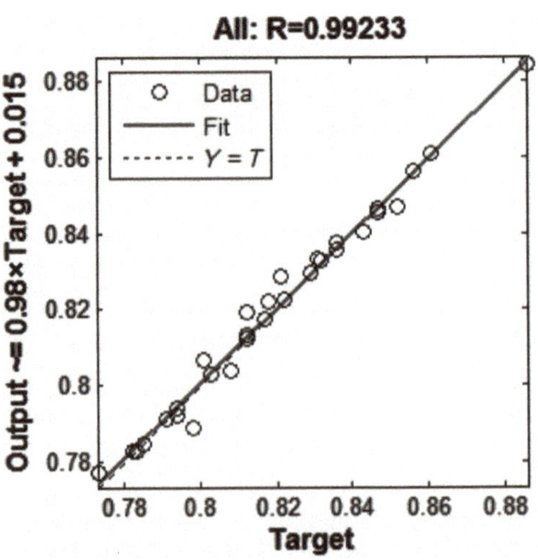

FIGURE 7.15 Comparison between experimental thermal conductivity data and ANN modeling. *From Kannaiyan, S., et al., Modeling of thermal conductivity and density of alumina/silica in water hybrid nanocolloid by the application of artificial neural networks. Chin. J. Chem. Eng., 2019. 27(3): p. 726–736. Copyright Elsevier (Lic# 5098701163385).*

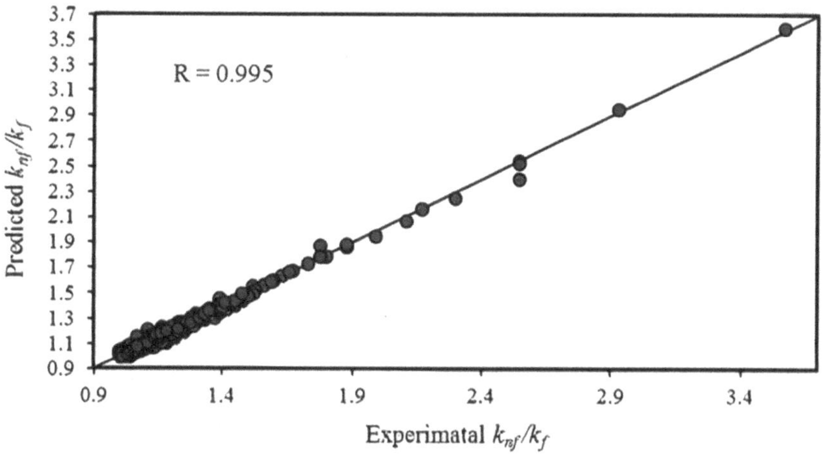

FIGURE 7.16 Predicted thermal conductivity values against experimental data. *From Ahmadloo, E. and S. Azizi, Prediction of thermal conductivity of various nanofluids using artificial neural network. Int. Commun. Heat. Mass. Transf., 2016. 74: p. 69–75. Copyright Elsevier (Lic# 5098701274789).*

increases with a volume concentration increase. ANN is one of the effective techniques to foresee the dynamic viscosity precisely. Comparable to numerical correlation, a suitable selection of variables as input alters the model precision. Esfe et al. [79] expected the viscosity of MWCNT-water nanofluids using ANN. Variables such as volume fraction and temperature were applied as input, and viscosity was applied as an output variable. It was reported that optimal ANN demonstrated a most significant variation margin of 0.28%. The outputs were compared with the experimental results as shown in Fig. 7.17, indicating better accuracy. Another study by Esfe et al. [80] predicted the viscosity of TiO_2–water nanofluids using the ANN model and compared it with experimental data. The regression coefficient obtained was 0.9998, indicating excellent precision with a simple structure. ANN modeling has been successful in estimating empirical data with high precision, as seen from Fig. 7.18. Meyer et al. [81] predicted the effective viscosity of Al_2O_3, CuO, TiO_2, and SiO_2, with H_2O as a host fluid using the FCM-ANFIS technique. It was reported that the projected FCM-ANFIS analysis fitted well with the experimental values.

Table 7.2 reviews the studies of different techniques of ML. ANN has been investigated in the research community in recent years. The table shows that the input parameters, such as temperature, particle concentration and size, shear rate, and Reynolds number, could be used. The output parameters, such as thermal conductivity, viscosity, and specific heat, are computed by internal learning algorithms. The regression coefficient (R^2) value in ML algorithms is greater than 0.99; a lower value of 0.9718 is observed by Osho and his coworkers [87]. To further enhance the prediction

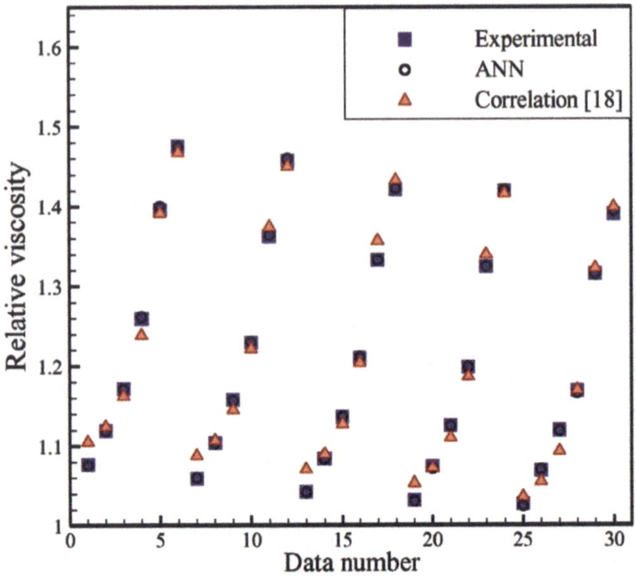

FIGURE 7.17 Comparison of artificial neural network with experimental data and correlation. *From Afrand, M., et al., Predicting the viscosity of multi-walled carbon nanotubes/water nanofluid by developing an optimal artificial neural network based on experimental data. Int. Commun. Heat. Mass. Transf., 2016. 77: p. 49–53. Copyright Elsevier (Lic# 5098701383011).*

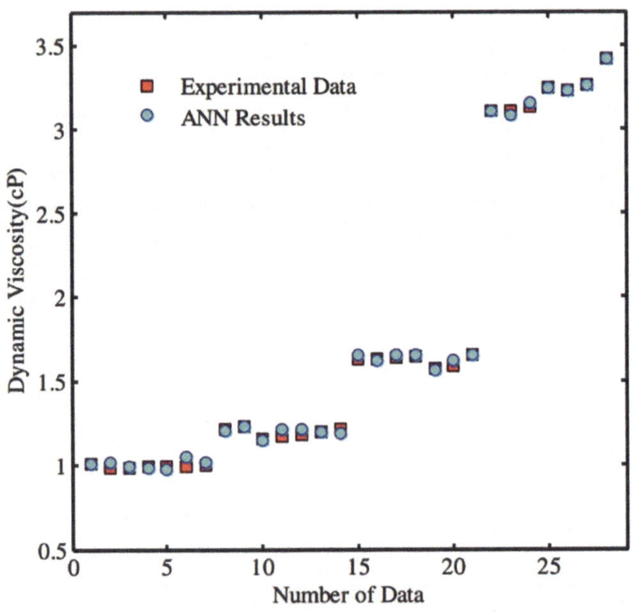

FIGURE 7.18 Comparison between experimental data and artificial neural network results. *From Hemmat Esfe, M., et al., Designing an artificial neural network to predict dynamic viscosity of aqueous nanofluid of TiO2 using experimental data. Int. Commun. Heat. Mass. Transf., 2016. 75: p. 192–196. Copyright Elsevier (Lic# 5098701480101).*

TABLE 7.2 Studies on the thermophysical properties of nanofluids using machine learning techniques.

Researchers	Technique	Nanofluid	Input parameters	Output parameters	Findings
Esfe et al. [82]	Artificial neural network (ANN)	Fe-EG	Temperature, particle loading, particle size	Thermal conductivity, dynamic viscosity	Maximal errors in thermal conductivity and dynamic viscosity are 2% and 2.5%, respectively.
Hassan et al. [83]	ANN	Molten salt	Temperature, particle concentration and size	Specific heat	R^2 value of 0.9998.
Alrashed et al. [84]	ANFIS; ANN	COOH-diamond, COOH-MWCNT in H_2O	Temperature, density, viscosity	Thermal conductivity, viscosity, density	Reported less errors as compared to other existing correlations.
Esfe et al. [85]	ANN;	Al_2O_3 + MWCNT-5W50	Temperature, particle concentration, shear rate	Viscosity	R^2 values of 0.998 and 0.982.
Longo et al. [75]	ANN	Oxide-water	Temperature, particle concentration, NP thermal conductivity	Thermal conductivity	R^2 values of 0.9910.
Esfe et al. [86]	ANN	MWCNT + SiO_2-EG	Temperature, volume concentration	Thermal conductivity	R^2 values of 0.9864 and 0.9977.
Osho et al. [87]	ANN; ANFIS; correlation	Al_2O_3 + ZnO-water	Temperature, volume concentration, mixture ratio of hybrid nanoparticles	Thermal conductivity	R^2 values of 0.9718, 0.9946, and 0.9928.

(Continued)

TABLE 7.2 (Continued)

Researchers	Technique	Nanofluid	Input parameters	Output parameters	Findings
Ma et al. [88]	ANN	$CuO + Al_2O_3$-water	Temperature, mixture ratio of hybrid nanoparticles	Thermal conductivity, viscosity	R^2 values of 0.9846 and 0.9755 for thermal conductivity and viscosity.
Alarifi et al. [89]	ANFIS-PSO; ANFIS-GA	Al_2O_3 + MWCNT-oil	Temperature and concentration	Viscosity	R^2 values of 0.9984 and 0.9967.
Baghban et al. [90]	ANFIS	TiO_2-water, Al_2O_3-water, SiO_2-water, CuO-water	Temperature, concentration, size, fluid viscosity, particle density	Viscosity	R^2 values of 0.9997.
Bahrami et al. [91]	ANN	Fe + CuO-EG + water	Temperature, concentration, shear rate	Viscosity	R^2 value of 0.99982.
Nguyen et al. [92]	RWLS-SVR	silicon carbide + graphene oxide-water	Temperature, particle concentration	Thermal conductivity	This technique provides scarcity and strength.

accuracy, some parameters such as the neural network structure, weight and threshold coefficients, number of hidden layers, and neurons must be devised based on the conditions.

7.5 Challenges and future opportunities

1. The outcomes predicted by ML models are only valid in their specific investigational scope. It is vital to develop significant databases for nanofluids' thermophysical properties and thermal performance to allow more extensive ML applications.
2. ML techniques to estimate performance and to optimize the energy system levels are necessary.
3. To enhance the precision, the neural network structure, weight and threshold coefficients, number of hidden layers, and neurons must be devised based on the particular dilemma.
4. The algorithms mentioned above so far have been restricted to the most standard ones. For example, a few optimization algorithms, like genetic algorithms, are analyzed in the ANN-based analysis, resulting in accurate and robust outputs. Additionally, different optimization methods like particle swarm optimization and genetic algorithms can be combined to lower the error of trained networks.

7.6 Conclusion

Nanofluids have drawn the attention of researchers. ML methods in nanotechnology research, as well as energy system assessment, are improving rapidly. This chapter presents an outline of ML techniques and nanofluids' thermophysical properties in heat transfer and solar energy systems. Models developed by various ML approaches, such as ANN, ANFIS, RBF, and LSSVM, are reviewed. The ANN approach is a promising approach for numerically demonstrating nanofluids' heat transfer performance and the MLP is the most studied type. The parameters such as thermal conductivity, viscosity, specific heat, exergy efficiency, Nusselt number, and pressure drop have been computed as outputs by learning algorithms. Investigations with ML techniques have shown remarkable prediction accuracy compared with the experimental data. In the final section, the challenges and future opportunities were discussed in terms of maximizing the accuracy of experimental data. ML promises a powerful and simple approach that predicts nanofluids' thermal performance with exceptional precision.

References

[1] P.K. Kanti, et al., Experimental investigation on thermal conductivity of fly ash nanofluid and fly ash-Cu hybrid nanofluid: prediction and optimization via ANN and MGGP model, Part. Sci. Technol. (2021) 1–14.

[2] Z. Said, et al., Thermophysical properties using ND/water nanofluids: an experimental study, ANFIS-based model and optimization, J. Mol. Liq. 330 (2021) 115659.
[3] S.L. Brunton, B.R. Noack, P. Koumoutsakos, Machine learning for fluid mechanics, Annu. Rev. Fluid Mech. 52 (1) (2020) 477–508.
[4] M. Raissi, A. Yazdani, G.E. Karniadakis, Hidden fluid mechanics: learning velocity and pressure fields from flow visualizations, Science 367 (6481) (2020) 1026–1030.
[5] K.T. Butler, et al., Machine learning for molecular and materials science, Nature 559 (7715) (2018) 547–555.
[6] M. Jamei, et al., On the specific heat capacity estimation of metal oxide-based nanofluid for energy perspective–a comprehensive assessment of data analysis techniques, Int. Commun. Heat. Mass. Transf. 123 (2021) 105217.
[7] Z. Said, et al., Optimizing density, dynamic viscosity, thermal conductivity and specific heat of a hybrid nanofluid obtained experimentally via ANFIS-based model and modern optimization, J. Mol. Liq. 321 (2021) 114287.
[8] A. Zendehboudi, X. Li, Robust predictive models for estimating frost deposition on horizontal and parallel surfaces, Int. J. Refrig. 80 (2017) 225–237.
[9] A.T.C. Goh, Back-propagation neural networks for modeling complex systems, Artif. Intell. Eng. 9 (3) (1995) 143–151.
[10] S. Walczak, Artificial neural networks, Advanced Methodologies and Technologies in Artificial Intelligence, Computer Simulation, and Human-Computer Interaction, IGI Global, 2019, pp. 40–53.
[11] J.R. Jang, ANFIS: adaptive-network-based fuzzy inference system, IEEE Trans. Syst. Man Cybern. 23 (3) (1993) 665–685.
[12] M. Buragohain, C. Mahanta, A novel approach for ANFIS modelling based on full factorial design, Appl. Soft Comput. 8 (1) (2008) 609–625.
[13] L.-C. Ying, M.-C. Pan, Using adaptive network based fuzzy inference system to forecast regional electricity loads, Energy Convers. Manag. 49 (2) (2008) 205–211.
[14] K.M. Tao, A closer look at the radial basis function (RBF) networks. in: Proceedings of Twenty- Seventh Asilomar Conference on Signals, Systems and Computers, IEEE, 1993.
[15] A. Najah, et al., Application of artificial neural networks for water quality prediction, Neural Comput. Appl. 22 (1) (2013) 187–201.
[16] A. Zendehboudi, A. Tatar, Utilization of the RBF network to model the nucleate pool boiling heat transfer properties of refrigerant-oil mixtures with nanoparticles, J. Mol. Liq. 247 (2017) 304–312.
[17] J.A.K. Suykens, J. Vandewalle, Recurrent least squares support vector machines, IEEE Trans. Circuits Syst. I: Fundam. Theory Appl. 47 (7) (2000) 1109–1114.
[18] T. van Gestel, et al., Benchmarking least squares support vector machine classifiers, Mach. Learn. 54 (1) (2004) 5–32.
[19] J.A.K. Suykens, J. Vandewalle, Training multilayer perceptron classifiers based on a modified support vector method, IEEE Trans. Neural Netw. 10 (4) (1999) 907–911.
[20] M. Ramezanizadeh, et al., A review on the utilized machine learning approaches for modeling the dynamic viscosity of nanofluids, Renew. Sustain. Energy Rev. 114 (2019) 109345.
[21] M.A. Ahmadi, M. Ebadi, Evolving smart approach for determination dew point pressure through condensate gas reservoirs, Fuel 117 (2014) 1074–1084.
[22] M.A. Ahmadi, et al., Evolving predictive model to determine condensate-to-gas ratio in retrograded condensate gas reservoirs, Fuel 124 (2014) 241–257.
[23] M. Ghodbane, et al., A numerical simulation of a linear Fresnel solar reflector directed to produce steam for the power plant, J. Clean. Prod. 231 (2019) 494–508.

[24] Z. Said, A. Mehmood, Standalone photovoltaic system assessment for major cities of United Arab Emirates based on simulated results, J. Clean. Prod. 142 (2017) 2722–2729.

[25] L.S. Sundar, et al., Energy, efficiency, economic impact, and heat transfer aspects of solar flat plate collector with Al2O3 nanofluids and wire coil with core rod inserts, Sustain. Energy Technol. Assess. 40 (2020) 100772.

[26] A.M. Hussein, et al., A review of forced convection heat transfer enhancement and hydrodynamic characteristics of a nanofluid, Renew. Sustain. Energy Rev. 29 (2014) 734–743.

[27] M. Raja, et al., Review on nanofluids characterization, heat transfer characteristics and applications, Renew. Sustain. Energy Rev. 64 (2016) 163–173.

[28] Z. Said, et al., Recent advances on nanofluids for low to medium temperature solar collectors: energy, exergy, economic analysis and environmental impact, Prog. Energy Combust. Sci. 84 (2021) 100898.

[29] M. Alhuyi Nazari, et al., Modeling thermal conductivity ratio of cuo/ethylene glycol nanofluid by using artificial neural network, Defect. Diffus. Forum 388 (2018) 39–43.

[30] L. Syam Sundar, et al., Heat transfer of rGO/CO3O4 hybrid nanomaterial-based nanofluids and twisted tape configurations in a tube, J. Therm. Sci. Eng. Appl. 13 (3) (2021) 031004.

[31] P.B. Maheshwary, C.C. Handa, K.R. Nemade, A comprehensive study of effect of concentration, particle size and particle shape on thermal conductivity of titania/water based nanofluid, Appl. Therm. Eng. 119 (2017) 79–88.

[32] M. Shamaeil, M. Firouzi, A. Fakhar, The effects of temperature and volume fraction on the thermal conductivity of functionalized DWCNTs/ethylene glycol nanofluid, J. Therm. Anal. Calorim. 126 (3) (2016) 1455–1462.

[33] J. Jeong, et al., Particle shape effect on the viscosity and thermal conductivity of ZnO nanofluids, Int. J. Refrig. 36 (8) (2013) 2233–2241.

[34] M. Gupta, et al., Up to date review on the synthesis and thermophysical properties of hybrid nanofluids, J. Clean. Prod. 190 (2018) 169–192.

[35] Z. Said, et al., Heat transfer, entropy generation, economic and environmental analyses of linear Fresnel reflector using novel rGO-Co3O4 hybrid nanofluids, Renew. Energy 165 (2021) 420–437.

[36] Y. Peng, et al., Develop optimal network topology of artificial neural network (AONN) to predict the hybrid nanofluids thermal conductivity according to the empirical data of Al_2O_3 – Cu nanoparticles dispersed in ethylene glycol, Phys. A: Stat. Mech. Appl. 549 (2020) 124015.

[37] A.K. Tiwari, et al., 3S (Sonication, surfactant, stability) impact on the viscosity of hybrid nanofluid with different base fluids: an experimental study, J. Mol. Liq. 329 (2021) 115455.

[38] Z. Said, et al., Stability, thermophysical and electrical properties of synthesized carbon nanofiber and reduced-graphene oxide-based nanofluids and their hybrid along with fuzzy modeling approach, Powder Technol. 364 (2020) 795–809.

[39] A.A. Hachicha, et al., A review study on the modeling of high-temperature solar thermal collector systems, Renew. Sustain. Energy Rev. 112 (2019) 280–298.

[40] M. Gupta, V. Singh, Z. Said, Heat transfer analysis using zinc Ferrite/water (hybrid) nanofluids in a circular tube: an experimental investigation and development of new correlations for thermophysical and heat transfer properties, Sustain. Energy Technol. Assess. 39 (2020) 100720.

[41] N.K. Cakmak, et al., Preparation, characterization, stability, and thermal conductivity of rGO-Fe_3O_4-TiO_2 hybrid nanofluid: an experimental study, Powder Technol. 372 (2020) 235–245.

[42] A. Kumar, A.K. Tiwari, Z. Said, A comprehensive review analysis on advances of evacuated tube solar collector using nanofluids and PCM, Sustain. Energy Technol. Assess. 47 (2021) 101417.

[43] Z. Said, et al., Recent advances on the fundamental physical phenomena behind stability, dynamic motion, thermophysical properties, heat transport, applications, and challenges of nanofluids, Phys. Rep. (2021).

[44] Z. Said, et al., Heat transfer enhancement and life cycle analysis of a Shell-and-Tube Heat Exchanger using stable CuO/water nanofluid, Sustain. Energy Technol. Assess. 31 (2019) 306–317.

[45] Q. Wang, et al., Prediction of heat transfer rates for shell-and-tube heat exchangers by artificial neural networks approach, J. Therm. Sci. 15 (3) (2006) 257–262.

[46] E.M.S. El-Said, M. Abd Elaziz, A.H. Elsheikh, Machine learning algorithms for improving the prediction of air injection effect on the thermohydraulic performance of shell and tube heat exchanger, Appl. Therm. Eng. 185 (2021) 116471.

[47] K. Sharifi, et al., A good contribution of computational fluid dynamics (CFD) and GA-ANN methods to find the best type of helical wire inserted tube in heat exchangers, Int. J. Therm. Sci. 154 (2020) 106398.

[48] J.D. Moya-Rico, et al., Characterization of a triple concentric-tube heat exchanger with corrugated tubes using Artificial Neural Networks (ANN), Appl. Therm. Eng. 147 (2019) 1036–1046.

[49] E. Rezaei, et al., Modeling the free convection heat transfer in a partitioned cavity using ANFIS, Int. Commun. Heat. Mass. Transf. 39 (3) (2012) 470–475.

[50] T.A. Tahseen, M. Ishak, M.M. Rahman, Performance predictions of laminar heat transfer and pressure drop in an in-line flat tube bundle using an adaptive neuro-fuzzy inference system (ANFIS) model, Int. Commun. Heat. Mass. Transf. 50 (2014) 85–97.

[51] S. Safarzadeh, et al., Energy and entropy generation analyses of a nanofluid-based helically coiled pipe under a constant magnetic field using smooth and micro-fin pipes: experimental study and prediction via ANFIS model, Int. Commun. Heat. Mass. Transf. 126 (2021) 105405.

[52] G. Sadeghi, et al., Energy and exergy evaluation of the evacuated tube solar collector using Cu2O/water nanofluid utilizing ANN methods, Sustain. Energy Technol. Assess. 37 (2020) 100578.

[53] O.N. Mensour, et al., Modeling of solar energy potential in Souss-Massa area-Morocco, using intelligence Artificial Neural Networks (ANNs), Energy Proc. 139 (2017) 778–784.

[54] S. Fischer, P. Frey, H. Drück, A comparison between state-of-the-art and neural network modelling of solar collectors, Sol. Energy 86 (11) (2012) 3268–3277.

[55] T.E. Boukelia, O. Arslan, M.S. Mecibah, Potential assessment of a parabolic trough solar thermal power plant considering hourly analysis: ANN-based approach, Renew. Energy 105 (2017) 324–333.

[56] S.A. Kalogirou, Prediction of flat-plate collector performance parameters using artificial neural networks, Sol. Energy 80 (3) (2006) 248–259.

[57] F. Nasirzadehroshenin, et al., Investigation of exergy of double-pipe heat exchanger using synthesized hybrid nanofluid developed by modeling, Int. J. Thermophys. 40 (9) (2019) 87.

[58] M. Saeedan, et al., CFD Investigation and neutral network modeling of heat transfer and pressure drop of nanofluids in double pipe helically baffled heat exchanger with a 3-D fined tube, Appl. Therm. Eng. 100 (2016) 721–729.

[59] H. Maddah, et al., Experimental and numerical study of nanofluid in heat exchanger fitted by modified twisted tape: exergy analysis and ANN prediction model, Heat. Mass Transf. 53 (4) (2017) 1413–1423.

[60] A.H.A. Al-Waeli, et al., Mathematical and neural network modeling for predicting and analyzing of nanofluid-nano PCM photovoltaic thermal systems performance, Renew. Energy 145 (2020) 963–980.

[61] H. Kalani, M. Sardarabadi, M. Passandideh-Fard, Using artificial neural network models and particle swarm optimization for manner prediction of a photovoltaic thermal nanofluid based collector, Appl. Therm. Eng. 113 (2017) 1170–1177.

[62] M. Hemmat Esfe, Designing a neural network for predicting the heat transfer and pressure drop characteristics of Ag/water nanofluids in a heat exchanger, Appl. Therm. Eng. 126 (2017) 559–565.

[63] A. Baghban, et al., Sensitivity analysis and application of machine learning methods to predict the heat transfer performance of CNT/water nanofluid flows through coils, Int. J. Heat. Mass. Transf. 128 (2019) 825–835.

[64] L.S. Sundar, et al., Properties, heat transfer, energy efficiency and environmental emissions analysis of flat plate solar collector using nanodiamond nanofluids, Diam. Relat. Mater. 110 (2020) 108115.

[65] A.A. Hachicha, et al., On the thermal and thermodynamic analysis of parabolic trough collector technology using industrial-grade MWCNT based nanofluid, Renew. Energy 161 (2020) 1303–1317.

[66] L.S. Sundar, et al., Combination of Co3O4 deposited rGO hybrid nanofluids and longitudinal strip inserts: thermal properties, heat transfer, friction factor, and thermal performance evaluations, Therm. Sci. Eng. Prog. 20 (2020) 100695.

[67] D. Xiaohong, et al., Statistical estimation the thermal conductivity of MWCNTs-SiO2/Water-EG nanofluid using the ridge regression method, Phys. A: Stat. Mech. Appl. 537 (2020) 122782.

[68] A. Vatani, P.L. Woodfield, D.V. Dao, A survey of practical equations for prediction of effective thermal conductivity of spherical-particle nanofluids, J. Mol. Liq. 211 (2015) 712–733.

[69] M. Levin, M. Miller, Maxwell a treatise on electricity and magnetism, Uspekhi Fizicheskikh Nauk. 135 (3) (1981) 425–440.

[70] D. Bruggeman, Calculation of various physical constants of heterogenous substances II. Dielectricity constants and conductivity of non regular multi crystal systems, Annal. Phys. 25 (7) (1936) 645–672.

[71] K. Leong, C. Yang, S. Murshed, A model for the thermal conductivity of nanofluids–the effect of interfacial layer, J. Nanopart. Res. 8 (2) (2006) 245–254.

[72] C.H. Chon, et al., Empirical correlation finding the role of temperature and particle size for nanofluid (Al 2 O 3) thermal conductivity enhancement, Appl. Phys. Lett. 87 (15) (2005) 153107.

[73] A.K. Tiwari, et al., Experimental comparison of specific heat capacity of three different metal oxides with MWCNT/water-based hybrid nanofluids: proposing a new correlation, Appl. Nanosci. (2020) 1–11.

[74] A. Akhgar, D. Toghraie, An experimental study on the stability and thermal conductivity of water-ethylene glycol/TiO2-MWCNTs hybrid nanofluid: developing a new correlation, Powder Technol. 338 (2018) 806–818.

[75] G.A. Longo, et al., Application of Artificial Neural Network (ANN) for the prediction of thermal conductivity of oxide–water nanofluids, Nano Energy 1 (2) (2012) 290–296.

[76] M. Hojjat, et al., Thermal conductivity of non-Newtonian nanofluids: experimental data and modeling using neural network, Int. J. Heat. Mass Transf. 54 (5) (2011) 1017–1023.

[77] S. Kannaiyan, et al., Modeling of thermal conductivity and density of alumina/silica in water hybrid nanocolloid by the application of artificial neural networks, Chin. J. Chem. Eng. 27 (3) (2019) 726–736.

[78] E. Ahmadloo, S. Azizi, Prediction of thermal conductivity of various nanofluids using artificial neural network, Int. Commun. Heat. Mass Transf. 74 (2016) 69–75.

[79] M. Afrand, et al., Predicting the viscosity of multi-walled carbon nanotubes/water nanofluid by developing an optimal artificial neural network based on experimental data, Int. Commun. Heat. Mass Transf. 77 (2016) 49–53.

[80] M. Hemmat Esfe, et al., Designing an artificial neural network to predict dynamic viscosity of aqueous nanofluid of TiO2 using experimental data, Int. Commun. Heat. Mass Transf. 75 (2016) 192–196.

[81] M. Mehrabi, M. Sharifpur, J.P. Meyer, Viscosity of nanofluids based on an artificial intelligence model, Int. Commun. Heat. Mass Transf. 43 (2013) 16–21.

[82] M. Hemmat Esfe, et al., Designing an artificial neural network to predict thermal conductivity and dynamic viscosity of ferromagnetic nanofluid, Int. Commun. Heat. Mass Transf. 68 (2015) 50–57.

[83] M.A. Hassan, D. Banerjee, A soft computing approach for estimating the specific heat capacity of molten salt-based nanofluids, J. Mol. Liq. 281 (2019) 365–375.

[84] A.A.A.A. Alrashed, et al., Effects on thermophysical properties of carbon based nanofluids: experimental data, modelling using regression, ANFIS and ANN, Int. J. Heat. Mass Transf. 125 (2018) 920–932.

[85] M.H. Esfe, et al., Modeling and prediction of rheological behavior of Al2O3-MWCNT/5W50 hybrid nano-lubricant by artificial neural network using experimental data, Phys. A Stat. Mech. Appl. 510 (2018) 625–634.

[86] M. Hemmat Esfe, S. Esfandeh, M. Rejvani, Modeling of thermal conductivity of MWCNT-SiO2 (30:70%)/EG hybrid nanofluid, sensitivity analyzing and cost performance for industrial applications, J. Therm. Anal. Calorim. 131 (2) (2018) 1437–1447.

[87] I. Wole-Osho, et al., An intelligent approach to predicting the effect of nanoparticle mixture ratio, concentration and temperature on thermal conductivity of hybrid nanofluids, J. Therm. Anal. Calorim. 144 (3) (2021) 671–688.

[88] M. Ma, et al., Statistical image analysis of uniformity of hybrid nanofluids and prediction models of thermophysical parameters based on artificial neural network (ANN), Powder Technol. 362 (2020) 257–266.

[89] I.M. Alarifi, et al., Feasibility of ANFIS-PSO and ANFIS-GA models in predicting thermophysical properties of Al2O3-MWCNT/Oil hybrid nanofluid, Materials 12 (21) (2019) 3628.

[90] A. Baghban, et al., Developing an ANFIS-based swarm concept model for estimating the relative viscosity of nanofluids, Eng. Appl. Comput. Fluid Mech. 13 (1) (2019) 26–39.

[91] M. Bahrami, et al., Develop 24 dissimilar ANNs by suitable architectures & training algorithms via sensitivity analysis to better statistical presentation: measure MSEs between targets & ANN for Fe–CuO/Eg–wWater nanofluid, Phys. A: Stat. Mech. Appl. 519 (2019) 159–168.

[92] A. Shahsavar, et al., Robust weighted least squares support vector regression algorithm to estimate the nanofluid thermal properties of water/graphene oxide–silicon carbide mixture, Phys. A Stat. Mech. Appl. 525 (2019) 1418–1428.

Chapter 8

Heat transfer enhancement with nanofluids in automotive

Xihui Wang[1], Kuo Huang[2], Yan Yuying[2] and Hailin Cen[1]
[1]*School of Mechanical and Automotive Engineering, South China University of Technology, GuangZhou, P.R. China,* [2]*Faculty of Engineering, The University of Nottingham, Nottingham, Nottinghamshire, United Kingdom*

8.1 Historical background

In the 1990s, with the development of nanotechnology, Professor Choi [1] proposed the concept of "nanofluids." As a new type of heat transfer medium, nanofluids are composed of metal or nonmetal nanoparticles that when added to traditional heat transfer fluid show better performance in heat transfer than the base fluid. Nanoparticles have the characteristics of being ultrafine and easy to flow. Compared with millimeter or micron-sized solid particles, nanoparticles are less likely to precipitate and cause bad results such as equipment wear and blockage. At present, the performance of heat exchangers is constantly improving, although the traditional heat transfer medium can not meet the requirements in most cases. Due to higher thermal conductivity than pure liquid and better stability than millimeter and micron-sized particle suspensions, nanofluids have been one of the candidates to solve this problem and have attracted extensive attention.

8.1.1 Applications of computational fluid dynamics in nanofluids studies

Computational fluid dynamics (CFD) is a technology that combines computer science, fluid dynamics, and numerical mathematics. Using the algorithm and fast operation ability of a computer to solve the governing equations, CFD can help us to get more expected information of fluid, thus allowing a comprehensive understanding of the fluid flow process and its phenomenon. For theoretical methods, CFD is easier to apply because it has less hypothetical limitations and a wider range of applications. For experimental methods, using CFD is short in time and low in cost, and there is no

Advances in Nanofluid Heat Transfer. DOI: https://doi.org/10.1016/B978-0-323-88656-7.00016-7

need to worry about the inevitable interference caused by measuring equipment to the real flow field. As a branch of fluid dynamics, CFD gradually became an independent discipline in the 1960s with the birth and development of computers. By 1995, when nanofluid technology was proposed for the first time, the study on the application of CFD in engineering and its application effect had achieved fruitful results and won the recognition of researchers. Meanwhile, currently widely used CFD software such as Fluent, CFX, PHOENICS, etc. appeared at that time, making CFD more convenient to be applied. Therefore, CFD has always been a common method and played an important part in the field of nanofluid research.

CFD steps include model building, domain discretization, governing equations discretization, algebraic equations solving, and postprocessing.

It is necessary to analyze the flow properties of the research object, to simplify the physical process of the flow phenomenon to build a physical model, and then convert it into a mathematical model with mathematical expressions (three conservation equations) and the definite conditions. The model will affect the computational amount and accuracy of CFD. Therefore, it is important to establish appropriate models for specific research problems. In general, according to the number of phases, the physical model of nanofluids can be divided into single-phase and two-phase models.

It can be known by definition that a nanofluid is a kind of two phase fluid, but some appropriate assumptions can be made to simplify it to a single-phase fluid. This simplification can shorten the time required for computer solution, but undoubtedly the precision is reduced, so it is mostly used in the research with low accuracy. The homogenization model is the simplest physical model of nanofluids. In this model, the slip between the nanoparticles and the base fluid is ignored, nanoparticles are considered to be ultrafine and uniformly distributed in the base fluid, and this two-phase fluid is considered to be in a state of hydrodynamic and thermal equilibrium, that is, the force and heat exchange between the solid–liquid phases is negligible, thus simplifying it to a single-phase continuum. In 2000, Yimin Xuan and Wilfried Roetzel [2] considered the influence of random movements of nanoparticles, and improved the homogenization model to the proposed thermal diffusion model, which had better consistency with experimental data [3–6]. In 2006, Buongiorno [7] believed that the Brownian diffusion and thermophoresis are the most important factors affecting slip in nanofluids, and established a new nonhomogeneous equilibrium model.

The two-phase model considers the relative motion and heat transfer between different phases, which is more in line with reality. The commonly used two-phase models include the mixture model, Euler model, and Euler–Lagrange model in nanofluids. The mixture model introduces the concept of drift velocity to allow the phases to move at different speeds. Most numerical studies of nanofluids have used the mixture model because of the higher precision and lower computational load in contrast with many other models [8]. The Eulerian model, often referred to as the two-fluid model, is the most complex two-phase flow model. In this model, both continuous and discrete phases are seen as a continuum, and the motion of the two phases is based on Eulerian coordinates. Also, the momentum equation and continuity equation are established for each phase, which can be solved through the coupling of the interaction force and interphase momentum exchange coefficient. Compared with the other two methods, the Eulerian model is less accurate in approaching nanofluids [8]. The Euler–Lagrangian model is also known as the discrete phase model. The fluid phase is regard as a continuous medium and solved by the N-S equation, while the discrete phase is modeled by calculating the motion of particles in the flow field. And there are exchanges of momentum, mass, and energy between fluid and particles. With the advantage of high precision, the Euler–Lagrangian model has been used a lot. However, using this method makes the computational cost much larger than using many other methods, because the track of each particle is calculated separately.

In order for the computer to numerically solve the established mathematical model, the partial differential equations must be discretized into algebraic equations. The following are three methods commonly used to study nanofluids: finite difference method (FDM), finite volume method (FVM), and finite element method (FEM).

In FDM, a finite number of grid nodes are used to replace the continuous solution domain, and the differential terms in the governing equations are discretized by the difference quotient of the values of the nodes. With an intuitive mathematical concept and simple expression, this method is an early and mature numerical method to solve flow and heat transfer problems. When solving the uncomplicated flow field, FDM has a high accuracy.

The FVM divides the solution region into successive control volumes, which are represented by the parameters of the central position of the control volume. The parameters on the surface of the control body are calculated by interpolation of the central parameters of two adjacent control volumes. The advantage of FVM is that the integral terms have clear physical meaning and complex grids can be dealt with easily. The disadvantage is that it is difficult to improve the precision.

The FEM divides the solution area into finite elements and finally solves the nonlinear algebraic equations obtained by using the weighted integrals of the governing equations on these elements. This method is flexible for dealing with complex structures but difficult for intricate equations.

The lattice Boltzmann method (LBM), derived from lattice gas automata in the 1980s [9,10], is a relatively new fluid simulation technique. The traditional CFD method is based on the continuum hypothesis and describes the fluid motion by the N-S equation, while the LBM method simulates fluid by using the discrete Boltzmann equation. Theoretically, the N-S equation is an approximation of the Boltzmann equation in a specific time and space scale. It means that LBM has a wider application range. In general, LBM is a mesoscale technique between macroscale and microscale. In this method, the macroscale movement is described by solving the velocity distribution function of the particles in the space. And it is convenient for LBM to deal with complex geometry and multicomponent fluids.

8.1.2 Applications of nanofluid in automotive system

Nanofluids have great potential in automotive applications due to their special properties. The function of the automobile engine cooling system is to keep the engine in the appropriate temperature range under all operating conditions. The overheating of the engine will lead to the deterioration of the working process, the strength of parts, oil deterioration, and wear of moving parts. Hence, the performance and life span of the engine are directly affected by the heat transfer efficiency of the cooling system. The thermal property of the coolant is one of the main factors affecting the heat exchange capacity of the engine cooling system. According to the literature [11], the idea of applying nanofluids as a coolant to the engine cooling system was first proposed by Choi et al. [12] in 2001. Their experiment found that the thermal conductivity of their homemade nanofluids was much higher than expected, followed by the suggestion that nanofluids could enhance the thermal performance of vehicles, and a large number of subsequent studies [13–19] have come to similar conclusions. In addition to ensuring better engine cooling, an increase in coolant heat transfer rate means that when maintaining the same level of cooling effect, the engine using nanocoolant stores less coolant and the overall size of the cooling system can be reduced (e.g., the reduction of the windward area of the radiator), thus realizing engine lightweight. It has also been found that the addition of nanoparticles improves the viscosity of the fluid and causes some degree of pressure drop [18–20]. As a result, the pump has to provide greater power to make up for the pressure drop. Nonetheless, with the same heat transfer effect, the nanofluids bring less pressure drop than the pure fluid, that is, the size of the pump can be reduced [20].

Nanofluids are also used as refrigerants in automobile air conditioners due to their excellent heat transfer ability. Previous studies [21–23] have shown that the thermal conductivity of nanorefrigerants is significantly improved compared with pure refrigerants. What's more, the addition of nanoparticles can improve the lubrication characteristics of frozen oil and reduce the input power of the compressor. With the improvement of people's environmental awareness, electric vehicles (EV) are becoming more and more popular in many countries because of the harmful emissions from traditional fuel vehicles. However, the short battery range of EV has always been criticized, limiting the spread of EV. The automobile air conditioning has a great impact on the increase of energy consumption of the vehicle. For limited battery energy, the application of nanorefrigerant can improve the efficiency of the air conditioning system and effectively extend the range of EVs.

Nanoparticles added into lubricating oil as additives can significantly improve the properties of lubricating oil, realizing antiwear and antifriction effects [24–26]. In the 1980s, tribological researchers were using nanoparticles in lubricants before the concept of nanofluids was proposed. In contrast to base oils, nanolubricants have better lubricating friction characteristics and stronger bearing capacity [27,28]. What's more, the excellent heat transfer characteristic of nanofluids can improve the heat dissipation of lubricating oils [13]. Therefore, nanolubricants are very suitable for lubrication and cooling of friction at high temperature and heavy load. The application of nanolubricants in the lubrication system of automobiles can not only enhance the lubrication effect for the reduction of mechanical loss, but also improve the heat dissipation of the powertrain, thus promoting the automotive power performance and economy [29].

The addition of nanoparticles into liquid fuel can significantly improve engine emissions and increase engine power and fuel economy [30,31]. The energy density of commonly used nanoparticles (copper, aluminum, cerium, copper oxide, cerium oxide, etc.) is much higher than that of diesel, gasoline, and other commonly used fuels in engines. It means that the addition of nanoparticles can increase the volume energy density of fuel. In addition, nanofuels can make combustion conditions better [32,33], such as shortening ignition delay, increasing combustion rate, and promoting complete combustion. As a result, the power performance and the economic performance of the engine can be enhanced. Many studies [34–37] have found that the application of nanofuels can significantly improve the exhaust pollutant emissions. What's more, some sort of nanoparticle additives can help achieve the regeneration of diesel particulate filters (DPF) [38,39]. For example, CeO_2 can reduce the ignition temperature of the PM in DPF, causing the PM to be burned into carbon dioxide at a low temperature, whereby DPF can maintain normal operation without being blocked.

In addition to the above aspects, nanofluids are also used to enhance the cooling efficiency of power cells and reduce the freezing point, and magnetic fluids composed of magnetic nanoparticles, base fluids, and surfactants are applied in vehicle dampers and brakes. In brief, the research of nanofluids applied in various aspects of automobiles has achieved good results, which is expected to bring further improvement to the performance of automobiles. However, as a new technology, we are still far from the practical application and popularization of nanofluids at present. For example, the sustainability, stability, and environmental impact of nanofluids in engineering equipment are still to be investigated, relevant standards have not been formulated, and the production cost is high, etc. Therefore, there are still many problems for researchers to overcome in nanofluid research.

8.2 Physical properties

Among various thermal properties, thermal conductivity and viscosity are of the greatest concerns. The thermal conductivity determines the heat transfer ability of the nanofluids, and the viscosity affects the pressure drop. This is followed by the discussion on these two features of nanofluids.

8.2.1 Thermal conductivity of nanofluids

The excellent heat transfer ability of nanofluids is expected to bring further optimization to the engineering thermal system, thus it has always been the focus of researchers. A sea of studies have been done around the thermal conductivity of nanofluids, but it is still a difficult problem to accurately reckon it in theory. Classic prediction models (Table 8.1) can accurately estimate the thermal conductivity of solid–liquid mixture flow when solid particles are micron-sized or larger, but often underestimates the thermal conductivity of nanofluids. For the sake of explaining this deviation, some hypotheses about the heat transfer mechanism were put forward, including Brownian motion, nanolayers, and clusters of particles. However, the reasons for the enhancement of the thermal conductivity of nanofluids are not clear at present. Therefore, researchers took different factors into consideration when establishing the thermal conductivity model of nanofluids, leading to various model forms (Table 8.2). Some correlations of thermal conductivity, based on experiments, are presented in Table 8.3.

8.2.2 Viscosity of nanofluids

Viscosity describes the internal resistance of a liquid to flow. The fluid with greater viscosity encounters greater resistance in the process of flow, resulting in a greater pressure drop. For the fluid to have the same pressure, the pump must provide more power to the fluid with greater viscosity. Therefore, heat transfer media with low viscosity are desired in thermal systems. However, studies have shown that the addition of nanoparticles not only increases the thermal conductivity of the fluid but also its viscosity. That is because the addition of nanoparticles into the base fluid enhances fluid friction and flow resistance. Table 8.4 lists some of the predicted models for the viscosity of nanofluids.

TABLE 8.1 Classic thermal conductivity models.

Researchers	Models	Remake
Maxwell [40]	$k_{nf} = k_{bf}\left[\frac{k_p + 2k_{bf} + 2\phi(k_p - k_{bf})}{k_p + 2k_{bf} - \phi(k_p - k_{bf})}\right]$	This model is suitable for spherical particles at low concentrations
Bruggeman [41]	$\frac{k_{nf}}{k_{bf}} = \frac{(3\phi - 1)\frac{k_p}{k_{bf}} + \{3(1-\phi) - 1\} + \sqrt{\Delta}}{4}$ $\Delta = \left[(3\phi - 1)\frac{k_p}{k_{bf}} + \{3(1-\phi) - 1\}\right]^2 + 8\frac{k_p}{k_{bf}}$	This model is suitable for spherical particles of various concentrations

TABLE 8.2 Modified theoretical models for thermal conductivity of nanofluids.

Researchers	Models	Remake
Yu and Choi [42]	$k_{nf} = k_{nf}\left[\frac{k_p + 2k_{bf} + 2\phi(k_p - k_{bf})(1+\beta)^3}{k_p + 2k_{bf} - \phi(k_p - k_{bf})(1+\beta)^3}\right]$	In this model, β is the ratio of the thickness of the nanolayer to radius of nanoparticles, which is normally calculated as 0.1. This model can be used when volume fraction is less than 1%
Xuan et al. [43]	$k_{nf} = k_{bf}\left[\frac{k_p + 2k_{bf} + 2(k_p - k_{bf})\phi}{k_p + 2k_{bf} - 2(k_p - k_{bf})\phi}\right] + \frac{1}{2}\rho_p C_{p,p}\phi\sqrt{\frac{\kappa_B T}{3\pi\mu_{bf} R_c}}$	Based on Maxwell model, the model considers the effect of Brownian motion and aggregation of nanoparticles. In this model, R_c is the average radius of the clusters, and κ_B is the Boltzmann constant
Koo and Kleinstreuer [44]	$k_{nf} = k_{bf}\left[\frac{k_p + 2k_{bf} + 2(k_p - k_{bf})\phi}{k_p + 2k_{bf} - 2(k_p - k_{bf})\phi}\right] + 5\times 10^4\theta\rho_{bf} C_{p,bf} f(T,\phi)\phi\sqrt{\frac{\kappa_B T}{\rho_p d_p}}$	The model comprehensively considering parameters such as particle size, volume fraction, temperature and base liquid properties and Brownian motion. In the formula, d_p is nanoparticle diameter, and θ and f are two empirical functions
Shukla et al. [45]	$\frac{k_{nf}}{k_{bf}} = \pi\left(\frac{6}{\pi}\right)^{\frac{1}{3}}\phi^{\frac{4}{3}}\left[\frac{1+0.5\left(\frac{6\phi}{\pi}\right)^{\frac{1}{3}}}{2}\left(\frac{k_{bf}}{k_p}\right) + \frac{\psi}{Nu}\right] + (1-\phi)$	This model takes into account convective heat transfer caused by the Brownian motion of the particles. In this model, ψ is particle sphericity
Zhang [46]	$\frac{k_{nf}}{k_{bf}} = \frac{1 + 2\phi\int_0^{\infty}\frac{(k_{cl}(r) - k_{bf})n(r)(1+\beta)^3}{k_{cl}(r) + 2k_{bf}}dr}{1 - \phi\int_0^{\infty}\frac{(k_{cl}(r) - k_{bf})n(r)(1+\beta)^3}{k_{cl}(r) + 2k_{bf}}dr} k_{cl} = (3\phi - 1)k_p + [3(1-\phi) - 1]k_{bf} + \sqrt{\Delta}$ $\Delta = (3\phi - 1)^2 k_p^2 + [3(1-\phi) - 1]^2 k_{bf}^2 + 2[2 + 9\phi(1-\phi)]k_p k_{bf}$	This model considers the influence of particle clusters distribution and the thickness of the interface between base fluid and nanofluids. And k_{cl} is the thermal conductivity of the clusters

TABLE 8.3 Experimental based models for the thermal conductivity of nanofluids.

Researchers	Models	Remark
Chiam et al. [47]	$\frac{k_{nf}}{k_{bf}} = 0.9683\left(1+\frac{\phi}{100}\right)^{11.13}\left(1+\frac{T}{70}\right)^{0.1676}(0.01+BR)^{0.00111}\left(\frac{d_p}{36}\right)^{0.0572}$	In this model, *BR* is of water and ethylene glycol. The model is suitable for concentration range of 0%–1.5%, temperature range of 20°C–70°C, particle diameter range of 13–36 nm and *BR* range of 0.4–0.6
Patel et al. [48]	$k_{nf} = k_{bf}\left(1 + 0.135 \times \left(\frac{k_p}{k_{bf}}\right)^{0.273} \times \varepsilon^{0.467} \times \left(\frac{T}{20}\right)^{0.547} \times \left(\frac{100}{d_p}\right)^{0.234}\right)$	This model is appropriate for the spherical diameter of 10–150 nm, volume concentration of 0.1%–0.3% and temperature of 20°C–50°C
Chon et al. [49]	$\frac{k_{nf}}{k_{bf}} = 1 + 64.7\phi^{0.7460}\left(\frac{d_f}{d_p}\right)^{0.3690}\left(\frac{k_p}{k_{bf}}\right)^{0.7476}\mathrm{Pr}_{bf}^{0.9955}\mathrm{Re}_p^{1.2321}$	The model is valid of Al-water nanofluids with nanoparticle sizes range from 11 to 150 nm at temperature range from 21°C to 71°C

TABLE 8.4 Experimental based models for viscosity of nanofluids.

Researchers	Models	Remark
Einstein [50]	$\mu_{nf} = \mu_{bf} \frac{1+0.5\phi}{(1-\phi)^2}$ When $\phi < < 1,; \mu_{nf} = \mu_{bf}(1 + 2.5\phi)$	This model is suitable for Newton solid–liquid two-phase suspension with spherical particles
Brinkman [51]	$\mu_{\mathrm{nf}} = \mu_{bf} \frac{1}{(1-\phi)^{2.5}}$	This model can be used when the volume fraction is less than 5%
Thomas [52]	$\mu_{nf} = \mu_{bf}\left(1 + 2.5\phi + 10.05\phi^2 + 0.00273e^{16.6\phi}\right)$	This viscosity model is suited for spherical particles with high particle volume fraction
Rea et al. [53]	$\mu_{nf}(\phi, T) = \mu_{bf}(T)\exp[4.91\phi(0.2092 - \phi)]$	This correlation is valid for Al_2O_3–water nanofluids with volume fractions less than 6% and temperatures range from 20°C to 80°C
Maiga et al. [54]	$\mu_{nf} = (123\phi^2 + 7.3\phi + 1)\mu_{bf}$ $\mu_{nf} = (306\phi^2 - 0.19\phi + 1)\mu_{bf}$	The former can be used for the viscosity of Al_2O_3–water and the latter is for Al_2O_3–EG nanofluids
Chiam et al. [47]	$\frac{\mu_{nf}}{\mu_{bf}} = \left(1 + \frac{\phi}{100}\right)^{32} \left(\frac{T}{70}\right)^{-0.001} (0.1 + BR)^{0.08}$	This model is based on the experimental data and considering the temperature, particle concentration and base liquid ratio and other parameters to conclude the viscosity calculation formula for Al_2O_3 nanofluid. In this model, *BR* is of water and ethylene glycol

8.3 The fundamental relation for computational fluid dynamics model

With the rapid development of computer science, CFD technology is very convenient to solve complex flow field or thermal systems problems, and the use of CFD is on the rise. This section introduced the CFD model of nanofluids, which can be classified into macroscopic models and nonmacroscopic models (including mesoscopic and microscopic models). Macroscopic models can be further divided into single-phase and two-phase models. For nonmacroscopic models, the LBM is shown; this is an increasingly widely used mesoscopic model.

8.3.1 Macroscopic models

8.3.1.1 Single-phase models

In single-phase models, a nanofluid can be simulated as an equivalent single-phase under certain assumptions. This assumption greatly reduces the cost of calculation but also reduces the accuracy. It is worth noting that if the model parameters are obtained from experiments, the accuracy of the model can be improved to some extent. The single-phase model is mainly classified into three approaches, including the homogenization model, thermal dispersion model, and Buongiorno model.

8.3.1.2 Two-phase models

In the two-phase models, the solid and liquid phases have different motion states and temperatures, which means there is relative slip and heat exchange between the solid and liquid phase. Considering that the nanofluids are actually two-phase flow, using the two-phase models for simulation is closer to practice, but the models are more complex and take a longer time to solve. The two-phase models are classified as volume of fluid, mixture, Eulerian models, and Lagrangian model. In the first three models, both two phases are treated as an interpenetrating continuum and are described by the Eulerian method. However, the solid phase is considered as a discrete phase and described by the Lagrange method in the Eulerian–Lagrange model, while the liquid phase is still described by the Eulerian method.

8.3.1.3 Control equation

Fluid flow should follow the laws of physics, and its governing equations include continuity equation, momentum equation, and energy equation, which correspond to the conservation law of mass, momentum, and energy in essence.

Continuity equation:

$$\nabla\cdot(\rho_{nf}\boldsymbol{u}) = 0 \tag{8.1}$$

Momentum equation:

$$\boldsymbol{F} - \nabla p + \nabla\cdot[\mu_{nf}(\nabla\boldsymbol{u} + \nabla\boldsymbol{u}^{\mathrm{T}})] = \rho_{nf}(\boldsymbol{u}\cdot\nabla)\boldsymbol{u} \tag{8.2}$$

Energy equation:

$$\nabla\cdot(\rho_{nf}c_{p,nf}\boldsymbol{u}T) = \nabla\cdot(k_{nf}\nabla T) + \Phi \tag{8.3}$$

where ρ_{nf}, $c_{p,nf}$, and k_{nf} are the density, specific heat capacity, and thermal conductivity of the nanofluid; $\boldsymbol{u}$ is the flow velocity; $\boldsymbol{F}$ is the external force; Φ is the source term.

8.3.2 Lattice Boltzmann method

As the study develops in depth, the conventional macro model has been unable to reveal the internal laws of the flow and energy transfer of nanofluids. Researchers began to further study the nanofluids at the mesoscopic and microscopic levels. The simulation in nanofluids is gradually transformed from the macroscopic models to the multiscale models. The LBM is one of the main mesoscopic methods used to study nanofluids. Based on molecular dynamics, this method adopts continuous method at the micro level while discrete method at the macro level, thus belonging to the mesoscopic simulation method. Compared with other models, LBM has the advantages of good physical model foundation, clear boundary conditions, and superior parallel performance. Therefore, this method has been applied to many fields, including multiphase flow, porous media flow, micron/nanometer scale flow, non-Newtonian fluid, turbulence, particle suspension flow, chemical reaction flow, magnetic fluid, and so on. With the gradual improvement of LBM theory, this method has been paid more and more attention in the study of nanofluids.

The discrete equation of LBM is expressed as:

$$f_\alpha(\mathbf{r} + \mathbf{c}_\alpha\Delta t, t + \Delta t) = f_\alpha(\mathbf{r}, t) + \frac{1}{\tau}\left[f_\alpha^{eq}(\mathbf{r}, t) - f_\alpha(\mathbf{r}, t)\right] \tag{8.4}$$

where f_α is the particle distribution function, f_α^{eq} is its corresponding equilibrium distribution function, $\mathbf{c}_\alpha$ is the velocity of the particle, $\mathbf{r}$ is the lattice position, and τ is the dimensionless single relaxation time. The angle α is the direction along which the particle moves.

The second term on the right of the equation is the BGK term.

According to Maxwell distribution, the equilibrium distribution function of D-dimensional space is given as:

$$f^{eq} = \rho\frac{1}{(2\pi R_g T)^{D/2}}\exp\left[-\frac{(\mathbf{c}-\mathbf{u})^2}{2R_g T}\right] \tag{8.5}$$

The exponential function on the right side of Eq. (8.5) is expanded by Taylor expansion.

$$f_\alpha^{eq} = \rho w_\alpha \left[1 + b\frac{\mathbf{c}_\alpha \cdot \mathbf{u}}{c^2} + e\frac{u^2}{c^2} + h\frac{(\mathbf{c}_\alpha \cdot \mathbf{u})^2}{c^4}\right] \tag{8.6}$$

where w_α, b, e, h are the undetermined coefficient, w_α is the weight factor, and c is the lattice speed for particles moving from site to site, expressed as $c = \Delta x/\Delta t$.

Associated with the macro equation, the equilibrium distribution function, density, and momentum must meet the following correlations:

$$\sum_\alpha f_\alpha^{eq} = \rho \tag{8.7}$$

$$\sum_\alpha f_\alpha^{eq}\mathbf{c}_\alpha = \rho\mathbf{u} \tag{8.8}$$

$$\sum_\alpha f_\alpha^{eq}\mathbf{c}_{\alpha i}\mathbf{c}_{\alpha j} = \rho u_i u_j + p\Delta ij \tag{8.9}$$

And then the concrete equilibrium distribution function can be inferred by solving Eqs. (8.6–8.9) based on the selected discrete velocity model.

The DnQm (n stands for dimension and m stands for the number of lattice chains in the discrete velocity model) series proposed by Qian et al. [55] is the basic model of LBM. The most common models are lattice D2Q9, D3Q15, and D3Q19.

The D2Q9 model is suitable for solving fluid flow problems, since it has relatively more motion directions and the velocity at the center node is 0, as shown in Fig. 8.1.

The discrete velocity is expressed as:

$$\mathbf{c}_\alpha = \begin{cases} (0,0)\ \alpha = 0 \\ c\{\cos[(\alpha-1)\pi/2], \sin[(\alpha-1)\pi/2]\}\ \alpha = 1,2,3,4 \\ \sqrt{2}c\{\cos[(2\alpha-1)\pi/4], \sin[(2\alpha-1)\pi/4]\}\ \alpha = 5,6,7,8 \end{cases} \tag{8.10}$$

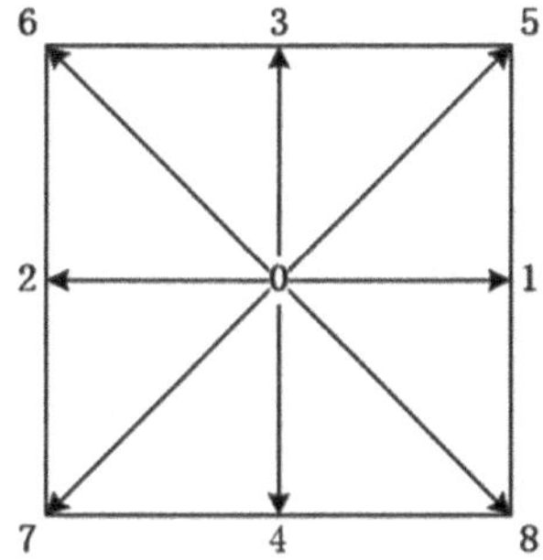

FIGURE 8.1 Lattice model for the D2Q9.

Substituting Eq. (8.10) into Eqs. (8.6–8.9), the values of the undetermined coefficients in Eq. (8.6) can be obtained as follows (the derivation steps in [10]):

$$b = 3, \quad e = -\frac{3}{2}, \quad h = \frac{9}{2},$$

$$w_0 = \frac{4}{9}, \quad w_1 = w_2 = w_3 = w_4 = \frac{1}{9}, \quad w_5 = w_6 = w_7 = w_8 = \frac{1}{36}$$

The D3Q15 and D3Q19 lattice is shown in Figs. 8.2 and 8.3, and their coefficients, velocity vectors, and weight factors are respectively given as:

– D3Q15

$$\mathbf{c} = \begin{Bmatrix} 0 & 1 & -1 & 0 & 0 & 0 & 0 & 1 & -1 & 1 & -1 & 1 & -1 & 1 & -1 \\ 0 & 0 & 0 & 1 & -1 & 0 & 0 & 1 & -1 & 1 & -1 & -1 & 1 & -1 & 1 \\ 0 & 0 & 0 & 0 & 0 & 1 & -1 & 1 & -1 & -1 & 1 & 1 & -1 & -1 & 1 \end{Bmatrix}$$

$$b = 3, \quad e = -\frac{3}{2}, \quad h = \frac{9}{2},$$

$$w_0 = \frac{2}{9}, \quad w_{1\sim 6} = \frac{1}{9}, \quad w_{7\sim 15} = \frac{1}{72}$$

– D3Q19

$$\mathbf{c} = \begin{Bmatrix} 0 & 1 & -1 & 0 & 0 & 0 & 0 & 1 & -1 & 1 & -1 & 1 & -1 & 1 & -1 & 0 & 0 & 0 & 0 \\ 0 & 0 & 0 & 1 & -1 & 0 & 0 & 1 & -1 & -1 & 1 & 0 & 0 & 0 & 0 & 1 & -1 & 1 & -1 \\ 0 & 0 & 0 & 0 & 0 & 1 & -1 & 0 & 0 & 0 & 0 & 1 & -1 & -1 & 1 & 1 & -1 & -1 & 1 \end{Bmatrix} b = 3, \quad e = -\frac{3}{2}, \quad h = \frac{9}{2},$$

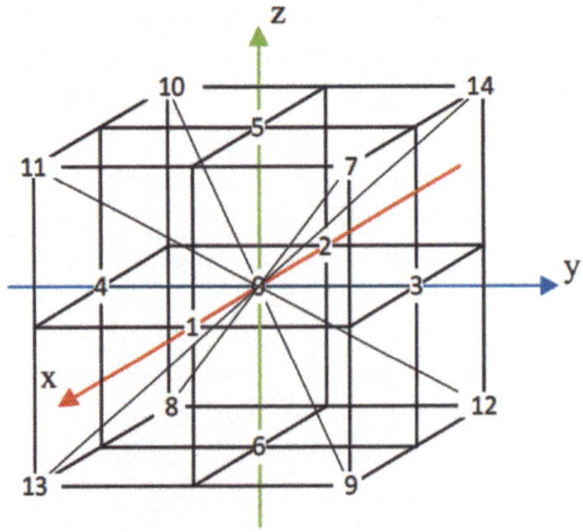

FIGURE 8.2 Lattice model for the D3Q15.

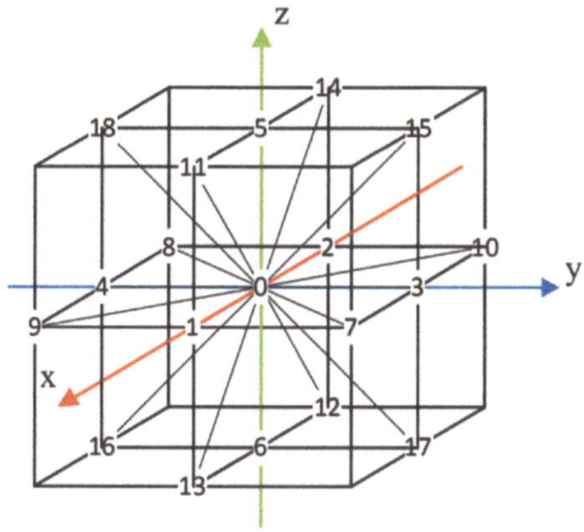

FIGURE 8.3 Lattice model for the D3Q19.

$$w_0 = \frac{1}{3}, \quad w_{1\sim6} = \frac{1}{18}, \quad w_{7\sim18} = \frac{1}{36}$$

8.4 Heat transfer enhancement with nanofluids in automotive

In this section, two simulation examples of nanofluids applied to vehicles are presented. One uses nanofluids in engine cooling systems [56] and the other in vehicle air conditioning systems [57].

8.4.1 Utilization in engine coolant

In this example, the cooling system performance of an in-line four-cylinder direct injection gasoline engine is studied after Fe_3O_4 is added to the coolant. Engine parameters are shown in Table 8.5.

8.4.1.1 Determination of nanofluid parameters

8.4.1.1.1 Thermal conductivity

Based on the regression analysis of a large number of experimental results, the Hrishikesh model can accurately simulate the thermal conductivity of metal and oxide nanofluids with a volume concentration of 1%–3% and a temperature of 20°C–50°C. However, the common operating temperature of engine coolant reaches 90°C, which exceeds the applicable temperature upper limit of the model. The simulated results are lower than those measured by experiments, and the influence of temperature on thermal

TABLE 8.5 Main parameters of the engine [56].

Name	Value
Type	In-line vertical, water-cooled, pressurized, four-stroke, direct injection
Cylinder diameter × stroke (mm)	82.5 × 84.2
Compression ratio	9.6:1
Piston total displacement (L)	1.799
Rated power (kW)	132.4
Rated speed (r/min)	5500
Maximum torque (N•m)	250
Speed at maximum torque (r/min)	1500–4500

conductivity is also underestimated. Therefore, Xu modified the Hrishikesh model for Fe_3O_4 nanofluids based on the experimental data, as follows:

$$k_{nf} = x(1) \times k_{bf} + x(2) \times (k_p/k_{bf})^{x(3)} \times \varphi^{b_2} \times (T/20)^{x(4)} \times (100/d_p)^{b_4} \quad (8.11)$$

where x(1), x(2), x(3), and x(4) are the matching parameters. Then, the thermal conductivity formulae for particle concentrations of 1%, 2%, and 5% are as follows:

$$\begin{aligned} k_{nf1} &= 0.754209 \times k_{bf} + 0.127133 \times (k_p/k_{bf})^{0.006995} \times \varphi^{0.467} \times (T/20)^{0.481224} \times (100/d_p)^{0.234}) \\ &= 0.3719 + 0.021677 \times T^{0.481224}; \end{aligned} \quad (8.12)$$

$$\begin{aligned} k_{nf2} &= 0.754209 \times k_{bf} + 0.127133 \times (k_p/k_{bf})^{0.006995} \times \varphi^{0.467} \times (T/20)^{0.481224} \times (100/d_p)^{0.234}) \\ &= 0.3719 + 0.029963 \times T^{0.481224}; \end{aligned} \quad (8.13)$$

$$\begin{aligned} k_{nf5} &= 0.754209 \times k_{bf} + 0.127133 \times (k_p/k_{bf})^{0.006995} \times \varphi^{0.467} \times (T/20)^{0.481224} \times (100/d_p)^{0.234}) \\ &= 0.3719 + 0.045965 \times T^{0.481224}; \end{aligned} \quad (8.14)$$

8.4.1.1.2 Density, specific heat capacity, and viscosity

The calculation formula of density and specific heat is shown in the following formula:

$$\rho_{\text{nf}} = (1 - \varphi)\rho_f + \varphi\rho_p \quad (8.15)$$

$$C_{nf} = \frac{(1-\varphi)(\rho C)_{bf} + \varphi(\rho C)_p}{(1-\varphi)\rho_{bf} + \varphi\rho_p} \tag{8.16}$$

The viscosity model purposed by Han [58] agreed well with the experimental result.

$$\mu_{\rm nf} = \mu_{nf}\left\{1 + 2.5\varphi\frac{\int (r_i+\delta)r^3 n(r)dr}{\int r_i^3 rn(r)dr} - 1.55\varphi^2\left[\frac{\int (r_i+\delta)r^3 n(r)dr}{\int r_i^3 rn(r)dr}\right]^2\right\} \tag{8.17}$$

$$n(r) = \frac{1}{r\sqrt{2\pi}\ln\sigma}\exp\left\{-\left[\frac{\ln\left(\frac{r}{\bar{r}}\right)}{\sqrt{2}\ln\sigma}\right]^2\right\} \quad r>0 \tag{8.18}$$

where $n(r)$ is the logarithmic normal distribution density function describing the spatial distribution of nanoparticle aggregates; r is the hydrodynamic radius of nanoparticle aggregates; $\bar{r}$ is its geometric average radius and approximately equal to the average radius of the nanoparticles r_i; σ is the root mean square deviation of r_i, generally taking the empirical value of 1.5; δ is the thickness of the surfactant, reflecting the effect of the thickness of the surfactant layer.

8.4.1.2 Computational fluid dynamics of water jackets

8.4.1.2.1 Geometric model and mesh generation

Since the water jackets contain many complex surfaces, the establishment of the water jackets model using CATIA combined with the reverse engineering technology. Fig. 8.4 is the geometric model of the cooling water jackets.

It is not easy work to generate mesh with high quality because of the complicated structure. Xu meshed the model through ICEM. A hexahedral mesh with high accuracy is used in most areas of the model, and a tetrahedral mesh with good adaptability is applied only in complex places and edge regions. Meanwhile, mesh encryption is carried out for the regions with severe flow. After that, the mesh is smoothed globally, as shown in Fig. 8.5. And Fig. 8.6 shows the mesh quality displayed in ICEM. The total number of grids in this model is about 15 million, among which the hexahedral grid accounts for 88%, and the minimum grid mass is 0.34.

8.4.1.2.2 Computational models and boundary conditions

The rated working condition (power 132.4 kW, speed 5500 rpm) with high working temperature and high thermal load was selected. The cooling medium was a mixture of water and ethylene glycol (30:7 volume fraction) and a Fe_3O_4 nanofluid coolant (1%, 2%, and 5% volume fraction). Since the volume of liquid varies little with temperature and pressure, the coolant is considered incompressible fluid.

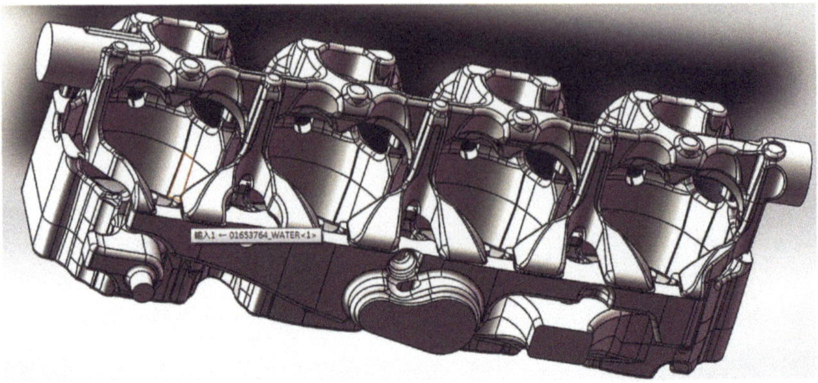

FIGURE 8.4 The model of the water jackets [56].

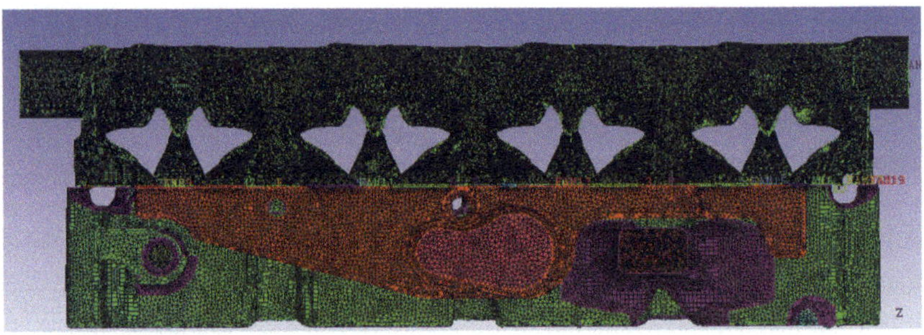

FIGURE 8.5 The mesh model [56].

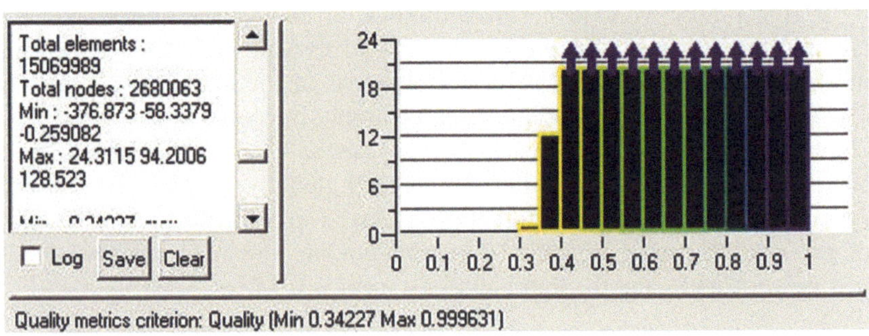

FIGURE 8.6 Mesh quality displayed in ICEM [56].

The Reynolds number of the coolant in the pipeline can be determined by the following

$$\mathrm{Re} = \rho v d / \mu \tag{8.18}$$

In which v is the characteristic velocity and characteristic velocity; d is the characteristic length. In this case, $v = 4.6$ m/s and $d = 9$ mm. ρ and μ are density and viscosity of fluid. According to the thermodynamic parameter table, when the average temperature of the coolant is 364K, the dynamic viscosity of the original coolant is 0.00056 kg/m × s, calculated as Re = 77,409.1, which means the coolant flows turbulent.

In engineering, the heat transfer process in the engine cylinder is generally regarded as the coupling effect of convection and radiation. And the radiation within the engine accounts for less than 10% of the total heat transfer in a work cycle [59]. When the engine operates under stable working conditions, though the gas conditions in the cylinder change periodically, these transient changes cannot affect the temperature at the dept of the cylinder liner wall surface due to the high engine speed. Therefore, the heat transfer process from the engine body to cooling liquid can be considered to be immutable under steady-state conditions of the engine [60]. Therefore, this example adopts steady state calculation and SIMPLEC method for numerical simulation of a water jacket in Fluent. The turbulence model is a k-epsilon model.

Table 8.6 shows the engine bench test results. On the basis of the results of the bench test, the inlet boundary is set as the speed inlet, and the flow rate can be converted to 0.46 m/s according to the pump flow. The outlet boundary is the pressure outlet, and the pressure is set as 20 kPa. The wall temperature combustion chamber is unknown before CFD calculation. The average heat transfer coefficients of cylinder liner and head are suggested to be about $3.863e^3$ and $1.0203e^4$ W/(m^2K), respectively [61]. In this example, Xu assumes that the temperature difference between the gas and the wall is 20K, then the initial heat flow values of the cylinder liner and cylinder head are set to 77,260 and 204,060 W/m^2. And the heat transfer process of the four cylinders under the same stroke is consistent. The third boundary condition is set between the gas and the inner wall of the cylinder liner. The

TABLE 8.6 Partial test data of engine bench test under rated operating conditions [56].

Name	Value
Coolant inlet pressure (kPa)	156.635
Coolant outlet pressure (kPa)	121.975
Coolant inlet temperature (K)	361
Coolant out temperature (K)	369
Coolant flow rate (L/min)	170

distribution law of convective heat transfer coefficient and gas temperature on the inner surface of cylinder liner in the axial direction is calculated according to the following formula:

$$a_m(h) = a_m(0)\cdot(1 + K_1\beta)\cdot e^{-\sqrt[3]{\beta}} \tag{8.19}$$

$$T_{\mathrm{res}}(h) = T_{res}(0)\cdot(1 + K_2\beta)\cdot e^{-\sqrt{\beta}} \tag{8.20}$$

$$a_m(0) = \frac{1}{4\pi}\int_0^{4\pi} a_g d\varphi \tag{8.21}$$

$$T_{res}(0) = \frac{\int_0^{4\pi} a_g T_g d\varphi}{\int_0^{4\pi} a_g d\varphi} \tag{8.22}$$

where weighted mean of heat transfer coefficient am, weighted mean of temperature T_{res}; $\beta = h/s$ $(0 \leq \beta \leq S)$; $K_1 = 0.5730\ (S/D)^{0.24}$, $K_2 = 1.45K_1$, T_g is the instantaneous temperature of the gas in the cylinder, ag is the instantaneous heat transfer coefficient of the gas in the cylinder, S is the stroke, and D is the cylinder diameter. From Eqs. 8.19 and 8.20, the temperature and heat transfer coefficient between the inner wall of the cylinder and the top of the cylinder in the axial direction can be obtained. By calculation, $K_1 = 0.5758$, $K_2 = 0.8349$.

In this study, a_g and T_g are calculated from the simulation of the engine under corresponding operating conditions in GT-SUITE. Limited by the article space, the simulation in GT-SUITE is not introduced here. Interested readers can refer to Chapter 2 and 5 of the original literature [56]. The curves of a_g and T_g changing with the crankshaft angle are shown in Figs. 8.7 and 8.8. From Eqs. 8.21 and 8.22, the weighted mean of heat transfer coefficient a_m (0) = 1092.46 W/(m^{2}K) and the weighted mean of temperature T_{res} (0) = 1149.98K were calculated.

8.4.1.2.3 Computational fluid dynamics calculation and result analysis

The convergence is determined from three aspects: residual curve, inlet and outlet mass flow rate, and temperature at the monitoring point. In this example, the residual curve is balanced without fluctuation. The difference between the inlet and outlet flow of coolant is 9.376e^{-12} kg/s, almost 0, which can be seen as the conservation of the system. The average temperature curves at the monitoring point and the center section of each cylinder tend to balance. In conclusion, it is considered that the calculation converges.

The result of temperature contour of different coolant is shown in Fig. 8.9. The maximum temperature of the original coolant, 1%, 2%, and 5% Fe_3O_4 nanofluids were 404.1, 401.5, 399.2, and 396.2K, respectively. The

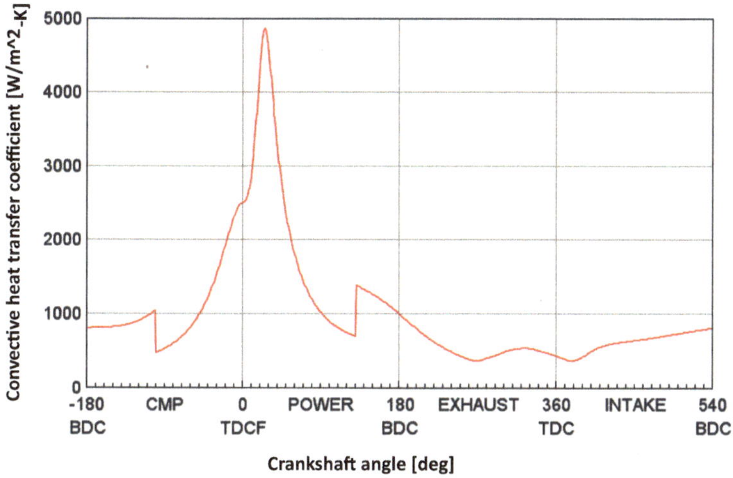

FIGURE 8.7 The convective heat transfer coefficient of gas in the engine cylinder at rated operating conditions [56].

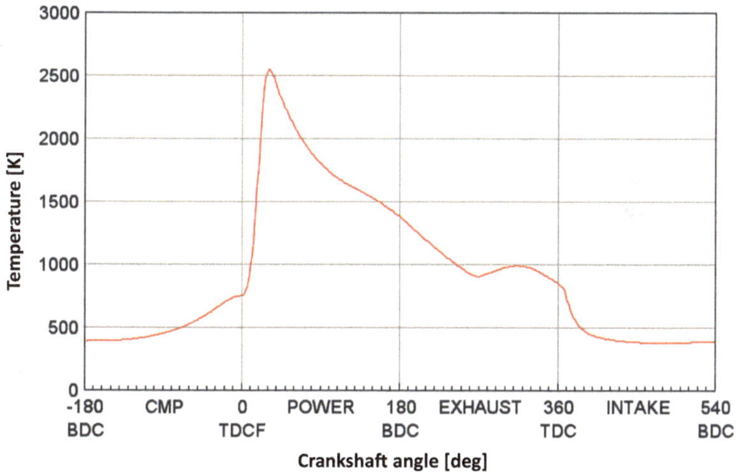

FIGURE 8.8 The temperature of the gas in the engine cylinder at rated operating conditions [56].

maximum temperature of 1%, 2%, and 5% nanofluids decreased by 2.6, 4.9, and 7.9K, respectively, compared with R134a, indicating a significant cooling effect of nanofluids (Fig. 8.10).

Under the same inlet speed, the difference between the average pressure of the inlet and outlet surface is the pressure loss of the water jacket. The pressure loss after using the pure R134a, 1%, 2%, and 5% nanofluid are 28.802, 31.179, 32.815, and 37.938 kPa. In engineering application, the pressure loss in a water jacket should be over 70kPa. Obviously, the application of Fe_3O_4 can meet this

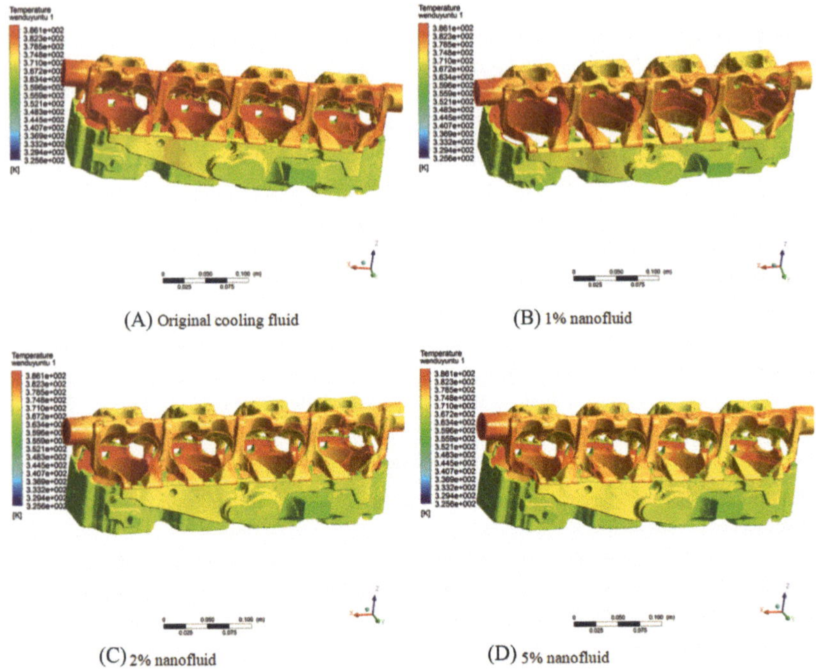

FIGURE 8.9 The temperature contours of different coolant [56].

requirement. Compared with the original coolant, the pressure drop of the cooling system after using 1%, 2%, and 5% nanofluid increases by 2.377, 4.013, and 9.136 kPa, respectively. But the heat transfer ability has been greatly improved, so the nanofluid has a broad application prospect.

8.4.2 Utilization in refrigerant of automotive air conditioning

In this example, Fe_3O_4 nanoparticles were combined with refrigerant R134a in automobile air conditioning, and the influence of nanorefrigerant on the performance of automobile air conditioning was numerically analyzed through LBM.

8.4.2.1 Determination of nanofluid parameters

There is a phase transition process for the nanorefrigerant working in air conditioning, so it is necessary to determine the thermodynamic parameters of Fe_3O_4-R134a nanofluid in gas and liquid forms.

8.4.2.1.1 Thermal conductivity

Liu found the physical property parameters of R134a under specific working conditions on REFPROP, and applied the Hrihikesh model to predict the increase of thermal conductivity of Fe_3O_4-R134a nanorefrigerant in gas and

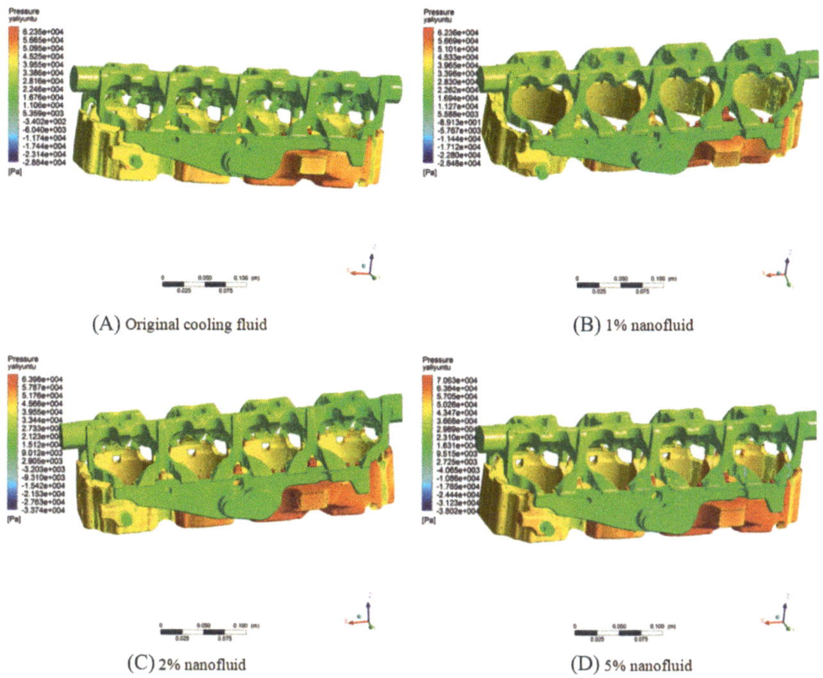

FIGURE 8.10 The pressure contour of different coolant [56].

liquid, respectively, compared with the original refrigeration. According to the variation law of thermal conductivity increase under the same pressure and different temperatures, Liu decided to use $y = A + BT + CT^2 + DT^3$ and $y = A + BT$ to match the relationship curve between thermal conductivity increase and temperature of liquid and gas nanofluids.

The fitting coefficients of four different concentrations (1%, 2%, 3%, and 5%) of the nanorefrigerant gas are shown in Table 8.7 under pressure of 0.1, 0.4, and 1.0 MPa, respectively.

Table 8.8 shows the fitting coefficients of the nanorefrigerant liquid at different concentrations and pressures of 0.4 and 1.0 MPa, respectively.

8.4.2.1.2 Viscosity

The estimation of viscosity is similar to that of thermal conductivity, and the Zuzovsky model [62] is used to calculate viscosity. Linear equations were used to fit the gas and liquid forms of nanocoolant.

$$\frac{\mu_{eff}}{\mu_f} = 1 + \frac{2.5\phi}{1 + 0.826\phi - 2.286\phi^{5/3}} \tag{8.23}$$

The fitting coefficients are shown in Tables 8.9 and 8.10.

TABLE 8.7 The fitting coefficients of thermal conductivity of Fe_3O_4-R134a (gas) [57].

Concentration	Pressure (MPa)	A	B ($\times 10^{-4}$)	C ($\times 10^{-6}$)	D ($\times 10^{-10}$)
1%	0.1 MPa	0.09019	0.27665	0.700151	6.25725
	0.4 MPa	0.08465	0.300753	−0.73085	6.31054
	1.0 MPa	0.05878	0.474004	−1.13967	9.62692
2%	0.1 MPa	0.12466	3.82415	−0.96773	8.64899
	0.4 MPa	0.11701	4.15713	−1.01021	8.72271
	1.0 MPa	0.08124	6.55187	−1.5753	13.3067
3%	0.1 MPa	0.15065	4.62136	−1.16952	10.452
	0.4 MPa	0.14140	5.02377	−1.22081	10.541
	1.0 MPa	0.09818	7.91779	−1.90372	16.081
5%	0.1 MPa	0.19123	5.086641	−1.4846	13.2679
	0.4 MPa	0.17949	6.37724	−1.54972	13.3811
	1.0 MPa	0.12463	10.1	−2.41658	20.4132

TABLE 8.8 The fitting coefficients of thermal conductivity of Fe_3O_4-R134a (liquid) [57]

Pressure	Concentration	A	B
0.4 MPa	1%	0.1058	2.12972
	2%	0.01388	3.01379
	3%	0.01443	3.9528
	5%	0.02499	4.2084
1.0 MPa	1%	0.00317	3.00141
	2%	0.00509	4.07546
	3%	0.00587	4.94736
	5%	0.00732	6.29398

8.4.2.1.3 Density and specific heat capacity

The density and specific heat are calculated by using Eqs. (8.15) and (8.16). The variation of density with temperature was fitted by $y = A + BT$. Cp was fitted by $y = A + BT + CT^2 + DT^3$.

TABLE 8.9 The fitting coefficients of viscosity of Fe_3O_4-R134a (gas) [57].

Pressure	Coefficient	R134a	1%	2%	3%	5%
0.1 MPa	A	−0.39	−0.40	−0.41	−0.42	−0.45
	B	0.043	0.044	0.045	0.047	−0.049
0.4 MPa	A	−0.97	−0.99	−1.02	−1.05	−1.11
	B	0.045	0.045	0.048	0.049	0.052
1.0 MPa	A	−1.66	−1.70	−1.74	−1.79	−1.89
	B	0.048	0.049	0.05	0.052	0.054

TABLE 8.10 The fitting coefficients of viscosity of Fe_3O_4-R134a (liquid) [57].

Pressure	Concentration	A	B
0.4 MPa	R134a	1534.03	−4.62
	1%	1573.35	−4.73
	2%	1614.57	−4.88
	3%	1657.72	−4.99
	5%	1749.76	−5.27
1.0 MPa	R134a	1359.41	−3.91
	1%	1394.24	−4.02
	2%	1430.77	−4.12
	3%	1469.01	−4.23
	5%	1550.57	−4.47

The fitting coefficients are shown in Tables 8.11–8.13.

8.4.2.2 The Boltzmann model of nanofluid

In general, thermal lattice Boltzmann models for nanofluids fall into two categories: mixing model and bidistribution function model. This case uses the bidistribution function model. The model has two distribution models: one is the density distribution function to simulate the velocity field, and the other

TABLE 8.11 The fitting coefficients of density of Fe3O4-R134a (gas) [57].

Pressure	Coefficient	R134a	1%	2%	3%	5%
0.1 MPa	A	8.3	60.01	111.73	183.45	266.88
	B	−0.0136	−0.0134	−0.0133	−0.0131	−0.0129
0.4 MPa	A	34.77	86.22	137.67	189.12	292.03
	B	−0.0571	−0.0566	−0.0561	−0.054	−0.0543
1.0 MPa	A	102.51	153.2	204.07	254.84	356.39
	B	−0.1785	−0.1767	−0.175	−0.1732	−0.1696

TABLE 8.12 The fitting coefficients of density of Fe_3O_4-R134a (liquid) [57].

Pressure	Concentration	A	B
0.4 MPa	R134a	2274.93	−3.64
	1%	2303.98	−3.61
	2%	2333.03	−3.57
	3%	2362.08	−3.53
	5%	2420.18	−3.46
1.0 MPa	R134a	2375.16	−4.04
	1%	2403.21	−4.01
	2%	2431.26	−3.96
	3%	2459.31	−3.92
	5%	2515.41	−3.84

TABLE 8.13 The fitting coefficients of specific heat of Fe_3O_4-R134a (liquid) [57].

Temperature (K)	R134a (J/kg-K)	1% (J/kg-K)	2% (J/kg-K)	3% (J/kg-K)	5% (J/kg-K)
243	780.4	778.8	777.2	775.6	772.5
253	815.3	813.3	811.4	809.5	805.6
263	853.7	851.4	849.1	846.8	842.2
273	896.6	893.8	891.1	888.3	882.8
283	944.7	941.5	938.2	935.0	928.6
293	999.8	996.0	992.2	988.4	980.9
303	1064.4	1060.0	1055.6	1051.1	1042.3
313	1143.2	1138.0	1132.8	1127.6	1117.1
323	1244.3	1238.1	1231.9	1225.6	1213.2
333	1384.3	1376.7	1369.1	1361.4	1346.2
343	1600.9	1591.1	1581.3	1571.5	1552.0
353	2003.4	1989.6	1975.8	1962.0	1934.3

is the internal energy distribution model to simulate the temperature field. The equation is:

$$g_\alpha(r + c_\alpha \Delta t, t + \Delta t) = g_\alpha(r, t) + \frac{\Delta t}{\tau_g}\left[g_\alpha^{eq}(r, t) - g_\alpha(r, t)\right] \quad (8.23)$$

The equilibrium distribution function should satisfy the following constraints:

$$\sum_\alpha g_\alpha^{eq} = \rho\varepsilon \quad (8.24)$$

$$\sum_\alpha c_{\alpha i} g_\alpha^{eq} = \rho\varepsilon u_i \quad (8.25)$$

$$\sum_\alpha c_{\alpha i} c_{\varepsilon i} g_\alpha^{eq} = \rho\varepsilon u_i u_j + \rho\varepsilon\delta_{ij} \quad (8.26)$$

So the equilibrium distribution function should be:

$$g_\alpha^{eq} = \rho\varepsilon w_\alpha\left[1 + \frac{\mathbf{c}_\alpha \cdot \mathbf{u}}{c_s^2} - \frac{u^2}{2c_s^2} + \frac{(\mathbf{c}_\alpha \cdot \mathbf{u})^2}{2c_s^4}\right] \quad (8.27)$$

In the derivation of macroscopic energy equation, the time derivative, space derivative, and distribution function are expanded by multiscale under D2Q9 model, and the derivation is compared with various coefficients:

$$\frac{\partial}{\partial t_1}(\rho\varepsilon) + \nabla \cdot (\rho u \varepsilon) = 0 \quad (8.28)$$

$$\frac{\partial}{\partial t_2}(\rho\varepsilon) = \Delta t\left(\tau_g - \frac{1}{2}\right)\frac{\partial}{\partial r_{1j}}\left(p\frac{\partial\varepsilon}{\partial r_{1j}}\right) \quad (8.29)$$

According to $\varepsilon = DR_gT/2$, the energy equation of the following form can be finally obtained:

$$\frac{\partial}{\partial t}(\rho\varepsilon) + \nabla \cdot (\rho u \varepsilon) = \nabla \cdot (\lambda \nabla T) \quad (8.30)$$

where thermal conductivity $\lambda = (\tau_g - 0.5)D_pR_gT/2$; macro temperature $T = \sum_\alpha g_\alpha$; and thermal diffusivity $\chi = c_s^2(\tau_T - \frac{1}{2})\Delta t$.

8.4.2.3 Computational fluid dynamics of nanorefrigerant based on lattice Boltzmann method

8.4.2.3.1 The physical model

The grid of the flow field is divided into 300×100, the upper and lower boundaries are nonsliding wall boundary, and the wall surface is constant temperature. The left boundary is the velocity inlet boundary with a constant velocity and temperature. The right boundary is an extended boundary with

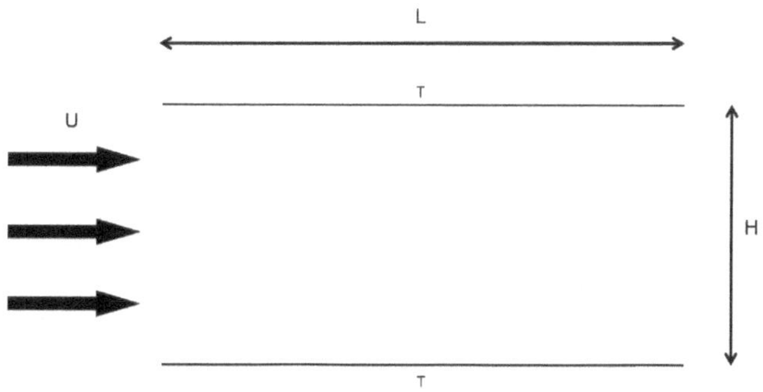

FIGURE 8.11 The model of flow in a tube [57].

no temperature gradient in the x direction. The initial conditions are that the nanoparticles are evenly distributed in the refrigerant and the inlet velocity is constant (Fig. 8.11).

The heat transfer of nanorefrigerant in the tube belongs to the forced convection heat transfer in the tube, and the classical Dittus–Boelter formula is used for the calculation of Nusselt number:

$$\mathrm{Nu} = 0.023\mathrm{Re}^{0.8}\mathrm{Pr}^{0.3} \tag{8.31}$$

Then the convective heat transfer coefficient is

$$h = \frac{\lambda}{d}\mathrm{Nu} \tag{8.32}$$

The heat transfer amount of the flow in the tube is calculated according to Newton's cooling formula:

$$\phi = hA\Delta t \tag{8.33}$$

where Δt is the logarithmic average temperature difference, and A is the heat transfer area.

The lattice model is D2Q9 model, and the boundary condition is standard bounce-back rule. The calculation was carried out under steady conditions, and the heat transfer of Fe_3O_4-R134A nanorefrigerant with different concentrations under different Reynolds numbers was compared. The final convergence condition was that the residual was less than 10^{-6}.

8.4.2.3.2 Numerical simulation results

Fig. 8.12 shows the temperature distribution contour of 1% concentration nanocoolant at Re = 800. The temperature boundary layer effect is very clear on the wall surface, indicating that the heat transfer is mainly concentrated near the flowing wall surface. But no significant difference was found in the temperature cloud maps at different concentrations, so the improvement of

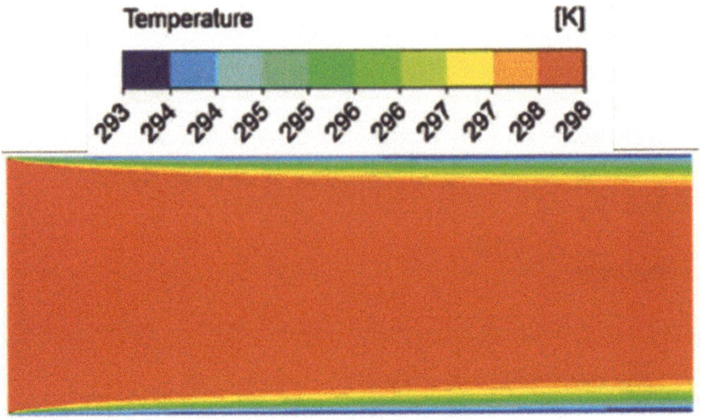

FIGURE 8.12 The temperature contour of the nanorefrigerant when Re = 800 [57].

TABLE 8.14 Heat transfer amount of the nanorefrigerant (liquid) [57].

Re	R134a (W)	1% (W)	2% (W)	3% (W)	5% (W)
800	42.83	43.51	44.06	44.42	45.23
1000	46.48	47.23	47.80	48.23	49.10
1200	51.21	52.09	52.70	53.25	54.22
1400	54.89	55.85	56.56	57.14	58.14
2000	64.37	65.55	66.40	67.03	68.31

TABLE 8.15 Heat transfer amount of the nanorefrigerant (gas) [57].

Re	R134a (W)	1% (W)	2% (W)	3% (W)	5% (W)
800	55.53	64.91	69.26	72.42	77.19
1000	61.07	71.76	76.32	79.83	85.24
1200	66.03	77.26	82.73	86.58	92.36
1400	70.67	82.81	88.61	92.77	99.03
2000	82.65	96.93	104.02	108.99	116.44

the heat transfer performance of the refrigerant by nanoparticles was reflected through the calculation of heat transfer. Tables 8.14 and 8.15 show the heat transfer of liquid and gas of nanorefrigerant respectively.

From Tables 8.14 and 8.15, we can find that with the increase of the concentration of Fe_3O_4 nanoparticles in the R134a, the heat transfer amount increased. Both for Fe_3O_4 nanorefrigerant liquid and gas, the maximum increases of heat transfer amount occurred when the concentration is 5% and the Reynolds number is 2000, which were 14.41% and 39.89%. That means that the addition of nanoparticles can improve the heat transfer performance of refrigerant significantly, especially on the heat transfer side of the gas.

Nomenclature

u	velocity (m/s)
k	thermal conductivity (W/m/K)
c_p	specific heat capacity (J/kg/K)
T	temperature (K)
m	mass (kg)
Nu	Nusselt number
Pr	Prandtl number
Re	Reynolds number
Greek symbols	
μ	viscosity (N m/s)
ϕ	volume fraction
ρ	density (kg/m^3)
bf	base fluid
p	nanoparticle
eff	effective
m	mixture
Subscripts	
nf	nanofluid

Problems

Although nanofluids have better performance in experiments compared with the base fluid, they are still far from being widely used in industry. One of the great challenges is the stability of nanoparticles dispersion. The excellent heat transfer ability of nanofluids is based on the good dispersion of nanoparticles in the base fluid, but actually nanoparticles cannot maintain a good dispersion state for a long time, resulting in the lack of reliability. And the high production cost is among the reasons that block the utility of nanofluids in industry. What's more, the higher viscosity after adding nanoparticles requires more comprehensive experimental data to select the optimal volume concentration of nanoparticles.

In terms of numerical simulation analysis, some problems concerning the simulation of nanofluids still exist, such as the lack of comprehensive understanding of some heat transfer phenomena in the actual flow, which make the model too simplified. And there is no reference standard for which model is best used for a certain problem.

References

[1] S.U.S. Choi, J.A. Eastman, Enhancing thermal conductivity of fluids with nanoparticles, ASME Int. Mech. Eng. Congr. Expo. 66 (1995) 99–105.

[2] Y. Xuan, W. Roetzel, Conceptions for heat transfer correlation of nanofluids, Int. J. Heat. Mass. Transf. 43 (2000) 3701–3707.

[3] S. Özerinç, A.G. Yazcoglu, S. Kakaç, Numerical analysis of laminar forced convection with temperature-dependent thermal conductivity of nanofluids and thermal dispersion, Int. J. Therm. Sci. 62 (2012) 138–148.

[4] S.Z. Heris, M.N. Esfahany, G. Etemad, Numerical investigation of nanofluid laminar convective heat transfer through a circular tube, Numer. Heat. Transf. A 52 (2007) 1043–1058.

[5] M. Bahiraei, S.I. Vasefi, A novel thermal dispersion model to improve prediction of nanofluid convective heat transfer, Adv. Powder Technol. 25 (2014) 1772–1779.

[6] F. Akbaridoust, M. Rakhsha, A. Abbassi, M. Saffar-Avval, Experimental and numerical investigation of nanofluid heat transfer in helically coiled tubes at constant wall temperature using dispersion model, Int. J. Heat. Mass. Transf. 58 (2013) 480–491.

[7] J. Buongiorno, Convective transport in nanofluids, J. Heat. Transf. 128 (2006) 240–250.

[8] O. Mahian, L. Kolsi, M. Amani, P. Estelle, G. Ahmadi, C. Kleinstreuer, et al., Recent advances in modeling and simulation of nanofluid flows-part I: fundamentals and theory, Phys. Rep. 790 (2019) 1–48.

[9] D. Tian, Theory and Applications of Lattice Boltzmann Method, Daqing Petroleum Institute, Daqing, Heilongjiang, China, 2009.

[10] A. Satoh, ISBN 978-0-12-385148-2 Introuduction to Practice of Molecular Simulation, Elsevier, 2011pp. 255–284.

[11] H.W. Xian, N.A.C. Sidik, G. Najafi, Recent state of nanofluid in automobile cooling systems, J. Therm. Anal. Calorim. 135 (2019) 981–1008.

[12] S.U.S. Choi, W. Yu, J.R. Hull, Z.G. Zhang, F.E. Lockwood, Nanofluids for vehicle thermal management, SAE Tech. Pap. (2001). 2001-1-1706.

[13] S.-C. Tzeng, C.-W. Lin, K.D. Huang, Heat transfer enhancement of nanofluids in rotary blade coupling of four-wheel-drive vehicles, Acta Mech. 179 (2005) 11–23.

[14] X. Zhong, X.-L. Yu, J. Wu, P.-Z. Jiang, Experimental investigation on alumina nanofluids in vehicle heat exchanger, Zhejiang Daxve Xuebao/J. Zhejiang Univ. Eng. Sci. 44 (2010) 761–764, 818.

[15] A.M. Hussein, R.A. Bakar, K. Kadirgama, K.V. Sharma, Heat transfer enhancement using nanofluids in an automotive cooling system, Int. Commun. Heat. Mass. Transf. 53 (2014) 195–202.

[16] S.Z. Heris, M. Shokrgozar, S. Poorpharhang, M. Shanbedi, S.H. Noie, Experimental study of heat transfer of a car radiator with CuO/ethylene glycol-water as a coolant, J. Dispers. Sci. Technol. 35 (2014) 677–684.

[17] X. Li, C. Zou, A. Qi, Experimental study on the thermo-physical properties of car engine coolant (water/ethylene glycol mixture type) based SiC nanofluids, Int. Commun. Heat. Mass. Transf. 77 (2016) 159–164.

[18] K.Y. Leong, R. Saidur, S.N. Kazi, A.H. Mamun, Performance investigation of an automotive car radiator operated with nanofluid-based coolants (nanofluid as a coolant in a radiator), Appl. Therm. Eng. 30 (2010) 2685–2692.

[19] W. Duangthonngsuk, S. Wongwises, An experimental study on the heat transfer performance and pressure drop of TiO_2-water nanofluids flowing under a turbulent flow regime, Int. J. Heat. Mass. Transf. 53 (2009) 334–344.

[20] R. Islam, B. Shabani, J. Andrews, G. Rosengarten, Experimental investigation of using ZnO nanofluids as coolants in a PEM fuel cell, Int. J. Hydrog. Energy 42 (2017) 19272–19286.

[21] K.-J. Park, D. Jung, Boiling heat transfer enhancement with carbon nanotubes for refrigerants used in building air-conditioning, Energy Build. 39 (2007) 1061–1064.

[22] J.K. Lee, J. Koo, H. Hong, Y.T. Kang, The effects of nanoparticles on absorption heat and mass transfer performance in NH3/H2O binary nanofluids, Int. J. Refrig. 33 (2010) 269–275.

[23] G. Jatinder, O.S. Ohunakin, D.S. Adelekan, O.E. Atiba, A.B. Daniel, J. Singh, et al., Performance of a domestic refrigerator using selected hydrocarbon working fluids and TiO_2-MO nanolubricant, Appl. Therm. Eng. 160 (2019) 114004.

[24] Y. Hou, L. Gan, H. Huang, J. Tu, C. Li, Tribological properties of lubrication oil with sheet graphite nanoparticles, J. East. China Univ. Sci. Technol. (Nat. Sci. Ed.) 31 (2009) 743–746.

[25] L.L. Zhang, J.P. Tu, H.M. Wu, Y.Z. Yang, WS2 nanorods prepared by self-transformation process and their tribological properties as additive in base oil, Mater. Sci. Eng. A 454 (2006) 487–491.

[26] R. Greenberg, G. Halperin, I. Etsion, R. Tenne, The effect of WS2 nanoparticles on friction reduction in various lubrication regimes, Tribol. Lett. 17 (2004) 179–186.

[27] H. Huang, J. Tu, L. Gan, T. Zou, C. Li, Preparation and tribological properties of graphite nanosheets as additive in liquid paraffin, Tribology 25 (2005) 312–316.

[28] D.-X. Peng, Y. Kang, C.-H. Chen, S.-K. Chen, F.-C. Shu, The tribological behavior of modified diamond nanoparticles in liquid paraffin, Ind. Lubr. Tribol. 61 (2009) 213–219.

[29] P. Shi, Y. Xu, Q. Liu, B. Xu, S. Ma, Preparation of nanoparicles self-repairing additive and its application in engines, China Surf. Eng. 17 (2004) 37–40.

[30] M. Hatami, M. Hasanpour, D. Jing, Recent developments of nanoparticles additives to the consumables liquids in internal combustion engines: Part I: Nano-fuels, J. Mol. Liq. 318 (2020) 114250.

[31] M.E.M. Soudagar, N.-N. Nik-Ghazali, M.A. Kalam, I.A. Badruddin, N.R. Banapurmath, N. Akram, The effect of nano-additives in diesel-biodiesel fuel blends: a comprehensive review on stability, engine performance and emission characteristics, Energ. Convers. Manage. 178 (2018) 146–177.

[32] G.A. Risha, J.L. Sabourin, V. Yang, R.A. Yetter, S.F. Son, B.C. Tappan, Combustion and conversion efficiency of nanoaluminum-water mixtures, Combust. Sci. Technol. 180 (2008) 2127–2142.

[33] H.A. Dhahad, S.A. Ali, M.T. Chaichan, Combustion analysis and performance characteristics of compression ignition engines with diesel fuel supplemented with nano-TiO2 and nano-Al2O3, Case Stud. Therm. Eng. 20 (2020) 100651.

[34] M. Mirzajanzadeh, M. Tabatabaei, M. Ardjmand, A. Rashidi, B. Ghobadian, M. Barkhi, et al., A novel soluble nano-catalysts in diesel-biodiesel fuel blends to improve diesel engines performance and reduce exhaust emissions, Fuel 139 (2015) 374–382.

[35] A.I. El-Seesy, H. Hassan, S. Ookawara, Effects of graphene nanoplatelet addition to jatropha biodiesel-diesel mixture on the performance and emission characteristics of a diesel engine, Energy 147 (2018) 1129–1152.

[36] D. Yuvarajan, N. Beemkumar, S. Ganesan, A comprehensive study on emission and performance characteristics of a diesel engine fueled with nanoparticle-blended biodiesel, Environ. Sci. Pollut. Res. 26 (2019) 10662–10672.

[37] A.K. Pandian, R.B.B. Ramakrishnan, Y. Devarajan, Emission analysis on the effect of nanoparticles on neat biodiesel in unmodified diesel engine, Environ. Sci. Pollut. Res. 24 (2017) 23273–23278.

[38] J. Meng, J. Liu, P. Sun, H. Feng, Analysis of oxidation characteristics on combustion of diesel fuels with CeO2 nano-catalyst particulates, Trans. CSAE 32 (2016) 93–99.

[39] S. Mohankumar, P. Senthilkumar, Particulate matter formation and its control methodologies for diesel engine: a comprehensive review, Renew. Sust. Energy Rev. 80 (2017) 1227–1238.

[40] J.C. Maxwell, A Treatise on Electricity and Magnetism, Oxford University Press, Cambridge, 1904, p. 2.

[41] D.A.G.N. Bruggema, Beechnung von verschiedener physikalischer Konstanten von Heterogenen Substanzen, Ann. Phys. 24 (1935) 636–664.

[42] W. Yu, S.U.S. Choi, The role of interfacial layers in the enhanced thermal conductivity of nanofluids: a renovated Maxwell model, J. Nanopart. Res. 5 (2003) 167–171.

[43] Y. Xuan, Q. Li, W. Hu, Aggregation structure and thermal conductivity of nanofluids, AIChE J. 49 (2003) 1038–1043.

[44] J. Koo, C. Kleinstreuer, A new thermal conductivity model for nanofluids, J. Nanopart. Res. 6 (2004) 577–588.

[45] K.N. Shukla, T.M. Koller, H.M. Rausch, A.P. Frba, Effective thermal conductivity of nanofluids—a new model taking into consideration Brownian motion, Int. J. Heat. Mass. Transf. 99 (2016) 532–540.

[46] S.B. Zhang, Experimental and Numerical Study on Heat Transfer Enhancement of Nanofluids, Zhejiang University, Hangzhou, China.

[47] H.W. Chiam, W.H. Azmi, N.A. Usri, R. Mamat, N.M. Adam, Thermal conductivity and viscosity of Al2O3 nanofluids for different based ratio of water and ethylene glycol mixture, Exp. Therm. Fluid. Sci. 81 (2017) 420–429.

[48] H.E. Patel, T. Sundararajan, S.K. Das, An experimental investigation into the thermal conductivity enhancement in oxide and metallic nanofluids, Int. Common. Heat. Mass. Transf. 49 (2013) 17–24.

[49] C.H. Chon, K.D. Kihm, S.P. Lee, S.U.S. Choi, Empirical correlation finding the role of temperature and particle size for nanofluid (Al2O3) thermal conductivity enhancement, Appl. Phys. Lett. 87 (2005) 1–3.

[50] A. Einstein, Investigation on the Theory of the Browmian Movement, Dover, New York, 1956.

[51] H.C. Brinkman, The viscosity of concentrated suspensions and solutions, J. Chem. Phys. 20 (1952) 571–766.

[52] D.G. Thomas, Transport characteristics of suspensions: VIII. A note on the viscosity of newtonian suspensions of uniform spherical particales, J. Colloid Sci. 20 (1965) 267–277.

[53] U. Rea, T. McKrell, L. wen Hu, J. Buongiorno, Laminar convective heat transfer and viscous pressure loss of alumina-water and zirconia-water nanofluids, Int. J. Heat. Mass. Transf. 52 (2009) 2042–2048.

[54] S. El Bécaye Maïga, S.J. Palm, C.T. Nguyen, G. Roy, Heat transfer enhancement by using nanofluids in forced convection flows, Int. J. Heat. Fluid Flow. 26 (2005) 530–546.

[55] Y.H. Qian, D. D'Humières, P. Lallemand, Lattice BGK models for navier-stokes equation, Europhys. Lett. 17 (1992) 479–484.

[56] M.J. Xu, Nanofluid Applied for Cooling DI Gasoline Engine, South China University of Technology, Guangzhou, China, 2017.

[57] Z.Y. Liu, Study of the Nanorefrigerant Applied in the Automotive Air Condition System, South China University of Technology, Guangzhou, China, 2018.

[58] Y.P. Han, Investigation of the Viscous Properties of Magnetic Fluids, Suzhou University, Suzhou, China, 2007.

[59] Z.D. Yan, Z.T. Liu, J. Chen, X. Yu, C.S. Cheung, Simulation of cylinder radiative heat transfer of diesel engine with Monte-carlo method, J. Zhejiang Univ. (Sci.) 3 (2000) 306–310.

[60] Z. Zhou, The Flow and Fluid-Solid Coupled Heat Transfer Analysis of a Diesel Engine Cooling System, Dalian University of Technology, Dalian, China, 2013.

[61] M. Duan, et al., CFD analysis on a four diesel engine cooling water jacket, Automobile Technol. 1 (2016) 69–71.

[62] M. Zuzovsky, P.M. Adler, H. Brenner, Spatially periodic suspensions of convex particles in linear shear flows. III. Dilute arrays spheres suspended Newtonian fluids, Phys. Fluids 26 (1983) 1714–1723.

Chapter 9

The use of nanofluids in solar desalination of saline water resources as antibacterial agents

Roya Mehrkhah[1], Mohammad Mustafa Ghafurian[2,3], Hamid Niazmand[3], Elaheh K. Goharshadi[1] and Omid Mahian[3,4]

[1]*Department of Chemistry, Faculty of Science, Ferdowsi University of Mashhad, Mashhad, Iran,* [2]*Department of Mechanical Engineering, Bozorgmehr University of Qaenat, Qaen, Iran,* [3]*Center for Nanotechnology in Renewable Energies, Ferdowsi University of Mashhad, Mashhad, Iran,* [4]*School of Chemical Engineering and Technology, Xi'an Jiaotong University, Xi'an, Shaanxi, P.R. China*

Nomenclature

C	dye concentration
CGCDs	chlorhexidine gluconate carbon dots
COD	chemical oxygen demand
CPS	cost of producing 1 g/s of steam
E. coli	*Escherichia coli*
GI	percentage of growth inhibition
GNP	graphene nanocomposite
GNP@MWCNT	nanofluid of graphene-multiwall carbon nanotubes
GPTMS	(3-glycidyloxypropyl) trimethoxysilane
k	pseudo-first-order rate constant
MIC	minimum inhibitory concentration
MWCNT	multiwall carbon nanotubes
NFs	nanofluids
NPs	nanoparticles
OD	optical density
PEG	polyethylene glycol
PVP	polyvinylpyrrolidone
RGO	reduce graphene oxide
S. aureus	*Staphylococcus aureus*
SDS	sodium dodecylsulfate

Advances in Nanofluid Heat Transfer. DOI: https://doi.org/10.1016/B978-0-323-88656-7.00009-X

SEM	scanning electron microscopy
SSG	solar steam generation
SVP	specific vapor productivity
TC	
Δm	mass loss (kg)
C_p	nanofluid specific heat capacity (kJ/kg/°C)
R	nanoparticle radius (nm)
h_{LV}	latent heat of water evaporation (kJ/g)
q	sunlight intensity (kW/m^2)
$\dot{m}$	evaporation flux (kg/m^2/h)
Δt	evaporation time (h)
$\dot{m}_r$	evaporation rate (g/s)
m	nanofluid total mass before evaporation (kg)
m_{NP}	nanoparticles mass in the nanofluid (g)
T	temperature (°C)
c	nanofluid concentration (wt.%)
ρ	water density (kg/m^3)
Q_{in}	total energy of the light irradiated to the solar receiver
Q_{eva}	heat dissipated
p	concentration of the NFs (g/L)
$\dot{V}$	steady–state volume

Greek symbols

Δ	variation
λ	wavelength (nm)
η	efficiency

Subtitle

abs	absorbed
LV	phase change

9.1 Harvesting solar energy by nanofluids

9.1.1 Introduction

As a renewable and green energy, solar energy is a promising solution to reduce the crisis of freshwater and energy shortage [1,2]. Solar steam generation (SSG) is beneficial for several applications, including water treatment [3–5], desalination [6,7], sterilization [8], and electricity generation [9–11]. Solar desalination for supplying freshwater is a process that extracts mineral components from saline water [12]. Owing to using energy with low efficiency, SSG as a straightforward method for achieving photothermal conversion is developing slowly. For efficient harvesting of the sunlight, the photothermal materials should possess several properties, such as ability of broadband absorption of the sunlight and efficient convertion of energy absorbed into the heat through reducing the heat loss occurring by radiation and conduction [13,14]. Several efforts have been performed for the enhancement of solar desalination devices. In this regard, various volumetric photoabsorbers have been used such as molten salts, gas–particle suspensions, and nanoparticles (NPs) dispersed in base fluids, namely nanofluids (NFs), with the aim of increasing light absorption and reducing the heat loss [15].

Choi et al. [16] first suggested NFs in 1995. NFs are suspensions formed by the dispersion of nanosized materials in base fluids, including water and glycol [17]. In comparison to traditional fluids, NFs have higher thermal conductivity (TC) [18]. The high absorption capacity for sunlight enables NFs to produce solar steam with high efficiency. NFs consisting of NPs of metals and metal oxides, including Au, Ag, and ZnO have been used to enhance the efficiency of SSG. Moreover, carbonaceous nanomaterials possess high TC and dark color and have the ability to achieve highly efficient SSG. Table 9.1 shows the performance of several NFs composed of metal and metal oxide NPs and carbonaceous nanomaterials in SSG.

9.1.2 Solar steam generation experiments by the nanofluids

Fig. 9.1 illustrates the experimental setup for SSG. The solar simulator with a xenon lamp is used for generating solar light. A solar receiver is a tube wrapped with a thermal insulation layer. Several sensors and a data-acquisition system are used for measuring and recording the increase of temperature. These sensors are placed at different heights in the solar collector [26]. An electric balance connected with a computer is utilized for measuring mass change. The light is absorbed and converted into heat by NF in the solar collector. By solar energy absorption via NF, the working fluid is evaporated [31].

The evaporation flux is calculated by Eq. (9.1) [31]:

$$\dot{m} = \frac{\Delta m}{A \Delta t} \tag{9.1}$$

where $\dot{m}$ (kg/m^{2}h) is the evaporation flux, Δm (kg) is the mass loss during a steady state, A (m^2) is the aperture area, and Δt (h) is the evaporation time.

The specific vapor productivity (SVP) is the capability of SSG device per unit mass per unit time. SVP is calculated using Eq. (9.2) for each NPs contributing in the evaporation of NF at steady state:

$$\text{SVP} = \frac{\dot{V}}{m_{\text{NP}}} \tag{9.2}$$

$$\dot{V} = \frac{\dot{m} A}{\rho_{\text{vapor}}} \tag{9.3}$$

TABLE 9.1 Evaporation efficiency of several nanofluids.

Reference	Nanofluids (NFs) type	Preparation method	Ultrasonic time (min)	NFs concentration	Solar intensity (kW/m^2 = sun)	Evaporation efficiency (%)
[19]	SWCNT	Two-step	120	19.04×10^{-4} vol.%	10	46.8
[20]	MWCNT	One-step	–	0.004 wt.%	3.5	44
[20]	GNP	One-step	–	0.004 wt.%	3.5	38
[20]	SWCNT	One-step	–	0.004 wt.%	3.5	34
[21]	MWCNT-OH	Two-step	20	0.002 wt.%	3.5	39
[22]	rGO@Fe_3O_4	Two-step	Not mentioned	1 m.m/L	1	70
[23]	Fe_3O_4@MWCNT	Two-step	120	0.5 g/L	1	43.8
[23]	Fe_3O_4@MWCNT	Two-step	120	0.5 g/L	3	23.3
[23]	Fe_3O_4@MWCNT	Two-step	120	0.5 g/L	10	60.3
[24]	TiO_2@Fe_3O_4	Two-step	Not mentioned	0.1 g/L	5	22
[25]	GO	Two-step	30	0.004 wt.%	1.5	36.54
[25]	GO	Two-step	30	0.004 wt.%	2.5	26.3
[25]	GO	Two-step	30	0.004 wt.%	3.5	21.7
[26]	rGO	Two-step	Not mentioned	10 ppm	1	46.17
[26]	rGO	Two-step	Not mentioned	10 ppm	3	36
[26]	rGO	Two-step	Not mentioned	10 ppm	5	42.82
[26]	rGO	Two-step	Not mentioned	10 ppm	7	47.4

[27]	Au	One-step	–	178 ppm	1	66.28
[27]	Au	One-step	–	178 ppm	3	54.45
[27]	Au	One-step	–	178 ppm	5	42.13
[27]	Au	One-step	–	178 ppm	7	39.1
[28]	Ag@TiO_2	Two-step	30	200 ppm	1	53.6
[28]	Ag@TiO_2	Two-step	30	200 ppm	3	45.9
[28]	Ag@TiO_2	Two-step	30	200 ppm	10	66.9
[29]	GO@Au	Two-step	30	15.6 wt.%	16.77	59.2
[30]	GNP@MWCNT	Two-step	120	0.01 wt.%	3.6	61.3

Source: Reproduced with permission M.M. Ghafurian, Z. Akbari, H. Niazmand, R. Mehrkhah, S. Wongwises, O. Mahian, Effect of sonication time on the evaporation rate of seawater containing a nanocomposite, Ultrason. Sonochem. 61 (2020) 104817. https://doi.org/10.1016/j.ultsonch.2019.104817. Copyright 2020, Elsevier.

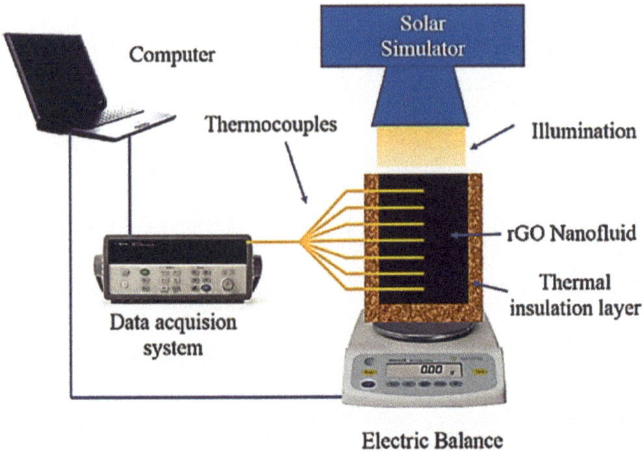

FIGURE 9.1 Schematic showing experimental setup for solar steam generation. *Reproduced with permission X. Liu, X. Wang, J. Huang, G. Cheng, Y. He, Volumetric solar steam generation enhanced by reduced graphene oxide nanofluid, Appl. Energy. 220 (2018) 302–312. https://doi.org/10.1016/j.apenergy.2018.03.097. Copyright 2018, Elsevier.*

where $\dot{V}$ (m^3) is the steady-state volume of SSG, ρ_{vapor} (kg/m^3) is the water density at 100°C, and m_{NP} (g) is the mass of NPs in NFs.

Q_{in} is the total energy of the light irradiated to the solar receiver and is calculated by Eq. (9.4):

$$Q_{in} = qA\Delta t \tag{9.4}$$

where q is (W/m^2) is the sunlight intensity. The heat dissipated is defined as Q_{eva} and explained by Eq. (9.5):

$$Q_{eva} = \dot{m}Ah_{LV}\Delta t \tag{9.5}$$

where Δt (h) is the total irradiation time, and h_{LV} (kJ/g) is the latent heat of water evaporation at 1 atm.

The evaporation efficiency of NFs, $\eta_{evaporation}$, is obtained by Eq. (9.6) [32]

$$\eta_{evaporation} = \frac{\dot{m}h_{Lv}}{q} \tag{9.6}$$

Efficiency of sensible heating is the consumed heat for heating up the fluid per the total irradiation Eq. (9.7): [15,29,33]

$$\eta_{sensible\ heating} = \frac{m \quad C_p\Delta T}{q \quad A} \tag{9.7}$$

where $m(kg)$ is the mass fluid before evaporation, C_p (kJ/kg/K) is the specific heat capacity of fluid, and ΔT stands for the fluid temperature rise during time t.

Total photothermal efficiency [26,34,35] is calculated by Eq. (9.8):

$$\eta_{\text{total}} = \eta_{\text{evaporation}} + \eta_{\text{sensible heating}} \tag{9.8}$$

9.1.3 Effective parameters on light harvesting capability of the nanofluids

9.1.3.1 Concentration of NFs

An increase in NFs concentration up to a certain concentration improves the sunlight harvesting. For example, Liu et al. [31] prepared graphene NFs with different concentrations of graphene (20, 40, and 80 ppm) in water and utilized them in SSG experiments. The results after 1 hour of evaporation are shown in Fig. 9.2A. As shown, the evaporation was improved with an increase in concentrations of graphene in NFs. More interestingly, the amount of the steam evaporated in the 20 ppm concentration was more than that of water only, showing that graphene plays an important role in sunlight absorption. The SVP representing NFs capability to produce steam and the evaporation rate of several graphene NFs are depicted in Fig. 9.2B. As illustrated, an increase in NFs concentration caused the evaporation rate to increase and the steam generation capacity per unit mass to decrease. This is because at low concentration of NFs, the graphene sheets on the surface are not able to absorb much solar energy. Hence, the sunlight can pass through the surface and the solar receiver underneath the graphene layer is capable of completely harvesting the light and converting it into the heat. In contrast, with the increase of concentration, light is completely absorbed by the graphene layer on the surface. However, the depth of the irradiation declines and only the graphene layer on the upper part of the solar receiver is exposed to the solar energy. This leads to a decrease of NFs capability for SSG at unit mass.

Furthermore, to assess the effect of NFs concentration on the solar evaporation, Ghafurian et al. [25] used GO NFs and investigated the effect of the concentration of these NFs on the temperature increase of NFs upper layers and mass of the evaporated water over time under 3.5 sun (1 sun = 1 kW/m^2). They also obtained the similar results to those of Liu et al. [31] and other researchers [19,36,37] so that the higher NPs concentration caused the higher evaporation flux to achieve. Among different concentrations of NFs ranging from 0.001% to 0.004%, the evaporation flux to the NF with a 0.004% concentration was the highest (1.25 $kg/m^2/h$) and 1.6 times more than that of water only. In another research, Ghafurian et al. [30] used GNP@MWCNT in seawater with various concentrations (ranging from 0.001 to 0.04 wt.%). They observed that a rise in NFs concentration enhanced the absorbance of sunlight (based on the Beer–Lambert law) and this was followed by an increase of the surface temperature of NF (Fig. 9.3A) and mass of the evaporated seawater (Fig. 9.3B).

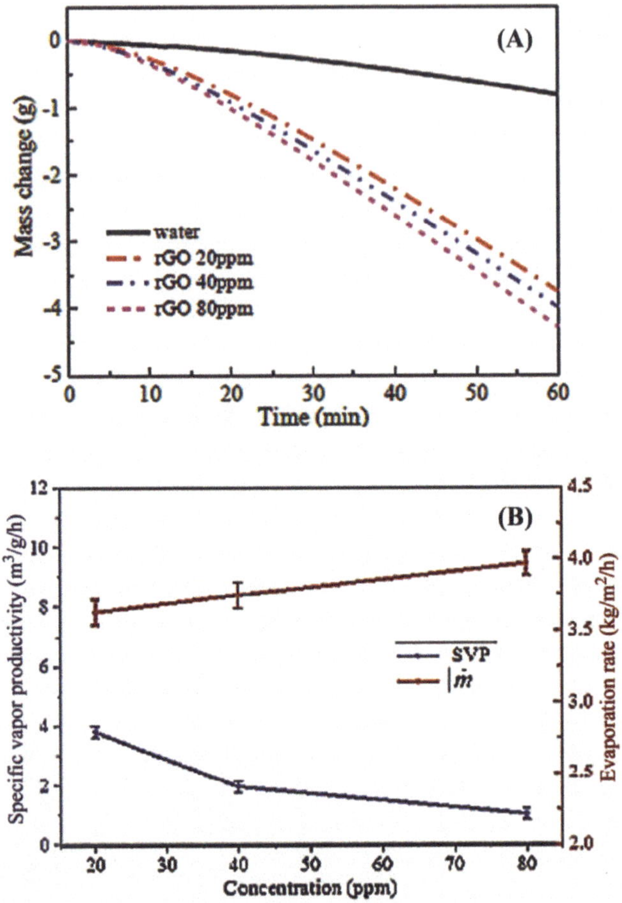

FIGURE 9.2 (A) Mass change versus time and (B) evaporation rate and specific vapor productivity of several rGO nanofluids. *Reproduced with permission X. Liu, J. Huang, X. Wang, G. Cheng, Y. He, Investigation of graphene nanofluid for high efficient solar steam generation, Energy Proc. 142 (2017) 350–355. https://doi.org/10.1016/j.egypro.2017.12.055. Copyright 2017, Elsevier.*

9.1.3.2 Stability and optical properties of NFs

The stability of NFs as a suspension plays a key role in a solar still [38], so that the low stability of NFs has negatively impact on the productivity of the solar still. Therefore, there is an urgent need to evaluate NFs stability in saline water.

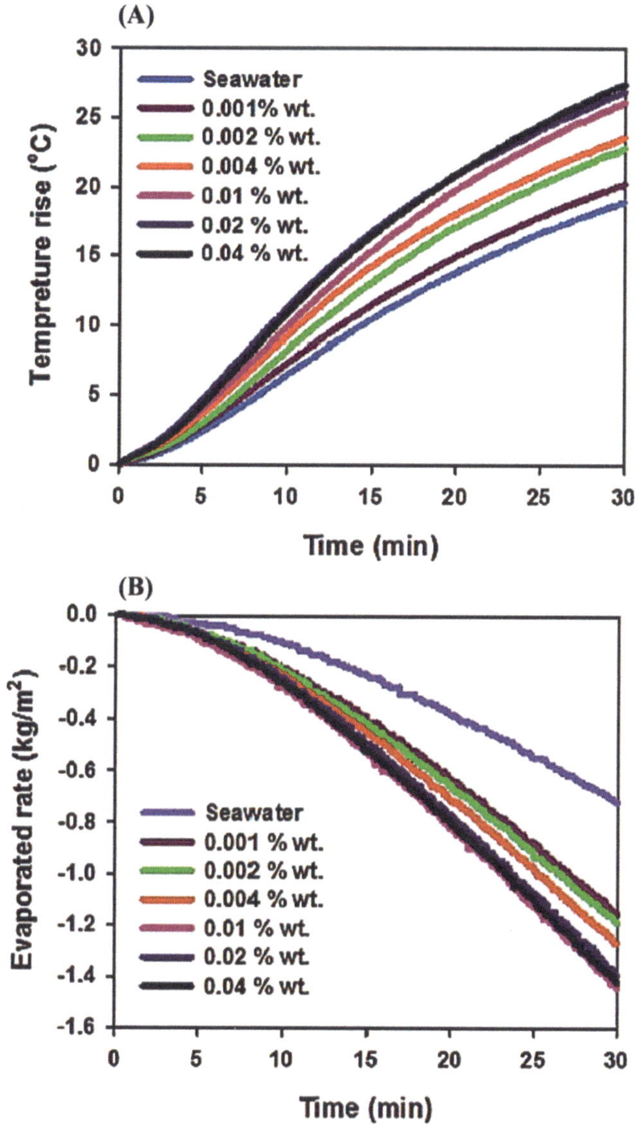

FIGURE 9.3 (A) The temperature increase of surface of several GNP@MWCNT nanofluids over time and (B) the mass change versus time for different concentrations of GNP@MWCNT nanofluids in comparison with the seawater under 3.5 sun. *Reproduced with permission M.M. Ghafurian, Z. Akbari, H. Niazmand, R. Mehrkhah, S. Wongwises, O. Mahian, Effect of sonication time on the evaporation rate of seawater containing a nanocomposite, Ultrason. Sonochem. 61 (2020) 104817. https://doi.org/10.1016/j.ultsonch.2019.104817. Copyright 2020, Elsevier.*

Liu et al. [26] exposed rGO NFs with concentrations of 1.25, 2.5, 5, and 10 ppm to solar simulated intensities and the results are shown in Fig. 9.4A. As observed, the rGO NFs are able to absorb a broadband light range. More importantly, an increase of rGO concentration caused solar energy harvesting

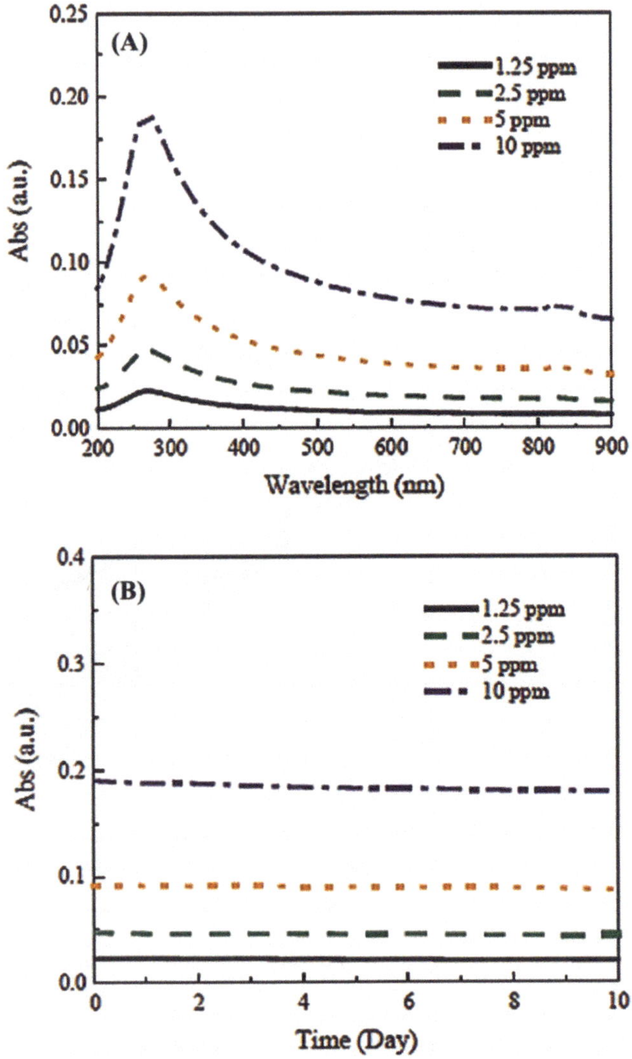

FIGURE 9.4 (A) absorbance spectra of UV−Vis and N-IR for several rGO nanofluids. (B) The change in the absorption peak for rGO nanofluids within time. *Reproduced with permission X. Liu, X. Wang, J. Huang, G. Cheng, Y. He, Volumetric solar steam generation enhanced by reduced graphene oxide nanofluid, Appl. Energy 220 (2018) 302–312. https://doi.org/10.1016/j.apenergy.2018.03.097. Copyright 2018, Elsevier.*

by the rGO NFs to increase in an accordance with Beer–Lambert law [39]. However, over time, precipitation of NPs leads to a reduction of the light absorbance. To prove the stability of rGO NFs, the changes in the UV–Vis absorbance by NFs were measured over 10 days (Fig. 9.4B) and a slight decrease in light absorption was observed during this time. This indicates a good stability of rGO NFs for using in long-term desalination.

The zeta potential is considered as another test for evaluating the stability of NFs. It is defined as a difference between the potential of the Stern layer adhered to NPs and the diffuse layer after this layer [40]. The electric charge of the Stern layer is opposite to that of NPs. Zeta potential is dependent on the value of pH so that by changing the pH, the zeta potential varies. Increasing the pH from that of the isoelectric point causes the zeta potential to increase and NFs stability enhances. In general, if the absoult value of Zeta potential is higher than 30 mV, NFs are stable. Ghafurian et al. [30] used NFs composed of a nanocomposite of graphene (GNP) and multiwall carbon nanotubes (MWCNT), namely GNP@MWCNT NFs, and showed that with the increase of the concentration from 0.001 to 0.04 wt.%, zeta potential reduces, leading to a reduction in NFs stability (Fig. 9.5).

To assess the stability of NFs, Chen et al. [41] also measured the zeta potential value of several MWCNT NFs with the concentrations of 0.0025, 0.005, 0.01, 0.02, and 0.04 wt.% in water and saline water with salinity of

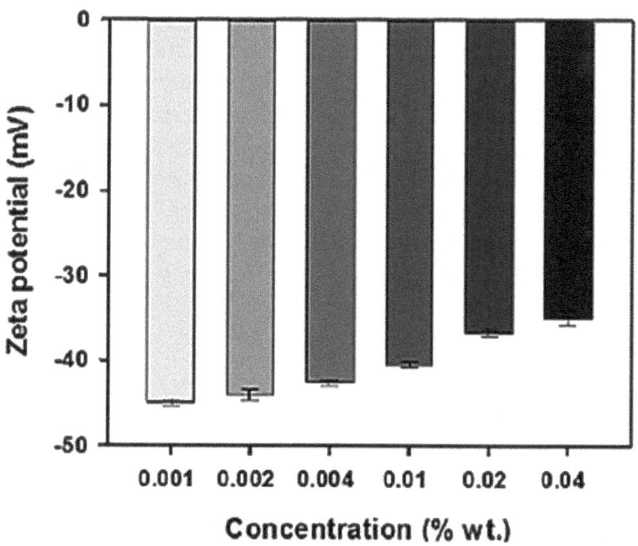

FIGURE 9.5 Zeta potential of several GNP@MWCNT nanofluids. *Reproduced with permission M.M. Ghafurian, Z. Akbari, H. Niazmand, R. Mehrkhah, S. Wongwises, O. Mahian, Effect of sonication time on the evaporation rate of seawater containing a nanocomposite, Ultrason. Sonochem. 61 (2020) 104817. https://doi.org/10.1016/j.ultsonch.2019.104817. Copyright 2020, Elsevier.*

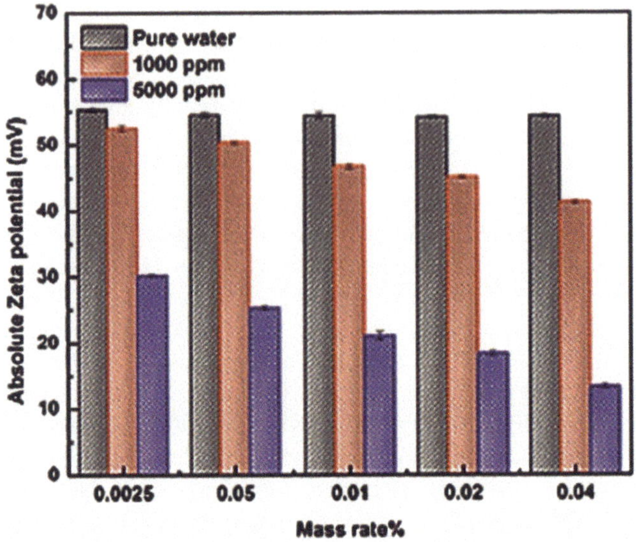

FIGURE 9.6 The zeta potential of several nanofluids at pure water, 1000, and 5000 ppm salty water. *Reproduced with permission W. Chen, C. Zou, X. Li, H. Liang, Application of recoverable carbon nanotube nanofluids in solar desalination system: an experimental investigation, Desalination (2019) 92–101. https://doi.org/10.1016/j.desal.2017.09.025. Copyright 2019, Elsevier.*

1000 and 5000 ppm and the results are depicted in Fig. 9.6. As shown, in water, zeta potential values of NFs are the same (above 50 mV) for all concentrations. The values of the zeta potential represent good stability of the NFs. In contrast, the values of zeta potential decrease in saline water so that by the increase in salt concentration from 1000 to 5000 ppm, the zeta potential reduces and, subsequently, the NFs become less stable. Zeta potential values of all NFs at 5000 ppm saline water are less than 30 mV. Although, this indicates the negative effect of the salt, NFs remain stable at 1000 ppm salt concentration because of the use of salt-resisting surfactant in the preparation of NFs.

9.1.3.3 Effect of solar illumination intensity

By increase in the solar power intensity, the evaporation efficiency increases. For instance, 10 ppm rGO NFs were exposed to several solar intensities ranging from 1 to 7 sun and the results are displayed in Fig. 9.7A and B. As shown in Fig. 9.7A, the evaporated mass increases with increasing the sunlight intensity, resulting in an improvement of evaporation flux ($\dot{m}$) and SVP achieved by the rGO NFs (Fig. 9.7B). In fact, the rise of the illumination intensity leads to an increase of incident total energy to the solar receiver and transmission capacity of the light. This enhances the evaporation capacity of the rGO NFs [26].

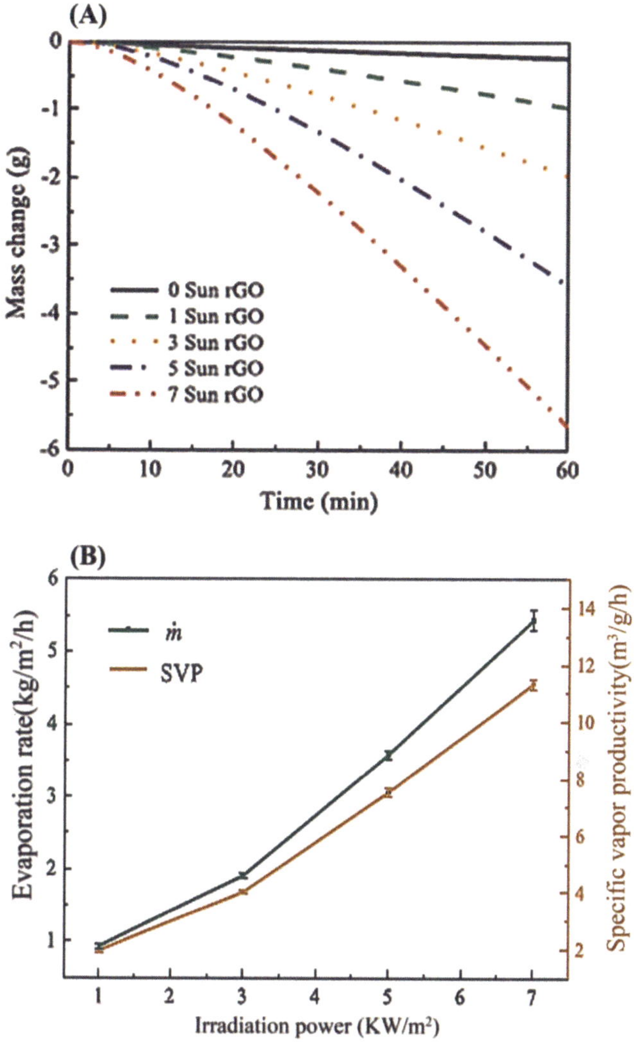

FIGURE 9.7 (A) The evaporation mass loss of the rGO NFs versus time under different light intensities. (B) The change of specific vapor productivity and ṁ achieved by the rGO NFs with increase of the sunlight intensity. *NFs*, Nanofluids. *Reproduced with permission X. Liu, X. Wang, J. Huang, G. Cheng, Y. He, Volumetric solar steam generation enhanced by reduced graphene oxide nanofluid, Appl. Energy. 220 (2018) 302–312. https://doi.org/10.1016/j.apenergy.2018.03.097. Copyright 2018, Elsevier.*

9.1.3.4 The effect of size of NPs

Ghafurian et al. [42] investigated the effect of different sizes of ZnO NPs (20 and 50 nm) on evaporation efficiency in ZnO NFs with the concentration of 0.004 wt.% in two volumetric and surface (ZnO-wood) approaches

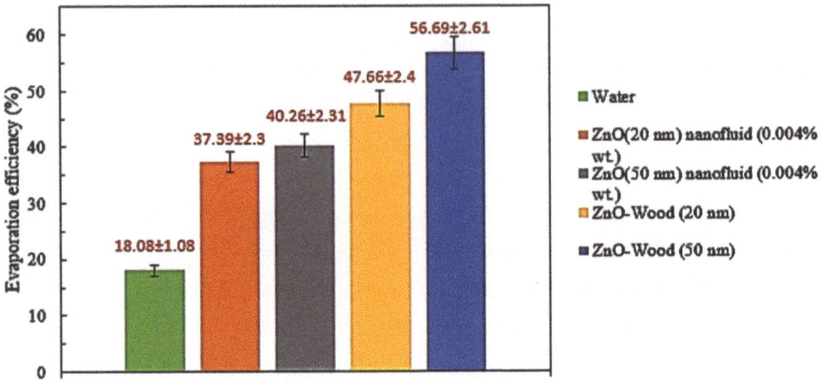

FIGURE 9.8 The comparison between the effect of different sizes of the ZnO NPs on evaporation efficiency by a 0.004% concentration of ZnO NFs in the volumetric and surface methods. *NFs*, Nanofluids. *Reproduced with permission M.M. Ghafurian, F.T. Dastjerd, A. Afsharian, F.R. Esfahani, H. Niazmand, H. Behzadnia, et al., Low-cost zinc-oxide nanoparticles for solar-powered steam production: superficial and volumetric approaches, J. Clean. Prod. 280 (2021) 124261. https://doi.org/10.1016/j.jclepro.2020.124261. Copyright 2021, Elsevier.*

(Fig. 9.8). As it can be observed, the higher NPs size, caused the greater evaporation efficiency to obtain, so that the highest efficiency belongs to the ZnO NPs with a 50 nm size in both approaches. This is due to the energy absorption increasing with the increase of the NPs size. Govorov et al. [43] stated that an increase in the surface temperature of NFs is proportional to $\frac{(2R)^2}{4}$, where R is radius of the NPs.

9.1.4 Effect of NFs on heat localization

By creating more concentrated hot area, the heat loss decreases and evaporation efficiency increases. Fig. 9.9 displays images from an infrared thermal camera of rGO NFs compared to water only. It can be seen that an increase of surface temperature in the presence of rGO NFs is greater than that of water only, showing that the photothermal efficiency of the rGO NFs is much higher than the water. Also, a temperature gradient during 30 min illumination and the existence of maximum temperature gradient at the top of bulk water resulted in the formation of hot area for evaporation which was similar to the evaporation in the presence of other NFs [15,19]. The results show that NFs are suitable for long-term SSG without heat loss [26].

9.1.5 Comparison of economic performance of NFs

Besides efficient performance in steam generation, a photothermal material should also be inexpensive in order to be suitable for cost-effective solar desalination devices [42]. The preparation cost to evaporation rate ratio is

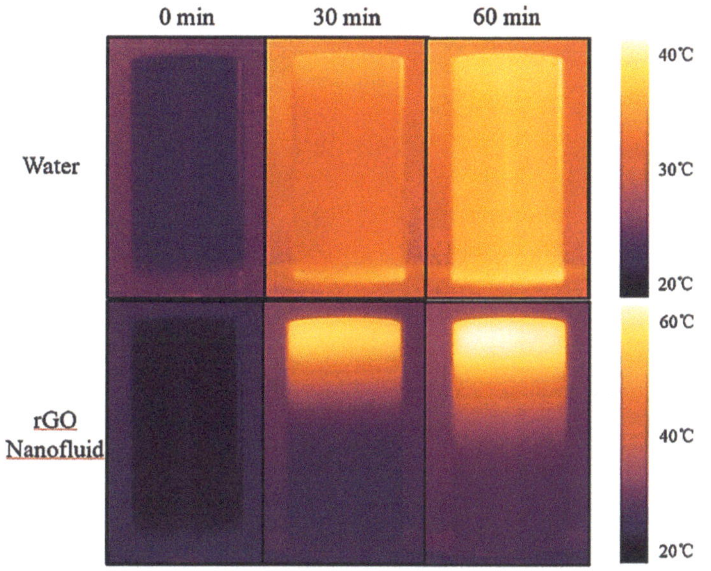

FIGURE 9.9 Infrared thermal image of water and rGO nanofluids at different times. *Reproduced with permission X. Liu, X. Wang, J. Huang, G. Cheng, Y. He, Volumetric solar steam generation enhanced by reduced graphene oxide nanofluid, Appl. Energy. 220 (2018) 302–312. https://doi.org/10.1016/j.apenergy.2018.03.097. Copyright 2018, Elsevier.*

calculated to compare economically NPs in SSG. The cost of producing 1 g/s of steam (CPS) by NPs is calculated by Eq. (9.9) [30,44,45]:

$$\mathrm{CPS}\left(\frac{\$}{\frac{g}{s}}\right) = \frac{C_{NP}}{\frac{\dot{m}_r}{v \cdot p}} \tag{9.9}$$

where C_{NP} is the cost of 1 g of NPs, $\dot{m}_r$ is the rate of the evaporation (g/s), v is NFs' volume (L), and p is concentration of NFs (g/L). Table 9.2 provides information on the calculation of the CPS of several NFs.

9.1.6 Photocatalytic experiments by NFs

The photocatalytic activity is considered as another application of NFs. It is assumed that the photocatalyst activity is based on several factors, including structure of each phase, capability of adsorption, efficient separation of electron–hole, and the NFs photostability [47]. Mahvelati et al. [48] investigated RB5 dye degradation under sunlight through the NFs consisting of ZnS and RGO-ZnS nanocomposites prepared with different weight ratios of 3, 5, and 10 of RGO to ZnS, so-called RGO3-ZnS, RGO5-ZnS, and RGO10-ZnS, respectively. The equilibrium of adsorption–desorption was established in the dark between the dye and surface of the photocatalyst by the addition of

TABLE 9.2 The CPS achieved by several nanofluids.

Nanoparticle	C_{NP} (\$/g)	$\dot{m}_r$ $\left(\frac{g}{s}\right)$ at 500 s	Vol (L)	p (g/L)	Sun (kW/m^2)	CSP $\left(\$/\frac{g}{s}\right)$	Reference
[a]MWCNT-OH	183	0.00028	0.06	0.019	3.5	772.7	[21]
[b]SWCNT	1290	0.00027	0.06	0.019	3.5	5446.7	[20,21]
[c]GO	8.03	0.00026	0.06	0.038	3.5	70.8	[25]
Au	549	0.0003	0.004	0.1	10	732	[44]
TiO_2	11	0.000352	0.06	0.038	3.5	78.3	[46]
ZnO (50 nm)	2.5	0.0006	0.06	0.019	3.5	4.7	[42]
ZnO (20 nm)	2.5	0.0005	0.06	0.019	3.5	5.7	[42]

[a]*Multiwall carbon nanotube-hydroxyl.*
[b]*Singlewall carbon nanotube-hydroxyl.*
[c]*Graphene oxide.*
Source: Reproduced with permission M.M. Ghafurian, F.T. Dastjerd, A. Afsharian, F.R. Esfahani, H. Niazmand, H. Behzadnia, et al., Low-cost zinc-oxide nanoparticles for solar-powered steam production: superficial and volumetric approaches, J. Clean. Prod. 280 (2021) 124261. https://doi.org/10.1016/j.jclepro.2020.124261. Copyright 2021, Elsevier.

photocatalyst to the solution containing dye within 30 min stirring. The photocatalytic tests were conducted on a sunny day for 4 h under natural sunlight and the results are shown in Fig. 9.10A and B. As shown, the RB5 structure is photodegraded through promising performance of RGO3-ZnS suspensions with respect to other NFs. As Fig. 9.10B illustrates, after 120 min, the intensity of UV–Vis absorption peaks related to the RB5 was

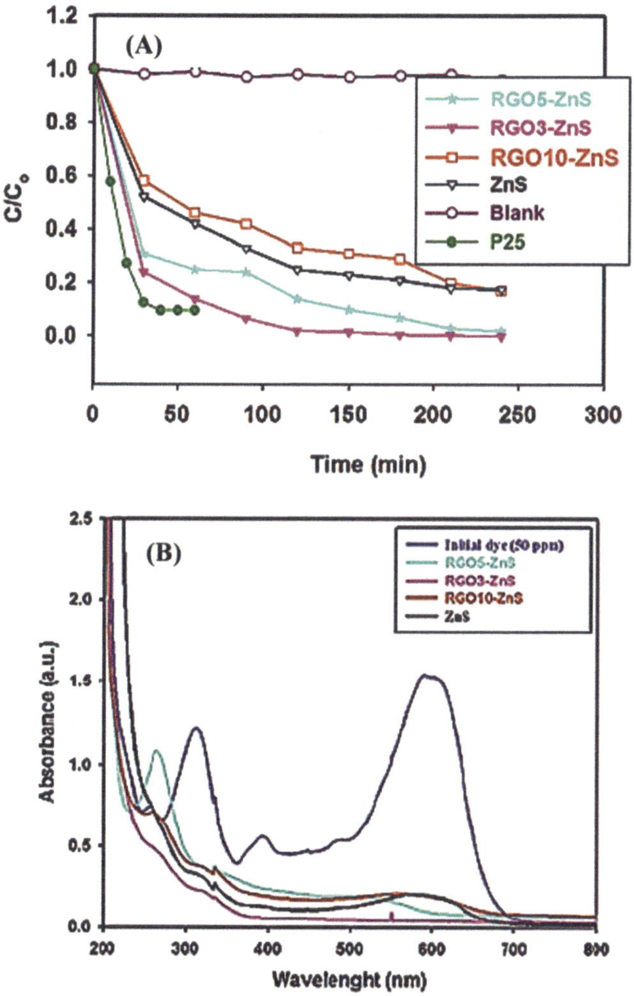

FIGURE 9.10 (A) The photocatalytic activity of several NFs under natural sunlight. (B) The UV–Vis spectra of RB5 after 120 min exposure to the different photocatalysts. *Reproduced with permission T. Mahvelati-Shamsabadi, E.K. Goharshadi, M. Shafaee, Z. Niazi, ZnS@ reduced graphene oxide nanocomposite as an effective sunlight driven photocatalyst for degradation of reactive black 5: a mechanistic approach, Sep. Purif. Technol. 202 (2018) 326–334. https://doi.org/10.1016/j.seppur.2018.04.001. Copyright 2018, Elsevier.*

disappeared by the RGO3-ZnS suspensions, indicating that there is an optimal ratio of RGO to ZnS NPs.

9.1.6.1 *Kinetic of photocatalytic activity of NFs*

To analyze the kinetics of photocatalytic activity of NFs, Mahvelati et al. [48] fitted the experimental data of RB5 dye degradation with the Langmuir–Hinshelwood model [49]

$$ln\frac{C_o}{C} = kt \qquad (9.10)$$

where C_o *is* dye concentration at $t = 0$ and C is dye concentration at the certain time of t. The initial concentration of the dye is 50 ppm. k is the pseudo-first-order rate constant of photodegradation (min^{-1}). The value of k was obtained by the curves slope of ln (C/C_o) versus irradiation time, for each nanofluid. As shown in Fig. 9.11A, among all NFs, the most value of k belonged to RGO3-ZnS suspension.

9.1.6.2 *The photocatalytic mineralization by NFs*

To investigate photocatalytic mineralization of the dye, the measurement of chemical oxygen demand (COD) value is necessary during the photodegradation process [50]. The efficiency of mineralization is calculated by Eq. (9.11):

$$\text{Efficiency of mineralization } (\%) = 1 - \frac{COD_t}{COD_0} \qquad (9.11)$$

As shown in Fig. 9.11B, the mineralization efficiency for RGO3-ZnS is 90% after 300 min photodegradation of RB5 dye ($C_o = 50$ ppm). This is mainly because a long time is needed for photodegradation of different parts of the dye structure.

9.2 Antibacterial activity of some NFs

9.2.1 Introduction

There are several diseases in human caused by bacterium of *Escherichia coli* (*E. coli*). *E. coli* is considered as the most important causal factor of urinary tract infection and diarrhea all over the world [51]. Furthermore, *E. coli* causes meningitis in infants. With insufficient defenses of the normal host, *E. coli* could enter the bloodstream and cause sepsis. Developing new, inexpensive, and environment-friendly antibacterial agents is of great importance owing to pathogenic bacterial resistance to the antibacterial agents available in the environment. Recently, nanostructures with antibacterial properties have attracted much attention worldwide [52,53]. Antibacterial agents are of great significance for several applications, including construction, decoration,

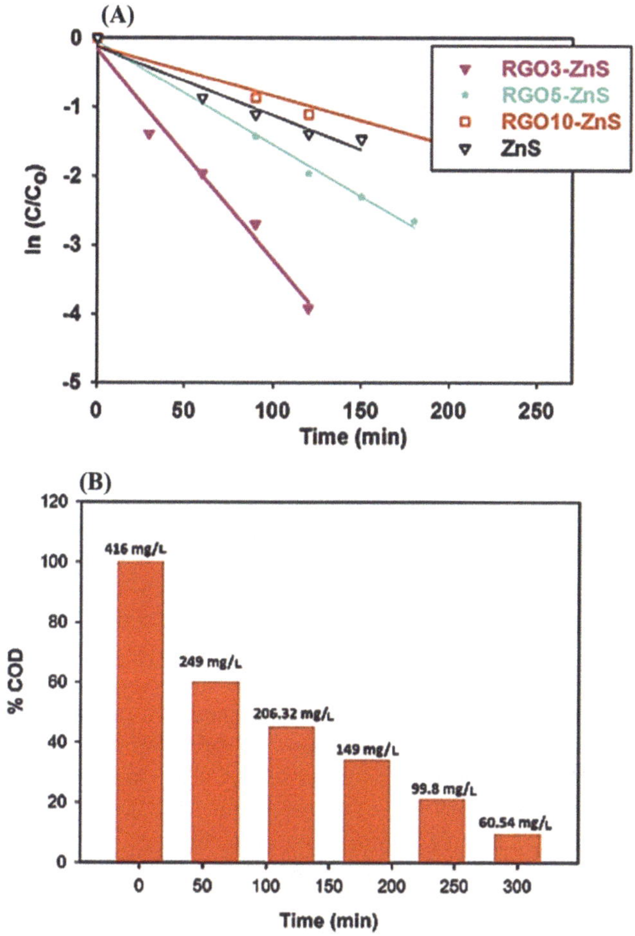

FIGURE 9.11 (A) The kinetic study of photodegradation reaction by several NFs (time exposure to the sunlight = 250 min, the NFs amount = 0.05 g). (B) COD measurement within degradation of RB5 (the NFs amount = 0.05 g, pH = 4.6). *Reproduced with permission T. Mahvelati-Shamsabadi, E.K. Goharshadi, M. Shafaee, Z. Niazi, ZnS@ reduced graphene oxide nanocomposite as an effective sunlight driven photocatalyst for degradation of reactive black 5: a mechanistic approach, Sep. Purif. Technol. 202 (2018) 326–334. https://doi.org/10.1016/j.seppur.2018.04.001. Copyright 2018, Elsevier.*

environmental, food, packaging, and synthetic textiles, as well as health and medical care. Antibacterial agents are usually divided into two types: organic and inorganic agents. In contrast to inorganic antibacterial agents, organic antibacterial materials have less stability, especially at high pressures and/or temperatures [54]. This operates as a hindering parameter for formulation of the product. Due to the capability to withstand harsh process conditions, inorganic materials, including metal and metal oxides, represent new opportunities [55–58]. Under harsh process conditions, metal oxides NPs such as Fe_2O_3, ZnO, and MgO NPs remain stable and are generally considered as safe materials to animals and human [57,59]. Furthermore, NFs composed of metal and metal oxide NPs [60] may have antibacterial activity.

9.2.2 Effect of NFs on prevention of *Escherichia coli* proliferation

ZnO NFs have showed much better antibacterial activity against *E. coli* with respect to the other metal oxides NPs. ZnO NFs cause *E. coli* to kill due to changing the configuration of *E. coli* cell membrane, especially the protein and lipid of its membrane [53]. ZnO NFs possess effective antibacterial performance against both Gram-positive and Gram-negative bacteria as well as to spores resistant to high temperature and high pressure [61–64]. Scanning electron microscopy (SEM) images of *E. coli* structure before treatment by ZnO NFs were taken by Zhang et al. [65]. After 5 h treatment by 0.2% ZnO NFs, they observed that the structure of *E. coli* has been deformed. Sun et al. [66] prepared different chlorhexidine gluconate carbon dots (CGCDs) suspensions, including s-CGCDs, m-CGCDs, and l-CGCDs, and investigated antibacterial activity of them against *E. coli* and *Staphylococcus aureus* (*S. aureus*). Fig. 9.12A and B depict the SEM images of structures of these bacteria before and after treatment by CGCDs NFs. As shown, after 6 h treatment by concentrations of 50 and 75 μg/mL of these suspensions, the structures of the bacteria have been damaged and deformed. Jalal et al. [53] prepared ZnO NFs through the dispersion of ZnO NPs in glycerol as a base fluid. The ammonium citrate was also used as dispersant. Fig. 9.12C and D show the images of *E. coli* on solid agar plates after bacteriological tests in the presence of ZnO NFs (0.5 g/dm^3) and in the absence of ZnO NPs, respectively. As Fig. 9.12C shows, a few colonies of *E. coli* are formed on the plate containing 0.5 g/dm^3 of ZnO NFs. This reveals that ZnO NFs are able to prevent *E. coli* colonies from proliferation. In contrast, in the absence of ZnO NPs, a lot of *E. coli* colonies appeared, as shown in Fig. 9.12D. Das et al. [67] assessed the antibacterial activity of the suspensions composed of GO/Ag nanocomposites against *E. coli*. As shown in Fig. 9.12E, a reduction was observed in the *E. coli* colonies after treated with different volumes of GO/Ag suspensions.

CuO NFs have antibacterial activity if the size of CuO NPs is less than the pore size of the bacterial cell membrane. Then, they easily pass through the membrane of any microorganism. CuO NFs are also very efficient in forming bonds with components of cell plasma like proteins and enzymes of bacteria and cause their death [68]. Mahapatra et al. [69] investigated the antibacterial activity of CuO NFs against enterobacteria families of bacteria, including *Salmonella*, *Klebsiella*, *Shigella*, and *Pseudomonas*. Bacterial cells are composed of cellular membranes with nanometer-sized pores. The CuO NFs had a size less than that of the bacterial pores, and hence they crossed the cell membrane without any hindrance, and after 30 h treatment, an inhibition zone was created. Ag_2O NFs have great antibacterial properties and are used for fabrics exposed to a lot of hand washing and laundering by machines [69]. The NFs composed of the Ag-RGO nanocomposite have also antibacterial properties. Moghayedi et al. [70] inhibited *E. coli* from proliferation effectively by the NFs composed of the Ag-RGO nanocomposites. By investigating the morphology of *E. coli* through SEM images, a disruption of *E. coli* cell integrity was examined before and after with Ag-RGO NFs treatment. Before treatment, cells of *E. coli* possess smooth membranes and a rodlike structure as illustrated in Fig. 9.13A. Fig. 9.13(B–D) shows bacteria morphology after 2 h treatment. The bacteria morphology changed significantly so that some *E. coli* cells were decomposed and their outer and inner membranes were damaged. The electrostatic attractions between the Ag-RGO NPs and the bacteria cause the inactivation of *E. coli* cells and finally they die (Fig. 9.13B) [71]. Fig. 9.13C shows bacteria morphology is deformed after treatment by Ag-RGO NFs. Fig. 9.13D shows that most of *E. coli* cells were killed with the increase of Ag-RGO concentration. As Fig. 9.13E and F show, after 24 h treatment, *E. coli* cells were completely damaged.

The antibacterial activity of hematite NPs, α-Fe_2O_3, against *E. coli* and *S. aureus* bacteria was investigated. Fig. 9.14A and B show inhibition zones in the investigated bacterial strains [72]. As illustrated, an increase in the NPs concentration caused the diameter of the inhibition zone increase for both bacteria [73]. Similar results have previously showed the dependence of bacterial inhibition on hematite NFs [74] and Ag NFs [75] concentrations. The hematite NPs act more effectively against Gram-positive *S. aureus* than the Gram-negative *E. coli*. Similar results were also reported by Ismail et al. [76].

9.2.3 Mechanisms of antibacterial activity of NFs

Several mechanisms were suggested to explain the antibacterial activity of inorganic NFs and the interaction between the NPs and the bacteria. Makhluf et al. [58] atributed the antibacterial activity of MgO NPs to several factors, including the formation of active oxygen species (ROS) interaction between the cell wall of the bacteria and MgO NPs, and MgO reformation inside the cell. Stoimenov et al. [59] suggested that the main reason is electrostatic interactions between the NPs and the bacteria surface. Zhang et al. [65] showed that due to interactions between the surface of the bacteria and ZnO NPs, the *E. coli* membrane wall is damaged in the presence of ZnO NPs. The ROS produced by ZnO NPs is the main inactivation mechanism. The presence of ROS were detected by Sawai et al. [64] and Yamamoto et al. [77]. Sawai et al. [64] reported that the concentration of H_2O_2 produced in ZnO slurry was proportional to the concentration of ZnO NPs. Similar

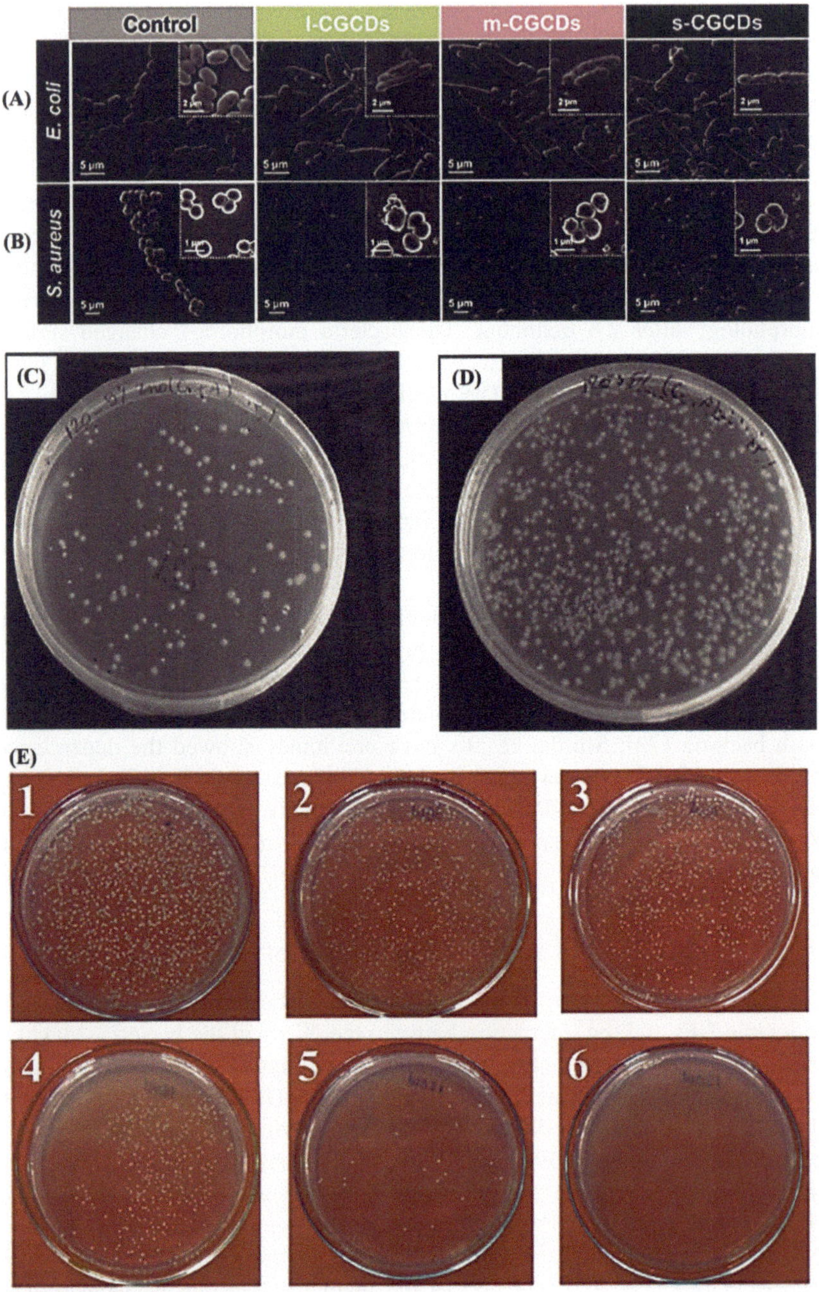

FIGURE 9.12 SEM images of (A) *Escherichia coli* and (B) *Staphylococcus aureus* before exposure to CGCDs suspensions and after 6 h treatment with CGCDs suspensions. Images of

(Continued)

findings were reported in literature [78–81]. In fact, the production of H_2O_2, OH^-, and O_2^{2-} from the surface of ZnO NPs is responsible for its antibacterial activity [53]. Electron–hole pairs are produced after ZnO activation by UV as well as visible light. OH^- and H^+ are created through splitting H_2O molecules from ZnO NFs by the holes. Then, superoxide radical anions, $^{\bullet}O^{2-}$, produced using dissolved oxygen, react with H^+ to produce $HO_2^{\bullet}$ radicals. After that, hydrogen peroxide anions (HO_2^-) are produced by subsequent collision with electrons. Finally, H_2O_2 is produced from HO_2^- reacting with hydrogen ions. By penetrating the bacteria cell, the produced H_2O_2 can kill them [81]. The number of produced H_2O_2 from the ZnO surface increases with the increase of NPs concentration and time [79]. The negatively charged hydroxyl radicals and superoxides are not able to penetrate into the membrane cell.

By capturing–killing process, the antibacterial activity of NFs composed of silver/graphene (Ag-RGO) nanocomposite was investigated [82]. This process may be induced by bacteria deposition and contact between Ag NPs and the bacteria cells in the presence of RGO. Due to strong contact of bacteria cells with RGO, Ag NPs, induce rupture of the cell wall of *E. coli*. In fact, RGO sheets prevent the agglomeration of Ag NPs. Therefore, RGO improves the antibacterial activity of Ag NPs [83]. Interaction between graphene-based materials and cells of bacteria takes place at the edges of graphene [83]. Ag NPs change the shape of bacteria cell, reduce the viability of cell, and eventually cause apoptosis and necrosis of the cell. Moreover, the release of Ag^+ ions and oxidative surface activity of Ag NPs leads to their strong antibacterial activity [84].

Rufus et al. [72] proposed that ROS, such as hydroxyl radicals, hydrogen peroxide (H_2O_2), and singlet oxygen (O_2^*), are produced during the interaction between the α-Fe_2O_3 NPs and bacteria [85]. With exposure to UV or visible light, the electron–hole pairs are produced through activation of α-Fe_2O_3 NPs. Splitting H_2O into H^+ and OH^- is done by holes.

bacteriological tests of *E. coli* on solid agar plates with (C) ZnO (0.5 g/dm^3) and (D) without ZnO NFs in glycerol and ammonium citrate. (E) Images of *S. aureus* colonies in the presence of (1) GO (control), (2) 50, (3) 80, (4) 100, (5) 120, and (6) 150 μL of GO/Ag NFs.

(A) and (B) reproduced with permission from S.U.S. Choi, J.A. Eastman, Enhancing Thermal Conductivity of Fluids with Nanoparticles, Argonne National Lab., IL, 1995. Copyright 2021, Elsevier; M. Karami, M.A. Akhavan-Behabadi, M.R. Dehkordi, S. Delfani, Thermo-optical properties of copper oxide nanofluids for direct absorption of solar radiation, Sol. Energy Mater. Sol. Cells 144 (2016) 136–142. Copyright 2013, Elsevier; (C) and (D) reproduced with permission F. Asadi, E.K. Goharshadi, M. Sadeghinia, Highly efficient solar-catalytic degradation of reactive black 5 dye using mesoporous plasmonic Ag/g-C 3 N 4 nanocomposites, ChemistrySelect. 5 (2020) 2735–2745. https://doi.org/10.1002/slct.201904730. Copyright 2010, Elsevier.

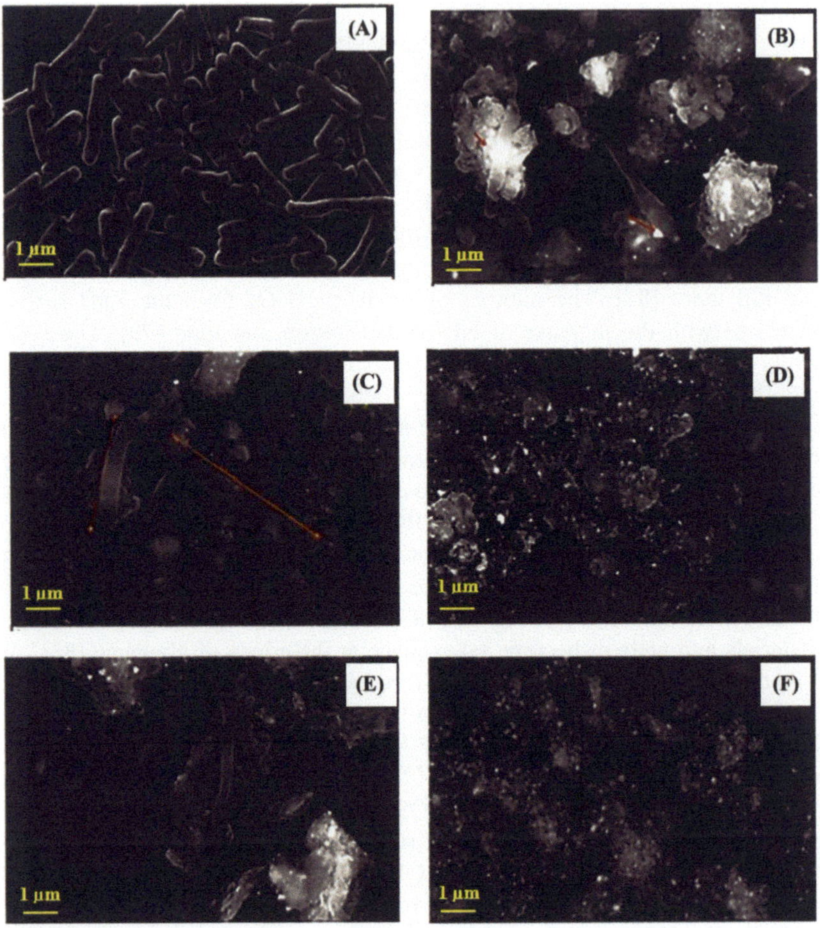

FIGURE 9.13 SEM images of (A) *Escherichia coli* before treatment; (B) Ag NPs anchored to *E. coli*; (C) deformed *E. coli* after treatment by Ag-RGO with concentration of (D) 40 μg/mL, (E) 20 μg/mL, (F) 40 μg/mL for 2, 24, and 24 hours, respectively. *Reproduced with permission M.M. Ghafurian, H. Niazmand, F.T. Dastjerd, O. Mahian, A study on the potential of carbon-based nanomaterials for enhancement of evaporation and water production, Chem. Eng. Sci. 207 (2019) 79–90. https://doi.org/10.1016/j.ces.2019.05.043. Copyright 2017, Elsevier.*

The dissolved oxygen is transformed to O_2^{-o} by electron addition. Polysaccharides are depolymerized in the reactions due to production of the free radicals of O_2^{-o} and OH^o. This results in the death of bacteria owing to enzymes inactivation and peroxidation of lipid [86,87]. The other proposed mechanism leading to the disruption of cellular function and membranes is the NPs binding to the bacteria cell membrane or proteins of the membrane via electrostatic interactions as well as the accumulation of the NPs in the periplasmic or cytoplasm [85,88].

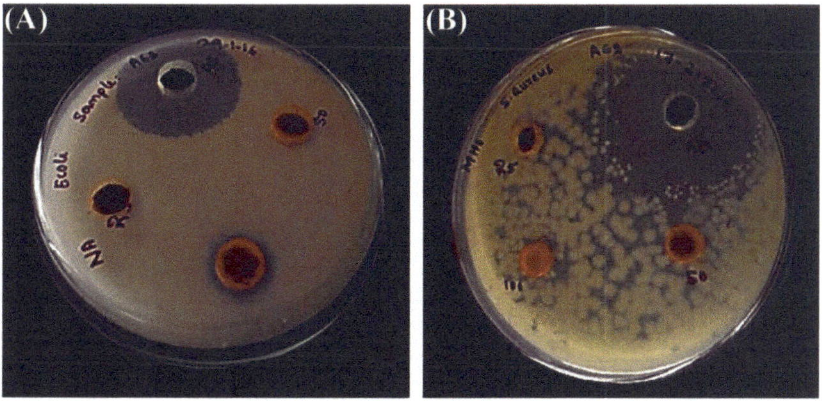

FIGURE 9.14 The inhibition zone for different concentrations of hematite NPs against organisms of (A) *Escherichia coli* and (B) *Staphylococcus aureus*. *Reproduced with permission X. Wang, G. Ou, N. Wang, H. Wu, Graphene-based recyclable photo-absorbers for high-efficiency seawater desalination, ACS Appl. Mater. Interfaces. 8 (2016) 9194–9199. https://doi.org/10.1021/acsami.6b02071. Copyright 2016, RSC Advance.*

9.2.4 Effective parameters on antibacterial activity of NFs

9.2.4.1 Effect of concentration of NPs

The antibacterial activity of a NF increases by increasing its concentration up to a certain concentration. Zhang et al. [65] prepared ZnO NFs with the concentrations of 0.1 and 0.25 g/L. Little growth of the bacteria was observed with the NF with concentration of 0.25 g/L within 8 h with respect to that of 0.1 g/L. Similar behavior was observed by Sawai et al. [54,63]. Moreover, the growth curve of *S. aureus* in the presence of several suspensions of GO/Ag nanocomposites with different concentrations of $AgNO_3$, including 0.0005, 0.001, 0.002, 0.004, and 0.008 mol/dm^3 (namely AgNP1, AgNP2, AgNP3, AgNP4, and AgNP5, respectively), showed that with an increase of concentration of the nanocomposites, the number of damaged bacteria increases (Fig. 9.15A) [67]. The antibacterial activity of several concentrations of ZnO NFs was tested. Fig. 9.15B represents curve of the survival ratio of the bacteria versus time was drawn for these concentrations. With increasing ZnO NPs concentrations, the survival ratio decreased, showing that with an increase in thr concentration of NPs, their antibacterial activity improves. When the ZnO NFs concentration was 0.125 g/dm^3, the survival ratio did not reduce during 120 min. In contrast, by increasing the NFs concentration to 0.25 or 0.5 g/dm^3, the survival ratio reduced and reached minimum values of 0.72 and 0.21 g/dm^3 after 40 and 120 min, respectively. The survival ratio for 0.5 g/dm^3 of NFs showed a greater decrease with respect to that for g/dm^3. The obtained results revealed that more antibacterial activity against *E. coli*

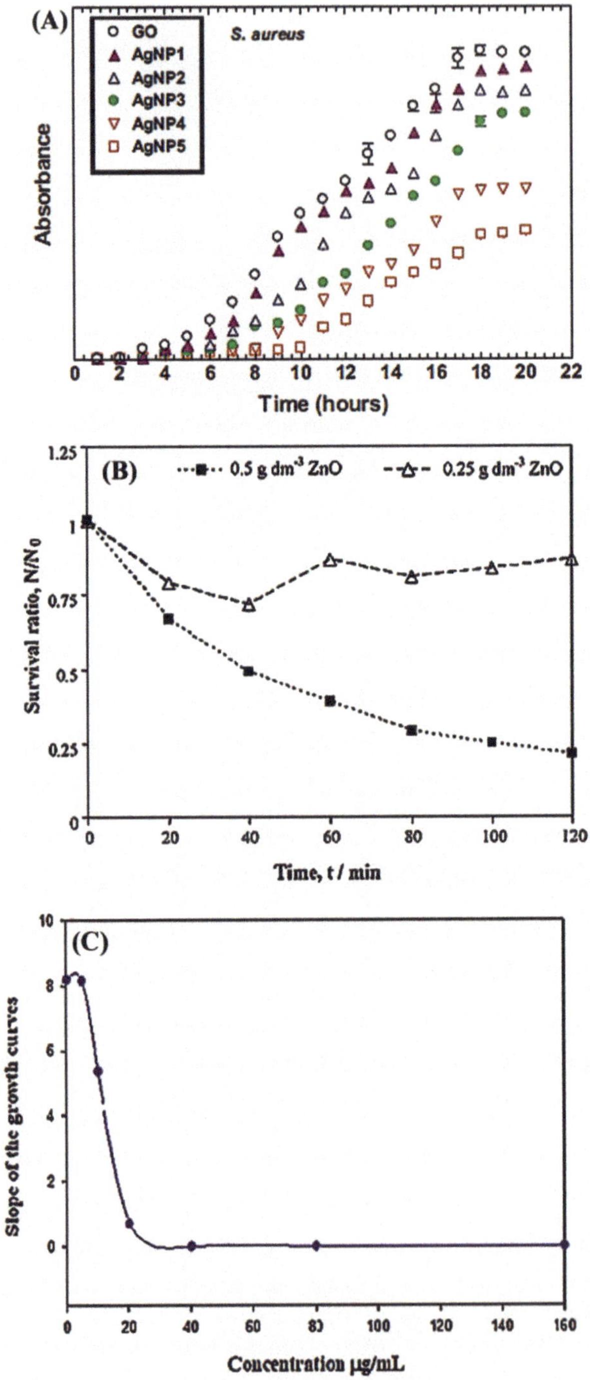
(A)
GO
AgNP1
AgNP2
AgNP3
AgNP4
AgNP5
S. aureus
Absorbance
0 2 4 6 8 10 12 14 16 18 20 22
Time (hours)
(B)
0.5 g dm⁻³ ZnO
0.25 g dm⁻³ ZnO
1.25 1 0.75 0.5 0.25 0
Survival ratio, N/N₀
0 20 40 60 80 100 120
Time, t / min
(C)
10 8 6 4 2 0
Slope of the growth curves
0 20 40 60 80 100 120 140 160
Concentration µg/mL

(Continued)

with the increase of concentration of NPs. The main reason is that with the increase of concentration of ZnO NFs and time, the concentration of H_2O_2 produced from the ZnO surface increases.

The minimum inhibitory concentration (MIC) of an antibacterial agent is the minimum value required to have the bacteriostatic effect. Moghayedi et al. [70] measured the MIC of Ag-RGO nanocomposite with different concentrations ranging from 5 to 40 μg/mL using the microdilution technique. The Ag-RGO NFs showed no antibacterial activity in the concentrations of 5 and 10 μg/mL. In contrast, the cellular viability decreased significantly in concentrations greater than 10 μg/mL so that in these concentrations the bacterial growth was totally prevented. Therefore, 20 μg/mL of Ag-RGO NFs was considered to be MIC.

The percentage of growth inhibition (GI %) is explained by:

$$\mathrm{GI}(\%) = \left[1 - \frac{OD_{630}(\text{sample})}{OD_{630}\quad(\text{control})}\right] \times 100 \tag{9.12}$$

GI % was calculated for Ag-RGO NFs with several concentrations at different times. It was found that an increase of concentration of Ag-RGO results in an increase of the GI rate. For concentrations of 5 and 10 μg/mL, GI % decreased sharply with time. The bacteria growth was completely inhibited until 2 h at a concentration of 20 μg/mL of Ag-RGO. Therefore, the bacteria growth was totally prevented during 6 h for 20, 40, and 80 μg/mL of the NFs. The slope of the growth curves of *E. coli* depended on the NFs concentration (Fig. 9.15C) so that the slope of the growth curves reduced sharply with increasing nanocomposite concentration up to 20 μg/mL. However, due to the complete inhibiton of the bacteria growth, the slopes did not change at higher concentrations.

FIGURE 9.15 (A) Curves of growth kinetics of *Staphylococcus aureus.* (B) Changes in survival ratio of *E. coli* through concentrations of 0.25 and 0.5 g/dm^3 of ZnO NFs. (C) The growth slopes of the bacteria versus concentrations of the NFs. *Reproduced with permission (A) Reproduced with permission from M. Karami, M.A. Akhavan-Behabadi, M.R. Dehkordi, S. Delfani, Thermo-optical properties of copper oxide nanofluids for direct absorption of solar radiation, Sol. Energy Mater. Sol. Cells. 144 (2016) 136–142. Copyright 2013, Elsevier. (B) Reproduced with permission from Asadi, E.K. Goharshadi, M. Sadeghinia, Highly efficient solar-catalytic degradation of reactive black 5 dye using mesoporous plasmonic Ag/g-C 3 N 4 nanocomposites, ChemistrySelect 5 (2020) 2735–2745. https://doi.org/10.1002/slct.201904730. Copyright 2010, Elsevier. (C) Reproduced with permission from M.M. Ghafurian, H. Niazmand, F.T. Dastjerd, O. Mahian, A study on the potential of carbon-based nanomaterials for enhancement of evaporation and water production, Chem. Eng. Sci. 207 (2019) 79–90. https://doi.org/10.1016/j.ces.2019.05.043. Copyright 2017, Elsevier.*

9.2.4.2 *Effect of size of NPs*

The influence of the size of the metal oxide NPs on their antibacterial activity was assessed in some reports. Yamamoto et al. [79] investigated the effect of size of ZnO NPs ranging from 100 to 800 nm. The electrical conductivity was used to assess the antibacterial ability. It was observed that with the decrease in particle size, the antibacterial ability increased. Makhluf et al. [58] prepared MgO NPs using micowave with several particle sizes (from 8 to 23 nm) to investigate their antibacterial activity. The results showed that a decrease in size of the MgO NPs with the concentration of 1 g/L results in increasing the antibacterial ability. Two ZnO NFs with the average particle size of 2417 nm (named as ZnO-4) and 230 nm (named as ZnO-5) were prepared with the concentration of 0.1 g/L by Zhang et al. [65]. The NPs with smaller size showed much better bacteriostatic activity. They found that ZnO NPs with a large average particle size of 2417 nm showed little antibacterial activity compared with the ZnO NPs with a size of 230 nm. The antibacterial activity of various CGCDs with different particle sizes, namely s-CGCDs, m-CGCDs, and l-CGCDs, respectively, were

FIGURE 9.16 Images of *Escherichia coli* and *Staphylococcus aureus* after treatment by CGCDs with different particle sizes. *Reproduced with permission S.U.S. Choi, J.A. Eastman, Enhancing Thermal Conductivity of Fluids with Nanoparticles, Argonne National Lab., IL, 1995. Copyright 2021, Elsevier.*

investigated (Fig. 9.16). The NPs with smaller size (s-CGCDs) showed much better bacteriostatic activity [66].

9.2.4.3 Effect of the presence of a stabilizer

The stabilizers can play an effective role in the antibacterial activity of NFs. Zhang et al. [65] prepared ZnO NFs with the concentration of 0.1 g/L and average particle size of 249, 293, and 314 nm, which were named as ZnO-1, ZnO-2, and ZnO-3, respectively. They used polyethylene glycol (PEG) and polyvinylpyrrolidone (PVP) as stabilizing agents. The results showed that both PEG and PVP have negligible effects on the antibacterial activity of ZnO NFs. However, PVP had a slight negative impact on the antibacterial ability of NFs. Cho et al. [89] investigated the effect of stabilizers of PVP and sodium dodecylsulfate (SDS) on the antibacterial activity of suspensions of Ag and Pt NPs. Their results revealed that although Ag and Pt NPs have excellent antibacterial activities, the SDS-stabilized Ag and Pt NPs have no antibacterial activity. Jalal et al. [53] prepared the ZnO NFs by dispersing ZnO NPs in glycerol as a base fluid in the presence of ammonium citrate. They investigated the trend in the survival ratio in several conditions: (1) without ZnO NPs, solvent, and dispersant (circle symbol) and (2) with solvent and dispersant with concentrations of 0.25 and 0.5 g/dm^3, shown by triangle and square symbols, respectively, in Fig. 9.17. Although the effect of

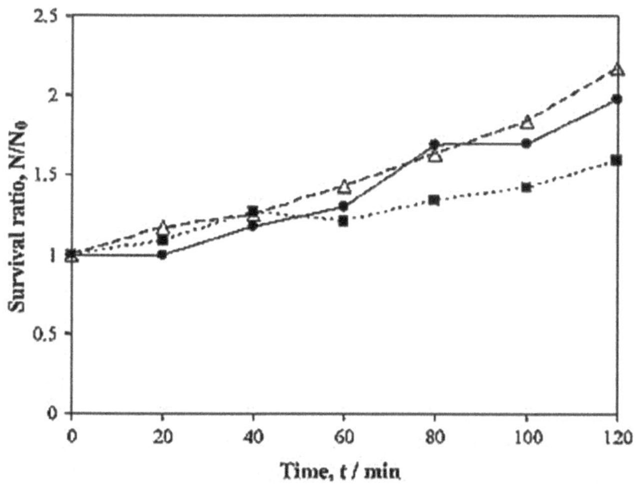

FIGURE 9.17 Changes in survival ratio of *Escherichia coli* without ZnO NPs without (circle symbol) and with solvent and 0.25 and 0.5 g/dm^3 of dispersant (triangle and square symbols, respectively). *Reproduced with permission F. Asadi, E.K. Goharshadi, M. Sadeghinia, Highly efficient solar-catalytic degradation of reactive black 5 dye using mesoporous plasmonic Ag/g-C 3 N 4 nanocomposites, ChemistrySelect 5 (2020) 2735–2745. https://doi.org/10.1002/slct.201904730. Copyright 2010, Elsevier.*

dispersant of ammonium citrate was negligible on GI of *E. coli* after 120 min, by 0.5 g/dm^3 of this dispersant, a 19.7% reduction was obserned in bacterial growth with respect to that in the absence of dispersant. This indicates that 0.5 g/dm^3 of this dispersant was more antibacterial activity than that of 0.25 g/m^3. The antibacterial activity of the citrate salts was reported in a variety of research in the literature [90,91]. It has been known that the antibacterial activity of ammonium citrate is better than PVP and PEG since long-chain polymers of PVP and PEG are adsorbed on the surface of NPs and act as a barrier between the wall of a bacterium and NPs.

9.2.4.4 Effect of pH, temperature, and pressure

pH and temperature are two important parameters affecting the the antibacterial activity of NFs. Saliani et al. [92] conducted several experiments at different pHs (4-10), incubation temperatures (25–42°C) in which the bacteria cells can normally grow, and ZnO concentrations (0.37–1.50 mg/mL). The MIC value of ZnO NPs was 1.5 mg/mL. The bacteria growth was determined by measuring the optical density (OD) every 2 h. Their results indicated that at concentrations below the MICs, with increasing pH from acidic to neutral, the GI percentage of *E. coli* decreased showing that the mechanism of antimicrobial properties of ZnO NPs is dependent on pH.

Investigation of temperature effect on antibacterial performance of ZnO NPs revealed that at concentrations less than the values of MICs, the antimicrobial activity of ZnO NPs increased by heat treatment. However, at acidic pH, ZnO NPs presented better antibacterial activity for a wide range of the temperatures.

Liu et al. [68] evaluated the effect of pressure on the antibacterial activity of ZnO NPs in the absence and presence of autoclave. They observe pressure has no effect on antibacterial activity of ZnO NPs.

9.2.5 Rate of antibacterial activity of NPs

The rate of antibacterial activity of NPs is calculated by the bacterial growth curve versus time. The grwoth curve of *E. coli in* the presence of Ag-RGO is shown in Fig. 9.18. It is obvious that the rate of bacteria growth decreases with increasing the nanocomposite concentration [70].

9.2.6 The effect of NFs on the bacteria morphology

Liu et al. [68] assessed the antibacterial activity of ZnO NPs against *E. coli*. They observed the bacteria growth was totally inhibited with the increase of ZnO concentration up to 12 mmol/L. The SEM images of the bacteria cells were taken before and after exposure to ZnO NPs. They observed that ZnO NPs adhered to the membrane of each bacterium cell. After 12 h treatment by ZnO NPs, no significant changes, including size, shape, and appear did

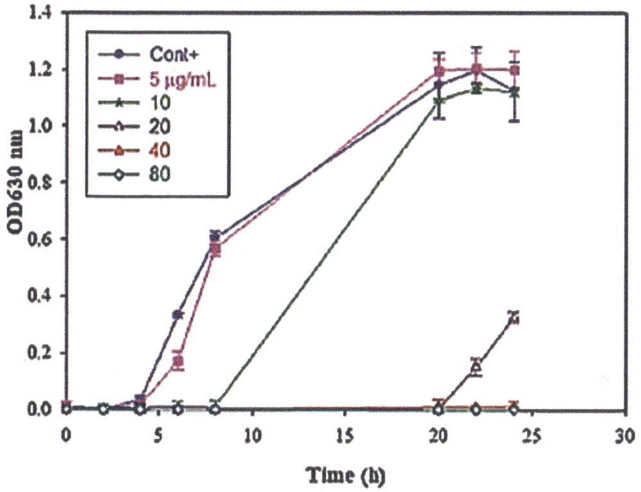

FIGURE 9.18 Growth curve of *Escherichia coli* cells exposed to the Ag-RGO NFs. *Reproduced with permission M.M. Ghafurian, H. Niazmand, F.T. Dastjerd, O. Mahian, A study on the potential of carbon-based nanomaterials for enhancement of evaporation and water production, Chem. Eng. Sci. 207 (2019) 79–90. https://doi.org/10.1016/j.ces.2019.05.043. Copyright 2017, Elsevier.*

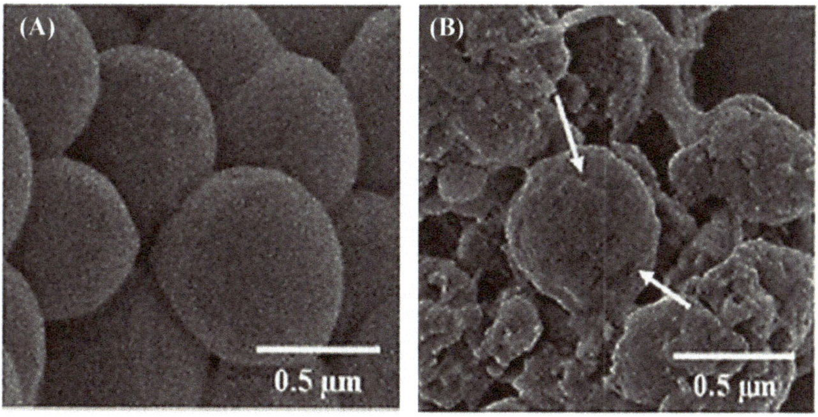

FIGURE 9.19 Surface morphology of *Staphylococcus aureus* shown by SEM images (A) without and (B) with 24 h treatment by GPTMS-ZnO NFs. *Reproduced with permission W. Chen, C. Zou, X. Li, H. Liang, Application of recoverable carbon nanotube nanofluids in solar desalination system: an experimental investigation, Desalination. (2019) 92–101. https://doi.org/10.1016/j.desal.2017.09.025. Copyright 2019, Elsevier.*

not change the morphology of *E. coli* cells. The SEM images of the bacteria cells of *S. aureus* were taken before and after exposure to (3-glycidyloxypropyl) trimethoxysilane-modified ZnO NPs (GPTMS-ZnO) (Fig. 9.19A and B). As shown in Fig. 9.19A, before treatment, the bacteria have a regular surface and spherical structure. After treatment, some holes were created in the cell wall of bacteria, as shown by white arrows. Due to cell death, the bacteria structure is deformed (Fig. 9.19B) [93].

9.3 Conclusion

In this chapter, the use of NFs in desalination of saline waters as well as antibacterial agent was investigated. Based on various reports, it was found that with an increase in NFs concentration, the performance of desalination devices improves. In this regard, the stability of NFs with time is of great importance. Also, by referring to several researches, it was found that the antibacterial activity of NFs increases with the increase of concentration and decrease of size of NPs. pH and temperature may play roles in he antibacterial activity of NFs.

References

[1] S.M. Sajadi, N. Farokhnia, P. Irajizad, M. Hasnain, H. Ghasemi, Flexible artificially-networked structure for ambient/high pressure solar steam generation, J. Mater. Chem. A 4 (2016) 4700–4705. Available from: https://doi.org/10.1039/C6TA01205A.

[2] Mohammad Mustafa Ghafuoryan, et al., Modeling and optimizing of a combined CHP system, compression chiller and reverse osmosis plant (CHP + C + W) in two strategies of connections with grid, Journal of the Brazilian Society of Mechanical Sciences and Engineering 6 (37) (2015) 1751–1763. Available from: https://doi.org/10.1007/s40430-015-0439-9.

[3] F. Asadi, E.K. Goharshadi, M. Sadeghinia, Highly efficient solar-catalytic degradation of reactive black 5 dye using mesoporous plasmonic Ag/g-C_3N_4 nanocomposites, Chem. Select 5 (2020) 2735–2745. Available from: https://doi.org/10.1002/slct.201904730.

[4] C. Chang, C. Yang, Y. Liu, P. Tao, C. Song, W. Shang, et al., Efficient solar-thermal energy harvest driven by interfacial plasmonic heating-assisted evaporation, ACS Appl. Mater. Interfaces 8 (2016) 23412–23418.

[5] Y. Liu, J. Lou, M. Ni, C. Song, J. Wu, N.P. Dasgupta, et al., Bioinspired bifunctional membrane for efficient clean water generation, ACS Appl. Mater. Interfaces 8 (2016) 772–779.

[6] M. Mustafa, Z. Akbari, H. Niazmand, R. Mehrkhah, Eff ect of sonication time on the evaporation rate of seawater containing a nanocomposite, Ultrason.—Sonochem. 61 (2020) 104817. Available from: https://doi.org/10.1016/j.ultsonch.2019.104817.

[7] H.M. Wilson, S.R. Ar, T. Garg, N. Jha, Recycling of hazardous diesel soot particles into a high performance solar evaporation device, Appl. Surf. Sci. 487 (2019) 951–961. Available from: https://doi.org/10.1016/j.apsusc.2019.05.080.

[8] P.W. Bohn, M. Elimelech, J.G. Georgiadis, B.J. Mariñas, A.M. Mayes, A.M. Mayes, Science and technology for water purification in the coming decades, Nanosci. Technol. A Collect. Rev. Nat. J. 452 (2009) 337–346. Available from: https://doi.org/10.1142/9789814287005_0035.

[9] Q. Zhang, S. Chen, Z. Fu, H. Yu, X. Quan, Temperature-difference-induced electricity during solar desalination with bilayer MXene-based monoliths, Nano Energy 76 (2020). Available from: https://doi.org/10.1016/j.nanoen.2020.105060.

[10] L. Zhu, M. Gao, C. Kang, N. Peh, X. Wang, Self-contained monolithic carbon sponges for solar-driven interfacial water evaporation distillation and electricity generation, Adv. Energy Mater. 1702149 (2018) 1–8. Available from: https://doi.org/10.1002/aenm.201702149.

[11] L. Zhu, T. Ding, M. Gao, C. Kang, N. Peh, G.W. Ho, Shape conformal and thermal insulative organic solar absorber sponge for photothermal water evaporation and thermoelectric power generation, Adv. Energy Mater. 1900250 (2019) 3–9. Available from: https://doi.org/10.1002/aenm.201900250.

[12] L. Zhu, M. Gao, C.K.N. Peh, G.W. Ho, Recent progress in solar-driven interfacial water evaporation: advanced designs and applications, Nano Energy 57 (2019) 507–518. Available from: https://doi.org/10.1016/j.nanoen.2018.12.046.

[13] Y. Ito, Y. Tanabe, J. Han, T. Fujita, K. Tanigaki, M. Chen, Multifunctional porous graphene for high-efficiency steam generation by heat localization, Adv. Mater. 27 (2015) 4302–4307.

[14] L. Zhang, B. Tang, J. Wu, R. Li, P. Wang, Hydrophobic light-to-heat conversion membranes with self-healing ability for interfacial solar heating, Adv. Mater. 27 (2015) 4889–4894. Available from: https://doi.org/10.1002/adma.201502362.

[15] G. Ni, N. Miljkovic, H. Ghasemi, X. Huang, S.V. Boriskina, C.-T. Lin, et al., Volumetric solar heating of nanofluids for direct vapor generation, Nano Energy 17 (2015) 290–301. Available from: https://doi.org/10.1016/j.nanoen.2015.08.021.

[16] S.U.S. Choi, J.A. Eastman, Enhancing Thermal Conductivity of Fluids with Nanoparticles, Argonne National Lab., IL, 1995.

[17] M. Karami, M.A. Akhavan-Behabadi, M.R. Dehkordi, S. Delfani, Thermo-optical properties of copper oxide nanofluids for direct absorption of solar radiation, Sol. Energy Mater. Sol. Cells 144 (2016) 136–142.

[18] A.K. Singh, Thermal conductivity of nanofluids, Def. Sci. J. 58 (2008) 600.

[19] X. Wang, Y. He, G. Cheng, L. Shi, X. Liu, J. Zhu, Direct vapor generation through localized solar heating via carbon-nanotube nanofluid, Energy Convers. Manag. 130 (2016) 176–183. Available from: https://doi.org/10.1016/j.enconman.2016.10.049.

[20] M.M. Ghafurian, H. Niazmand, F.T. Dastjerd, O. Mahian, A study on the potential of carbon-based nanomaterials for enhancement of evaporation and water production, Chem. Eng. Sci. 207 (2019) 79–90. Available from: https://doi.org/10.1016/j.ces.2019.05.043.

[21] M.M. Ghafurian, H. Niazmand, E. Ebrahiminia-Bajestan, Improving steam generation and distilled water production by volumetric solar heating, Appl. Therm. Eng. 158 (2019) 113808. Available from: https://doi.org/10.1016/j.applthermaleng.2019.113808.

[22] X. Wang, G. Ou, N. Wang, H. Wu, Graphene-based recyclable photo-absorbers for high-efficiency seawater desalination, ACS Appl. Mater. Interfaces 8 (2016) 9194–9199. Available from: https://doi.org/10.1021/acsami.6b02071.

[23] L. Shi, Y. He, Y. Huang, B. Jiang, Recyclable Fe_3O_4@CNT nanoparticles for high-efficiency solar vapor generation, Energy Convers. Manag. 149 (2017) 401–408. Available from: https://doi.org/10.1016/j.enconman.2017.07.044.

[24] L. Shi, Y. He, X. Wang, Y. Hu, Recyclable photo-thermal conversion and purification systems via Fe_3O_4@TiO2 nanoparticles, Energy Convers. Manag. 171 (2018) 272–278. Available from: https://doi.org/10.1016/j.enconman.2018.05.106.

[25] M.M. Ghafurian, H. Niazmand, E. Ebrahimnia-Bajestan, H. Elhami Nik, Localized solar heating via graphene oxide nanofluid for direct steam generation, J. Therm. Anal. Calorim. 135 (2019) 1443–1449. Available from: https://doi.org/10.1007/s10973-018-7496-0.

[26] X. Liu, X. Wang, J. Huang, G. Cheng, Y. He, Volumetric solar steam generation enhanced by reduced graphene oxide nanofluid, Appl. Energy 220 (2018) 302–312. Available from: https://doi.org/10.1016/j.apenergy.2018.03.097.

[27] X. Wang, Y. He, X. Liu, L. Shi, J. Zhu, Investigation of photothermal heating enabled by plasmonic nanofluids for direct solar steam generation, Sol. Energy 157 (2017) 35–46. Available from: https://doi.org/10.1016/j.solener.2017.08.015.

[28] H. Li, Y. He, Z. Liu, Y. Huang, B. Jiang, Synchronous steam generation and heat collection in a broadband Ag@TiO_2 core–shell nanoparticle-based receiver, Appl. Therm. Eng. 121 (2017) 617–627. Available from: https://doi.org/10.1016/j.applthermaleng.2017.04.102.

[29] Y. Fu, T. Mei, G. Wang, A. Guo, G. Dai, S. Wang, et al., Investigation on enhancing effects of Au nanoparticles on solar steam generation in graphene oxide nanofluids, Appl. Therm. Eng. 114 (2017) 961–968. Available from: https://doi.org/10.1016/j.applthermaleng.2016.12.054.

[30] M.M. Ghafurian, Z. Akbari, H. Niazmand, R. Mehrkhah, S. Wongwises, O. Mahian, Effect of sonication time on the evaporation rate of seawater containing a nanocomposite, Ultrason. Sonochem. 61 (2020) 104817. Available from: https://doi.org/10.1016/j.ultsonch.2019.104817.

[31] X. Liu, J. Huang, X. Wang, G. Cheng, Y. He, Investigation of graphene nanofluid for high efficient solar steam generation, Energy Proc. 142 (2017) 350–355. Available from: https://doi.org/10.1016/j.egypro.2017.12.055.

[32] M. Aziznezhad, E. Goharshadi, M. Namayandeh-Jorabchi, Surfactant-mediated prepared VO_2 (M) nanoparticles for efficient solar steam generation, Sol. Energy Mater. Sol. Cells 211 (2020). Available from: https://doi.org/10.1016/j.solmat.2020.110515.

[33] H. Jin, G. Lin, L. Bai, A. Zeiny, D. Wen, Steam generation in a nanoparticle-based solar receiver, Nano Energy 28 (2016) 397–406. Available from: https://doi.org/10.1016/j.nanoen.2016.08.011.

[34] X. Wang, Y. He, X. Liu, J. Zhu, Enhanced direct steam generation via a bio-inspired solar heating method using carbon nanotube films, Powder Technol. 321 (2017) 276–285.

[35] X. Wang, Y. He, X. Liu, G. Cheng, J. Zhu, Solar steam generation through bio-inspired interface heating of broadband-absorbing plasmonic membranes, Appl. Energy 195 (2017) 414–425.

[36] M.M. Ghafurian, H. Niazmand, E. Ebrahimnia Bejestan, Performance evaluation of multi-wall carbon nanotube in solar fresh water production, Amirkabir J. Mech. Eng. (2018).

[37] M.S. Zielinski, J.-W. Choi, T. La Grange, M. Modestino, S.M.H. Hashemi, Y. Pu, et al., Hollow mesoporous plasmonic nanoshells for enhanced solar vapor generation, Nano Lett. 16 (2016) 2159–2167.

[38] W. Chen, C. Zou, X. Li, L. Li, Experimental investigation of SiC nanofluids for solar distillation system: stability, optical properties and thermal conductivity with saline water-based fluid, Int. J. Heat. Mass. Transf. 107 (2017) 264–270.

[39] S.K. Kohl, J.D. Landmark, D.F. Stickle, Demonstration of absorbance using digital color image analysis and colored solutions, J. Chem. Educ. 83 (2006) 644.

[40] D. Dey, P. Kumar, S. Samantaray, A review of nanofluid preparation, stability, and thermo-physical properties, Heat. Transf. Res. 46 (2017) 1413–1442.

[41] W. Chen, C. Zou, X. Li, H. Liang, Application of recoverable carbon nanotube nanofluids in solar desalination system: an experimental investigation, Desalination (2019) 92–101. Available from: https://doi.org/10.1016/j.desal.2017.09.025.

[42] M.M. Ghafurian, F.T. Dastjerd, A. Afsharian, F.R. Esfahani, H. Niazmand, H. Behzadnia, et al., Low-cost zinc-oxide nanoparticles for solar-powered steam production: superficial and volumetric approaches, J. Clean. Prod. 280 (2021) 124261. Available from: https://doi.org/10.1016/j.jclepro.2020.124261.

[43] A.O. Govorov, H.H. Richardson, Generating heat with metal nanoparticles, Nano Today 2 (2007) 30.

[44] A. Zeiny, H. Jin, G. Lin, P. Song, D. Wen, Solar evaporation via nanofluids: a comparative study, Renew. Energy 122 (2018) 443–454.

[45] M.M. Ghafurian, H. Niazmand, E. Ebrahimnia-Bajestan, R.A. Taylor, Wood surface treatment techniques for enhanced solar steam generation, Renew. Energy 146 (2020) 2308–2315.

[46] M.M. Ghafurian, H. Niazmand, A.E. Moallemi, F. Tavakoli Dastjerd, Experimental investigation of nanofluid based on titanium dioxide nanoparticles in absorption sunlight and steam generation, Modares Mech. Eng. 20 (2020) 57–65.

[47] Y. Wang, R. Shi, J. Lin, Y. Zhu, Enhancement of photocurrent and photocatalytic activity of ZnO hybridized with graphite-like C_3N_4, Energy Environ. Sci. 4 (2011) 2922–2929.

[48] T. Mahvelati-Shamsabadi, E.K. Goharshadi, M. Shafaee, Z. Niazi, ZnS@ reduced graphene oxide nanocomposite as an effective sunlight driven photocatalyst for degradation of reactive black 5: a mechanistic approach, Sep. Purif. Technol. 202 (2018) 326–334. Available from: https://doi.org/10.1016/j.seppur.2018.04.001.

[49] A.M. Golsheikh, H.N. Lim, R. Zakaria, N.M. Huang, Sonochemical synthesis of reduced graphene oxide uniformly decorated with hierarchical ZnS nanospheres and its enhanced photocatalytic activities, RSC Adv. 5 (2015) 12726–12735.

[50] M. Movahedi, A.R. Mahjoub, S. Janitabar-Darzi, Photodegradation of congo red in aqueous solution on ZnO as an alternative catalyst to TiO_2, J. Iran. Chem. Soc. 6 (2009) 570–577.

[51] S. Zhan, D. Zhu, S. Ma, W. Yu, Y. Jia, Y. Li, et al., Highly efficient removal of pathogenic bacteria with magnetic graphene composite, ACS Appl. Mater. Interfaces 7 (2015) 4290–4298. Available from: https://doi.org/10.1021/am508682s.

[52] K. Krishnamoorthy, M. Veerapandian, L.-H. Zhang, K. Yun, S.J. Kim, Antibacterial efficiency of graphene nanosheets against pathogenic bacteria via lipid peroxidation, J. Phys. Chem. C 116 (2012) 17280–17287.

[53] R. Jalal, E.K. Goharshadi, M. Abareshi, M. Moosavi, A. Yousefi, P. Nancarrow, ZnO nanofluids: green synthesis, characterization, and antibacterial activity, Mater. Chem. Phys. 121 (2010) 198–201.

[54] J. Sawai, Quantitative evaluation of antibacterial activities of metallic oxide powders (ZnO, MgO and CaO) by conductimetric assay, J. Microbiol. Methods 54 (2003) 177–182.

[55] Y.L. Wang, Y.Z. Wan, X.H. Dong, G.X. Cheng, H.M. Tao, T.Y. Wen, Preparation and characterization of antibacterial viscose-based activated carbon fiber supporting silver, Carbon NY 36 (1998) 1567–1571.

[56] C.J. Hewitt, S.R. Bellara, A. Andreani, G. Nebe-von-Caron, C.M. McFarlane, An evaluation of the anti-bacterial action of ceramic powder slurries using multi-parameter flow cytometry, Biotechnol. Lett. 23 (2001) 667–675.

[57] G. Fu, P.S. Vary, C.-T. Lin, Anatase TiO2 nanocomposites for antimicrobial coatings, J. Phys. Chem. B 109 (2005) 8889–8898.

[58] S. Makhluf, R. Dror, Y. Nitzan, Y. Abramovich, R. Jelinek, A. Gedanken, Microwave-assisted synthesis of nanocrystalline MgO and its use as a bacteriocide, Adv. Funct. Mater. 15 (2005) 1708–1715.

[59] P.K. Stoimenov, R.L. Klinger, G.L. Marchin, K.J. Klabunde, Metal oxide nanoparticles as bactericidal agents, Langmuir 18 (2002) 6679–6686.

[60] A. Rana, A review on nanofluids: preparation and application 5, 2019, 12–18.

[61] J. Sawai, H. Igarashi, A. Hashimoto, T. Kokugan, M. Shimizu, Effect of particle size and heating temperature of ceramic powders on antibacterial activity of their slurries, J. Chem. Eng. Jpn. 29 (1996) 251–256.

[62] J. Sawai, I. Saito, F. Kanou, H. Igarashi, A. Hashimoto, T. Kokugan, et al., Mutagenicity test of ceramic powder which have growth inhibitorory effect on bacteria, J. Chem. Eng. Jpn. 28 (1995) 352–354.

[63] J. Sawai, H. Igarashi, A. Hashimoto, T. Kokugan, M. Shimizu, Effect of particle size and heating temperature of ceramic powders on antibacterial activity of their slurries, J. Chem. Eng. Jpn. 29 (1996) 251–256. Available from: https://doi.org/10.1252/jcej.29.251.

[64] J. Sawai, E. Kawada, F. Kanou, H. Igarashi, A. Hashimoto, T. Kokugan, et al., Detection of active oxygen generated from ceramic powders having antibacterial activity, J. Chem. Eng. Jpn. 29 (1996) 627–633.

[65] L. Zhang, Y. Jiang, Y. Ding, M. Povey, D. York, Investigation into the antibacterial behaviour of suspensions of ZnO nanoparticles (ZnO nanofluids), J. Nanopart. Res. 9 (2007) 479–489. Available from: https://doi.org/10.1007/s11051-006-9150-1.

[66] B. Sun, F. Wu, Q. Zhang, X. Chu, Z. Wang, X. Huang, et al., Insight into the effect of particle size distribution differences on the antibacterial activity of carbon dots, J. Colloid Interface Sci. 584 (2021) 505–519. Available from: https://doi.org/10.1016/j.jcis.2020.10.015.

[67] M.R. Das, R.K. Sarma, S.C. Borah, R. Kumari, R. Saikia, A.B. Deshmukh, et al., The synthesis of citrate-modified silver nanoparticles in an aqueous suspension of graphene oxide nanosheets and their antibacterial activity, Colloids Surf. B Biointerfaces 105 (2013) 128–136. Available from: https://doi.org/10.1016/j.colsurfb.2012.12.033.

[68] Y. Liu, L. He, A. Mustapha, H. Li, Z.Q. Hu, M. Lin, Antibacterial activities of zinc oxide nanoparticles against *Escherichia coli* O157: H7, J. Appl. Microbiol. 107 (2009) 1193–1201.

[69] O. Mahapatra, M. Bhagat, C. Gopalakrishnan, K.D. Arunachalam, Ultrafine dispersed CuO nanoparticles and their antibacterial activity, J. Exp. Nanosci. 3 (2008) 185–193. Available from: https://doi.org/10.1080/17458080802395460.

[70] M. Moghayedi, E.K. Goharshadi, K. Ghazvini, H. Ahmadzadeh, L. Ranjbaran, R. Masoudi, et al., Kinetics and mechanism of antibacterial activity and cytotoxicity of Ag-RGO nanocomposite, Colloids Surf. B Biointerfaces 159 (2017) 366–374. Available from: https://doi.org/10.1016/j.colsurfb.2017.08.001.

[71] J. Shen, T. Li, M. Shi, N. Li, M. Ye, Polyelectrolyte-assisted one-step hydrothermal synthesis of Ag-reduced graphene oxide composite and its antibacterial properties, Mater. Sci. Eng. C 32 (2012) 2042–2047.

[72] A. Rufus, N. Sreeju, D. Philip, Synthesis of biogenic hematite (α-Fe2O3) nanoparticles for antibacterial and nanofluid applications, RSC Adv. 6 (2016) 94206–94217. Available from: https://doi.org/10.1039/c6ra20240c.

[73] R. Sonohara, N. Muramatsu, H. Ohshima, T. Kondo, Difference in surface properties between *Escherichia coli* and *Staphylococcus aureus* as revealed by electrophoretic mobility measurements, Biophys. Chem. 55 (1995) 273–277.

[74] E.N. Taylor, T.J. Webster, The use of superparamagnetic nanoparticles for prosthetic biofilm prevention, Int. J. Nanomed. 4 (2009) 145.

[75] J.S. Kim, E. Kuk, K.N. Yu, J.-H. Kim, S.J. Park, H.J. Lee, et al., Antimicrobial effects of silver nanoparticles, Nanomed. Nanotechnol. Biol. Med. 3 (2007) 95–101.

[76] R.A. Ismail, G.M. Sulaiman, S.A. Abdulrahman, T.R. Marzoog, Antibacterial activity of magnetic iron oxide nanoparticles synthesized by laser ablation in liquid, Mater. Sci. Eng. C 53 (2015) 286–297.

[77] O. Yamamoto, J. Sawai, T. Sasamoto, Change in antibacterial characteristics with doping amount of ZnO in MgO–ZnO solid solution, Int. J. Inorg. Mater. 2 (2000) 451–454.

[78] J. Sawai, S. Shoji, H. Igarashi, A. Hashimoto, T. Kokugan, M. Shimizu, et al., Hydrogen peroxide as an antibacterial factor in zinc oxide powder slurry, J. Ferment. Bioeng. 86 (1998) 521–522.

[79] O. Yamamoto, Influence of particle size on the antibacterial activity of zinc oxide, Int. J. Inorg. Mater. 3 (2001) 643–646.

[80] T. Ohira, O. Yamamoto, Y. Iida, Z. Nakagawa, Antibacterial activity of ZnO powder with crystallographic orientation, J. Mater. Sci. Mater. Med. 19 (2008) 1407–1412.

[81] N. Padmavathy, R. Vijayaraghavan, Enhanced bioactivity of ZnO nanoparticles—an antimicrobial study, Sci. Technol. Adv. Mater. 9 (2008) 35004.

[82] A.F. de Faria, D.S.T. Martinez, S.M.M. Meira, A.C.M. de Moraes, A. Brandelli, A.G. Souza Filho, et al., Anti-adhesion and antibacterial activity of silver nanoparticles supported on graphene oxide sheets, Colloids Surf. B Biointerfaces 113 (2014) 115–124.

[83] L. Shahriary, R. Nair, S. Sabharwal, A.A. Athawale, One-step synthesis of Ag–reduced graphene oxide–multiwalled carbon nanotubes for enhanced antibacterial activities, N. J. Chem. 39 (2015) 4583–4590.

[84] T. Zhang, L. Wang, Q. Chen, C. Chen, Cytotoxic potential of silver nanoparticles, Yonsei Med. J. 55 (2014) 283.

[85] Y. Yamamoto, N. Imai, R. Mashima, R. Konaka, M. Inoue, W.C. Dunlap, [3] Singlet oxygen from irradiated titanium dioxide and zinc oxide, Methods Enzymol. 319 (2000) 29–37.

[86] M. Fang, J.-H. Chen, X.-L. Xu, P.-H. Yang, H.F. Hildebrand, Antibacterial activities of inorganic agents on six bacteria associated with oral infections by two susceptibility tests, Int. J. Antimicrob. Agents 27 (2006) 513–517.

[87] N. Jones, B. Ray, K.T. Ranjit, A.C. Manna, Antibacterial activity of ZnO nanoparticle suspensions on a broad spectrum of microorganisms, FEMS Microbiol. Lett. 279 (2008) 71–76.

[88] G. Applerot, A. Lipovsky, R. Dror, N. Perkas, Y. Nitzan, R. Lubart, et al., Enhanced antibacterial activity of nanocrystalline ZnO due to increased ROS-mediated cell injury, Adv. Funct. Mater. 19 (2009) 842–852.

[89] K.-H. Cho, J.-E. Park, T. Osaka, S.-G. Park, The study of antimicrobial activity and preservative effects of nanosilver ingredient, Electrochim. Acta 51 (2005) 956–960.

[90] Y. Lee, T. Cesario, J. Owens, E. Shanbrom, L.D. Thrupp, Antibacterial activity of citrate and acetate, Nutrition 18 (2002) 665–666.

[91] M.C. Weijmer, Y.J. Debets-Ossenkopp, F.J. Van De Vondervoort, P.M. ter Wee, Superior antimicrobial activity of trisodium citrate over heparin for catheter locking, Nephrol. Dial. Transplant. 17 (2002) 2189–2195.

[92] M. Saliani, R. Jalal, E.K. Goharshadi, Effects of pH and temperature on antibacterial activity of zinc oxide nanofluid against Escherichia coli O157: H7 and Staphylococcus aureus, Jundishapur J. Microbiol. 8 (2015) 1–6. Available from: https://doi.org/10.5812/jjm.17115.

[93] B. Lallo da Silva, B.L. Caetano, B.G. Chiari-Andréo, R.C.L.R. Pietro, L.A. Chiavacci, Increased antibacterial activity of ZnO nanoparticles: influence of size and surface modification, Colloids Surf. B Biointerfaces 177 (2019) 440–447. Available from: https://doi.org/10.1016/j.colsurfb.2019.02.013.

Chapter 10

Application of nanofluids in combustion engines with focusing on improving heat transfer process

Mohammad Hemmat Esfe[1], Saeed Esfandeh[1,2] and Mohammad Hassan Kamyab[1,3]

[1]*Department of Mechanical Engineering, Imam Hossein University, Tehran, Iran,* [2]*Department of Mechanical Engineering, Jundi-Shapur University of Technology, Dezful, Iran,* [3]*Department of Mechanical Engineering, University of Kashan, Kashan, Iran*

Nomenclature

Abbreviation

ACEA	European Automobile Manufacturers Association (Europe)
AD	Anno Domini
API	American Petroleum Association (United states)
B	constant.
BSFC	brake specific fuel consumption
BTE	brake thermal efficiency
C	constant.
CI	compression ignition
CO	carbon monoxide
COF.	friction coefficient
EGT	exhaust gas temperature
EGR	exhaust gas recirculation
HC	hydrocarbon
ILSAC	International Organization for Standardization and Approval of Committee (Asia)
NOX	nitric oxide
PAO	poly-alpha- olefin
PM	particulate matter
R	universal gas constant
SI	spark-ignition
SAE	Society of Automobile Engineers

Advances in Nanofluid Heat Transfer. DOI: https://doi.org/10.1016/B978-0-323-88656-7.00008-8

TDC top dead center
UHC unburned hydrocarbon

Greek letters

$\triangle$**T (°C)**	temperature difference
$\triangle$**x**	distance of difference
μ **(N.s/m²)**	dynamic viscosity
υ **(m² /s)**	gas kinematic viscosity
ω_P**(rad/s)**	rotation of the paddle wheel
ρ **(kg/m³)**	density

Latin letters

A (m²)	cylinder area (surface area)
C_p**(J/mol.K)**	specific heat at constant pressure
g(m/s²)	gravity acceleration
h (W/m² K)	heat transfer coefficient
h_c**(W/m² K)**	convective heat transfer coefficients
k (W/(m K)	gas thermal conductivity
L (m)	engine stroke
L (m)	cylinder characteristic length
m (kg)	mass
M (kg/mol)	molecular mass
N(m/s)	engine speed
Nu	Nusselt number
P (kpa)	instantaneous cylinder pressure
P_m**(k pas)**	motored cylinder pressure at the same crankangleas p
Pr	working fluid pressure
Pr	Prantel number
Q (W/m²)	overall heat transfer
$\dot{q}$**(r) (kj)**	heat flux for each radius
Re	Reynolds number
$\tilde{R}$**(J/k.mol)**	special gas constant
$\tilde{S}$ **p(m/s)**	mean piston speed
$\tilde{T}$ **g (°c)**	average temperature of the cylinder
Tcool (°C)	coolant temperature
Tgas (°C)	effective gas temperature
TW (°C)	water temperature
Tr (°C)	temperature
u (m/s)	characteristic gas velocity
V (m³)	volume
V_d**(m³)**	displaced volume
ν_s**(m/s)**	swirl velocity
V_r	volume

10.1 Introduction

In this chapter, the heat transfer in combustion engines is discussed, which is an important issue for having suitable engine performance. About one third of the total energy in the fuel that enters the internal combustion engines converts to the useful work of the crankshaft. Another one third of total energy is released by engine exhaust gases. These values indicate that the remaining 30%, that is, remained one third of the total energy entering the engine, should be released to the environment by one of the heat transfer methods. 2700K is the temperature that can be reached inside the combustion chamber of engines. The materials used in the engine can not withstand such temperatures, and if proper heat transfer does not occur, they are quickly destroyed by high heat. Heat dissipation is critical to protecting the engine and engine lubricating fluids from damage. Nanofluids have much better thermal properties than conventional lubricants or conventional cooling fluids. The purpose of using nanofluids is to achieve the highest possible thermal properties at the lowest possible concentration and volume percentage of nanoparticles in the fluids and lubricants flowing in ignition engines. Nanofluids contain nanoparticles with the size of 1–100 nm dispersed in base fluid. Recently research has been conducted to improve the thermal conductivity and thermal behavior of nanofluids in combustion engines [1–4]. The Massachusetts Institute of Technology, the University of Leeds, the Royal

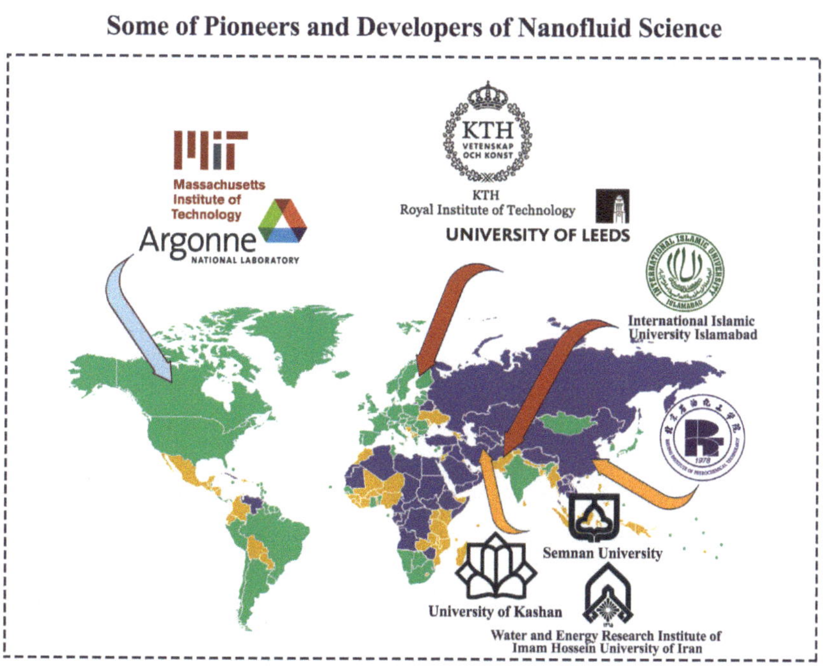

FIGURE 10.1 Some of pioneers and developers in the world.

Swedish Institute of Technology, and the Water and Energy Research Institute of Imam Hossein University of Iran [5–9] are among the leading nanofluid research groups in this field. Fig. 10.1 shows some of pioneers and developer universities in the world in the field of nanofluids.

10.1.1 History of heat transfer in combustion engines

Around 1800 CE, researchers turned their attention to building internal combustion engines that generated power by burning fuel. During the 1800s a number of engines were built, none of which were actually successful, until finally the use of compression in the engine cycle and Barnett's modification [10] on the fuel ignition system in 1838 (Fig. 10.2). The manufacturing of a commercial internal combustion engine was achieved by Jean-Joseph Lenoir [11] in 1860, and internal combustion engines were born. With the development of the engine industry, the need of the society for this industry increased in order to transport, trade, and generate more power. But due to the friction between the moving parts of the engine, a large part of the energy produced by the engine is always converted into thermal energy. Sometimes the heat produced in the engine is so high that it causes the parts

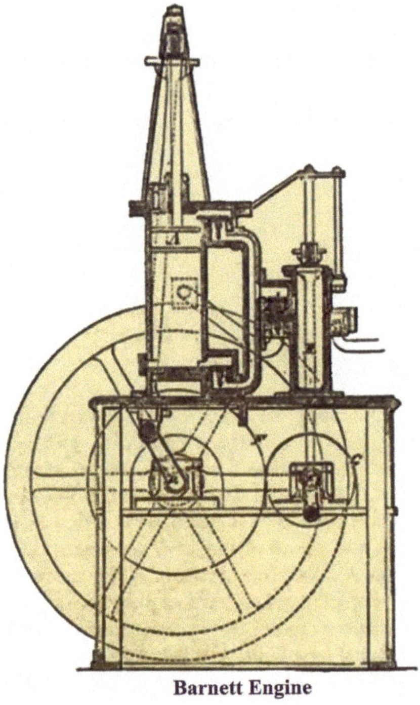

FIGURE 10.2 Invented engine by Barnett [10].

to melt and deform rapidly. Heat transfer in the engines became an important issue and in order to prevent the engines from overheating, a cooling system was used; the cooling operation was done by water and some others worked with air. Therefore, two methods of cooling with water and cooling with air, were used to cool the internal combustion engines. The body of the water-cooled engine is restricted by a water jacket, which is filled with a cooling fluid that circulates inside the engine, whereas the air-cooled engine has surface fins on the body of the engine and the air flows across them.

10.2 Parameters affecting the heat transfer of combustion engines

Heat transfer in engines depends on a variety of variables and parameters, so it is difficult to study all the effective parameters in this chapter. Among the effective parameters on heat transfer of engines the following are noteworthy: engine size, engine speed, spark timing, inlet air temperature, engine materials, engine cooling fluid temperature, and compression ratio [12]. These are described in detail below.

10.2.1 Engine size and dimension

Engines performance depends on too many variables; some of them are working temperature, air–fuel ratio, fuel types, engine material, engine load, etc. In order to analyze the effect of engine size on the performance of the engine all the other effective parameters on engine performance need to considered as constants. In this condition the larger engine has higher heat loss and simultaneously greater thermal efficiency. The higher heat losses of the larger engine is because of its higher surface area. In fact the heat loss per unit of surface area is the same for both small and large engines. Therefore, what is needed to be considered to have good thermal efficiency is the larger ratio of volume to surface area of the engine combustion chamber. This is one of the reasons why the new generation internal combustion engines with overhead valves are more efficient than the L-cylinder engines (flathead engine) with a larger combustion chamber surface area. Fig. 10.3 shows two overhead and flathead engines and their difference.

10.2.2 Engine speed

The higher speed of the working engine will result in turbulent flow and an increase in the heat transfer coefficient. Increasing the engine speed causes an increase in the heat transfer rate at the start of the compression stroke and also in the gas in and out steps. On the other hand in the combustion and power stages the engine speed will not affect gas velocity in the cylinder. So in two abovementioned strokes the engine speed will not affect the heat

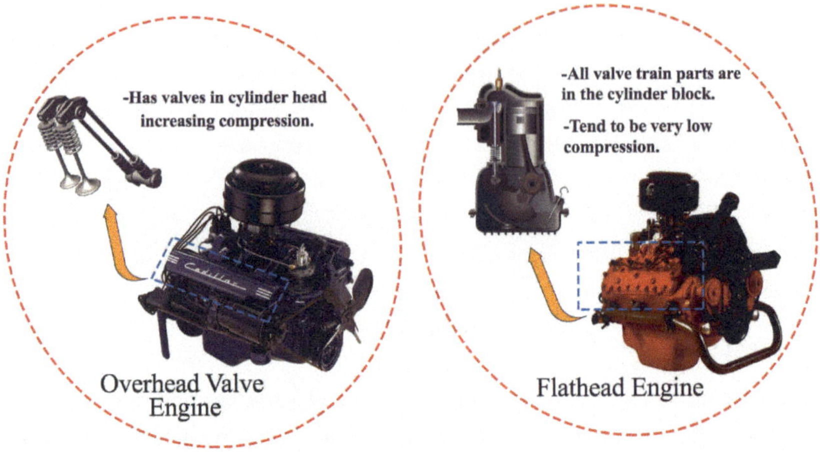

FIGURE 10.3 Overhead and flathead engines.

transfer rate and the heat transfer rate is constant during the combustion. Overall the higher engine speed will result in shorter strokes and less heat transfer time, which improves the thermal efficiency of engines at higher speeds.

10.2.3 Engine load

When the engine load increases, the mass rate of air and fuel at a certain engine speed should be increased with increasing load. The heat transfer in these conditions increases according to the Eq. (10.1):

$$\dot{Q} = hA\triangle T \tag{10.1}$$

In the above formula the heat transfer coefficient is related to the Reynolds number as follows: $h \propto Re^c$.

The coefficient c is usually 0.8 and the Reynolds number and mass rate are proportional. Therefore, the heat transfer rate increases in proportion to $m^{0.8}$. On the other hand, the amount of fuel entering the engine and the amount of engine load are proportional. At light loads, a lower amount of fuel enters and burns and as a result a lower stable temperature is expected. For the heavy loads the condition is the opposite of the aforementioned. Combustion with a richer mixture in heavy loads also produces large amounts of soot and solid carbon, and since solid carbon is a good radiator, it increases radiant heat transfer. So the amount of heat transfer increases as the load on internal combustion engines increases.

10.2.4 Inlet air temperature

An increase in inlet air temperature will cause inappropriate results. It increases the heat loss because of the increased temperature in the engine cycle. Based on Pulkrabek's study [12] a 100°C increase in inlet air temperature may result in a 10%–15% heat loss increase. It also increases the knocking possibility. Super and turbocharged type engines may encounter higher air inlet temperatures more than other engine types. So they need aftercooling systems to reduce the air temperature.

10.2.5 Coolant temperature

Increasing the temperature of the engine cooling fluid leads to an increase in the temperatures of the parts that need to be cooled. Therefore, choosing a fluid with more heat transfer power can prevent engine damage, such as the knocking phenomenon, and can even improve engine power and efficiency.

10.2.6 Engine materials

Engine material type affects the allowed operating temperature of engine parts like the cylinder and piston. For example, in equilibrium conditions, aluminum pistons with higher thermal conductivity are generally 30°C–80°C cooler than cast iron pistons. Fig. 10.4 shows the aluminum piston and its high thermal conductivity.

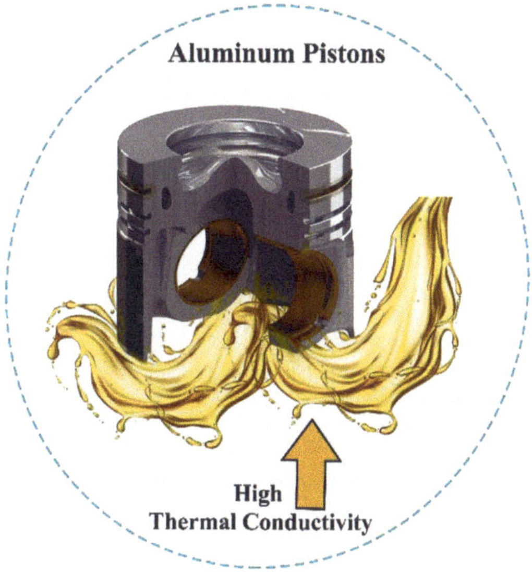

FIGURE 10.4 Aluminum piston.

10.2.7 Friction coefficient

There are many moving parts that come into contact with each other as they move inside internal combustion engines. Any contact of moving parts produces frictional force. The power of the force relates to the collision and engagement of engine parts rate. Friction force is one of the causes of energy loss in the engine. By lubricating the surfaces involved, the amount of these friction losses can be greatly reduced. The most frictional losses are in the friction parts between the piston rings with the cylinder wall, the crankshaft bearings, the valve control mechanism, and the camshaft. Most of the losses are related to the valve mechanism and the camshaft. The internal combustion engine parts engaged with friction are depicted in Fig. 10.5.

Common methods for measuring the friction factor in combustion engines are divided into performance measurement methods, motoring friction measurement, firing/motoring tests, and direct friction measurement. The Williance line and the Morse test methods belong to the firing/motoring test category [13–15].

FIGURE 10.5 View of engine parts engaged with possible frictions.

10.2.8 Heat transfer coefficient

To calculate average heat transfer from the gas to the cylinders' coolant, convective heat transfer can be used [Eq. (10.2)]:

$$\frac{Q}{A} = h_{c,o}\left(\tilde{T}_g - \tilde{T}_c\right) \tag{10.2}$$

It is clear that the heat transfer coefficient relates to engine geometry and engine speed. With the aim of measuring and estimating the heat transfer various correlations are proposed. In the following some of these correlations are presented for the prediction of heat transfer. For example, in Eq. (10.3) the Reynolds number–Nusselt number correlation is proposed for heat transfer coefficient prediction.

Nusselt # = a (Reynolds #) ^{m}

$$Nu = CRe^m Pr^n, \frac{hL}{k} = C\left(\frac{\rho u L}{\mu}\right)^m \left(\frac{\mu c_p}{k}\right)^n \tag{10.3}$$

"u" is the velocity of the gas, but usually the piston velocity will be considered for it. However, there are other various opinions about the gas velocity estimation by using mean velocity or special combustion speed. The parameter of "C" and "m" can be obtained from the moto test. Table 10.1 gives information about three ways of calculating the heat transfer coefficient.

The amount of momentary and average heat transfer coefficient are not equal and the peak values of the momentary are higher. Based on C.F. Taylor's theory [16], which uses piston velocity as a characteristic velocity, for a certain point in time and space the average heat transfer coefficient will be used.

In determining the heat transfer coefficient in internal combustion engines, scientists have pointed to various relationships. For example, the instantaneous spatial coefficient [17] [Eqs. (10.4) and (10.5)]:

TABLE 10.1 Average heat transfer methods.

Heat transfer coefficient	Description
$h(x,t)$	Average heat transfer coefficient as a function of time and space
$h(x,t)$	Momentary and average as a function of space
$h(x,t)$	Momentary, local

$$\left(\frac{h_c B}{\mu}\right) = a\left(\frac{\rho \tilde{S}_P B}{\mu}\right)^b \tag{10.4}$$

The coefficient "a" is strongly dependent on the load and the design of the motor. For the normal combustion, $b = 0.7$ and "a" is between 0.35 and 0.8.

The properties of the gas are evaluated at the average temperature of the cylinder.

$$\tilde{T}_g = \frac{PVM}{m\tilde{R}} \tag{10.5}$$

Also, for the Woschni [18] average spatial-temporal heat transfer coefficient, the average gas velocity is considered equal to the piston velocity [Eqs. (10.6)–(10.13)].

$$Nu = 0.035Re^m, \textbf{assuming } k \propto T^{0.75}, \mu \propto T^{0.62}, P = \rho RT \tag{10.6}$$

$$h_c = CB^{m-1}p^m w^m T^{0.75-1.62m} \tag{10.7}$$

$$w = \left[C_1\tilde{S}_P + C_2\frac{V_d T_r}{p_r V_r}(p - p_m)\right] \tag{10.8}$$

Considering $m = 0.8$

$$h_c = 3.26B(m)^{-.02}p(kpa)^{0.8}T(K)^{-0.55}w(m/s)^{0.8} \tag{10.9}$$

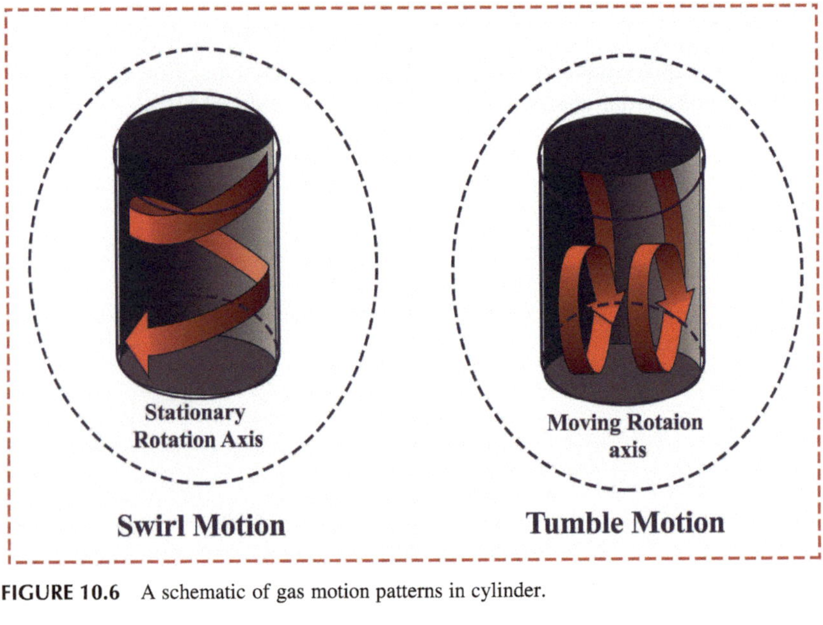

FIGURE 10.6 A schematic of gas motion patterns in cylinder.

$$\tilde{S}_P = 2LN \tag{10.10}$$

$C_1 = 6.18, C_2 = 0$, for the gas ex change period,
$C_1 = 2.28, C_2 = 0$, for the compression period,
$C_1 = 6.18, C_2 = 3.24 \times 10^{-3}$, for the combustion and expansion period.
In engines with swirl motions (Fig. 10.6) the C_1 is as below:
For gas exchange period in engine cycle:

$$C_1 = 6.18 + 0.417 \frac{\nu_S}{\tilde{S}_P} \tag{10.11}$$

For other strokes:

$$C_1 = 2.28 + 0.308 \frac{\nu_S}{\tilde{S}_P} \tag{10.12}$$

where

$$\nu_s = B\omega_P/2 \tag{10.13}$$

and ω_p is the rotation of the paddle wheel that can be used to measure the swirl velocity.

For direct injection diesel engines with swirl motion, the local-temporal heat transfer coefficient was proposed by LeFeuvre et al. [19] and Dent-Sulaiman [20] [Eqs. (10.14) and (10.15)]:

$$\left(\frac{h_c l}{k}\right) = 0.036 \left(\frac{\rho \nu l}{\mu}\right)^{0.8} \left(\frac{\mu c_p}{k}\right)^{0.333}, l = 2\pi r, \nu = rw \tag{10.14}$$

The heat flux for each radius is obtained with the *Pr* number of 0.73:

$$\dot{q}(r) = 0.023 \frac{k}{r} \left(\frac{wr^2}{\nu}\right)^{.8} \left[\tilde{T}_g(r) - T_w(r)\right] \tag{10.15}$$

The measurement of a heat transfer coefficient in internal combustion engines is based on the assumption of considering the heat transfer process inside a cylinder as quasiconstant, in which the heat transfer speed can be considered to be in proportion to the temperature difference between the working fluid and the metal surface at any time. Although this method has been accepted, as Annand stated in 1963 [21], the process of quasifixed analysis cannot be justified by any exact theory or empirical method. The reason for this statement is the fact that heat transfer in a reciprocating engine is different in terms of gas motion (swirl, tumble, etc.).

10.2.9 Conduction heat transfer coefficient

One of the parameters affecting the working fluid in internal combustion engines is the thermal conductivity. Hence, a lot of research is being done to

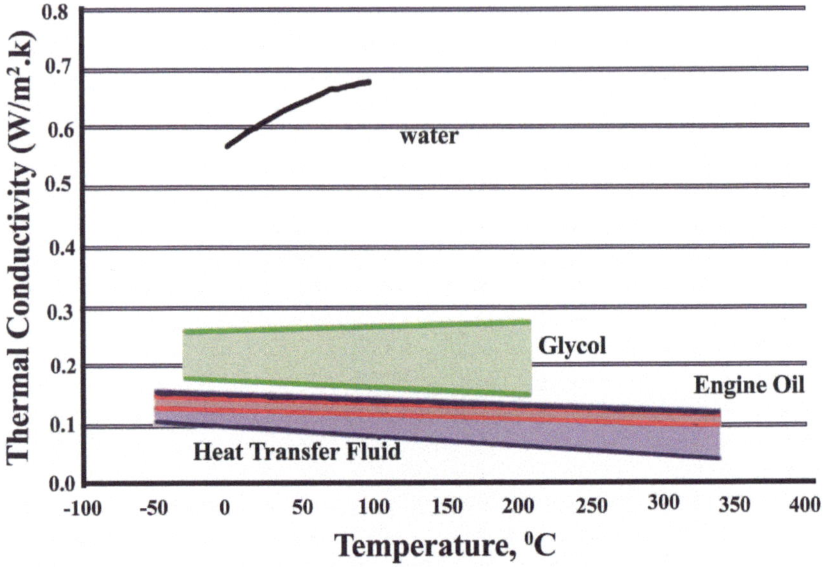

FIGURE 10.7 Thermal conductivity variation with temperature [24].

TABLE 10.2 Thermal conductivity coefficient of various materials.

Material	K (W/m.K)
Copper	400
Aluminum	240
Cast iron	80
Water	0.61
Air	0.026

increase the transport capacity of engine oil [22] to make the engine more efficient. Studies by Wrenick et al. [23] show that the thermal conductivity of oil in internal combustion engines is a function of temperature, and they state that oil with higher heat conduction transfers heat energy more effective. In addition, their studies show the change in thermal conductivity for different fluids with respect to temperature, as shown in Fig. 10.7, which clearly shows that as the engine oil temperature rises, its thermal conductivity decreases

Fourier found a relation between the heat transfer per unit area Q/A (W/m^2) and the temperature gradient dT/dx. In order to change the proportionality to equality, the coefficient of the heat conduction "k" is used. The

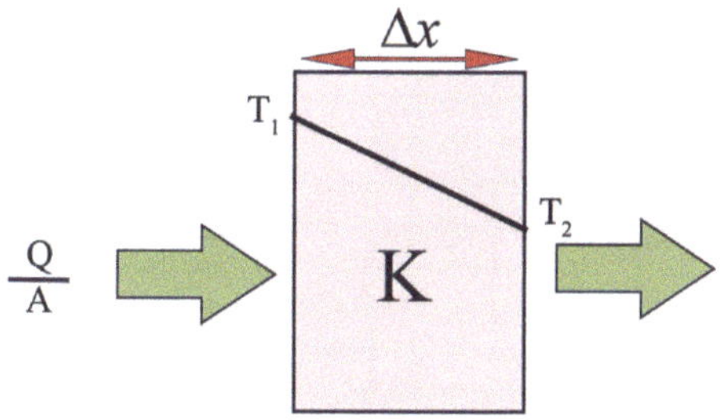

FIGURE 10.8 Conduction through piston cylinder wall.

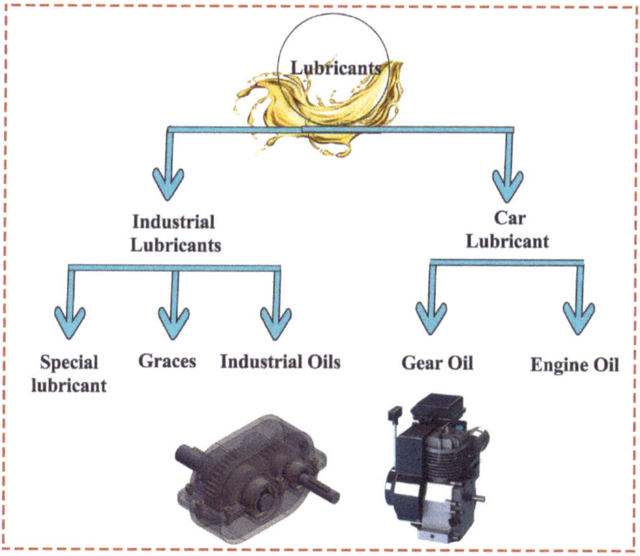

FIGURE 10.9 Various types of lubricants.

"k" coefficient depends on the properties of the material and does not depend on geometry and other parameters. For example, different materials used in engines have the following thermal conductivity coefficient (Table 10.2) (W/m.K)

$$\frac{Q}{A} = -k\frac{dT}{dx} \tag{10.16}$$

For example, the calculation of heat transfer by thermal conductivity between the piston ring and the cylinder wall for a cast iron cylinder block of 0.012 m in steady state is simply calculated as follows (Fig. 10.8):

FIGURE 10.10 Standardization references for engine oils.

$$\frac{Q}{A} = -k\frac{dT}{dx} = \frac{80(100-300)}{0.012} = 1.3 \text{ Mw/m2}$$

10.3 Type of lubricants

Oil is a fluid with various applications and in result there are many types of oils. Lubricants create a film with suitable thickness between moving parts of engine and lower the friction and abrasion while working. This cause energy saving and improve the efficiency of internal combustion engines. The classification of lubricants is depicted in Fig. 10.9.

The heat transfer created by sliding lubricant particles should be removed from the lubrication positions (e.g., bearings) without having an undesirable effect on surfaces with relative motion. For this purpose, various additives such as nanoparticles are used to increase the speed of heat transfer.

10.3.1 Lubrication of combustion engines

Lubricants used in the market today can be divided into four categories: liquid lubricants (mineral oils and synthetic oils), gas lubricants, semisolid lubricants (greases), and solid lubricants (polymer, silver, and gold). A very large percentage of lubricating oils are based on petroleum oils that are used to lubricate internal combustion engines in the transportation. The level of oil quality in internal combustion engines is determined by three standardization references (Fig. 10.10):

1. (API) American Petroleum Association (United States)
2. (ACEA) European Automobile Manufacturers Association (Europe)
3. (ILSAC) International Organization for Standardization and Approval of Committee (Asia).

TABLE 10.3 Temperature range for summer single grade oils.

Summer types	Temperature range (°C)	
	Minimum	Maximum
SAE20	-7	27
SAE30	0	38
SAE40	5	50

TABLE 10.4 Temperature range for winter single grade oils.

Winter types	Temperature range (°C)	
	Minimum	Maximum
SAE20W	−1	27
SAE15W	−38	18
SAE10W	−50	29
SAE5W	−40	0

These standards recommend the use of special oils for different engines. Engine type, fuel type, and engine design year are factors that affect the selection of oil type.

10.3.1.1 Single-grade engine oils

Single -grade oils do not use polymeric additives to improve their viscosity index. These oils are designed and used for warm and balanced weather conditions. Single-grade engine oils are oils that have only one viscosity. These oils have the ability to work at low or high temperatures. In the Society of Automotive Engineering (SAE) system, single-grade oils are divided into two groups. The group with the letter W are also called winter grade oils. These oils are designed to work at low temperatures. The second group, which has no letters and is also known as summer oils, is designed to work at high temperatures. At present, single-grade oil is not used in variable geographical conditions [25]. Tables 10.3 and 10.4 show the appropriate temperature range for the performance of various engine oils based on the SAE standard.

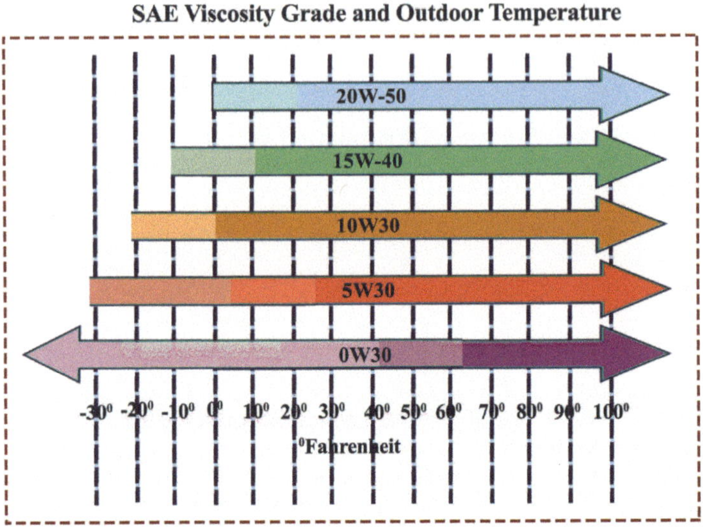

FIGURE 10.11 Society of automotive engineering viscosity grade and outdoor temperature [26].

TABLE 10.5 Multigrade engine oils temperature range.

	Temperature range (°C)	
	Minimum	Maximum
SAE5W20	−35	16
SAE10W40	−24	38
SAE15W50	−10	49

10.3.1.2 Multigrade engine oils

Multigrade oils are obtained by mixing a low-viscosity oil with viscosity index improver polymers. For example, when enough polymers are added to an oil with a SAE 15W viscosity to achieve a SAE40 viscosity, the resulting oil acts like a 15W oil at low temperatures, while at high temperatures it also has the properties of a SAE40 oil. The use of multigrade oils has advantages over single-grade oils; some of them are stated below:

1. It is easier to start the engine, which will reduce engine wear and increase starter and battery life.
2. Fuel consumption reduction.
3. Can be used in all seasons and temperatures with best performance.

4. Oil burning reduction [25].

Multigrade oils, also known as four-season oils, are light oils that by means of viscosity index improvers create the conditions necessary for lubrication at high temperatures. At low temperatures, this oil has the viscosity of W oil and at high temperatures it has the properties of W-free oil, for example, 10W30 SAE oil, which has the viscosity of 10W oil at low temperatures and SAE30 oil characteristics at high temperatures.

The SAE classification system is used only to select the appropriate engine oil grade according to the ambient temperature and engine temperature conditions (Fig. 10.11). This classification system does not provide any information for the consumer about the efficiency and usability of different oils for various engines [27]. Table 10.5 shows the minimum and maximum temperature range for suitable performance of multigrade engine oils.

10.3.2 Basic engine oils and challenges

Today, in addition to lubrication and reducing the wear of engine parts, engine oils have functions such as sealing the gaps between parts, washing and cleaning parts from soot and compounds resulting from the decomposition of oil and sediments, protecting surfaces against rust and chemical corrosion, shock, waste suspensions, and preventing their deposition on the parts. Since liquid lubricants are not able to provide all of the above required properties in an optimal way, engine oil additives are used to achieve or improve many of the above properties. All lubricants (mineral or synthetic based) need compounds or additives that can create new properties or enhance the existing properties to provide optimal performance.

The engine plays a major role in the car and performing this role requires continuous movements of the various engine parts in contact with each other. So the oil is used as a lubricant to reduce friction and wear between them. Lubrication of components in internal combustion engines allows an increase in the efficiency and power of the engine and consequently reduces fuel consumption. However, in the beginning of the engine history, engine oil was only used to lubricate engine moving parts.

10.3.2.1 Engine oils additives to improve lubrication process

Oil additives are often used for reasons such as protection of lubricated surfaces (cleaners, suspenders, antiwear and anticorrosion agents), increase of lubricant efficiency (viscosity index improvers), and protection of lubricants (antifoam materials and antioxidation), and the quality of the oils is measured depending on the type of additives in it.

The increasing use of nanotechnology in various industries has also affected lubricant products. Nanooil additives are used with the aim of

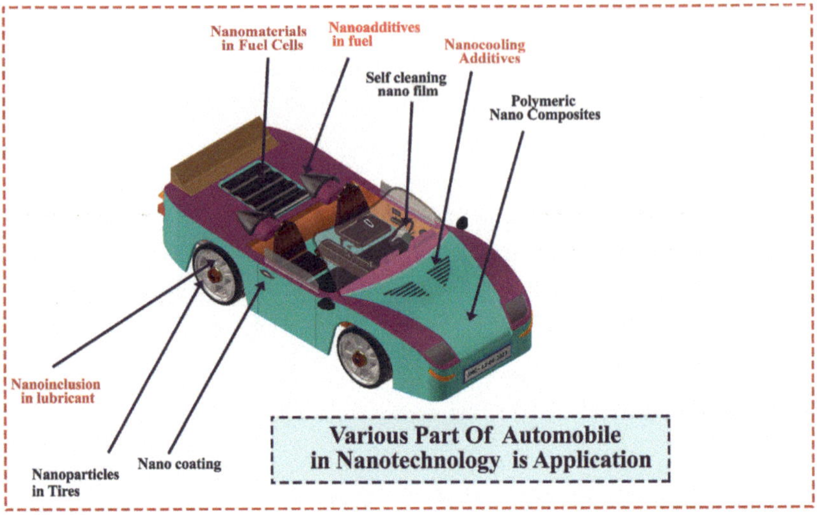

FIGURE 10.12 Application of nanotechnology in various parts of automobile.

reducing the coefficient of friction and wear on engine components, repairing surfaces and improving surface properties, improving thermal properties, improving sealing, reducing engine noise and maintenance costs, increasing engine efficiency and fuel savings, and other goals (Fig. 10.12).

Nanoadditives generally improve and enhance the properties of the base liquid lubricant by four functional mechanisms including the conversion of sliding friction to rolling friction, very high heat transfer capability, friction surface repair, and the formation of a protective film on friction surfaces. In this regard, over the past years, various researchers and companies [28–30] have also introduced and presented various engine oil additives, such as additives containing fullerene, nanodiamond, nanofluorine, and other similar compounds.

10.4 Using nanoparticles in internal combustion engines

Investigating the effects of using nanoparticles in internal combustion engines is one of the most beloved research areas among researchers. In addition to the application of nanoparticles and nanoscience in the automotive paint industry, nanosensors for cars, superhydrophobic car glass and mirrors, and nanoenriched tires, the most important applications of nanoscience and nanoparticles in car industries could be: (1) nanooils for optimizing lubrication purposes; (2) nanofuels as the consumable fluid in cars; and (3) nanocoolants as enriched working fluids in the radiator. In this section, the effect of nanoparticles on each of the abovementioned cases is investigated. Also, the possible side effects of using nanoparticles, such as the effect of

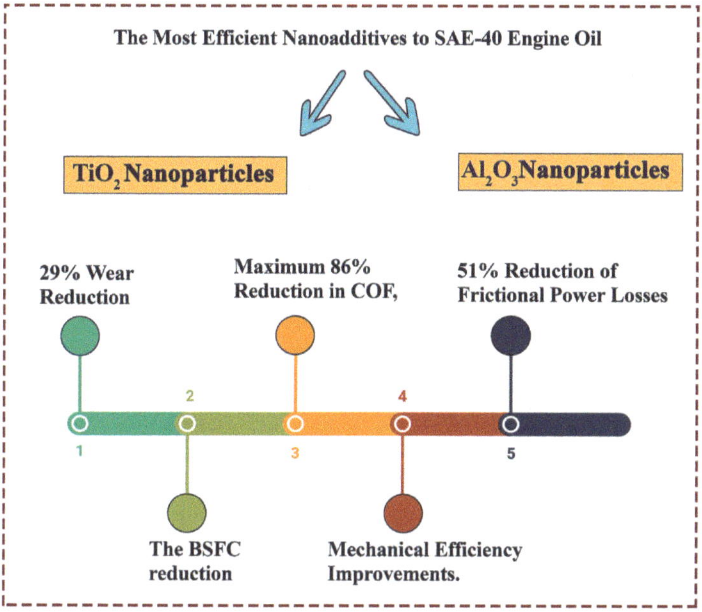

FIGURE 10.13 Best nanoparticles as SAE40 engine oil additives [35].

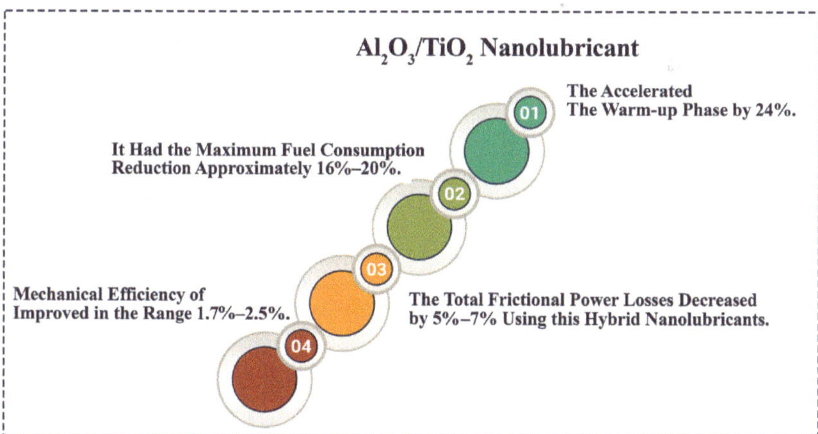

FIGURE 10.14 Al_2O_3/TiO_2 enriched nanolubricants as hybrid nanofluid [35].

nanoparticles on friction and wear between engine parts, engine power, amount of pollutants and pollution caused by engine operation, etc. have been investigated.

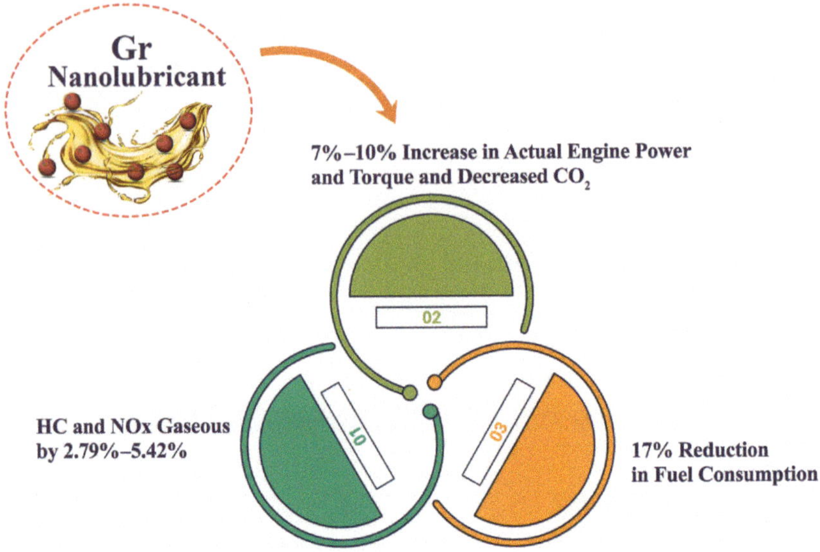

FIGURE 10.15 Graphene-enriched nanolubricants [35].

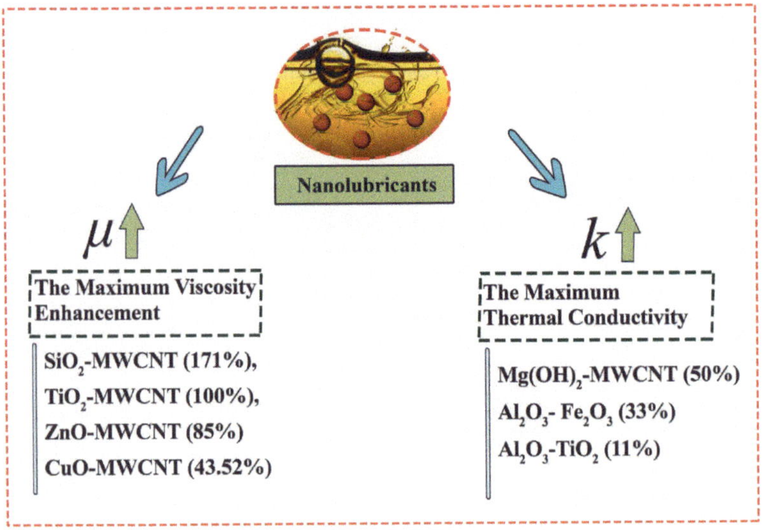

FIGURE 10.16 Maximum thermal conductivity and viscosity after adding nanomaterials [35].

10.4.1 Adding nanoparticles to engine oils (nano oil)

Oils and lubricants are used for various purposes in the car. Some of these oils include engine oil, gearbox oil, differential oil, hydraulic oil, and

TABLE 10.6 Main possible improvements after adding nanoparticles to the oil.

Base oil	Nanoparticles	Type	Main effect or possible improvement
Polyalpha-olefin 4 (PAO) [36]	MoS_2	Simple nanofluid	• Torque reduction between 40% to 50%
Zinc-free 5W30 [32]	MoS_2	Simple nanofluid	• 0.9% reduction in fuel consumption
5W-30 [37]	Al_2O_3, TiO_2	Simple nanofluid	• Investigated part: piston ring • TiO_2 additive effect: 21% reduce in abration rate • Al_2O_3 additive effect: 29% reduce in abration rate
5W-30 [38]	Al_2O_3/TiO_2	Hybrid nanofluid	• Enhancement of heat transfer characteristics by 9%–14%
5W-30 [39]	MoS_2	Simple nanofluid	• Emissions or pollutant reduction
SAE 20W40 [31]	Cu TiO_2	Simple nanofluid	• BSFC improvement that means a reduce in the ratio of fuel consumption rate to generated power
SAE 5W30 [32]	MoS_2	Simple nanofluid	• Positive effect on CO_2 reduction • Positive effect on NOx reduction • Fuel consunmption optimizer
5W30 [34]	Gr	Simple nanofluid	• 17% fuel saving • 3%–5% reduction in exhaust pollutants • Optimizing real engine power for about 7%–10%
5W30 [40]	Al_2O_3/TiO_2	Hybrid nanofluid	• About 16%–20% fuel saving
SAE15W40 [41]	Al_2O_3, SiO_2	Simple nanofluid	• The Al_2O_3 showed better performance in comparison to SiO_2 nanoparticle • The Al_2O_3 reduced frictional wasted power • The Al_2O_3 lowered the BSFC • Wear decreased and engine performance improved at low volume fractions

brake oil. In this section, the effect of adding nanoparticles to some of the mentioned oils is investigated. Various engine types like motorbike [31], diesel [32,33], and gasoline four-cylinder engines [34] have been tested in the literature. Hatami et al. [35] conducted an extensive study on the recent developments using nanosized additives in internal combustion engine oils. Based on their study TiO_2 and Al_2O_3 are the most effective additives for SAE40 engine oil (Fig. 10.13). Also based on the literature the combination of TiO_2 and Al_2O_3 nanoparticles was introduced as one of the best additives in order to synthesize a hybrid nano oil (Fig. 10.14) and graphene enriched nanolubricants were presented as another valuable combination of nano oil (Fig. 10.15). Because of the importance of thermal conductivity coefficient and viscosity of nanolubricants, the percent of thermal conductivity and viscosity increase for some nanoadditive combinations are presented in

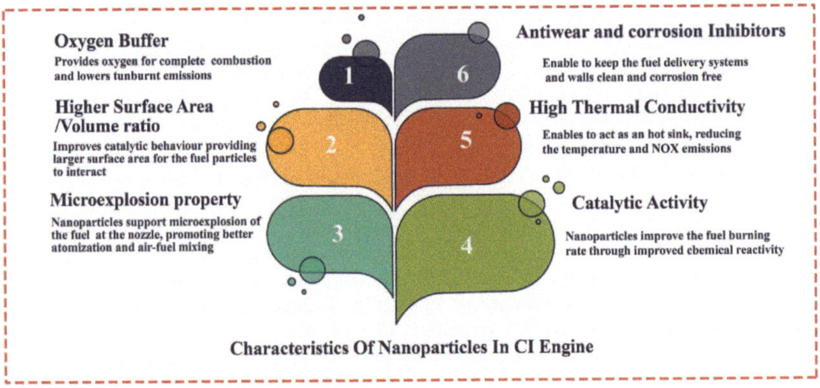

FIGURE 10.17 Nanoparticle function in fuel [42,43].

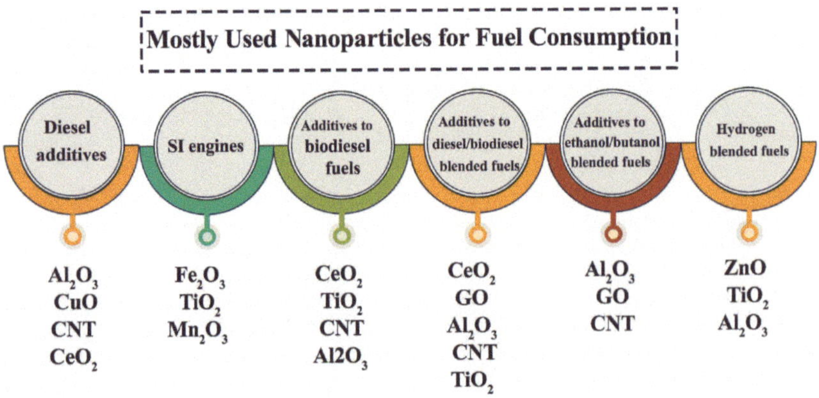

FIGURE 10.18 The most used nanoparticles as fuel additives [44].

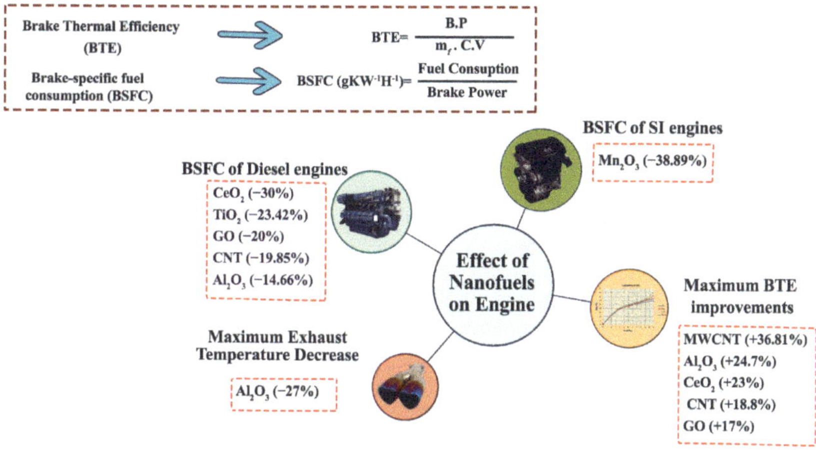

FIGURE 10.19 Best nanoparticle additives based on BTE and BSFC criterions [44].

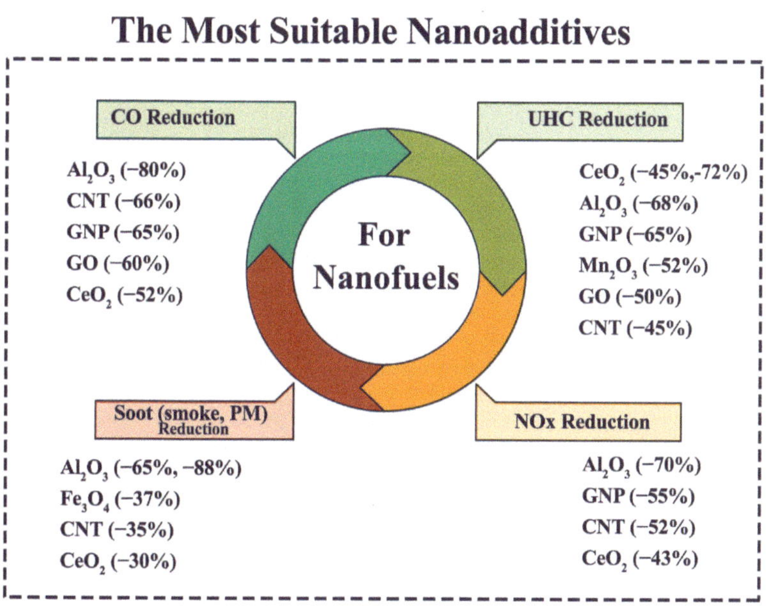

FIGURE 10.20 Best nanoadditives for fuel based on highest CO reduction, UHC reduction, soot reduction, and NOx reduction [44].

Fig. 10.16. Having lower viscosity and higher thermal conductivity after adding nanoparticle combinations is very desirable. Table 10.6 shows the main possible improvements in engine performance after adding nanoparticles to pure oil.

TABLE 10.7 Effect on engine performance of adding nanoparticles to various engine fuels.

Base fuel	Nanoparticle	type	Main effect
Diesel [45]	CNT, MoO_3	Simple nanofluid	• CNT-diesel worked better than MoO_3 nanoparticles based on emissions reduction level
Diesel [46]	Graphite, iron oxide	Simple nanofluid	• Graphite worked better than Iron oxide in terms of viscosity decrease • Power optimization • Decrease in BSFC and NOx reduction
Diesel [47]	Water droplets in nanodimensions	Simple nanofluid	• 14.2% improve in BTE (brake thermal efficiency) • 30.6% reduction in the NOx emissions
Diesel [48]	Cu	Simple nanofluid	• 7%–14% soot reduction
Diesel [49]	TiO_2, copper(II) nitrate, triacetoxycerium hydrate	Simple nanofluid	• TiO_2 and triacetoxycerium hydrate caused pollutants emissions reduction
Diesel [50]	Al_2O_3, CNT, and SiO_2	Simple nanofluid	• SiO_2 acts better than Al_2O_3 as fuel additive • Fuels containing Al_2O_3 and SiO_2 are more stable than those enriched with CNTs • Carbon nanotube blends reduced NOx emissions better than other two blends
Diesel [51]	Al_2O_3, CuO	Simple nanofluid	• Better performance of aluminum oxide nanoparticles in emissions reduction of CO, HC and NOx in comparison to copper oxide
Diesel [52]	MnO and CuO	Simple nanofluid	• More engine power saving • Pollutants controlled
Gasoline [53]	Fe_2O_3 and TiO_2	Simple nanofluid	• Optimization of engine efficiency • Pollutants reduction
Gasoline [54]	Hydrogen nanobubble	Simple nanofluid	• 4% power improvement • About 7.5% BSFC improvement

(Continued)

TABLE 10.7 (Continued)

Base fuel	Nanoparticle	type	Main effect
Biofuel [55]	CeO_2	Simple nanofluid	• 1.75% improve in thermal performance
Biodiesel [56]	Ag_2O	Simple nanofluid	• Improvement in ignition characteristics • BTE enhancement • reduction in BSFC

10.4.2 Adding nanoparticles to fuel (nanofuel)

The specific properties of nanoparticles make them suitable for use as nanofuel additives (Figs. 10.17 and 10.18). In fact another important application of nanofluids is their application as fuel additives, which has been extensively studied by researchers. Nanofuels are conventional fuels containing nanoadditives with the aim of improving combustion properties. Coventional fuels include diesel, gasoline, biofuels, alcohols, and hybrids. There are many criteria that help researchers and industries to select the best nanoadditives. Two of these criteria are brake thermal efficiency (BTE) and brake specific fuel consumption (BSFC). Fig. 10.19 introduces the best nanoadditives based on the two abovementioned criteria [44]. Also some of the nanoadditives for fuels with the highest effect on CO, unburned hydrocarbon (UHC), soot, and NOx reduction are presented in Fig. 10.20. Table 10.7 provides useful information about adding nanoparticles to various fuels and the effect on engine and automobile performance.

10.4.3 Effects of nanoparticles additives on friction

When two objects are in relative motion and in contact with each other, the two phenomena of friction and wear will occur. Abrasion will eventually cause the geometric shape to be lost and the surfaces to be uneven, which will reduce engine life. The easiest solution is lubrication in order to reduce the effect of friction and wear.

The base lubricant separates the moving parts by creating a layer of oil on the friction surfaces and removes excess heat and abrasion particles. There are several methods to improve the lubricating properties of base oil. The use of various additives with unique properties to enhance the physical, chemical, and mechanical properties of base lubricants is one of these ways. The combination of nanoparticles with pure oils will result in nanolubricants. Nanolubricants are a type of engineering lubricant made of nanoparticles, dispersants, and base lubricants that have improved lubrication and heat

TABLE 10.8 Main possible friction improvements after adding nanoparticles to pure oils.

Base oil	Nanoparticles	Type	Friction and other main effects
SN 500 mineral oil [64]	Ni-MoS_2	Simple nanofluid	• About 40%–50% improving antiwear properties • About 20% reduction in friction coefficient
motor oil [65]	Multilayer graphene	Simple nanofluid	• 43% reduction in the friction coefficient • 63% reduction in wear
5W30 [40]	Al_2O_3/TiO_2	Hybrid nanofluid	• About 2% improve in engines' mechanical performance • 53% decrease in friction coefficient
Calcium grease [66]	TiO_2/CNTs	Hybrid nanofluid	• Coefficient of friction reduction by 60% • Wear reduction by 72%
Polyalkylene glycol [67]	TiO_2, SiO_2, and Al_2O_3	Simple nanofluid	• Friction coefficient reduction by 23.8%, 15.8% and 2.3% with adding SiO_2, TiO_2, and Al_2O_3 nanoparticles to the base oil
5W-30 [39]	Al_2O_3 and TiO_2	Simple nanofluid	• Friction power loss reduced by 50% for TiO_2 and 45% for Al_2O_3
SAE 20W50 [68]	Hexagonal BN, WS_2 and graphite	Simple nanofluid	• Reduction of coefficient of friction (COF): 56%–84% reduction by Hexagonal boron nitride 37%–80% reduction by tungsten disulfide 2.5%–49% reduction by graphite
20W50 [69]	Nanoflakes	Simple nanofluid	• 70% and 23% enhancement in heat transfer rate and thermal conductivity respectively • 21% coefficient of friction reduction

transfer properties compared to conventional lubricants [57]. Many researchers have reported the positive effect of nanolubricants on reducing friction and wear as well as increasing heat transfer capacity. They have used different types of nanoparticles made of polymeric, metallic, organic, and inorganic materials to make nanolubricants [58–62].

TABLE 10.9 Effect on friction of adding nanoparticles to various fuels.

Base fuel	Nanoparticles	Power engine	Reference
Diesel	Graphite	+9.6%	[46]
5% Biodiesel–95% diesel 10% Biodiesel–90% diesel	Alumina	+5.36%	[70]
20% Biodiesel–80% diesel	Grapheme oxide	+14.3%	[71]
Gasoline	Mn_2O_3	+19.56%	[72]
Palm oil	Titania	+3%	[73]
Diesel	Al_2O_3	+3.69%	[74]
Diesel	Al_2O_3	+3.28%	[51]
Diesel + biodiesel	CNT-S	+2%	[75]
B15 (diesel containing 15% biodiesel)	Quantum dot nanoparticles + containing 5% water	+21%	[76]
20% Biodiesel–80% diesel	MWCNTs-CeO_2	+7.81%	[77]
Diesel	CuO	+1%	[51]

Although there is no clear information about the corrosion effect on engine oils because of using nanofuel, some researchers have claimed that enriched fuel with nanoparticles stops and lowers iron and carbon depositions in cylinder. In fact it causes a reduction in the friction between engine parts and then decreases the corrosion and increases the pure engine power [63]. Table 10.8 gives classified information about the effect of adding nanoparticles to different automobile oils on friction parameters.

10.4.4 Effects of nanoparticles additives on engines power

The addition of nanoparticles to the types of oils used in the car, as well as the car fuel, has various effects on the performance of the car, including changes in the useful power of the car engine. Researchers have conducted extensive research to study these changes, some of which are mentioned below. Also other parameters like BSFC, EGT, and BTE are affected by nanoadditives. In the following the effect of adding some of nanoparticles to base oils and fuels are presented. Table 10.9 details the effect on the friction parameter of adding nanoparticles to different fuels.

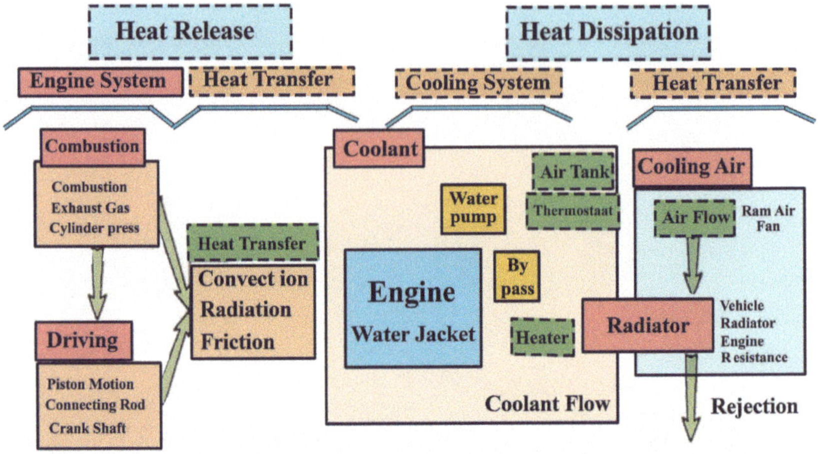

FIGURE 10.21 Engine and the various parts of the vehicles' cooling system [78].

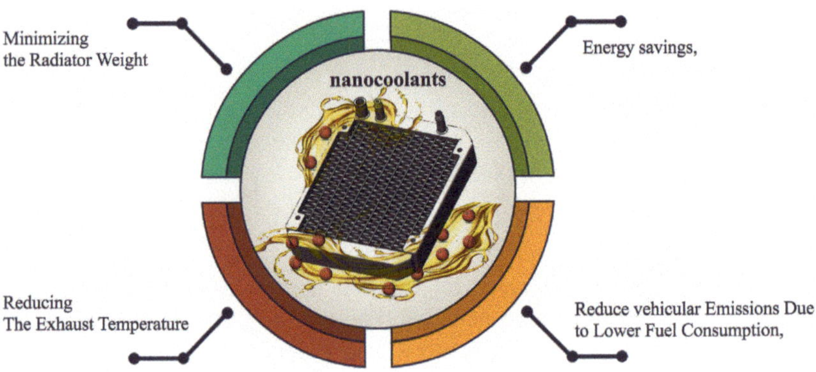

FIGURE 10.22 Nanocoolant benefits.

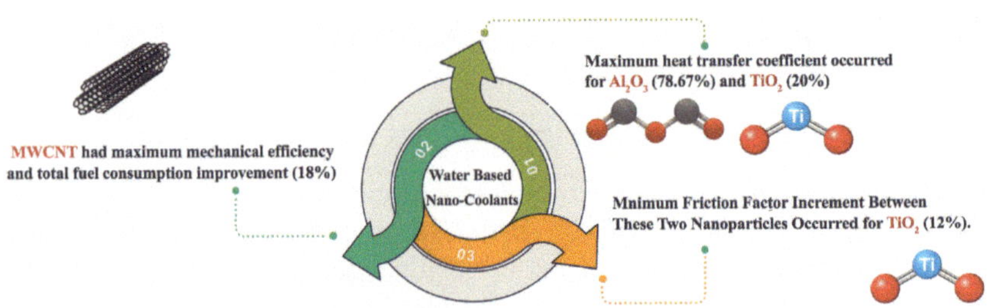

FIGURE 10.23 Best water-based nanocoolant [78].

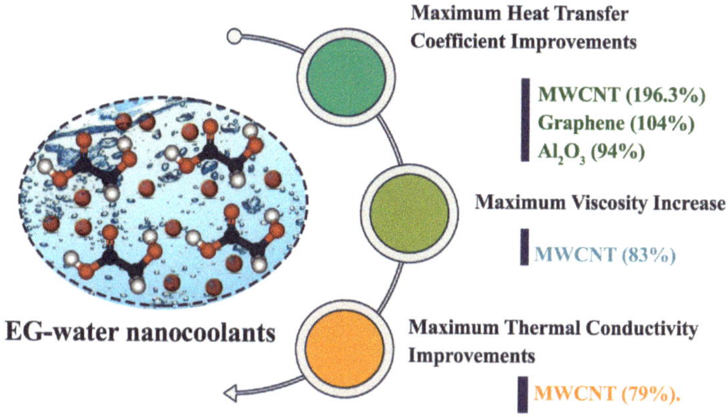

FIGURE 10.24 Best EG–water-based nanocoolants [78].

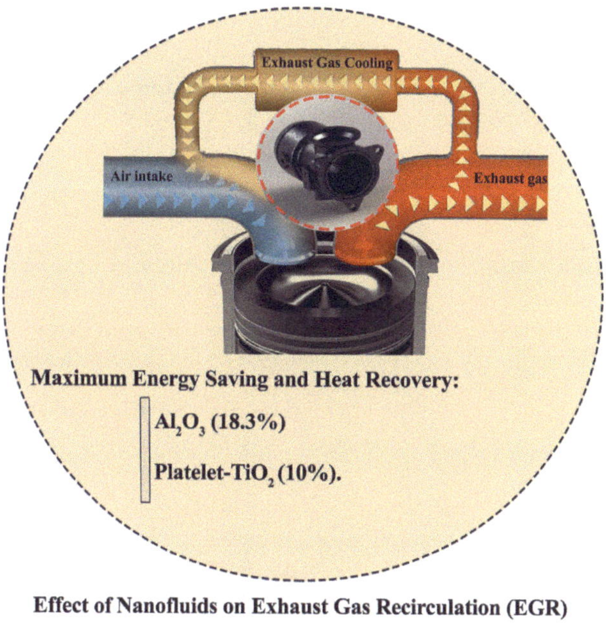

FIGURE 10.25 Best nanoparticles for exhaust gas recirculation systems [78].

10.4.5 Effects of nanoparticle additives on the cooling cycle of combustion engines

Nanocoolants are another application of nanoscience in engines. The combination of nanoparticles with the working fluid of cooling systems can improve the efficiency of coolants and cooling systems. The nanocoolants

TABLE 10.10 Effect of nanoparticle dispersion in conventional working fluids of engine cooling systems.

Host fluid	Nanoadditive	Main results
Water [80]	Al_2O_3	44.29% enhancement in heat transfer
Water [81]	Al_2O_3	About 45% heat transfer improvement
Water [82]	CuO	8% enhancement in heat transfer
Water [83]	ZnO	46% enhancement in heat transfer
EG [84]	MgO-FMWCNTs	21.3% enhancement of thermal conductivity
EG [85]	Al_2O_3	13% improvement in Nu number
EG [86]	Cu	3.8% heat transfer enhancement
Water–EG [87]	Al-multiwalled carbon nanotube	45% improve in heat transfer of cooling system
Water–EG [88]	Gr nanoplatelets	51% improvement in "h" coefficient
Water–EG [89]	Al_2O_3	40% heat transfer improvement
Water [90]	TiO_2	47% improvement in car radiator performance
Water–EG [91]	TiO_2	Improved heat transfer rate by 37%
Water–EG [92]	MWCNT	About 196% increment for "h"
Water–EG [93]	Gr and Ag	Between 2% to 9% improvement in thermal conductivity by using of three types of Gr sample 4.7% improve in heat transfer rate by using of silver nanoparticles

can be used in the new generation of car radiators. Fig. 10.21 also schematically shows the general relationship between the engine and the various parts of the vehicles' cooling system. To cool the engine and other components, the cooling fluid is circulated by the pump throughout the cooling system. The coolant is usually a mixture of water, ethylene glycol, and other additives (such as nanoparticles, etc.). In Figs. 10.22–10.24, the benefits of using nanocoolants in internal combustion engines, highly efficient water-based nanocoolant, and also the best EG–water-based nanocoolants are presented, respectively. Also Fig. 10.25 proposes the best combinations of nanoparticles for exhaust gas recirculation (EGR) with the maximum ability in terms of energy saving and heat recovery [78].

Nanofluid technology is a new endeavor for automotive lubrication and cooling systems. Nanoparticles not only improve the behavior of nanofluids and reduce friction between components, but also increase the amount of heat transferred by these fluids [79]. The reason for the increase in heat transfer rate in nanofluids is due to the following two factors (1) type of nanoparticles; and (2) increasing the surface area-to-volume ratio. The smaller the particle diameter, the higher the surface-to-volume ratio, which increases heat transfer. Design and construction of a cooling system with maximum efficiency and minimum power consumption has always been a goal for car manufacturers, to prevent the engine temperature from rising and can be in an ideal temperature and conditions. Table 10.10 details the main effects of nanoparticles addition on the performance of engine cooling systems. Some of the presented cooling systems are car radiators and some others are laboratory test setups.

10.5 Conclusion on threats and opportunities of applying nanoscience in combustion engines

This chapter discusses the positive effects of using nanoparticles in various parts of an internal combustion engine. For example, the positive effects of using nanoparticles as additives in fuel, cooling systems, and car lubrication systems are undeniable in many studies. Nanoparticles have clearly improved the process of burning car fuel and made combustion a cleaner process with the least amount of pollutants. Also, the antifriction effect and the help in improving the heat transfer process provided by different nanoparticles have been proven in many studies. But an important point to note, along with all these positives, is the environmental impact that nanoparticles may pose as a threat to human health and the environment. The small size of nanoparticles has the potential to spread around roads and the surrounding space and enter the tissues of plants and the human body, which is a major threat.

But this is not the end for nanoparticles. With the right orientation, in addition to using the capacity of nanoparticles, the risk of using them can be minimized. One of the innovative applications of nanoparticles is their use in fuel cells as a replacement for internal combustion engines. Certainly, nanoparticles act as catalysts in the chemical reactions of fuel cells and accelerate the production of electricity for the car. Also, battery systems engineered with nanotechnology will have lighter batteries, higher energy densities, and the ability to hold a charge for a longer period of time. In short, the use of nanoparticles in fuel cells will pave the way for the expansion of the electric vehicle industry and its popularity. This will minimize the environmental concerns of using nanoparticles in internal combustion vehicles, as well as address the common environmental pollution caused by the use of fossil fuels in internal combustion engines.

References

[1] L. Yang, M. Mao, J.N. Huang, W. Ji, Enhancing the thermal conductivity of SAE 50 engine oil by adding zinc oxide nano-powder: an experimental study, Powder Technol. 356 (2019) 335–341.

[2] M.H. Esfe, A.A.A. Arani, S. Esfandeh, M. Afrand, Proposing new hybrid nano-engine oil for lubrication of internal combustion engines: preventing cold start engine damages and saving energy, Energy 170 (2019) 228–238.

[3] M.T. Sulgani, A. Karimipour, Improve the thermal conductivity of 10w40-engine oil at various temperature by addition of Al_2O_3/Fe_2O_3 nanoparticles, J. Mol. Liq. 283 (2019) 660–666.

[4] M.H. Esfe, S. Esfandeh, M. Afrand, M. Rejvani, S.H. Rostamian, Experimental evaluation, new correlation proposing and ANN modeling of thermal properties of EG based hybrid nanofluid containing ZnO-DWCNT nanoparticles for internal combustion engines applications, Appl. Therm. Eng. 133 (2018) 452–463.

[5] M.H. Esfe, S. Esfandeh, A.A.A. Arani, Proposing a modified engine oil to reduce cold engine start damages and increase safety in high temperature operating conditions, Powder Technol. 355 (2019) 251–263.

[6] M.H. Esfe, A.A.A. Arani, S. Esfandeh, Improving engine oil lubrication in light-duty vehicles by using of dispersing MWCNT and ZnO nanoparticles in 5W50 as viscosity index improvers (VII), Appl. Therm. Eng. 143 (2018) 493–506.

[7] M.H. Esfe, S. Wongwises, A. Naderi, A. Asadi, M.R. Safaei, H. Rostamian, et al., Thermal conductivity of Cu/TiO_2–water/EG hybrid nanofluid: experimental data and modeling using artificial neural network and correlation, Int. Commun. Heat. Mass Transf. 66 (2015) 100–104.

[8] M.H. Esfe, A.A.A. Arani, M. Rezaie, W.M. Yan, A. Karimipour, Experimental determination of thermal conductivity and dynamic viscosity of Ag–MgO/water hybrid nanofluid, Int. Commun. Heat. Mass Transf. 66 (2015) 189–195.

[9] M.H. Esfe, W.M. Yan, M. Akbari, A. Karimipour, M. Hassani, Experimental study on thermal conductivity of DWCNT-ZnO/water-EG nanofluids, Int. Commun. Heat. Mass. Transf. 68 (2015) 248–251.

[10] Barnett, W., Obtaining motive power from inflammable gases by compression and explosion UK Patent 7615 of 1838.

[11] Jean-Joseph Étienne Lenoir, Lenoir Engine. Scientific American. 22 September 1860. p. 193.

[12] W.W. Pulkrabek Engineering Fundamentals of the Internal Combustion Engine, 2004.

[13] D.E. Richardson, Review of power cylinder friction for diesel engines, J. Eng. Gas. Turbines Power 122 (4) (2000) 506–519.

[14] D.E. Winterbone, D.W.H. Tennant, The variation of friction and combustion rates during diesel engine transients, SAE Paper 810339, 1981.

[15] S.N.M. Haines, S.A. Shields, The determination of diesel engine friction characteristics by electronic cylinder disablement, J. Power Eng. 203 (1989) 203–204.

[16] T.C. Fayette, The Internal Combustion engine in Theory and Practice, 1985.

[17] M.S. Chitti, N.K. Anand, An analytical model for local heat transfer coefficients for forced convective condensation inside smooth horizontal tubes, Int. J. Heat. Mass Transf. 38 (4) (1995) 615–627.

[18] G. Woschni, J. Fieger, Determination of local heat transfer coefficients at the piston of a high speed diesel engine by evaluation of measured temperature distribution, SAE Trans. (1979) 2807–2815.
[19] E.J. Le Fevre J.W. Rose A theory of heat transfer by dropwise condensation, in: Proceedings of the International Heat Transfer Conference Digital Library, Begel House Inc., 1966.
[20] J.C. Dent, S.J. Sulaiman, Convective and radiative heat transfer in a high swirl direct injection diesel engine, SAE Trans. (1977) 1758–1783.
[21] Thermodynamics and Fluid Mechanics Group, W.J.D. Annand, Heat transfer in the cylinders of reciprocating internal combustion engines, Proc. Inst. Mech. Eng. 177 (1) (1963) 973–996.
[22] S.C. Vishweshwara, O.K.H. Al Badi, Density and thermal conductivity changes in engine oil during its life cycle: an experimental study, Int. J. Mutidiscip. Sci. Eng. 6 (10) (2015) 24–28.
[23] S. Wrenick, P. Sutor, H. Pangilinan, E.E. Schwarz, January. Heat transfer properties of engine oils, in: Proceedings of the World Tribology Congress, Vol. 42010, 2005, pp. 595–596.
[24] G.K. Jaiswal, et al., Design of a smart automotive ventilation system for a parked car, Int. J. Theor. Appl. Res. Mech. Eng. I (I) (2012) 83–88.
[25] J. Denis, G. Parc (Eds.), Rheological Properties of Lubricants, Institute Francias Du petrole, 1989.
[26] Available from: http://www.lubescience.com.
[27] Review of the used oil management program in New York State, legislative commission on toxic substance and hazardous wastes, 1986.
[28] M. Ivanov, O. Shenderova, Nanodiamond-based nanolubricants for motor oils, Curr. Opin. Solid. State Mater. Sci. 21 (1) (2017) 17–24.
[29] M.G. Ivanov, D.M. Ivanov, S.V. Pavlyshko, I. Petrov, A. Vargas, G. McGuire, et al., Nanodiamond-based nanolubricants, Fuller., Nanotub. Carbon Nanostruct. 20 (4–7) (2012) 606–610.
[30] Q.L. Yan, F.Q. Zhao, K.K. Kuo, X.H. Zhang, S. Zeman, L.T. DeLuca, Catalytic effects of nano additives on decomposition and combustion of RDX-, HMX-, and AP-based energetic compositions, Prog. Energy Combust. Sci. 57 (2016) 75–136.
[31] P.K. Sarma, V. Srinivas, V.D. Rao, A.K. Kumar, Experimental study and analysis of lubricants dispersed with nano Cu and TiO_2 in a four-stroke two wheeler, Nanoscale Res. Lett. 6 (1) (2011) 233.
[32] M.F. Sgroi, M. Asti, F. Gili, F.A. Deorsola, S. Bensaid, D. Fino, et al., Engine bench and road testing of an engine oil containing MoS_2 particles as nano-additive for friction reduction, Tribol. Int. 105 (2017) 317–325.
[33] M.K. Ahmed Ali, P. Fuming, H.A. Younus, M.A.A. Abdelkareem, F.A. Essa, A. Elagouz, et al., Fuel economy in gasoline engines using Al_2O_3/TiO_2 nanomaterials as nanolubricant additives, Appl. Energy 211 (2018) 461–478.
[34] M.K. Ahmed Ali, H. Xianjun, M.A.A. Abdelkareem, M. Gulzar, A.H. Elsheikh, Novel approach of the graphene nanolubricant for energy saving via anti-friction/wear in automobile engines, Tribol. Int. 124 (2018) 209–229.
[35] M. Hatami, M. Hasanpour, D. Jing, Recent developments of nanoparticles additives to the consumables liquids in internal combustion engines: part II: nano-lubricants, J. Mol. Liq. (2020) 114156.

[36] M.R. Ripoll, A.M. Tomala, V. Totolin, M. Remškar, Performance of nanolubricants containing MoS_2 nanotubes during form tapping of zinc-coated automotive components, J. Manuf. Process. 39 (2019) 167–180.

[37] M.K. Ahmed Ali, H. Xianjun, L. Mai, C. Qingping, R.F. Turkson, C. Bicheng, Improving the tribological characteristics of piston ring assembly in automotive engines using Al_2O_3 and TiO_2 nanomaterials as nanolubricant additives, Tribol. Int. 103 (2016) 540–554.

[38] M.K. Ahmed Ali, H. Xianjun, Improving the heat transfer capability and thermal stability of vehicle engine oils using Al_2O_3/TiO_2 nanomaterials, Powder Technol. 363 (2020) 48–58.

[39] S.J. Castillo Marcano, S. Bensaid, F.A. Deorsola, N. Russo, D. Fino, Nanolubricants for diesel engines: related emissions and compatibility with the after-treatment catalysts, Tribol. Int. 72 (2014) 198–207.

[40] M.K. Ahmed Ali, P. Fuming, H.A. Younus, M.A.A. Abdelkareem, F.A. Essa, A. Elagouz, et al., Fuel economy in gasoline engines using Al_2O_3/TiO_2 nanomaterials as nanolubricant additives, Appl. Energy 211 (2018) 461–478.

[41] A. Kotia, S. Borkakoti, S.K. Ghosh, Wear and performance analysis of a 4-stroke diesel engine employing nanolubricants, Particuology 37 (2018) 54–63.

[42] K. Nanthagopal, R. Susanth Kishna, A.E. Atabani, H. Al-Muhtaseb Ala'a, G. Kumar, B. Ashok, A compressive review on the effects of alcohols and nanoparticles as an oxygenated enhancer in compression ignition engine, Energy Convers. Manag. 203 (2020) 112244.

[43] V.W. Khond, V.M. Kriplani, Effect of nanofluid additives on performances and emissions of emulsified diesel and biodiesel fueled stationary CI engine: a comprehensive review, Renew. Sust. Energ. Rev. 59 (2016) 1338–1348.

[44] M. Hatami, M. Hasanpour, D. Jing, Recent developments of nanoparticles additives to the consumables liquids in internal combustion engines: part I: nano-fuels, J. Mol. Liq. (2020) 114250.

[45] D. Mei, L. Zuo, D. Adu-Mensah, X. Li, Y. Yuan, Combustion characteristics and emissions of a common rail diesel engine using nanoparticlediesel blends with carbon nanotube and molybdenum trioxide, Appl. Therm. Eng. 162 (2019) 114238.

[46] A. Ahmed, A.N. Shah, A. Azam, G.M. Uddin, M.S. Ali, S. Hassan, et al., Environment-friendly novel fuel additives: investigation of the effects of graphite nanoparticles on performance and regulated gaseous emissions of CI engine, Energy Convers. Manag. 211 (2020) 112748.

[47] W.M. Yang, H. An, S.K. Chou, S. Vedharaji, R. Vallinagam, M. Balaji, et al., Emulsion fuel with novel nano-organic additives for diesel engine application, Fuel 104 (2013) 726–731.

[48] A. La Rocca, A. Ferrante, E. Haffner-Staton, A. Cairns, A. Weilhard, V. Sans, et al., Investigating the impact of copper leaching on combustion characteristics and particulate emissions in HPCR diesel engines, Fuel 263 (2020) 116719.

[49] A. Yaşar, A. Keskin, Ş. Yıldızhan, E. Uludamar, Emission and vibration analysis of diesel engine fuelled diesel fuel containing metallic based nanoparticles, Fuel 239 (2019) 1224–1230.

[50] A.F. Chen, M. Akmal Adzmi, A. Adam, M.F. Othman, M.K. Kamaruzzaman, A.G. Mrwan, Combustion characteristics, engine performances and emissions of a diesel engine using nanoparticle-diesel fuel blends with aluminium oxide, carbon nanotubes and silicon oxide, Energy Convers. Manag. 171 (2018) 461–477.

[51] S. Gumus, H. Ozcan, M. Ozbey, B. Topaloglu, Aluminum oxide and copper oxide nano-diesel fuel properties and usage in a compression ignition engine, Fuel 163 (2016) 80–87.

[52] M.A. Lenin, M.R. Swaminathan, G. Kumaresan, Performance and emission characteristics of a DI diesel engine with a nanofuel additive, Fuel 109 (2013) 362–365.

[53] M. Valihesari, V. Pirouzfar, F. Ommi, F. Zamankhan, Investigating the effect of Fe_2O_3 and TiO_2 nanoparticle and engine variables on the gasoline engine performance through statistical analysis, Fuel 254 (2019) 115618.

[54] S.H. Oh, S.H. Yoon, H. Song, J.G. Han, J.-M. Kim, Effect of hydrogen nanobubble addition on combustion characteristics of gasoline engine, Int. J. Hydrogenenergy 38 (2013) (14849 e14853).

[55] B. Dhinesh, Y. Maria Ambrose Raj, C. Kalaiselvan, R. KrishnaMoorthy, A numerical and experimental assessment of a coated diesel engine powered by highperformance nano biofuel, Energy Convers. Manag. 171 (2018) 815–824.

[56] Y. Devarajan, D.B. Munuswamy, A. Mahalingam, Investigation on behavior of diesel engine performance, emission, and combustion characteristics using nano-additive in neat biodiesel, Heat Mass Transf. 55 (6) (2019) 1641–1650.

[57] C.G. Lee, Y.J. Hwang, Y.M. Choi, J.K. Lee, C. Choi, J.M. Oh, A study on the tribological characteristics of graphite nano lubricants, Int. J. Precis. Eng. Manuf. 10 (1) (2009) 85–90.

[58] Y. Wu, W. Tsuia, T. Liub, Experimental analysis of tribological properties of lubricating oils with nanoparticle additives, Wear 262 (7–8) (2007) 819–825.

[59] J. Lee, S. Cho, Y. Hwang, C. Lee, S. Kim, Enhancement of lubrication properties of nano-oil by controlling the amount of fullerene nanoparticle additives, Tribol. Lett. 28 (2) (2007) 203–208.

[60] B.C. Ku, Y.C. Han, J.E. Lee, J.K. Lee, S.H. Park, Y.J. Hwang, Tribological effects of fullerene (C60) nanoparticles added in mineral lubricants according to its viscosity, Int. J. Precis. Eng. Manuf. 11 (4) (2010) 607–611.

[61] X. Li, D. Zhu, X. Wang, Experimental investigation on viscosity of Cu-H_2O nanofluids, J. Wuhan. Univ. Technol.-Mater 24 (1) (2009).

[62] S. Harish, K. Ishikawa, E. Einarsson, S. Aikawa, S. Chiashi, J. Shiomi, et al., Enhanced thermal conductivity of ethylene glycol with single-walled carbon nanotube inclusions, Int. J. Heat. Mass Transf. 55 (2012) 3885–3890.

[63] S.S. Hoseini, G. Najafi, B. Ghobadian, M.T. Ebadi, R. Mamat, T. Yusaf, Performance and emission characteristics of a CI engine using graphene oxide (GO) nanoparticles additives in biodiesel-diesel blends, Renew. Energy 145 (2020) 458e465.

[64] N. Rajendhran, S. Palanisamy, P. Periyasamy, R. Venkatachalam, Enhancing of the tribological characteristics of the lubricant oils using Ni-promoted MoS_2 nanosheets as nano-additives, Tribol. Int. 118 (2018) 314–328.

[65] E.D. Ramón-Raygoza, C.I. Rivera-Solorio, E. Giménez-Torres, D. Maldonado-Cortés, E. Cardenas-Alemán, R. Cué-Sampedro, Development of nanolubricant based on impregnated multilayer graphene for automotive applications: analysis of tribological properties, Powder Technol. 302 (2016) 363–371.

[66] A. Mohamed, S. Ali, T.A. Osman, B.M. Kamel, Development and manufacturing an automated lubrication machine test for nano grease, J. Mater. Ress Technol. 9 (9) (2020) 2054–2062.

[67] S.S. Sanukrishna, S. Vishnu, T.S. Krishnakumar, M. Jose Prakash, Effect of oxide nanoparticles on the thermal, rheological and tribological behaviours of refrigerant compressor oil: an experimental investigation, Int. J. Refrig. 90 (2018) 32–45.

[68] M.S. Charoo, M. Hanief, Improving the tribological characteristics of a lubricating oil by nano sized additives, Mater. Today: Proc. 28 (2020) 1205–1209.
[69] A.K. Rasheed, M. Khalid, A. Javeed, W. Rashmi, T.C.S.M. Gupta, A. Chan, Heat transfer and tribological performance of graphene nanolubricant in an internal combustion engine, Tribol. Int. 103 (2016) 504–515.
[70] S.H. Hosseini, A. Taghizadeh-Alisaraei, B. Ghobadian, A. Abbaszadeh-Mayvan, Effect of added alumina as nano-catalyst to diesel-biodiesel blends on performance and emission characteristics of CI engine, Energy 124 (2017) 543–552.
[71] S.S. Hoseini, G. Najafi, B. Ghobadian, R. Mamat, M.T. Ebadi, T. Yusaf, Novel environmentally friendly fuel: the effects of nanographene oxide additives on the performance and emission characteristics of diesel engines fuelled with Ailanthus altissima biodiesel, Renew. Energy 125 (2018) 283e294.
[72] M. Amirabedi, S. Jafarmadar, S. Khalilarya, Experimental investigation the effect of Mn_2O_3 nanoparticle on the performance and emission of SI gasoline fueled with mixture of ethanol and gasoline, Appl. Therm. Eng. 149 (2019) 512–519.
[73] V. Velmurugan, S.M. Aathif Akmal, V. Paramasivam, S. Thanikaikarasan, Prediction of vibration and exhaust gas emission characteristics using palm oil with nano particle diesel fuel, Mater. Today: Proc. 21 (2020) 800–805.
[74] A.K. Hasannuddin, W.J. Yahya, S. Sarah, A.M. Ithnin, S. Syahrullail, N.A.C. Sidik, et al., Nano-additives incorporated water in diesel emulsion fuel: fuel properties, performance and emission characteristics assessment, Energy Convers. Manag. 169 (2018) 291–314.
[75] M. Ghanbari, G. Najafi, B. Ghobadian, T. Yusaf, A.P. Carlucci, M. Kiani Deh Kiani, Performance and emission characteristics of a CI engine using nanoparticles additives in biodiesel-diesel blends and modeling with GP approach, Fuel 202 (2017) 699–716.
[76] E. Ettefaghi, B. Ghobadian, A. Rashidi, G. Najafi, M.H. Khoshtaghaza, M. Rashtchi, et al., A novel bio-nano emulsion fuel based on biodegradable nanoparticles to improve diesel engines performance and reduce exhaust emissions, Renew. Energy 125 (2018) 64e72.
[77] M. Mirzajanzadeh, M. Tabatabaei, M. Ardjmand, A. Rashidi, B. Ghobadian, M. Barkhi, et al., A novel soluble nanocatalysts in diesel–biodiesel fuel blends to improve diesel engines performance and reduce exhaust emissions, Fuel 139 (2015) 374–382.
[78] M. Hatami, M. Hasanpour, D. Jing, Recent developments of nanoparticles additives to the consumables liquids in internal combustion engines: part III: nano-coolants, J. Mol. Liq. (2020) 319.
[79] S.U.S. Choi, Enhancing thermal conductivity of fluid with nanoparticles, in: D.A. Siginer and H.P. Wang (Eds.), Developments and Applications of Non- Newtonian Flows, FED, V. 231/MD,V. 66, 1995, 99 pp.
[80] P. Chaurasia, A. Kumar, A. Yadav, P.K. Rai, V. Kumar, L. Prasad, Heat transfer augmentation in automobile radiator using Al_2O_3–water based nanofluid, SN Appl. Sci. 1 (2019) 257.
[81] S.M. Peyghambarzadeh, S.H. Hashemabadi, M. Seifi Jamnani, S.M. Hoseini, Improving the cooling performance of automobile radiator with Al_2O_3/water nanofluid, Appl. Therm. Eng. 31 (2011) (1833 e 1838).
[82] M. Naraki, S.M. Peyghambarzadeh, S.H. Hashemabadi, Y. Vermahmoudi, Parametric study of overall heat transfer coefficient of CuO/water nanofluids in a car radiator, Int. J. Therm. Sci. 66 (2013) 82e90.

[83] H.M. Ali, H. Ali, H. Liaquat, H.T.B. Maqsood, M.A. Nadir, Experimental investigation of convective heat transfer augmentation for car radiator using ZnO-water nanofluids, Energy 84 (2015) 317e324.

[84] M. Afrand, Experimental study on thermal conductivity of ethylene glycol containing hybrid nano-additives and development of a new correlation, Appl. Therm. Eng. 110 (2017) 1111–1119.

[85] K. Goudarzi, H. Jamali, Heat transfer enhancement of Al_2O_3-EG nanofluid in a car radiator with wire coil inserts, Appl. Therm. Eng. 118 (2017) 510–517.

[86] K.Y. Leong, R. Saidur, S.N. Kazi, A.H. Mamun, Performance investigation of an automotive car radiator operated with nanofluid-based coolants (nanofluid as a coolant in a radiator), Appl. Therm. Eng. 30 (2010) 2685e2692.

[87] K.S. Raju Jadar, S.R. Shashishekar, Manohara, Performance evaluation of Al-MWCNT based automobile radiator, Mater. Today: Proc. 9 (2019) 380–388.

[88] C. Selvam, D. Mohan Lal, S. Harish, Enhanced heat transfer performance of an automobile radiator with graphene based suspensions, Appl. Therm. Eng. 123 (2017) 50–60.

[89] S.M. Peyghambarzadeh, S.H. Hashemabadi, S.M. Hoseini, M. Seifi Jamnani, Experimental study of heat transfer enhancement using water/ethylene glycol based nanofluids as a new coolant for car radiators, Int. Commun. Heat. Mass Transf. 38 (2011) 1283–1290.

[90] S.A. Ahmed, M. Ozkaymak, A. Sözen, T. Menlik, A. Fahed, Improving car radiator performance by using TiO_2-water nanofluid, Eng. Sci. Technol. Int. J. 21 (2018) 996–1005.

[91] S. Devireddy, C.S.R. Mekala, V.R. Veeredhi, Improving the cooling performance of automobile radiator with ethylene glycol water based TiO_2 nanofluids, Int. Commun. Heat. Mass Transf. 78 (2016) 121–126.

[92] B. M'hamed, N.A.C. Sidik, M.F.A. Akhbar, R. Mamat, G. Najafi, Experimental study on thermal performance of MWCNT nanocoolant in Perodua Kelisa 1000cc radiator system, Int. Commun. Heat. Mass Transf. 76 (2016) 156–161.

[93] E.M.C. Contreras, G.A. Oliveira, E.P.B. Filho, Experimental analysis of the thermohydraulic performance of grapheme and silver nanofluids in automotive cooling systems, Int. J. Heat. Mass Transf. 132 (2019) 375–387.

Chapter 11

Applications of nanofluids in solar energy collectors focusing on solar stills

Saeed Esfandeh[1,2], Mohammad Hemmat Esfe[2] and Mohammad Hassan Kamyab[2,3]
[1]*Department of Mechanical Engineering, Jundi-Shapur University of Technology, Dezful, Iran,* [2]*Department of Mechanical Engineering, Imam Hossein University, Tehran, Iran,* [3]*Department of Mechanical Engineering, University of Kashan, Kashan, Iran*

Nomenclature

Abbreviation

A	Euro/g ratio
BCE	before common era
CE	common era
CPV	concentrated photovoltaic
L	length
NPs	nanoparticles
USD	United States dollar
W	width
WWII	World War II

Latin letters

h_{H2O} **(W/m^2 K)**	thermal convection
I_{solar} **(t,x,y)**	light intensity
S/I	freshwater cost
T **(°C)**	temperature
T_{air} **(°C)**	roatation of the paddle wheel
x **(m)**	distance
Y_L **(m)**	height

Advances in Nanofluid Heat Transfer. DOI: https://doi.org/10.1016/B978-0-323-88656-7.00010-6

11.1 History of solar energy collectors

The thought and idea of using the capabilities of the sun for heating and lighting goals dates back to ancient times. The ancient Greeks were the first to change their home design with the aim of capturing sunlight and heat in the 5th century BCE. As the second important step, Romans improved sun heat capture by creating south facing windows on roof tops to trap and save the heat in daylight hours. The third important step with more scientific process was conducted by a Swiss scientist, Horace de Saussure, in the 1760s. He designed a box with glass material to heat the water inside which can be considered to be the first prototype in the history of the solar energy collectors [1].

The 19th century was an important period in the history of solar energy collectors' improvement. Some of the most important improvements followed the discovery of the photovoltaic effect by a young physicist, Edmund Becquerel. Also solar-powered steam engine and setting up the first solar steam-powered plant using parabolic dish collectors are other highlights of the 19th century [2,3].

The low cost and wide availability of fossil fuels was a strong barrier against solar energy collector developments. But efforts for the promotion of using solar energy didn't stop. For this goal in the last decade of the 19th century many companies started designing and manufactured hot water heaters [4,5].

The next step in development was planned and started by William Adams in the 1870s. William Adams gathered much information about previous studies on solar energy collectors and also published some of his discoveries. He provided high-pressure steam to run modern steam engines using concentrated solar energy. Also he spent a long time improving the mentioned process, although governments didn't support him in the next steps of his studies because of the lack of economic justifications. So again he started his studies and wrote books about solar energy. Between his studies he faced with an important issue that was the scarcity of drinking water in some areas of the world and its expensive price because of the distilling process from seawater sources. Engineers developed his apparatus to produce potable water in hot regions with no freshwater sources. It can be nominated as one of the solar energy victories in competition with other energy sources, although this victory did not last long with the discovery of cheap oil resources after WWII [6].

But a close look at the cost trend of fossil fuels shows an increase in the price of these fuels with reducing resources. In addition, with the reduction of oil and gas resources, there is a possibility of international political and economic crises [7]. Another issue that has attracted attention in the last two decades is the risk of environmental pollution. The most probable cause of environmental problems can be considered to be a combination of several factors caused by the increase of human activities. These factors are population growth, energy consumption and economic activities. Achieving solutions for environmental problems that human beings face today requires extensive activities. An effective energy solution is renewable and sustainable energy sources [8]. As a conclusion, the highlights of ancient history of using solar energy can be summarized in Table 11.1.

TABLE 11.1 History of solar energy: the highlight points [9–12].

History	Event
Three centuries BCE	Using mirrors to light torches in ancient Greece and Rome
Two centuries BCE	Archimedes used the reflection of sunlight from a bronze shield and concentrated sunlight to set fire to enemy wooden ships.
20 years BCE	The Chinese used mirrors to light their torches
1th century to 4th century CE	The Romans designed the baths of their houses to use sunlight to heat water.
1767–1860 CE	• A Swiss scientist builds the first solar collector. • Scottish Minister Robert Sterling patented his heat engine preheater. His invention was later used to generate electricity by heating solar energy. • Working on solar steam engine by August Mouchet as a French mathematician. Twenty years later, he and his assistant, Abel Pifre, built motors whose modern models are now used in linear parabolic collectors.
1860–1900 CE	• Willoughby Smith discovers the ability to conduct selenium light. • Generating electricity by using of selenium under radiation of sunlight (a metod that introduced by William Grylls Adams and Richard Evans Day). • Charles Fritts, an American, thinks of making solar cells from selenium wafers. • The first solar water heater is patented by American Clarence.
1900–1950 CE	• Sensitivity of Cu and CuO to light that discovered by Wilhelm Hallwachs. • Designing a collector with copper coil and insulating box by William J. Bailey. • Scientists discover an electron strap in a photovoltaic device. • Scientists experimentally prove the photoelectric effect. • The first commercial solar water heater manufactured.
1950–1980 CE	• The first silicon photovoltaic cell developed by scientists with the ability of providing power for electronic systems. • New age of photovoltaic cells for space missions. • NASA launches the first 470-W satellite powered by photovoltaic cells.

(Continued)

TABLE 11.1 (Continued)

History	Event
	• A solar oven is built in Odeillo company, France, using eight parabolic mirrors. • Elliott Berman and Exxon Crepe design cheaper solar cells, and this factor led to the widespread use of photovoltaic cells. • Delaware University establishes the Energy Conversion Institute, the world's first laboratory for the research and development of photovoltaic cells. The following year, the institute built a hybrid photovoltaic thermal system called the Solar One. • NASA's Lewis Research Center first installs 83 photovoltaic systems around the world, used to illuminate clinics, pump water, classroom televisions, and more. • US Government established the Solar Energy Research Institute
1980–2008 CE	• The first solar-powered aircraft is flown from France to the United Kingdom. • An Australian first solar car travels 2,800 miles from Sydney to Perth. • California launches the largest solar thermal power plant. • The first solar parabolic plate is connected to the grid using a free-piston Stirling engine. • The photovoltaic film is made. • California's largest rooftop solar system is installed. • The largest solar park in Germany is launched by Thin Film Systems.

11.2 Classification of Solar energy collectors

Solar collectors are a special type of heat exchanger that convert energy, solar radiation, into heat. Solar collectors differ in several ways compared to conventional heat exchangers. In conventional heat exchangers two fluids exchange the heat with each other and also solar radiation has no effect on the performance of heat exchangers. On the other hand in solar collectors the energy will transfer from a heat radiation source to the working fluid [13].

11.2.1 Classification based on solar energy application

Solar radiation can be convert into different types of energy. The mains are (1) electrical, (2) mechanical, (3) chemical, and (4) heat energy. The chemical type can be found in the mechanism of action of plants and trees, so solar radiation is a main parameter in the clean air cycle. The thermal energy obtained from solar energy can be found in solar-based water heating or building heating applications. Also electrical energy generation and mechanical power energy generation, are other converted forms of solar energy that can be accessible by using solar thermodynamic engines. For example, the output of a solar collector that concentrates the solar radiation can be applied and coupled to a pump with the goal of water desalination or to a Stirling cycle during solar irradiation, etc.

11.2.2 Classification based on concentration level ability

A solar collector can convert a specific amount of the sun's radiant energy to heat based on its flexibility, geometry, and structures and also environmental conditions. This amount of energy conversion depends on the ability of the solar collector to concentrate the mentioned radiant energy. Fig. 11.1 shows the classification of solar collectors based on concentration ability and the accessible temperature of solar collector. It is shown that the flat plate and evacuated tubes solar collectors have lower efficiency in comparison to concentrated types of solar collectors.

11.2.3 Classification based on tracking or nontracking solar collectors

Tracking the sun radiation by solar collectors is one of the most important capabilities that really improves the efficiency of solar collectors in the way of absorbing solar energy. Nontracking or fixed solar collectors, one-axis

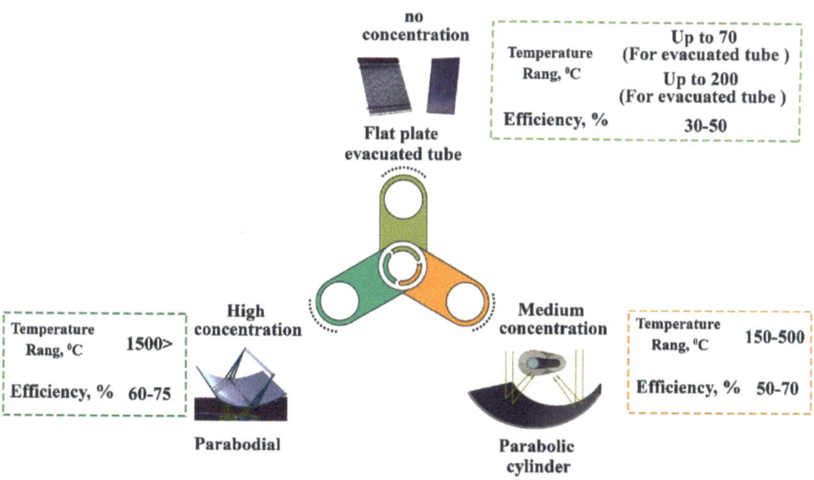

FIGURE 11.1 Classification based on concentration level ability [14].

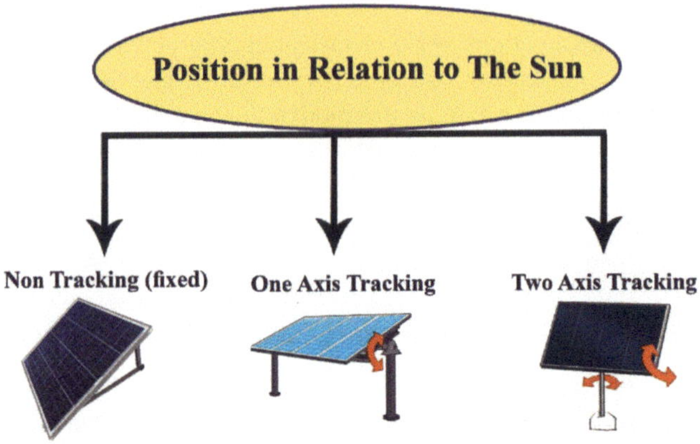

FIGURE 11.2 Tracking or nontracking solar collectors.

tracking, and two-axis tracking are the three groups of solar collectors [15]. Fig. 11.2 shows non-tracking, one-, and two-axis tracking solar collectors.

11.2.4 Passive and active solar systems

Solar energy systems could be divided into two groups: passive and active. Active refers to solar systems with moving parts or systems equipped with electronic intelligent based packages. Heating the liquid or any working fluid in solar collectors is the basis of active solar energy systems. The mentioned heat energy will transfer to heat exchangers or any equipment like that and heats the target liquid phase fluid or heats the air in an air conditioner system where both of them help to heat a home or any other place. Having pumps to push the fluid in the cycle of a solar system or using fans to move warm air in central heating systems makes these types of systems active solar systems.

On the other hand the passive solar system doesn't use any moving parts or other electronic controls. They use natural absorption methods for solar radiation saving, using thermal mass as heat preservative materials. Some of the heat preservative materials are water, concrete, clay, earth, rock, and wood. When solar radiation enters a passive building the heat could be stored by thermal mass materials and could be released over time, helping to keep the building warm or cool all day long.

11.2.4.1 Benefits and drawbacks of active solar systems

The benefits of active solar systems are (1) having higher capacity than passive systems, (2) existing buildings could be equipped by active systems, (3) more efficient than passive systems because of having moving elements, and (4) they are clean and eco-friendly. The drawbacks of active systems are

(1) the high cost of active systems, (2) needing space inside and outside of the building, and (3) needing periodical overhaul

11.2.4.2 Benefits and drawbacks of passive solar systems

Also the benefits of passive systems are (1) clean and good for the environment and do not produce greenhouse gases, (2) cheaper than active systems because it doesn't have any moving part, (3) no need to repair and maintenance. On the other hand drawbacks are (1) passive solar is less efficient than active solar technology, (2) only new buildings are able to be equipped with passive solar systems, (3) possible cutting down of trees around the buildings to obtain maximum sunlight.

11.2.5 Main types of solar collectors

The main types of solar collectors can be divided into three main groups. The groups are flat plates, evacuated tube collectors, and concentrating collectors.

11.2.5.1 Flat plate solar collectors

Flat plate collectors are the most common type of solar collectors that are in use for domestic water heating. Their simplicity and effectiveness make them a benchmark for evaluating the performance quality of other types of solar collectors. Based on season and location on the Earth, flat plate collectors should be placed at suitable angle to have maximum absorbing energy. With regard to the best direction for solar flat plate installation, south-facing panels are suitable for installing in northern hemisphere areas and north-facing solar collectors are suitable for installing in southern hemisphere areas. A schematic of a flat plate solar collector is given in Fig. 11.3.

The main parts of flat plate solar collectors are some sort of transparent covers that are placed on the top of coated absorber plates. Absorber plates are usually dark colored to absorb the maximum of incident radiation. The transparent glazed cover prevents the generated heat loss and will trap the heat between the absorber plate and the cover [16]. The distance between the transparent glazed cover and absorbing plate needs to be optimized to have the minimum energy loss. The absorbing plate is usually made from material with high conductive heat transfer like copper, aluminum, etc. in order to transfer the maximum heat to the target fluid. The pipes are another main part of solar flat plates; they are connected to the absorber plate and are responsible for transferring the heated fluid for consumption or storage goals [17] (Fig. 11.4).

Flat plate collectors could be classified into two categories: water or liquid heaters in which sunlight works on heating the water, and air heaters on which the sunlight focuses on heating the entered cool air. The liquid heater

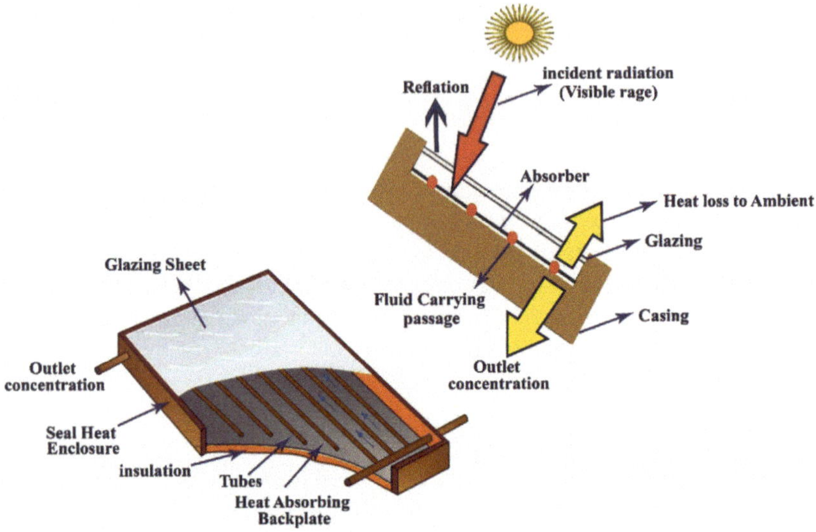

FIGURE 11.3 Schematic of flat plate solar collector and its heat transfer mechanism.

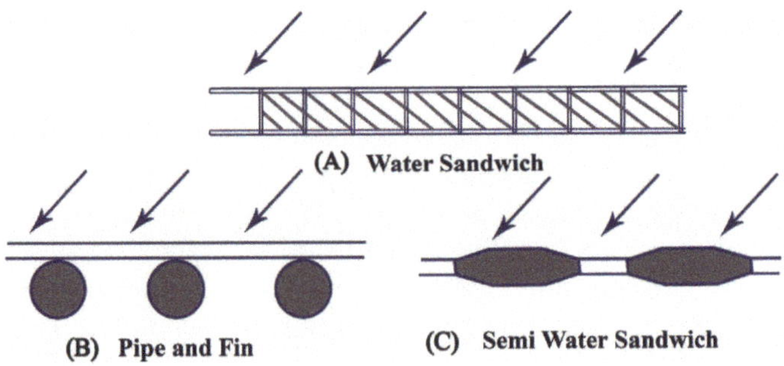

FIGURE 11.4 Structure of flat plate solar collectors.

group is divided into three groups: water sandwich type, semi sandwich type, and pipe and fin type, based on their extent of wetted surface area in comparison to the absorbing surface area. In pipe and fin type the capacity of liquid and wetted area are low. In full sandwich type both liquid capacity and wetted area are at the maximum level. In the semi sandwich types the liquid capacity and wetted area are in an intermediate level between the two other types mentioned above. Fig. 11.5 shows the detailed classification of liquid and air heater type of flat plate collectors. The pipe and fin type is divided into four groups: (1) tube in flat plate, (2) tubes bonded to upper surface of flat plate, (3) tubes fastened to lower surface of plate, and (4) tubes that are fitted in grooved plates (Fig. 11.6).

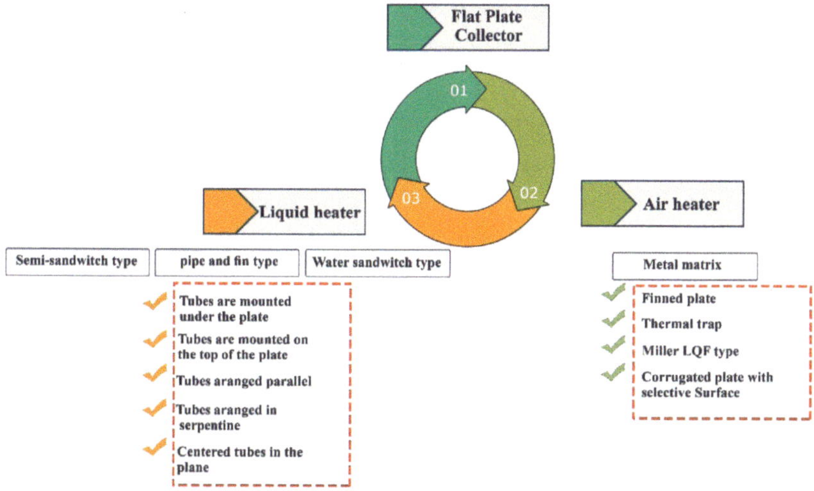

FIGURE 11.5 Classification of liquid and air heater type of flat plate collectors.

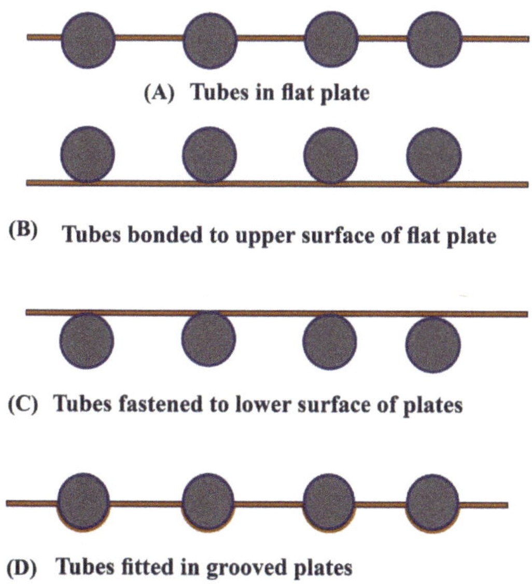

FIGURE 11.6 Various pipe and fin type of flat plate collectors.

11.2.5.2 Evacuated tube collectors

Evacuated tube solar collectors are one of the most common and simplest forms of the solar technology field. The mechanism of working of an evacuated tube solar collector is like leaving a jar exposed to radiation and letting its liquid contents heat up, although the evacuated tube solar collector works in a more complex and also more efficient way than the stated mechanism. This solar collector has various types. Fig. 11.7 shows a schematic of evacuated tube collector and its parts.

- **Heat pipe evacuated tubes**

The heat pipe evacuated tube is the most common type of evacuated tube collectors. It works according to the following steps: (1) the heat pipe fills with a special liquid, (2) the special liquid will heat by the sunlight and its phase will change to vapor, (3) the vapor will transfer to the top of the heat pipe which is known as a condenser, (4) the cold water or the heat transfer fluid flows through the header pipe, (5) the water or heat transfer fluid will heat by contact with the condenser, (6) the vapor will change to special liquids again after heat exchange between heat transfer fluid and special liquid, (7) the steps 1 to 6 are repeated. Fig. 11.8 shows the mechanism of working in above mentioned type of evacuated tube.

- **U-pipe evacuated tube**

Another form of evacuated tube solar collector is the U-pipe solar collector and this is one of the most popular evacuated tubes. This collector has distinguishing characteristics in comparison to the other types of evacuated tubes. Its working process is based on the following steps: (1) water (heat transfer fluid) flows through the header pipe, (2) the water is guided toward the copper U-pipe that is surrounded by the evacuated tube, and (3) the heat transfer fluid absorbs the energy of sunlight while it goes through the U-pipe. Fig. 11.9 shows the mechanism of working in the above mentioned type of evacuated tube.

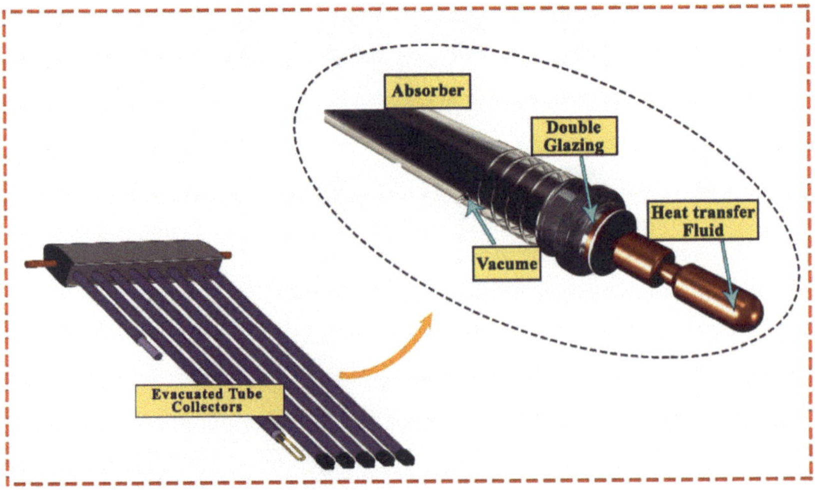

FIGURE 11.7 Evacuated tube collectors and its parts.

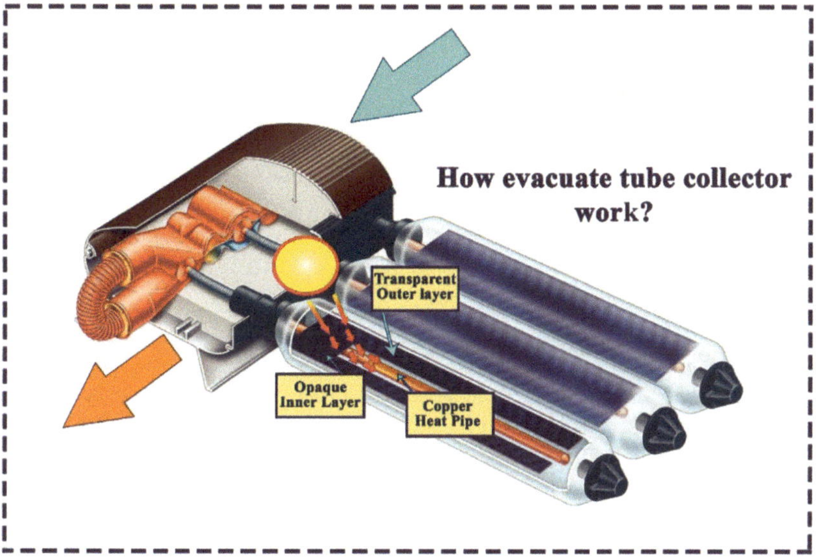

FIGURE 11.8 Mechanisms of working in heat pipe evacuated tube collectors.

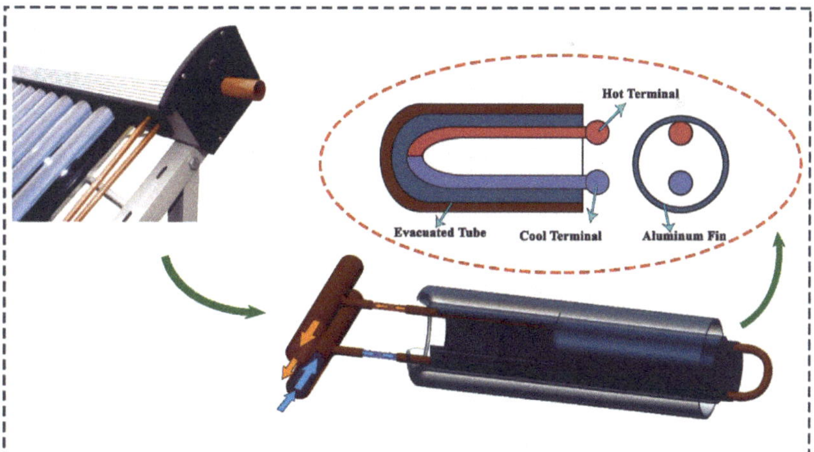

FIGURE 11.9 Mechanisms of working in U-pipe evacuated tube collectors.

- **Thermosyphon evacuated tube**

Thermosyphon evacuated tube could be categorized in the group of affordable price evacuated solar collectors in comparison to other types. The following steps are the working process of Thermosyphon evacuated tube: (1) the round shaped storage tank and each of the evacuated tubes fill with heat transfer fluid or water, (2) cold water goes downward to the bottom of the tubes and is heated up by the sunlight, and (3) the heated water rises up to the storage tank and because of its lowest density in comparison to cold water, the heated water gathers on the top of the cold water in the storage tank. Fig. 11.10 shows the working mechanism in the above mentioned type of evacuated tube.

The main types of evacuated tube solar collectors are described in the above sections. Other types of evacuated solar collectors are listed in Fig. 11.11. Adding reflectors, applying metal fins, and using insulation and covers are the ways to improve the performance of evacuated tube solar collectors.

- **Comparisons of flat plate and evacuated tube collectors**

In this section flat plate and evacuated tube collectors have been compared from different aspects like cost, performance, design, installation, etc. Table 11.2 shows the characteristics of evacuated plate and flat plate collectors.

11.2.5.3 Concentrating collectors

Concentrating solar collectors use mirrors, lenses, or parabolic surface reflectors to guide and concentrate the sunlight to a receiver tube. The receiver tube usually transfers working fluid, which could be synthetic oil, water, or

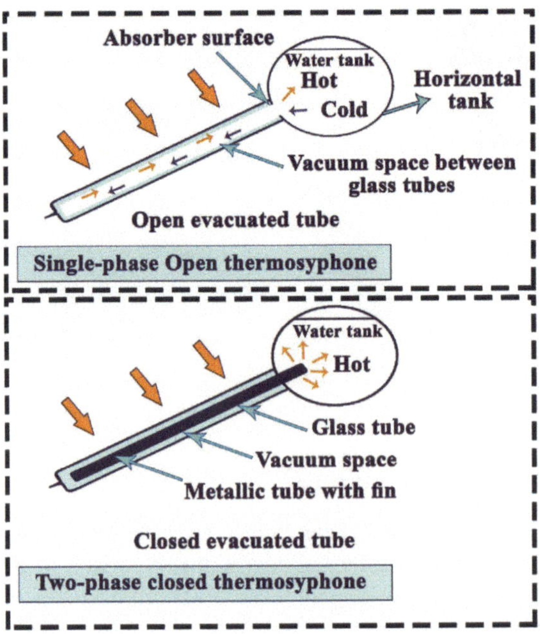

FIGURE 11.10 Mechanisms of working in thermosyphon evacuated tube collectors.

FIGURE 11.11 Other types of evacuated tube collectors.

TABLE 11.2 Comparison of flat plate and evacuated tube collectors.

Flat plate collector	Evacuated tube collector
Slow heat generation	Fast heat generation
Heat loss is high	Heat loss is low
Low efficiency in high temperature	Higher efficiency in high temperature
Long system life	Long system life
Maintenance cost is low	Maintenance cost is high
Temperature range: 60°C–80°C	Temperature range: 60°C–120°C
The collector may be damaged by freezing of water	Suitable performance even in cold condition
Cannot be cleaned manually	Can be cleaned manually
Hard and expensive to replace glass sheet	Easy to replace glass tube
Snow may decrease the performance because of covering the absorbing surface.	Snow falls through the gaps of collector tubes

nanoenriched water. The concentrated solar light will increase the working fluid temperature and change its phase to steam. This steam can be used as a driver of a steam turbine in order to run an electrical generator. There are many types of concentrating solar collectors: (1) parabolic trough collector, (2) power tower receiver, (3) parabolic dish collector, and (4) Fresnel lens collector. The most popular type is the parabolic trough collector of which its working steps are described above. To have the maximum efficiency the parabolic trough collectors usually can rotate with the changing sun position during the day. It rotates about the axis of the absorber pipe or receiver tube. The parabolic trough collectors are more in use in comparison to dish type collectors because they have low cost and a simpler sun tracking system. Figs. 11.12–11.15 show four types of concentrating collectors.

In power tower receivers, the receiver can be seen at the top of the apparatus. Large numbers of moving flat mirrors are scattered around the central receiver that is located at the top of the tower. The scattered mirrors usually track the sun in two directions and reflect the radiation to the c. 400-meter tall receiver tower. Exactly like the previous solar collector the receiver will transfer the concentrated heat to a working fluid that will increase its temperature. The phase of heated working fluid will change to steam that can be used as fuel for a conventional steam power plant in order to generate electricity. The steam temperature in power tower receivers can reach 500°C–600°C.

The parabolic dish reflectors have a focal point as their receiver. A low amount of working fluid can be heated in the receiver and because of this reason, this collector can be coupled with small generators of a few kW.

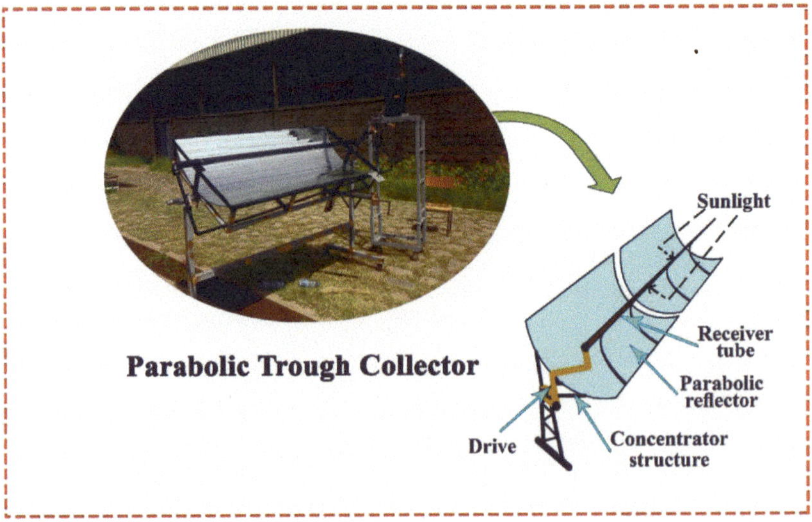

FIGURE 11.12 Parabolic trough collector.

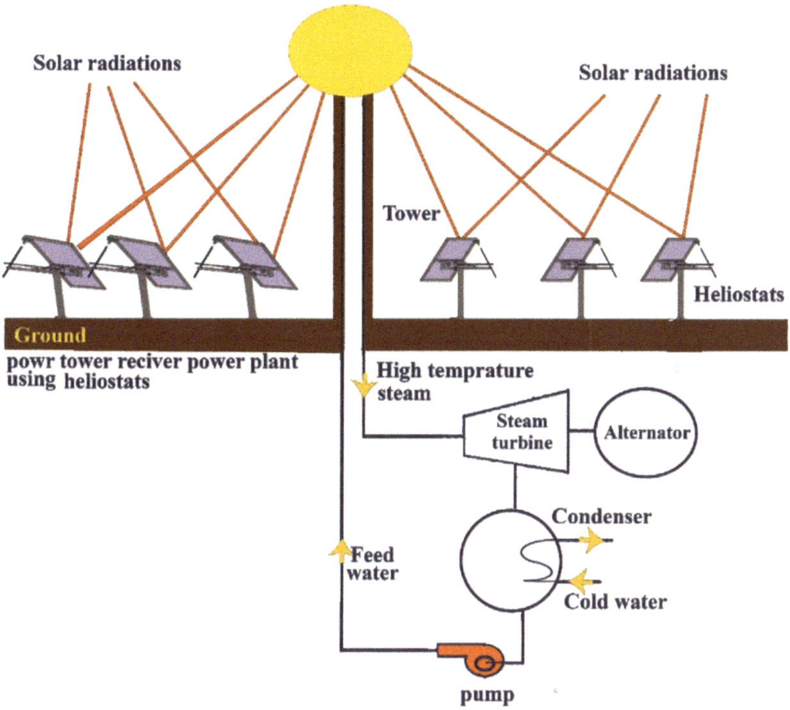

FIGURE 11.13 Power tower receiver.

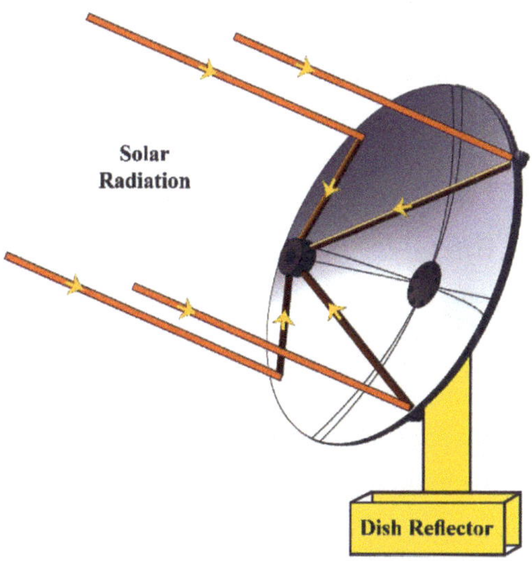

FIGURE 11.14 Parabolic dish collector.

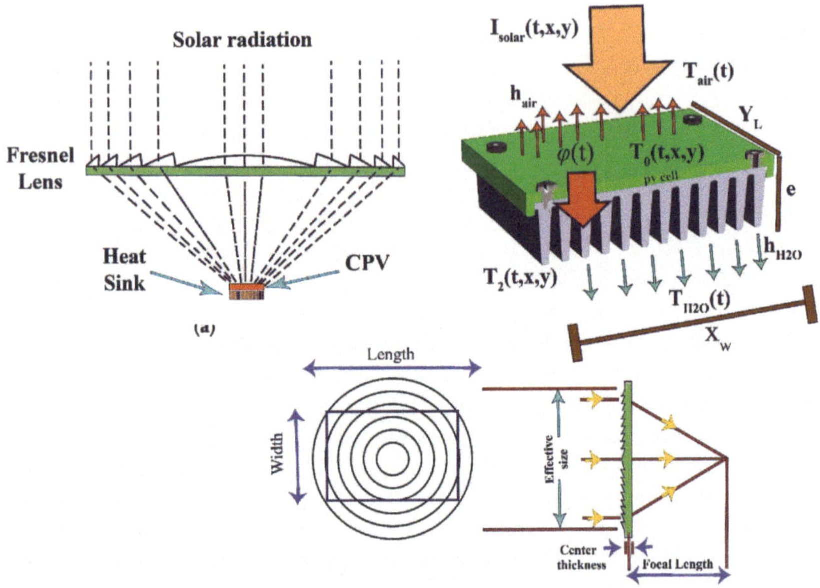

FIGURE 11.15 Power tower receiver.

This collector usually has a two-axis tracking system and can increase the temperature of working fluid to about 3000°C.

In the last type of concentrated solar collector we have the Fresnel solar collector. In this collector the sun radiates on a Fresnel lens that has fine and linear grooves on a fragile surface with optical quality and a flat surface with no groove on the other side. The sun radiates to the lens normally and light refraction will occur. The refracted light will be focused on an absorber or receiver tube and it will result in increasing the temperature of the working fluid that is passing through the tube. Fig. 11.16 shows a more complete classification of concentrating solar collectors.

11.3 Effective parameters on solar still performance

There are three categories of parameters affecting the performance of solar desalination plants: climatic parameters, design parameters, and operational parameters. The subsections of each parameter are shown in Fig. 11.17.

11.3.1 Various design parameters

In this section some of the effective designing parameters on solar still performance are studied. Using external or internal condensers in solar stills, water depth in the solar stills' basin, and applying external water heaters are the parameters that are focused upon in this section.

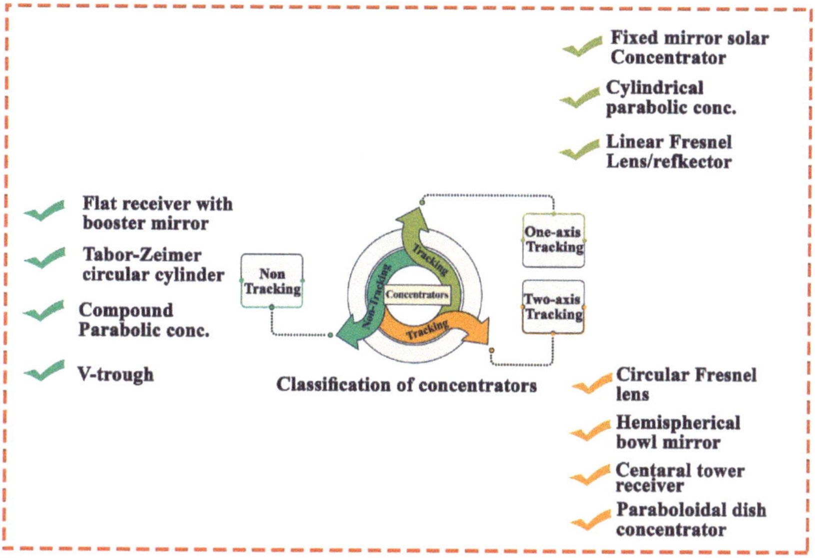

FIGURE 11.16 Classification of concentrating collectors.

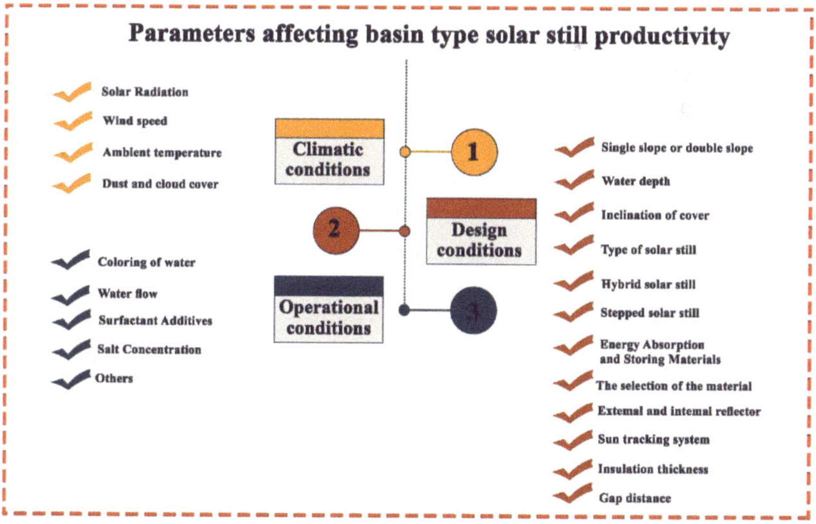

FIGURE 11.17 Parameters affecting basin type solar still productivity level.

- **Water depth**

Water depth in the basin of the solar still is one of the most effective parameters in terms of solar still performance. Many researchers have confirmed that the lesser depth of saline water in solar still will improve its productivity.

Also the convective heat transfer coefficient between the cover of solar still (the condensing cover) and water depends on the available water in the solar still [18–23]. According to studies, the performance of dual-basin desalination plants is more dependent on the water depth in the lower basin. Of course, like single-basin desalination plants, double-basin desalination plants will be more efficient if the available water depth is less [24]. Stepped and basin type solar still act as two solar water desalinators based on the greenhouse phenomena. They trap energy and sunlight, so cause energy accumulation and heat exchange. By evaporating saline water and condensing it on glass surfaces, they obtain freshwater. As mentioned in solar distillation technology, the depth of saline water is of special importance and experiments have shown that the lower the depth of saline water in desalination water, the higher the efficiency and production. Therefore, applying a stepped solar still can be very effective in the way of reducing water depth by increasing the surface area and path of the saline water flow [25]. In fact the production rate of freshwater in stepped solar stills is higher than the basin type.

- Installing heat exchangers

As discussed in previous sections about the types of active and inactive solar collectors, desalination plants are divided into active and inactive categories. In the inactive category, the water in the pond is evaporated by direct sunlight and the water desalination process takes place. But on the other hand, in the active category, external apparatuses like solar collectors are used to absorb as much thermal energy and to accelerate the evaporation process. Active solar collectors use various apparatuses like solar heaters, solar concentrators, waste heat recovery systems, and heat exchangers to make the evaporation process faster more and more [26–30].

Badran et al. coupled a solar still with a flat plate collector and improved the productivity by about 53% [31]. In another study Zeinab and Ashraf [32] coupled the solar still with both a parabolic trough collector and a heat exchanger, and based on their study an 18% improvement in the amount of distilled water was reported in comparison to conventional passive solar stills. Kumar and Tiwari [33] designed a hybrid solar still coupled with a photovoltaic system and reported 3.5 times improvement in productivity in comparison to passive solar stills. In another study Yadav [34] studied the effect of connecting a solar still to a flat plate collector and the result was a 30%–35% enhancement in distilled water production compared to a simple solar still. El-Sebaii et al. [35] used a serpentine heat exchanger in a water storage tank and a shallow solar pond next to a one-slope solar still and according to their results the efficiency of the water production improved by about 55%. Karuppusamy [36] augmented a conventional solar still with an evacuated tube collector. Based on their results the productivity of the system increased 50% in comparison to simple solar stills. Fig. 11.18 shows the solar still equipped with heat exchanger.

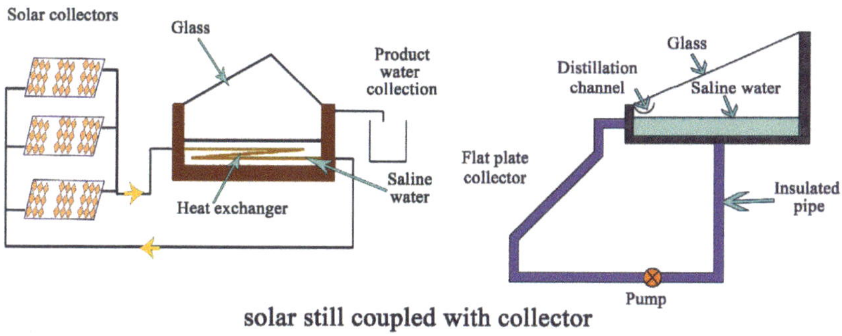

FIGURE 11.18 Solar still with heat exchanger.

- External/internal condenser

The condensation that occurs because of temperature difference is not limited to the glass surface. Side walls can also be used to produce water by applying internal condensers on the side walls (one or all four walls) of solar still [37] or external condensers [38]. The four side walls can be cooled by attached tubes on the walls with the aim of low temperature water circulation. Kumar and Bai [39] performed an extensive experimental study on the performance of solar stills equipped with enhanced condensation. They used different water types such as tap water, seawater, and dairy industry effluent in their experiment. Based on Kumar and Bai's study the efficiency of a solar still is doubled by using enhanced condensation Figs. 11.19 and 11.20 show a schematic of internal and external condensers in solar stills, respectively.

11.3.2 Climatic parameters

- Wind velocity

The wind velocity or wind speed is one of the important parameters in solar stills efficiency level [40,41]. The higher wind speed will result in the higher efficiency of solar still. Soliman [42] studied a roof type solar still and applied a forced convection to the apparatus to check the effect of wind speed on the performance of solar still. Based on his study the evaporation rate in solar still will be enhanced with increasing the velocity speed. Also El-Sebaii [43,44] analyzed the effect of wind velocity on various active and passive solar stills and reported its positive effect on solar still productivity. In another study, Nafey et al. [45] had an opposite opinion based on his study results and stated that by increasing the wind velocity from 1 to 9 m/s the total solar still performance reduced by about 13%. Reddy and Reddy [46] conducted more complete studies on the effect of wind velocity on various heat transfer coefficients that are determinant in heat flow of solar stills. They assumed other parameters like ambient temperature and water temperature in basin to be constant. Based on their study results

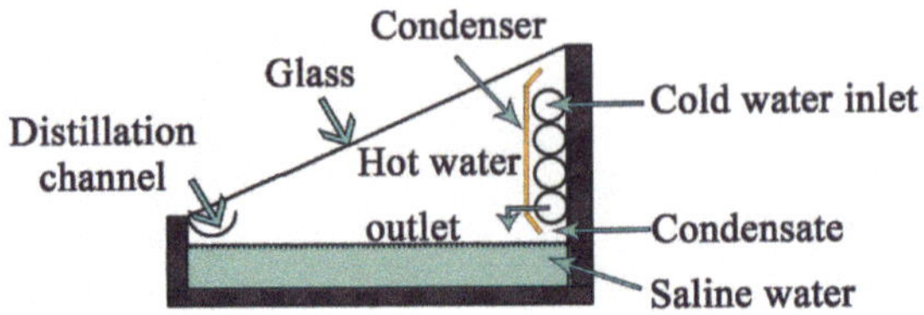

FIGURE 11.19 Schematic of internal condenser in solar still.

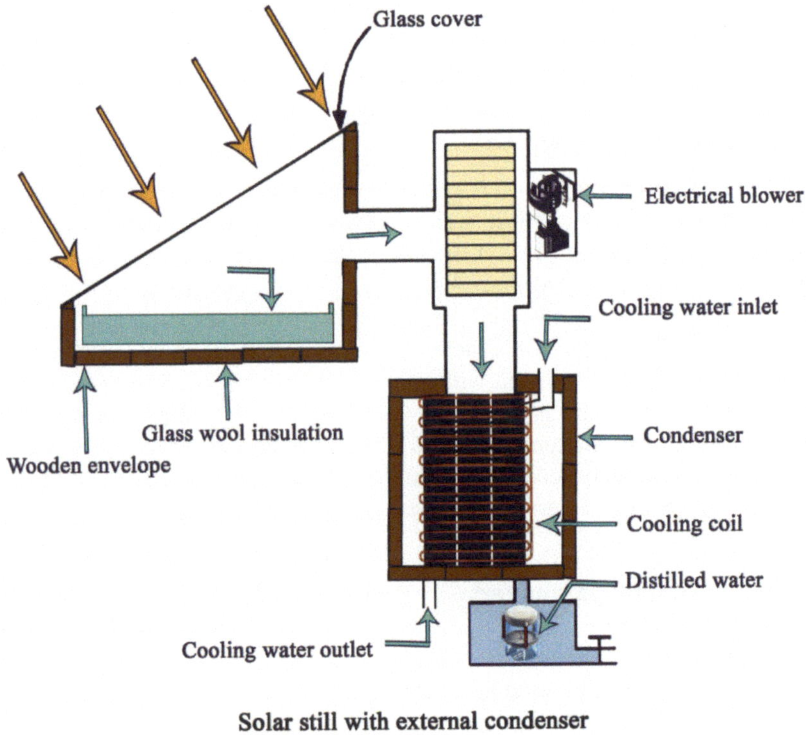

FIGURE 11.20 Schematic of external condenser in solar still.

wind speed has indirect, neutral, and indirect relations with radiative heat transfer coefficient between water and glass, evaporative heat transfer coefficient between water and glass, and the radiative heat transfer coefficient between glass and sky, respectively.

- Ambient temperature

Many researchers have conducted experiments on the effect of ambient temperature on the performance of solar stills. Based on Tiwari Anil and Tiwari's [23] findings, increasing the ambient temperature will result in a slight increase in solar still productivity. Rahbar and Esfahani [47] stated that a higher ambient temperature will give higher water productivity. Also Nafey et al. [45] showed that in order to enhance the performance of a solar still by about 3%, a 5°C increase in ambient temperature could be the solution. In a similar study, Hinai et al. [48] stated that a 10°C rise in ambient air temperature could enhance the solar still performance by about 8%. Also Boukar and Harmim [49] conducted their experiments on uncoupled and coupled simple solar still with a flat plate solar collector and concluded that the performance and productivity of both examined solar stills is highly dependent on solar radiation intensity and ambient temperature. Next to all the aforementioned experimental studies, many researchers have used theoretical models [50] to study the effect of ambient temperature on solar still productivity level.

- Solar radiation

Based on many search results one of the most effective parameters on solar still productivity is the amount and intensity of solar radiation [45,51]. Although some other researchers focused on the negative aspects of radiation increases and stated that more solar radiation heats the liquid more, so it increases the heat loss in a solar still system [52]. In another study Morse and his colleagues [53] studied the effect of wind speed, intensity of solar radiation, and ambient temperature, etc. on productivity of solar stills. Based on their analytical study the intensity of solar radiation was the most important compared to other parameters. Also Aburideh et al. [54], Rahbar and Esfahani [47], and Kamal [40], based on experimental studies, concluded that the intensity of incidental solar radiation is the most important parameter compared with other climate conditions.

11.4 Application of nanofluids in solar stills

Due to the limitations of the thermophysical properties of heat transfer fluids such as water or ethylene glycol, researchers have always sought to improve these properties, the most important of which is the use of nanoparticles as an additive to fluids [55].

Nanofluids at laboratory and semi industrial scale have been widely used in various equipment, such as heat exchangers [56], solar collectors [57], nuclear industries [58], and waste energy recovery systems [59]. In this section, the feasibility and effect of using nanoparticles in small-size desalination systems (solar stills) are discussed.

11.4.1 Solar stills without nanofluids

How a typical solar desalination plant works is very simple. The saline water enters the desalination tank and is separated from the outside environment by a transparent cover. Exposure of saline water to sunlight gradually increases its temperature and thus increases the kinetic energy of water molecules. This causes some of the salt water to evaporate, thus separating its molecules from the salt. Certainly not all evaporated particles have enough kinetic energy to reach the transparent cover solar still. So after evaporation, they condense again and return to the saline water. Of course, on the other hand, particles with higher kinetic energy levels move toward the transparent cover and after reaching it, they lose their latent heat and will condense. This condensation will produce freshwater that can be collected by gravity in an external or internal reservoir tank [60–62]. Fig. 11.21 shows a schematic of performance of a conventional solar still.

11.4.2 Solar stills with nanofluids

The performance of nanoparticle-enriched solar still is not much different from conventional solar desalination plants. So all descriptions about solar still without nanofluid in the previous section are the same. The main difference is the use of nanoparticles in solar still cycle directly and indirectly. This section will address these two issues briefly.

- Direct use of nanofluids in solar stills

In the method of using nanoparticles directly, the nanoparticles are dispersed in the saline water available in the basin (Fig. 11.22). The better

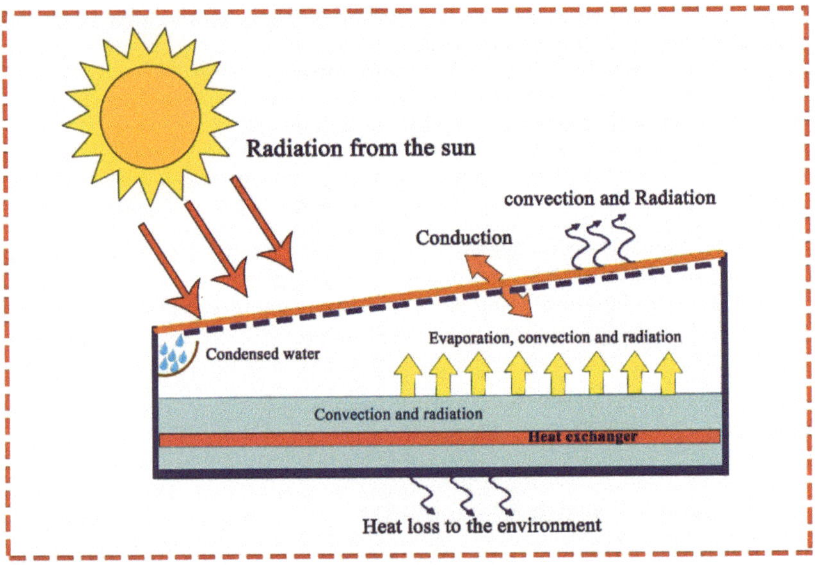

FIGURE 11.21 Schematic of performance of a conventional solar still.

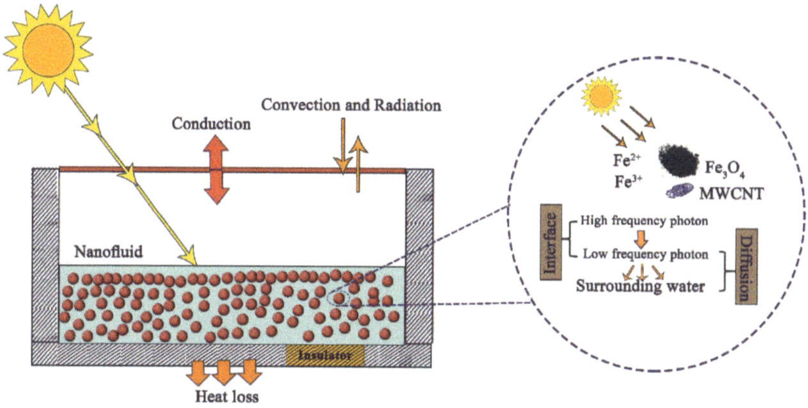

FIGURE 11.22 Schematic of direct using of nanofluids in solar stills.

thermal properties of nanoparticles, including better thermal conductivity, as well as helping to improve the overall heat transfer of saline water in the basin, have motivated researchers to disperse nanoparticles into water. This causes the saline water to absorb more heat at the same intensity and time of irradiation, and will cause more saline water to evaporate.

Of course, it should be noted that in addition to these positive effects, the environmental aspects and toxicity of some nanoparticles should not be neglected to maintain the health of the environment and humans.

- Indirect use of nanofluids in solar stills

Because of the potential problems described in the previous section, the researchers came up with the idea of using nanoparticles in a separate cycle, such as an external heat exchanger coupled with a solar still, to avoid the physical and chemical interactions of water and nanoparticles. These desalination plants actually use nanofluids in lateral heat cycles and use them to exchange heat between saline water and nanofluid available in coupled heat exchangers (Fig. 11.23). This removes the restrictions on the use of some nanoparticles and nanofluids due to their potential toxicity. Of course, limitations such as sedimentation and deposition of nanoparticles in closed heat exchanger cycles as well as the stability of nanofluids exposed to radiation and high heat exchange are some of the things that should be considered in the selection and synthesis and production of nanofluids.

11.5 Most applied nanoparticles in solar stills

As mentioned above, the use of nanoparticles and nanofluids in desalination cycles can increase the efficiency and amount of freshwater produced by this equipment. In Table 11.3, we detail the effects of using each of the

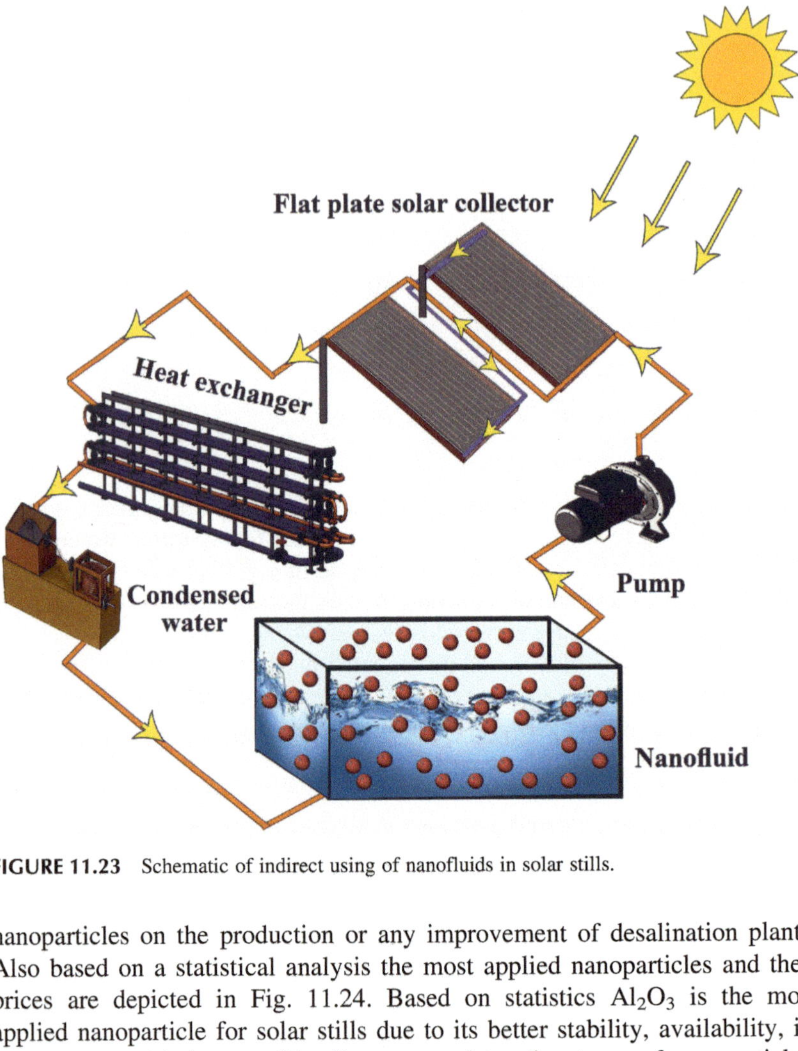

FIGURE 11.23 Schematic of indirect using of nanofluids in solar stills.

nanoparticles on the production or any improvement of desalination plants. Also based on a statistical analysis the most applied nanoparticles and their prices are depicted in Fig. 11.24. Based on statistics Al_2O_3 is the most applied nanoparticle for solar stills due to its better stability, availability, its lower cost, and being eco-friendly compared to other types of nanoparticles. Also a lack of studies on some of cheap nanoparticles like Fe_3O_4 can be seen. Next to all these issues studying and working on introducing recoverable nanoparticles can remove the limitations against using other nanoparticles with higher prices.

11.6 Challenges of nanofluid application in solar collectors

Next to various application of nanotechnology in water distillation process, like the new generation of absorbing materials that can be used in many parts of solar stills, the use of nanoparticles as working fluid improvers in solar

TABLE 11.3 Effect of using nanoparticles in solar stills.

Nanoparticles type	Type of nanofluid	Type of solar stills	Improvement (type and level of improvement)	Researchers
Al_2O_3 + SDBS	Single	Conventional solar still	About 20% increase in performance of solar still	[63]
SiC	Single	Solar desalination system	6% improvement in thermal conductivity	[64]
ZnO	Single	Tubular solar stills	38% improvement in efficiency and productivity	[65]
Aluminum oxideTitanium oxideCopper oxide	Single	Double slope solar still	Thermal conductivity improvementAbout 29%, 19% and 13% for three nanoparticles respectively	[66]
Aluminum oxideCopper oxide	Single	Conventional solar stillNew age of corrugated wick solar still equipped with mirror walls and fans	Increase in performance:By using of CuO: about 280%By using of Al_2O_3: about 250%	[67]
CuO	Single	Conventional solar still	About 50% improvement in productivity	[68]
Modified MWCNTs by using of Fe_3O_4 nanoparticles (Recoverable MWCNT)	Single (modified nanoparticles)	Conventional solar still	Increase in evaporation rate between 25%–75% for various volume fraction of nanoparticles dispersed in saline water	[69]
Al_2O_3	Single	Single basin solar still + external condense system + flat plate solar collector	About 65% improvement in productivity of solar still after using nanofluid	[70]

(Continued)

TABLE 11.3 (Continued)

Nanoparticles type	Type of nanofluid	Type of solar stills	Improvement (type and level of improvement)	Researchers
SiO_2 Cu	Single	Solar still + heat exchanger	All yield, thermal efficiency, and exergy efficiency improved About 10% by using of Cu NPsLow effect of using SiO_2investigated	[71]
MWCNT + SDS	Single	Modified single basin solar still	The efficiency improved by 150%	[72]
CuO	Single	Conventional solar still + water cooling system for the glass	Efficiency increase: about 55%	[73]
Al_2O_3	Single	Solar still (cascade type)	22% improvement in production	[74]

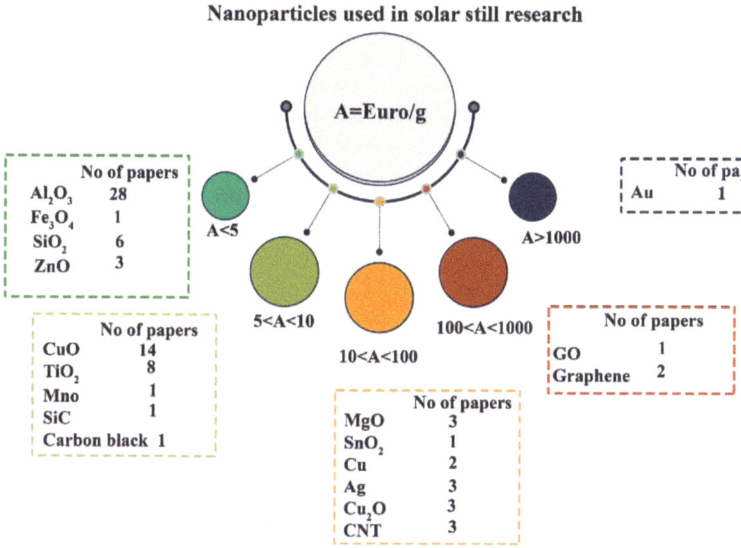

FIGURE 11.24 Frequency of use of nanoparticles in solar stills [75].

stills is certainly a promising way to improve the overall performance of this equipment. But some of specific possible challenges of using nanofluids should be studied and answered before industrial-scale use. The main possible challenges are stability of nanofluids, reusability, environmental issues, and economic analysis.

- Stability of nanofluids

Nanofluids samples preparation can be done in two ways: one-step or two-step methods. In the one-step preparation method the synthesis and dispersion of nanoparticles in the host fluid occur simultaneously. On the other hand in the two-step method dispersion and synthesis are done separately. Based on experimentation history the chance of having suitable dispersion of nanoparticle in a host fluid in one-step method is greater than in the two-step method. But the problem is the higher costs of one-step method in comparison to the two-step method. Although some physical methods are available to get a more stable nanofluid, that is, ultrasound vibration, magnet stirrer, etc. The chemical methods like using surfactant are also effective for the stability life of nanofluid. Despite all of the above mentioned, achieving economically optimized nanofluids with high ability to maintain stability over time is crucial.

- Reusability of nanofluids

As mentioned, in addition to the possibility of using nanoparticles in heat exchangers coupled with solar desalination plants, which is known as the indirect method, nanoparticles can also be used directly in the basin of solar stills. Regardless of the environmental effects, nanoparticles that are used directly must be able to withstand the salinity of the water in the pond as well as the very high temperature of available water in the basin. Agglomeration of nanoparticles and changing properties of nanoparticles could be possible observations. Therefore, the selection of nanoparticles with these capabilities is one of the important principles in providing the possibility of reuse and long-term use of nanoparticles in the water desalination cycle.

- Environmental issues

As mentioned in the previous sections, the use of nanoparticles in solar stills can be direct or indirect. This section deals with the environmental effects of using nanoparticles in a direct way and the possible problems caused by it. Of course, in direct use there are two possible modes. In the first mode, after the production of freshwater, the used nanoparticles in the remaining water of the pond are discharged into the environment, which will certainly have irreparable risks, and therefore it is not wise to do so.

The second mode is the use of equipment and methods to separate nanoparticles from the remaining water available in the pool, including centrifugation or filtration methods that require financial credit and energy consumption that are not beloved [76,77].

On the other hand, using magnetic nanoparticles and then collecting them using magnet-based apparatuses is another way to separate nanoparticles from the saline water. As a safer way, bio-based or organic-based nanosized materials can be used to minimize the damage of nanoparticles to the environment [78].

- Economic analysis

As stated in previous sections, using nanoparticles in a solar still cycle has a positive effect on its performance but there are other parameters that should be checked before selecting a specific type of nanoparticle in solar-based water distillation. Some of the main parameters are life cycle cost, payback period, and net present value. Another important note is the cost of the produced freshwater. Based on the results of Kabeel et al. [70,77], Elango et al. [79], Gupta et al. [73], Abdullah et al. [80], Modi et al. [81], Balachandran et al. [82], Sathyamurthy et al. [83], and Sharshir et al. [84], the price of distilled water by nanoenriched solar stills could be in the range of US$ 0.013 for Cu_2O and US$ 0.27 for CuO per liter. The greatest competitor to nanoenriched solar stills in water desalination process is the RO system with a cost of 0.0008–0.0013 US$/L for large-scale pants and 0.0032 US$/L for small plants. So nanoenriched solar stills should be optimized economically. Fig. 11.25 shows the NPs cost fraction and fresh water cost per liter for some of the common nanoparticles.

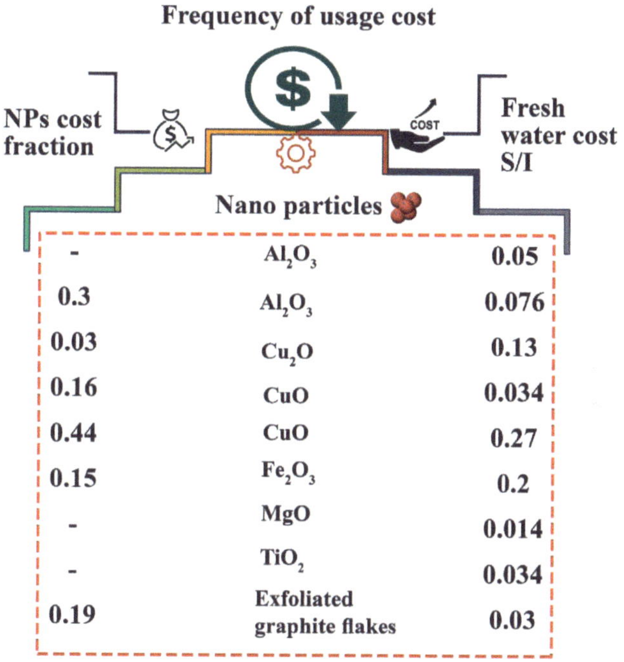

FIGURE 11.25 Price of distilled water by nanoenriched solar stills.

References

[1] <http://www.californiasolarcenter.org/history_solarthermal.html> (accessed 4.04.12).

[2] W.W. Anderson, Y.G. Chai, Becquerel effect solar cell, Energy Convers. 15 (3–4) (1976) 85–94.

[3] J. Puddicombe, Solar thermal electric power, Sci. Creat. Quart. 6 (2011).

[4] A. Madrigal, Powering the Dream. The History and Promise of Green Technology, Da Capo, Cambridge, MA, 2011, pp. 84–86.

[5] <http://www.seia.org/galleries/FactSheets/Factsheet_solar%20history.pdf> (accessed 20.05.11).

[6] W. Adams, Solar Heat: A Substitute for Fuel in Tropical Countries, Bombay, 1878. Cited Kryza, Power, p. 95.

[7] M. Thirugnanasambandam, S. Iniyan, R. Goic, A review of solar thermal technologies, Renew. Sustain. Energy Rev. 14 (2010) 312–322.

[8] I. Dincer, Energy and environmental impacts: present and future perspectives, Energy Sources 20 (1998) 427–453.

[9] <http://solar-module-panels.com/pv/history-of-photovoltaic-pv/>

[10] <http://www.eere.energy.gov>

[11] <http://www.britannica.com>

[12] <http://www.iranrea.com>

[13] J.A. Duffie, W.A. Beckman, Solar Engineering of Thermal Process, John Wiley and Sons, New York, 2013.

[14] Direct Thermal Conversion and Storage, U.S. Department of Energy, Office of Scientific and Technical Information <http://www.osti.gov/accomplishments/pdf/DE06877213/10.pdf>

[15] B. Ramlow, B. Nusz, Types of solar collectors <http://oikos.com/library/solarwaterheating/collector_types.html>.

[16] Flasolar, Flat plate solar collectors <http://www.flasolar.com/active_dhw_flat_plate.htm>, August 10, 2015

[17] G.N. Tiwari, Solar Energy. Fundamentals, Design, Modeling Applications, Alpha Science International Ltd, New Delhi, 2006, p. 525.

[18] N. Khalifa Abdul Jabbar, Hamood, M. Ahmad, Effect of insulation thickness on the productivity of basin type solar stills: an experimental verification under local climate, Energy Convers. Manag. 50 (2009) 2457–2461.

[19] M.K. Phadatare, S.K. Verma, Influence of water depth on internal heat and mass transfer in a plastic solar still, Desalination 217 (2007) 267–275.

[20] Anil K.R. Tiwari, G.N. Tiwari, Thermal modeling based on solar fraction and experimental study of the annual and seasonal performance of a single slope passive solar still: the effect of water depths, Desalination 207 (2007) 184–204.

[21] R. Tripathi, G.N. Tiwari, Effect of water depth on intemal heat and mass transfer for active solar distillation, Desalination 173 (2005) 187–200.

[22] R. Tripathi, G.N. Tiwari, Thermal modeling of passive and active solar stills for different depths of water by using the concept of solar fraction, Sol. Energy 80 (2006) 956–967.

[23] Anil K.R. Tiwari, G.N. Tiwari, Effect of water depths on heat and mass transfer in a passive solar still: in summer climatic condition, Desalination 195 (2006) 78–94.

[24] G.N. Tiwari, Sumegha Chetna, Y.P. Yadav, Effect of water depth on the transient performance of a double basin solar still, Energy Convers. Manag. 32 (1991) 293–301.

[25] A.E. Kabeel, Z.M. Omara, M.M. Younes, Techniques used to improve the performance of the stepped solar still—a review, Renew. Sustain. Energy Rev. 46 (2015) 178–188.

[26] K. Voropoulos, E. Mathioulakis, V. Belessiotis, Experimental investigation of a solar still coupled with solar collectors, Desalination 138 (2001) 103–110.

[27] K. Voropoulos, E. Mathioulakis, V. Belessiotis, A hybrid solar desalination and water heating system, Desalination 164 (2004) 189–195.

[28] C. Tiris, M. Tiris, Y. Erdalli, Sohmen, Experimental studies on a solar still coupled with a flat-plate collector and a single basin still, Energy Convers. Manag. 39 (1998) 853–856.

[29] Y.P. Yadav, B.P. Yadav, Transient analytical solution of a solar still integrated with a tubular solar energy collector, Energy Convers. Manag. 39 (1998) 927–930.

[30] A. Ali, Badran. Inverted trickle solar still: effect of heat recover, Desalination 133 (2001) 167–173.

[31] A.A. Badran, A.A. Al-Hallaq, I.A. Eyal Salman, M.Z. Odat, A solar still augmented with a flat-plate collector, Desalination 172 (2005) 227–234.

[32] S. Abdel Rehim Zeinab, Lasheen Ashraf, Experimental and theoretical study of a solar desalination system located in Cairo, Egypt, Desalination 217 (2007) 52–64 [8] G.N. Tiwari, S. Sinha, Parametric studies of active regenerative solar still, Energy Convers. Manag. 34 (3) (1993) 209–218.

[33] Kumar Shiv, Tiwari Arvind, An experimental study of hybrid photovoltaic thermal (PV/T) active solar still, Int. J. Energy Res. 32 (2008) 847–858.

[34] Y.P. Yadav, Analytical performance of a solar still integrated with a flat plate solar collector: thermosyphone mode, Energy Convers. Manag. 31 (1991) 255–263.

[35] A.A. El-Sebaii, S. Aboul-Enein, M.R.I. Ramadan, A.M. Khallaf, Thermal performance of an active single basin solar still (ASBS) coupled to shallow solar pond (SSP), Desalination 280 (2011) 183–190.

[36] Sampath Kumar Karuppasamy, An experimental study on single basin solar still augumented with evacuated tubes, Therm. Sci. 16 (2012) 573–581.
[37] A.-J.N. Khalifa, A.S. Al-Jubouri, M.K. Abed, An experimental study on modified simple solar stills, Energy Convers. Manag. 40 (1999) 1835–1847.
[38] G.N. Tiwari, H.N. Singh, Rajesh Tripathi, Present status of solar distillation, Sol. Energy 75 (2003) 367–373.
[39] K. Vinoth kumar, R. Kasturi bai, Performance study on solar still with enhanced condensation, Desalination 230 (2008) 51–61.
[40] W. Kamal, A theoretical and experimental study of the basin-type solar still under the arabian gulf climatic conditions, Sol. Wind. Technol. 5 (2) (1988) 147–157.
[41] A.A. El-Sebaii, Effect of wind speed on some designs of solar stills, Energy Convers. Manag. 41 (2000) 523–538.
[42] S.H. Soliman, Effect of wind on solar distillation, Sol. Energy 13 (1972) 403–415.
[43] A.A. El-Sebaii, Effect of wind speed on active and passive solar stills, Energy Convers. Manag. 45 (2004) 1187–1204.
[44] A.A. El-Sebaii, On effect of wind speed on passive solar still performance based on inner/outer surface temperatures of the glass cover, Energy 36 (2011) 4943–4949.
[45] A.S. Nafey, M. Abdelkader, A. Abdelmotalip, A.A. Mabrouk, Parameters affecting solar still productivity, Energy Convers. Manag. 41 (2000) 1797–1809.
[46] R.M. Reddy, K.H. Reddy, Upward heat flow analysis in basin type solar still, J. Min. Metall. 45 (2009) 121–126.
[47] N. Rahbar, J.A. Esfahani, Experimental study of a novel portable solar still by utilizing the heat pipe and thermoelectric module, Desalination 284 (2012) 55–61.
[48] H. Al-Hinai, M.S. Al-Nassri, B.A. Jubran, Effect of climatic, design and operational parameters on the yield of a simple solar still, Energy Convers. Manag. 43 (2002) 1639–1650.
[49] M. Boukar, A. Harmim, Effect of climatic conditions on the performance of a simple basin solar still: a comparative study, Desalination 137 (2001) 15–22.
[50] M.A.S. Malik, G.N. Tiware, A. Kumar, M.S. Sodha, Solar Distillation, Pergamon Press, Oxford, 1982.
[51] H.E.S. Fath, Solar thermal energy storage technologies: technical note, Renew. Energy 14 (1998) 35–40.
[52] C.E. Okeke, S.U. Egarievwe, A.O.E. Anmalu, Effects of coal and charcoal on solar still performance, Energy 15 (11) (1990) 1071–1073.
[53] R.N. Morse, W.R. Read, Sol. Energy 11 (1968) 5–17.
[54] Hanane Aburideh, Adel Deliou, Brahim Abbad, Fatma Alaoui, Djilali Tassalit, Zahia Tigrine, An experimental study of a solar still: application on the sea water desalination of Fouka, Proced. Eng. 33 (2012) 475–484.
[55] S.U. Choi, J.A. Eastman. Enhancing thermal conductivity of fluids with nanoparticles (No. ANL/MSD/CP-84938; CONF-951135-29). Argonne National Lab., IL, 1995.
[56] A.H. Pordanjani, S. Aghakhani, M. Afrand, B. Mahmoudi, O. Mahian, S. Wongwises, An updated review on application of nanofluids in heat exchangers for saving energy, Energy Convers. Manag. 198 (2019) 111886.
[57] Q. Xiong, A. Hajjar, B. Alshuraiaan, M. Izadi, S. Altnji, S.A. Shehzad, State-of-the-art review of nanofluids in solar collectors: a review based on the type of the dispersed nanoparticles, J. Clean. Prod. (2021) 127528.
[58] T. Iqbal, M. Zafar, M. Ijaz, Use of nano fluids in nuclear technology: a review, Pak. J. Sci. & Ind. Res. Ser. A: Phys. Sci. 64 (2) (2021) 149–160.

[59] A.G. Olabi, K. Elsaid, E.T. Sayed, M.S. Mahmoud, T. Wilberforce, R.J. Hassiba, et al., Application of nanofluids for enhanced waste heat recovery: a review, Nano Energy (2021) 105871.

[60] H. Zheng, Chapter 4—traditional solar desalination units, in: H. Zheng (Ed.), Solar Energy Desalination Technology, Elsevier, Amsterdam, 2017, pp. 259–321.

[61] S. Rashidi, N. Karimi, O. Mahian, J.A. Esfahani, A concise review on the role of nanoparticles upon the productivity of solar desalination systems, J. Therm. Anal. Calorim. 135 (2019) 1145–1159.

[62] Chapter 3—desalination and drying, in: H. Messel, S.T. Butler (Eds.), Solar Energy, Pergamon, 1975, pp. 256–274.

[63] A. Singh, H. Singh, Performance evaluation of solar still with and without nanofluid, Int. J. Sci. Eng. Technol. 3 (2015) 1093–1101.

[64] W. Chen, C. Zou, X. Li, L. Li, Experimental investigation of SiC nanofluids for solar distillation system: stability, optical properties and thermal conductivity with saline water-based fluid, Int. J. Heat. Mass. Tran. 107 (2017) 264–270.

[65] S.M. Saleh, A.M. Soliman, M.A. Sharaf, V. Kale, B. Gadgil, Influence of solvent in the synthesis of nano-structured ZnO by hydrothermal method and their application in solar-still, J. Environ. Chem. Eng. 5 (2017) 1219–1226.

[66] L. Sahota, G. Tiwari, Effect of nanofluids on the performance of passive double slope solar still: a comparative study using characteristic curve, Desalination 388 (2016) 9–21.

[67] Z. Omara, A. Kabeel, F. Essa, Effect of using nanofluids and providing vacuum on the yield of corrugated wick solar still, Energy Convers. Manag. 103 (2015) 965–972.

[68] S. Sharshir, G. Peng, L. Wu, N. Yang, F. Essa, A. Elsheikh, et al., Enhancing the solar still performance using nanofluids and glass cover cooling: experimental study, Appl. Therm. Eng. 113 (2017) 684–693.

[69] W. Chen, C. Zou, X. Li, H. Liang, Application of recoverable carbon nanotube nanofluids in solar desalination system: an experimental investigation, Desalination 451 (2019) 92–101.

[70] A. Kabeel, Z. Omara, F. Essa, Enhancement of modified solar still integrated with external condenser using nanofluids: an experimental approach, Energy Convers. Manag. 78 (2014) 493–498.

[71] O. Mahian, A. Kianifar, S.Z. Heris, D. Wen, A.Z. Sahin, S. Wongwises, Nanofluids effects on the evaporation rate in a solar still equipped with a heat exchanger, Nanomater. Energy 36 (2017) 134–155.

[72] G.M. Koilraj, K.P. Senthil, G. Jemilda, J.S. Sherin, Effect of nanofluids in a modified vacuum single basin solar still, Int. J. Sci. Eng. Res. 3 (2012) 1–7.

[73] B. Gupta, A. Kumar, P.V. Baredar, Experimental investigation on modified solar still using nanoparticles and water sprinkler attachment, Front. Mater. 4 (2017) 23.

[74] S. Rashidi, M. Bovand, N. Rahbar, J.A. Esfahani, Steps optimization and productivity enhancement in a nanofluid cascade solar still, Renew. Energy 118 (2018) 536–545.

[75] A. Iqbal, M.S. Mahmoud, E.T. Sayed, K. Elsaid, M.A. Abdelkareem, H. Alawadhi, et al., Evaluation of the nanofluid-assisted desalination through solar stills in the last decade, J. Environ. Manag. 277 (2021) 111415.

[76] S. Shanmugan, S. Palani, B. Janarthanan, Productivity enhancement of solar still by PCM and nanoparticles miscellaneous basin absorbing materials, Desalination 433 (2018) 186–198.

[77] A. Kabeel, E.M. El-Said, Applicability of flashing desalination technique for small scale needs using a novel integrated system coupled with nanofluid-based solar collector, Desalination 333 (2014) 10–22.

[78] F.A. Essa, A.H. Elsheikh, A.A. Algazzar, R. Sathyamurthy, M.K.A. Ali, M. Abd Elaziz, et al., Eco-friendly coffee-based colloid for performance augmentation of solar stills, Process. Saf. Environ. Prot. 136 (2020) 259–267.

[79] T. Elango, A. Kannan, K. Kalidasa Murugavel, Performance study on single basin single slope solar still with different water nanofluids, Desalination 360 (2015) 45–51.

[80] A. Abdullah, F. Essa, Z. Omara, Y. Rashid, L. Hadj-Taieb, G. Abdelaziz, et al., Management, Rotating-drum solar still with enhanced evaporation and condensation techniques, Compr. Study 199 (2019) 112024.

[81] K.V. Modi, H.K. Jani, I.D. Gamit, Impact of orientation and water depth on productivity of single-basin dual-slope solar still with Al_2O_3 and CuO nanoparticles, J. Therm. Anal. Calorim. (2020) 1–15.

[82] G.B. Balachandran, P.W. David, R.K. Mariappan, A.E. Kabeel, M.M. Athikesavan, R. Sathyamurthy, Improvising the efficiency of single-sloped solar still using thermally conductive nano-ferric oxide, Environ. Sci. Pollut. Res. (2019).

[83] R. Sathyamurthy, A.E. Kabeel, E.S. El-Agouz, D. Rufus, H. Panchal, T. Arunkumar, et al., Experimental investigation on the effect of MgO and TiO_2 nanoparticles in stepped solar still, Int. J. Energy 43 (2019) 3295–3305.

[84] S.W. Sharshir, A.H. Elsheikh, Y.M. Ellakany, A.W. Kandeal, E.M.A. Edreis, R. Sathyamurthy, et al., Improving the performance of solar still using different heat localization materials, Environ. Sci. Pollut. Res. (2020) 1–13.

Chapter 12

Utilization of nanofluids (mono and hybrid) in parabolic trough solar collector: a comparative analysis

Muhammad Sajid Khan[1,2], Mi Yan[1] and Hafiz Muhammad Ali[3,4]
[1]*Institute of Energy and Power Engineering, College of Mechanical Engineering, Zhejiang University of Technology, Hangzhou, P.R. China,* [2]*Department of Mechanical Engineering, Mirpur University of Science & Technology (MUST), Mirpur, AJK, Pakistan,* [3]*Mechanical Engineering Department, King Fahd University of Petroleum and Minerals, Dhahran, Saudi Arabia,* [4]*Interdisciplinary Research Center for Renewable Energy and Power Systems (IRC-REPS), King Fahd University of Petroleum and Minerals, Dhahran, Saudi Arabia*

12.1 Introduction

The exploitation of sustainable energy is a feasible way to overcome the energy dilemma and perform an important role to save the environment from pollution and hazardous gas emissions. Solar energy with photovoltaic or thermal collectors is a key energy source in the class of renewable energy sources, that is, eco-friendly and inexhaustible, which can be easily harnessed to useful energy [1]. Solar energy is used for the production of electricity, heating/cooling, hydrogen production, water desalination, and other purposes [2,3]. Collectors are the most important feature of solar energy systems that utilize heat transfer fluids (HTFs) circulating in the pipes and that are able to absorb and convert incidence solar radiation to useful heat [4,5].

A parabolic trough collector (PTC) is a better and sophisticated technology as compared to other concentrated solar power technologies, contributing almost 85% of total plant capacity, from an economically vital and installation point of view [6]. PTC consists of a parabolic mirror and the receiver lies on its focal line. Solar radiation strikes the aperture and will reflect to the receiver. The receiver will get heated and transfers this heat to the working fluid circulating inside the receiver [7]. The operating temperature range of PTC is approximately 323K–700K. The useful heat gain from

Advances in Nanofluid Heat Transfer. DOI: https://doi.org/10.1016/B978-0-323-88656-7.00004-0

PTC can be applied to various applications: industrial heating, refrigeration and air-conditioning, pumping irrigation water, space heating, domestic water heating, and desalination. Absorber tube material must have higher absorption and less emittance properties for better performance, while the efficiency of the system depends on two main factors: thermal and optical efficiency of PTC. Firstly, the optical efficiency of PTC is influenced by several factors, as explained in Fig. 12.1 [6]. The type of mirror used, material, coating, and smoothness influence the optical efficiency of the reflector. Other influencing factors are the incidence angle, which depends on location, day, and time, and the tracking system, which is related to the orientation and control. Secondly, thermal efficiency is affected by HTF type (fluid flow rate, fluid type, and thermophysical properties), absorber tube geometry (smooth, wavy, inserts, material length, size and deformations), and the amount of heat losses (orientation, operation, and weather conditions).

A lot of research has been carried out on PTC with different HTFs. The utilization of Syltherm or thermal oils, Therminol D12, Therminol VP-1,

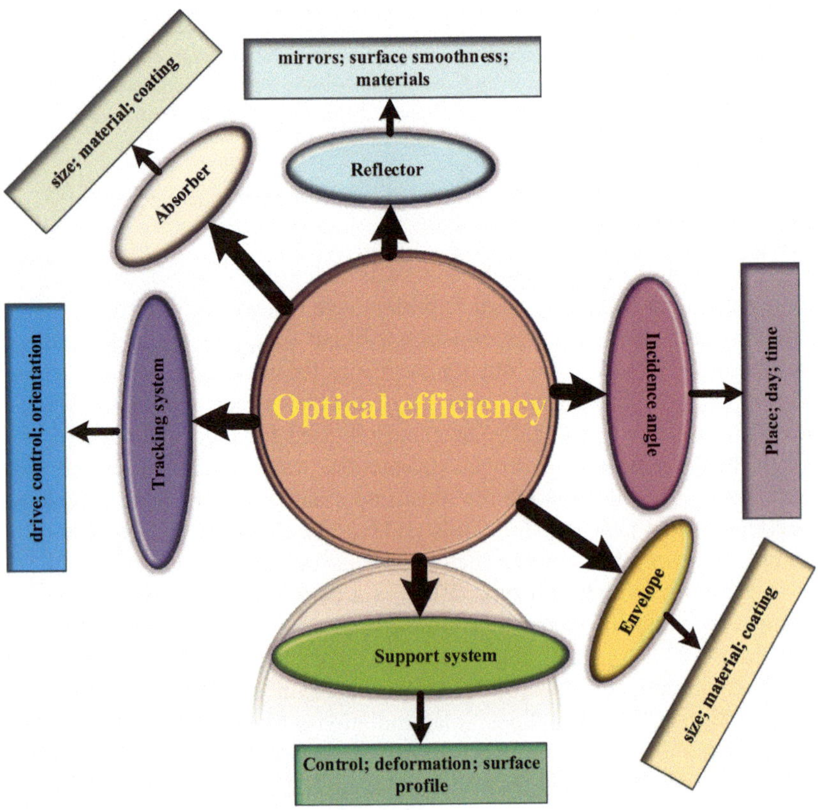

FIGURE 12.1 Effect of various factors on the optical efficiency of PTC [6].

Sandotherm 59, and Dowtherm A is common for indirect systems for heat generation [8]. However, the maximum temperature range for thermal oils is limited to 673K [9]. On the other hand, molten salts are able to operate up to 873K approximately, delivering a higher allowance for electricity conversion [10]. Although these HTFs can be utilized in PTCs and storage media for CSP plants, they face a solidification problem if temperature drops to 473K [8]. Numerous techniques have been investigated in the literature to improve the efficiency of PTC [11]. The aim of these techniques is to increase the coefficient of heat transfer between HTFs and the absorber tube, while reducing the circumferential temperature. The first class of these techniques concentrates on the geometrical modifications: use of vortex generators, internal fins, twisted tape inserts, metallic foams, etc. [5,12,13]. In addition, a huge portion of the literature investigates the dispersion of nanoparticles in the HTFs, mainly to increase the thermal conductivity of the HTFs. Such fluids are named nanofluids, a terminology put forward by Choi [14]. Al, Al_2O_3, CuO, Cu, TiO_2, Fe, ZnO, Au, and CNTs are the most common nanoparticles and their volumetric concentration is generally low: 0.1%–6% [15]. Nanoparticles are immersed in base fluids to become a homogeneous solution with superb heat transportability and absorbing characteristics. Nanofluids have higher density and thermal conductivity compared to the respective base fluid, which makes it possible to increase the heat transfer coefficient and finally thermal efficiency [16].

Ghasemi et al. [17] examined the performance of PTC by utilizing water-based CuO and Al_2O_3 nanofluids. They found the enhancement in heat transfer coefficient of almost 35% and 28% with CuO and Al_2O_3, respectively. Coccia et al. [18] numerically examined the application of different H_2O-based nanofluids with different volumetric concentrations and simulation results were validated by an experimental setup using water as the HTF. The results showed that some nanofluids presented slight improvements at low concentrations as compared to the water, while there was no improvement obtained by increasing the nanoparticles concentration. Mwesigye et al. [19] analyzed the Al_2O_3/Syltherm 800-based nanofluid in PTC up to 8% concentration and observed an optimum Reynolds number close to 10^5 to reduce the entropy generation. The thermal efficiency of the PTC receiver improved to approximately 7.6% with nanofluids utilization. In addition to this, it was found that by increasing the volume fraction, the optimal Reynolds number decreased. Bellos et al. [20] examined the PTC utilizing Al_2O_3/oil-based nanofluids and a wavy absorber tube, and found nearly 4.25% mean enhancement in thermal efficiency using nanofluids. However, 4.55% enhancement in the efficiency was observed with a convergent–divergent absorber tube. Mwesigye et al. [21] explored the optimal thermodynamic and thermal operating conditions of PTC with different concentrations of Ag, Cu, and Al_2O_3. Results showed that Ag/therminol-VP1 was the best nanoparticle followed by Cu and Al_2O_3. The thermal efficiency of the system was enhanced to nearly 13.9%, 12.5%, and 7.2% with Ag-Therminol-VP1, Cu-Herminol-VP1, and Al_2O_3-Therminol-VP1, respectively.

Hybrid nanofluids are another class of nanofluids, synthesized by dissolving two or more different nanoparticles in the base fluid to become a homogenous solution [22]. Such nanoparticles change the heat transfer characteristics, modify the base fluid evaporative rate, and finally enhance the performance of solar collectors. These are the potential fluids that offer excellent performance and thermophysical properties compared to the commonly used mono nanofluids and resolve the imperfections that result from the utilization of mono nanofluids: higher viscosity [23]. The concept of hybrid nanofluids to enhance the thermal and heat transfer performance has been utilized and applied in different solar-thermal collectors. The utilization of hybrid nanofluids has been investigated in heat exchangers [24,25], cooling electronic systems [26,27], chillers, and solar thermal applications [28–30]. The benefit of using hybrid nanofluids as a replacement to unitary nanofluids is their superior thermal and rheological properties [22].

Hybrid nanofluids can be prepared using a one-step or two-step method [6,31], while thermophysical properties, challenges, and applications are discussed in detail [32,33]. Gupta et al. [32] presented a summary of the synthetization of HNFs from different nanoparticles as well as the thermophysical properties of various hybrid nanofluids briefly. Physical vapor deposition (PVD) and liquid chemical method are the two main one-step techniques to synthesize hybrid nanofluids. Initially, a one-step direct evaporation method was proposed and later on a modified VEROS technique (physical method) was suggested. On the other hand, nanopowders are mixed in the base fluid to synthesize nanofluids in a two-step method. The nanopowders can be manufactured by various physical, chemical, and mechanical processes and are available commercially. Furthermore, numerous stability measurement methods are available including zeta potential method, sedimentation method, centrifugation method, dynamic light scattering method, and spectral absorbance method [6]. Also, some stability enhancement methods have been characterized: surfactant addition method, pH value control method, ball milling method, ultrasonic vibration method, magnetic stirring method, and steric stabilization method.

Moreover, Fig. 12.2 shows the effect of different factors on the properties of hybrid nanofluids [6]. Menbari et al. [34] analyzed the utilization of hybrid Al_2O_3 and CuO water-based HNFs in a direct absorption PTC. The optical properties of PTC with HNFs were higher as compared to the mono nanofluids. A conventional LS 2 PTC was investigated using Syltherm 800-based Al_2O_3 and CuO mono and HNFs [35]. Total volumetric concentration for mono nanofluids was set to 3%, whereas for the hybrid case 1.5% TiO_2–1.5% Al_2O_3 was selected to make a comparison suitable at the same operating conditions. Results showed that thermal efficiency enhancement with HNFs was almost 1.8%, while 0.7% has been found with mono

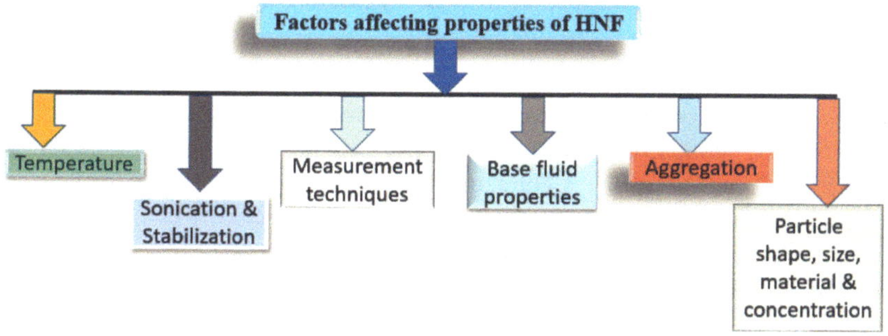

FIGURE 12.2 Effect of various factors on the properties of hybrid nanofluids [6].

nanofluid. In addition, Nusselt number enhancement with HNFs was 2.2% more than the respective base fluid.

Minea and El-Maghlany [36] presented an updated review report on the utilization and behavior of HNFs in solar PTCs. Their conclusion stated that the research on HNFs in solar system was limited to only preparation methods and certain thermophysical properties. The highest enhancement in Nusselt number was found to be almost 2% for Cu-MgO HNF. In addition to this, Khan et al. [7] investigated and compared convergent–divergent and smooth absorber tubes of PTC utilizing therminol-VP1-based mono and hybrid nanofluids. The performance of the system was evaluated at a ample range of operating temperature considering 1.5% TiO_2 and 1.5% MWCNTs as HNFs, whereas 3% MWCNTs and 3% TiO_2 were taken for both mono nanofluids. The results demonstrated that convergent–divergent tube with HNFs offers the highest thermal efficiency as compared to the other examined cases. Furthermore, the utmost improvement in thermal efficiency was approximately 5.27% for a convergent–divergent tube with HNFs with the penalty of higher-pressure drop (PD) and pumping work. Al-Oran et al. [37] investigated Syltherm 800-based various mono and hybrid NFs in LS-2 PTC with 4% volumetric fraction. Apart from energy and exergy analysis, some other parameters are considered (Nusselt number, PD, and heat transfer coefficient). Findings presented by the authors stated that Al_2O_3-CuO HNFs had the highest exergy and energy enhancements, nearly 1.03% and 1.09%, respectively. Ekiciler et al. [38] compared three different Syltherm 800-based hybrid nanofluids (Ag-ZnO, Ag-TiO_2, and Ag-MgO) with 1% to 4% volumetric fractions and Reynolds number from 10,000 to 80,000. Results showed that the highest convective heat transfer enhancement was approximately 31% for Ag-MgO/Syltherm 800 at 4% volume concentration. Xiaoke et al. [39] examined the stability, solar-thermal conversion performance, and optical properties of ethylene glycol-based SiC-MWCNTs HNF in direct absorption solar collectors. Results showed that 0.5 wt.% MWCNTs-SiC HNFs absorbed 99.9% energy at 10 mm path length. The highest value of solar thermal conversion efficiency was observed to be approximately 97.3% at 1 wt.% SiC-MWCNTs nanofluid and it was 48.6% more than the pure ethylene glycol. Graphene oxide (GO) and titanium oxide (TiO_2) hybrid nanofluids were utilized for thermal energy storage application by Vakaa et al. [40]. Their findings stated that the specific heat was improved by 19.6% by using hybrid nanoparticles s compared to the mono nanoparticles at the same operating conditions. Okonkwo et al. [41] investigated and presented the thermodynamic evaluation of flat plate solar collector with mono and hybrid nanofluids. Nanofluids were experimentally synthesized with 0.05%, 0.1%, and 0.2% nanoparticle concentrations. In addition, parametric analysis and single objective optimization of the system was conducted. Results showed that an enhancement in exergetic efficiency was found to be almost 6.9% using hybrid nanofluids. A summary of the nanofluids utilization in PTC after 2015 is presented in Table 12.1.

By reviewing the available literature, it can be seen that numerous studies are available on the thermal enhancement techniques of PTC utilizing nanofluids and different inserts or modifications inside the absorber tube. Moreover, the application of hybrid nanofluids in different cooling components and heat sinks is carried out in the past. However, exceptional thermophysical properties of hybrid nanofluids and their application in PTC is still rare. In addition, a combination of MWCNT with other nanofluids, especially CuO, CeO_2, and TiO_2 in the literature has not been investigated broadly. Therefore, the present work bridges this gap by considering different combinations of MWCNTs/Syltherm 800-based HNFs in PTC at a wide range of operating temperatures. In addition, enhancements at different temperature levels will be presented, and finally performance evaluation criteria or thermohydraulic performance will be carried out for all investigated mono and hybrid nanofluids.

12.2 System description and thermodynamic modeling

In this section, detail will be provided about the investigated PTC module size, the utilized HTFs, and the mathematical relations used for the thermodynamic modeling of PTCs and HTFs (mono and hybrid NFs). Fig. 12.3 shows the cross-sectional view of PTC; the sun's rays are incident on the aperture and some are reflected and strike the absorber tube [7]. The absorber consists of a stainless steel tube and thermal losses are decreased by employing a selective coating to enhance its absorptivity. HTFs circulate inside the tube that transfers its heat to the working fluid, and the outlet temperature of the fluid is relatively higher than the inlet temperature. The lower or bottom side of the absorber tube experiences the maximum radiation compared to the upper side, which causes nonuniform flux distribution on the PTC. In our present work, uniform flux of 1000 W/m^2 is considered for the

TABLE 12.1 Summary review of nanofluids on PTC.

Sr. no.	Ref.	Nanofluids	Nanoparticle size (nm), volume, and concentration	Results
1	[20]	Al_2O_3/water	$D_p = 20nm; \phi = 1 - 2\%$	$\eta_{mean} = \uparrow 4.25\%$;
2	[42]	Al_2O_3/Syltherm 800	$D_p = 10nm - 20nm; \phi = 0 - 4\%$	$\eta = \uparrow 10\%\ at\ \phi = 4\%$
3	[21]	Al_2O_3, Cu, and Ag/ therminol-VP1	$D_p < 100nm; \phi = 0 - 6\%$	$\eta = \uparrow 13\%$ higher with Ag/Therminol-VP1
4	[43]	Al_2O_3-SWCNTs/ Therminol and Syltherm800	$D_p < 100nm; \phi = 0 - 5\%$	Highest efficiency and saving belongs to Cu-Syltherm 800, which is 3.1% and 2.4%, respectively.
5	[44]	CuO, SiO_2, TiO_2, Al_2O_3/oil	$D_p = 10nm - 50nm; \phi = 0.02 - 0.056\%$	$\eta = \uparrow$ if $\phi = \uparrow$; CuO/oil has the highest performance
6	[45]	Fe_2O_3 and Al_2O_3/ water	$D_{Al_2O_3} = 20nm; D_{Fe_2O_3} = 20nm - 40nm; \phi = 0.2\%, 0.25\%, 0.3\%$	$\eta = \uparrow 17\%$ and $\uparrow 13\%$ for Al_2O_3 and Fe_2O_3, respectively.
7	[35]	Al_2O_3-TiO_2/Sytherm 800	$\phi = 1.5\%, 3\%$	$\eta = \uparrow 1.8\%$ and $h = 178\%\uparrow$
8	[34]	Al_2O_3 and CuO/water	$D_{Al_2O_3} = 40nm; D_{CuO} = 100nm; \phi = 0.05\%,\ 0.1\%,\ 0.2\%$	$\eta = \uparrow 48\%@\phi = \uparrow$
9	[46]	Cu/Therminol VP-1	$D_p = 100nm; \phi = 0 - 6\%$	$\eta = \uparrow 12.5\%$
10	[46]	Nanosilica and MWCNT/EG	$D_{NS} = 15nm; D_{MWCNT} = 10nm \phi = 0 - 0.6\%$	$\eta_{NS} = 70.9\%$ at $\phi = 0.4\%$ $\eta_{MWCNT} = 80.7\%$ at $\phi = 0.55\%$
11	[47]	Fe_3O_4/therminol 66	$D_p = 10nm;\ \phi = 1\% - 4\%$	Collector efficiency, local-heat transfer coefficient increase as magnetic field intensity increases.

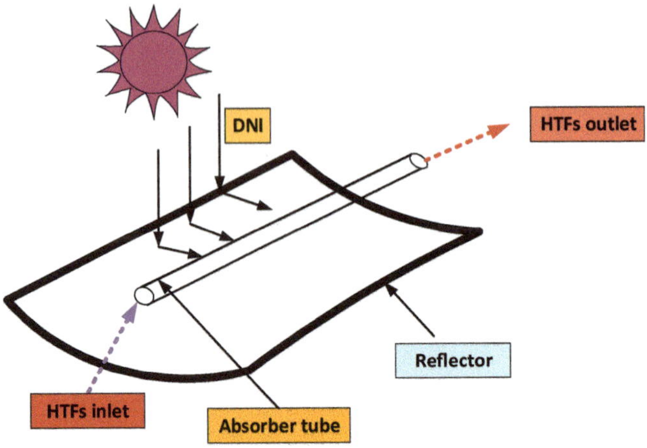

FIGURE 12.3 Schematic of investigated PTC model [7].

TABLE 12.2 PTC design input parameters [37].

Module dimensions	**7.8 × 5 m**	**solar irradiation, G_b**	**1000 W/m^2**
$D_{r,o}$	0.07 m	$D_{r,i}$	0.04 m
$D_{c,o}$	0.115 m	$D_{c,i}$	0.072 m
Aperture area, A_{ap}	38.1 m^2	T_{amb}	300K
Optical efficiency	0.755	Emittance of cover	0.86
Sun temperature	5600	Mass flow rate of HTF in PTC	0.7 kg/s
Concentration ratio	22.21	Base fluid	Syltherm-800

analysis. PTC design parameters are listed in Table 12.2. The following eight cases are studied for the PTC absorber tube:

1. Utilization of Syltherm 800 is regarded as base or reference case.
2. Utilization of MWCNTs/Syltherm 800 mono nanofluid.
3. Utilization of TiO_2/Syltherm 800 mono nanofluid.
4. Utilization of CuO/Syltherm 800 mono nanofluid.
5. Utilization of CeO_2/Syltherm 800 mono nanofluid.
6. Utilization of MWCNTs-TiO_2/Syltherm 800 hybrid nanofluid.
7. Utilization of MWCNTs-CuO/Syltherm 800 hybrid nanofluid.
8. Utilization of MWCNTs-CeO_2/Syltherm 800 hybrid nanofluid.

The mathematical relations used for the thermodynamic analysis of Syltherm 800-based mono and hybrid nanofluids in PTC will be given here. The examined system is modeled in engineering equation solver [48]. Modeling is conducted on the theory of steady state and uniform flux conditions at 300K ambient temperature and 101-kPa pressure.

12.2.1 Energy and exergy analysis of parabolic trough collector

The useful heat gain described by energy balance is a vital parameter in the solar collector system, and it is required to calculate the collector thermal efficiency [7].

$$Q_u = \dot{m}C_p(T_{out} - T_{in})$$

where $\dot{m}$ is the mass flow rate of HTFs, and T_{out} and T_{in} are the temperatures of the HTFs at the exit and inlet of the absorber tube, respectively.

The available solar energy is the product of incidence beam radiation and the aperture area of the collector and is defined by the following relation.

$$Q_{sun} = G_b A_{ap}$$

The parameters mentioned in the above equations are used to calculate the thermal efficiency of PTC collector [20].

$$\eta_{th} = Q_u / Q_{sun} \tag{12.1}$$

In addition, the concentration ratio is the ratio of aperture area to the receiver area, which are given as:

$$A_{ap} = L\left(W - D_{c,o}\right) \tag{12.2}$$

$$A_r = \pi L D_{r,o} \tag{12.3}$$

$$C_{rat} = \frac{A_{ap}}{A_r} \tag{12.4}$$

To calculate the Nusselt number and convection coefficient of wind, it is necessary to calculate the Reynolds number first [20,35].

$$R_e = \frac{\rho V D_{c,o}}{\mu} \tag{12.5}$$

For $1000 < R_e < 50{,}000$

$$N_{U,air} = 0.3 R_e^{0.6} \tag{12.6}$$

And for $0.1 < R_e < 1000$

$$N_{U,air} = 0.4 + 0.54 R_e^{0.52} \tag{12.7}$$

The overall heat loss coefficient can be explained mathematically as:

$$U_L = \left[\frac{A_r}{\left(h_{c,ca} + h_{r,ca}\right) A_c} + \frac{1}{h_{r,cr}} \right]^{-1} \tag{12.8}$$

The convective wind heat transfer for the receiver is defined as:

$$h_{c,ca} = \left(\frac{N_{u,air} K_{air}}{D_{co}} \right) \tag{12.9}$$

where K_{air} is the thermal conductivity of the air.

The heat transfer from the ambient to the glass cover due to the radiation coefficient can be evaluated as:

$$h_{r,ca} = \sigma \epsilon_{cv} (T_c + T_a)({T^2}_c + {T^2}_a) \tag{12.10}$$

ϵ_{cv} and σ are the glass cover emissivity and Stefan Boltzman constant, respectively.

Eq. (12.11) is used to determine the coefficient of radiation heat exchange between glass cover and receiver.

$$h_{r,cr} = \frac{\sigma(T_c + T_{r,av})({T_c}^2 + {T^2}_{r,av})}{\frac{1}{\epsilon_r} + \frac{A_r}{A_c}\left(\frac{1}{\epsilon_{cv}} - 1\right)} \tag{12.11}$$

Furthermore, the heat removal factor and heat loss coefficient are defined by the following equations [12].

$$F_R = \frac{\dot{m}_r Cp_r}{A_r U_L} \left[1 - \exp\left(- \frac{A_r U_L F_1}{\dot{m}_r Cp_r} \right) \right] \tag{12.12}$$

$$U_L = \frac{\dot{Q}_{loss}}{A_r (T_r - \mathrm{T}_a)} \tag{12.13}$$

The useful energy gain from collector to glass cover is calculated as:

$$\dot{Q}_u = A_{ap} F_R \left(\mathrm{S} - \frac{A_r}{A_{ap}} U_L (T_{ri} - \mathrm{T}_0) \right) \mathrm{t}_r \tag{12.14}$$

S is absorbed solar radiation, given as:

$$S = \eta_0 G_b \tag{12.15}$$

The overall coefficient of heat transfer for the collector can be evaluated by following equation.

$$U_o = \left[\frac{1}{U_L} + \frac{D_{r,o}}{h_{c,r,in D_{r,i}}} + \left(\frac{D_{r,o}}{2 \; K_r} ln \frac{D_{r,o}}{D_{r,i}} \right) \right]^{-1} \tag{12.16}$$

Apart from energy analysis, exergy is another potential and powerful tool that has qualities that make it handy and suitable as a quantifier for the sustainability of the process [49]. It is an extensive property and its value is calculated by the parameters of the system under consideration and the reference environment. Exergy efficiency depends on the useful collector exergy and the available solar exergy, defined as:

$$\eta_{X,PTSC} = \frac{\dot{E}_{X,Col}}{\dot{E}_{X,solar}} \tag{12.17}$$

Exergy from the sun and the collector can be computed by the following equations.

$$\dot{E}_{x,solar} = G_b.A_a.\eta_{pet} \tag{12.18}$$

$$\dot{E}_{X,Col} = \dot{Q}_{prod} - \dot{m}.c_p.T_{amb}.\ln\left[\frac{T_{out}}{T_{in}}\right] \tag{12.19}$$

Pattela efficiency is defined by the following equation:

$$\eta_{pet} = 1 - \frac{4T_0}{3T_{sun}} + \frac{1}{3}\left(\frac{T_0}{T_{sun}}\right)^4 \tag{12.20}$$

12.2.2 Heat transfer analysis

The useful heat gain between the receiver surface and HTFs can be calculated using the equation:

$$Q_u = h_{conv}.A_{ri}(T_r - T_{avg}) \tag{12.21}$$

The convective heat transfer coefficient h_{conv} in the above equation can be determined from the Nusselt number.

$$h_{conv} = \frac{Nu_{nf}.k_{,nf}}{D_{ri}} \tag{12.22}$$

The associated Reynolds and Prandtl numbers can be computed as:

$$Re_{nf} = \frac{4.\dot{m}}{D_{ri}.\mu.\pi} \tag{12.23}$$

$$Pr = \frac{\mu_{nf}.C_{p,nf}}{k_{nf}} \tag{12.24}$$

The Nusselt number can be calculated in the turbulent region for $3000 \le R_e \le 5 \times 10^6$ and $0.5 \le P_r \le 2000$; a correlation developed by Gnielinski [50] is:

$$N_u = \frac{\left(\frac{f}{8}\right)P_r(R_{eD} - 1000)}{1 + 12.7\left(\frac{f}{8}\right)^{\frac{1}{2}}\left(P_r^{\frac{2}{3}} - 1\right)} \tag{12.25}$$

Friction factor is calculated by using the Petukhov correlation [51].

$$f = (0.790 \ln \mathrm{Re} - 1.64)^{-2} \tag{12.26}$$

In addition, PD and pumping power can be evaluated as:

$$\Delta P = f \frac{L}{D_{r,i}} \left(\frac{1}{2} . \rho u^2 \right) \tag{12.27}$$

$$\dot{W}_p = \frac{\dot{m}}{\rho} . \Delta P \tag{12.28}$$

12.2.3 Hybrid and mono nanofluids specifications

The aim of the present section is to outline the mathematical equations to define and conclude the thermal properties of various mono and hybrid nanofluids with different symbols. In our work, Syltherm 800 is used as the base fluid (bs), along with mono and hybrid nanoparticles of MWCNTs (φ_1), while TiO_2, CuO, and CeO_2 nanoparticles are denoted by φ_2 [2,52]. The performance of PTC is evaluated using 3% volumetric concentration of each of the mono nanofluids, whereas 1.5% is taken for each individual nanoparticle for the hybrids (1.5% WCNT: 1.5% TiO_2), (1.5% WCNT: 1.5% CuO), and (1.5% WCNT: 1.5% CeO_2). In this way a suitable comparison is performed at the same operating conditions. A summary of the tested nanoparticles is given in Fig. 12.4.

Total volumetric concentration of nanoparticles can be explained by the equation:

$$\varphi = \varphi_1 + \varphi_2 \tag{12.29}$$

The following mathematical relations are used to calculate the thermophysical properties of nanoparticles.

$$\rho_{np} = \frac{\varphi_1 . \rho_{np-1} . + \varphi_2 . \rho_{np}}{\varphi} \tag{12.30}$$

$$C_{p,np} = \frac{\varphi_1 . \rho_{np-1} . C_{p,np-1} + \varphi_2 . \rho_{np-2} . C_{p,np-2}}{\varphi . \rho_{np}} \tag{12.31}$$

$$k_{np} = \frac{\varphi_1 . k_{np-1} + \varphi_2 . k_{np-2}}{\varphi} \tag{12.32}$$

ρ_{np}, $C_{p,np}$, and k_{np} are the density, specific heat capacity, and thermal conductivity of nanoparticles, respectively.

Moreover, specific heat capacity (C_p), density, and thermal conductivity (k) of nanofluids can be evaluated as:

$$C_{p,nf} = \frac{\rho_{bf} . (1 - \varphi)}{\rho_{nf}} . C_{p,bf} + \frac{\rho_{np} . \varphi}{\rho_{nf}} . C_{p,np} \tag{12.33}$$

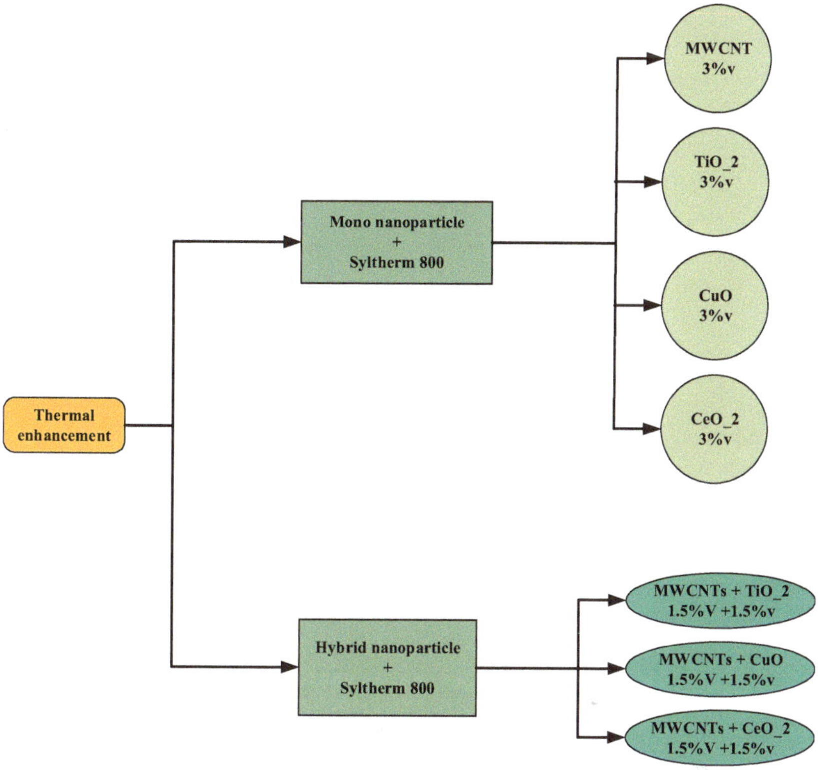

FIGURE 12.4 Investigated mono and hybrid nanofluid cases.

$$\rho_{nf} = \rho_{bf}.(1-\varphi) + \rho_{np}.\varphi \tag{12.34}$$

$$k_{nf} = \frac{k_{np} + 2.k_{bf} - 2.\varphi.(k_{bf} - k_{np})}{\frac{k_{np}}{k_{bf}} + 2 + \frac{k_{bf} - k_{np}}{k_{bf}}.\varphi} \tag{12.35}$$

Nanofluids dynamic viscosity is given by the Batchelor model [53] and Brinkman model [54], respectively:

$$\mu_{nf} = \mu_{bf} \times (1 + 2.5 \times \varphi + 6.5 \times \varphi^2) \tag{12.36}$$

$$\mu_{nf} = \frac{\mu_{nf}}{(1-\varphi)^{2.5}} \tag{12.37}$$

In order to evaluate and determine the Nusselt number of selected mono nanofluids, the Pak and Cho model [55], used by researchers in their previous studies, can be given as:

$$Nu = 0.021.Re^{0.8}.Pr^{0.5}(\varphi \leq 3\%) \tag{12.38}$$

TABLE 12.3 Thermophysical properties of the tested nanoparticles at 300K.

Nanoparticles	Ref.	ρ (kg/m^3)	k (W/mK)	C_p (J/kgK)
MWCNTs	[2]	1600	2000	796
TiO_2	[5]	4250	8.95	686
CuO	[37]	6000	77	551
CeO_2	[37]	7220	12	460
Syltherm-800		884.5	0.1243	1706

In addition, the Minea relation [56] gives the Nusselt number of hybrid nanofluids as:

$$Nu = 0.0074.Re^{0.9}.Pr^{0.67}.(100.\varphi)^{0.063} \tag{12.39}$$

Thermophysical properties of the examined nanoparticles are given in Table 12.3.

12.2.4 Performance evaluation criteria

It is already stated above that the utilization of nanofluids (mono and hybrid) increases the heat transfer and simultaneously the PD increase inside the absorber tube. Therefore, it is imperative to figure out the applicability of nanofluids for process applications using performance evaluation criteria [57,58]. In these criteria, the absorber tube operating with pure Sytherm-800 is taken as a reference case, whereas the tubes with hybrid and mono nanofluids are compared with the base or reference case according to the following three conditions.

- PEC > 1: it indicates that driving power is less than the heat transfer efficiency and nanofluid is suitable as a HTF to boost the heat transfer process.
- PEC = 1: it shows that heat transfer efficiency and driving power are equal and utilization of nanofluid will not have substantial impact on heat transfer rate enhancement.
- PEC < 1: it means that driving power is greater than the heat transfer efficiency and nanofluids cannot be utilized to improve the heat transfer rate.

The following relation describes the PEC.

$$\eta_I = \frac{[Nu/Nu_0]}{[f/f_0]^{1/3}} \tag{12.40}$$

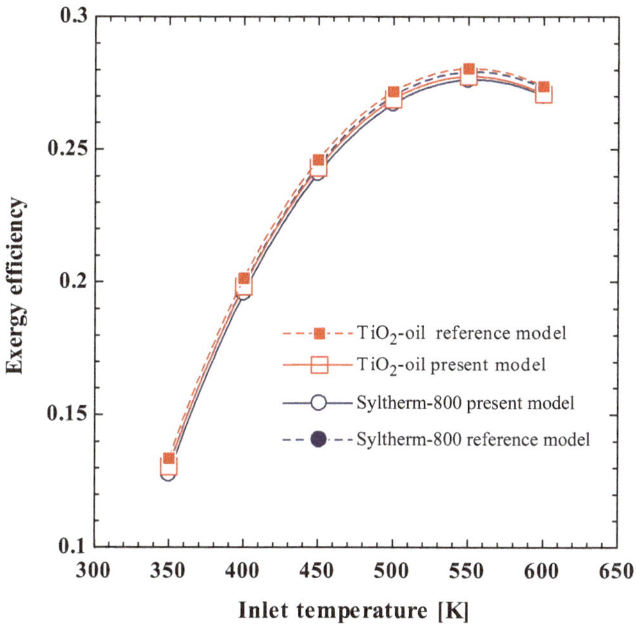

FIGURE 12.5 Validation of the present and reference model [35].

12.2.5 Validation of parabolic trough collector model

Validation results of exergy efficiency for the investigated PTC system have been compared with the reference case [35] for pure Syltherm-800 and TiO_2/Syltherm-800. The results in Fig. 12.5 show that a mean deviation of almost 1.31% is found for pure Syltherm-800, while there is a 1.29% deviation for TiO_2/Syltherm-800. The exergetic efficiency results presented high accuracy behavior, which indicates that our system is quite reliable.

12.3 Results and discussion

In the present work, a parabolic trough solar collector has been analyzed and compared using different Syltherm oil-based mono and hybrid nanofluids. Therefore, it is worth investigating the thermophysical properties of utilized HTFs. Different thermophysical properties of HTFs are presented before debating and evaluating the main performance parameters.

Fig. 12.6A–D shows the influence of inlet temperature of working fluids ranging from 350K to 600K on various properties like specific heat capacity, thermal conductivity, density, and dynamic viscosity. The specific heat capacity (6a) of Syltherm oil is higher compared to the other investigated HTFs; almost 5.7%, 8.24%, and 10.27% higher than MWCNTs-TiO_2,

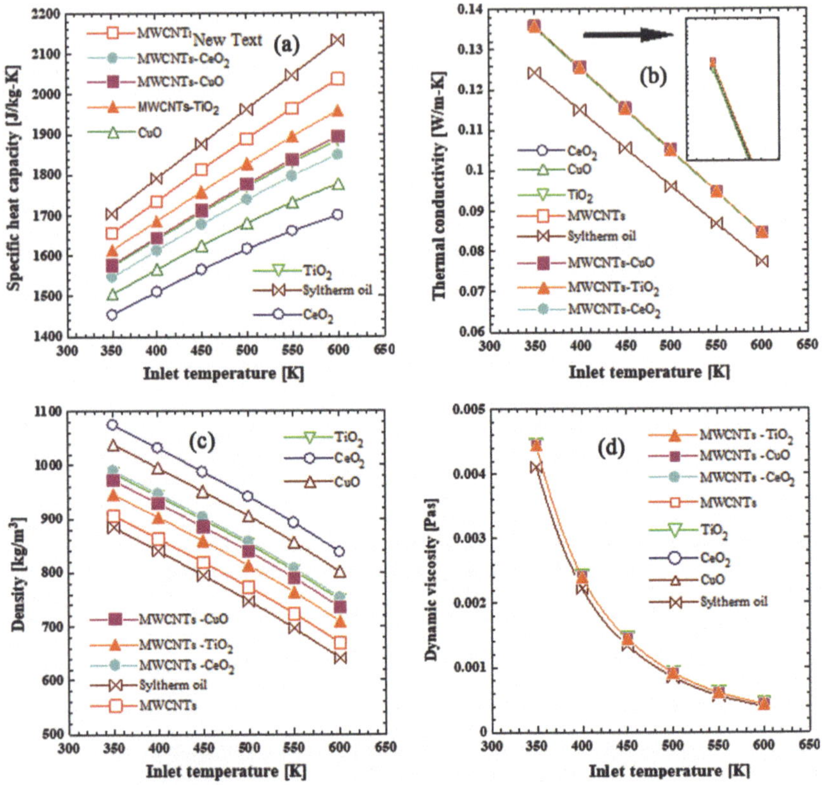

FIGURE 12.6 Thermophysical properties of HTFs: (a) specific heat capacity (C_p); (b) thermal conductivity (k); (c) density (ρ); and (d) dynamic viscosity.

MWCNTs-CuO, and MWCNTs-CeO_2 hybrid nanofluids, respectively. However, this property of HTFs increases linearly with the rise in the temperature. In addition, thermal conductivities (Fig. 12.6B) of hybrid and mono nanofluids are approximately 9.25% higher than the Syltherm oil and it reduces for all investigated HTFs as the inlet temperature increases. Fig. 12.6C shows that as the density of investigated fluids decreases, the temperature will rise. The highest density is observed for CeO_2/Syltherm oil-based nanofluid, which is nearly 1075 kg/m^3, followed by CuO/Syltherm oil and MWCNTs/Syltherm oil mono nanofluids, while pure Syltherm oil has the lowest density (884.5 kg/m^3) at 350K. MWCNTs-TiO_2/Syltherm oil has almost 6.91% higher density than pure oil but is 13.67% lower density than CeO_2/Syltherm oil-based mono nanofluid. Finally, the dynamic viscosity of hybrid nanofluids is approximately 8.10% higher than the respective base fluid. It is important to state that the total concentration of hybrid nanofluids is considered 3% (1.5% for MWCNTs and 1.5% for others), and 3% is taken for all mono nanofluids to make comparison suitable.

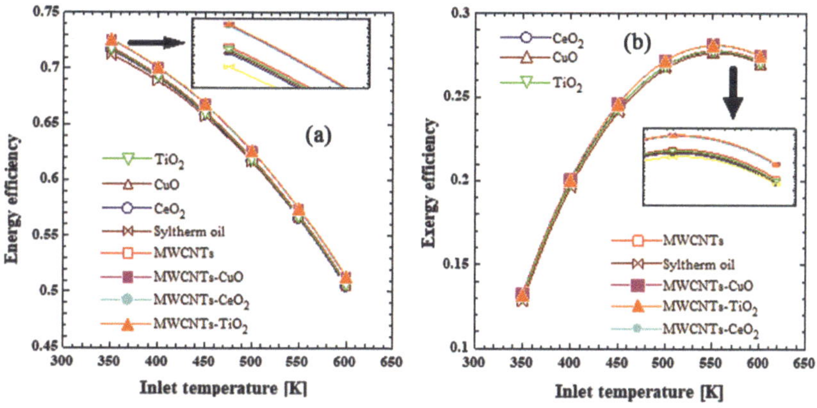

FIGURE 12.7 Comparison of (a) energy efficiencies, and (b) exergetic efficiencies at different temperature levels.

Fig. 12.7A and B are the graphical representation of solar PT receiver energy and exergy efficiencies against inlet temperature of HTFs. Thermal or energy efficiency tends to be reduced, as the temperature will rise due to the greater heat losses from the surface of the absorber tube. The thermal efficiency of a receiver utilizing MWCNTs-TiO_2 hybrid nanofluid is the highest compared to the other HTFs. It is 0.04%, 0.08%, and 2% more efficient than MWCNTs-CuO, MWCNTs-CeO_2 HNFs, and pure Syltherm oil, respectively. Furthermore, the thermal efficiency of mono nanofluids is better than Syltherm oil but less than hybrid nanofluids. Thermal efficiency of TiO_2, CuO, MWCNTs, and CeO_2 mono nanofluids are 71.76%, 71.7%, 71.85%, and 71.64%, respectively.

In addition, the exergy efficiency of the solar collector with hybrid nanofluids is higher than the mono nanofluids and the respective base fluid and it increases with an increas in the temperature. Efficiency of MWCNTs-TiO_2 HNF increases from 13.18% to 27.47%, while MWCNTs-CuO and MWCNTs-CeO_2 increase from 13.23% to 27.46% and 13.26% to 27.44%, respectively. These are almost 3.21%, 3.6%, and 3.83% more than pure base fluid, respectively at 350K. Exergy efficiency of mono nanofluids are higher than pure Syltherm oil (CeO_2 is 3.4%, MWCNTs is 1.6%, while CuO and TiO_2 are 2.9% and 2.3%, respectively at 350K).

Fig. 12.8A and B evaluates the comparative analysis and results of heat transfer coefficient and Nusselt number of various HTFs at different temperature levels. The heat transfer coefficient of MWCNTs-TiO2 HNF is 1.6% and 2.85% higher than MWCNTs-CuO and MWCNTs-CeO_2 HNFs, accordingly, and 135.6% more than pure Syltherm oil. It increases from 821.4 to 1371 W/m^2-K for MWCNTs-TiO_2/Syltherm-800 as temperature changes between 350 and 600K. For mono nanofluids, MWCNT has the better heat transfer coefficient as compared to the others in its class and Syltherm oil.

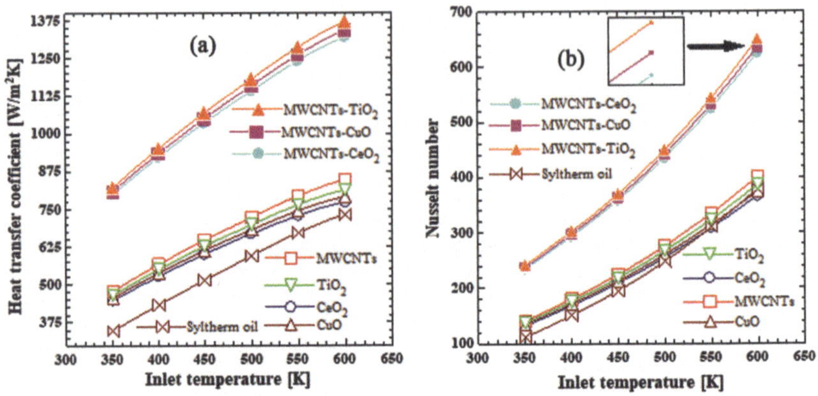

FIGURE 12.8 Temperature effect on (a) heat transfer coefficient, and (b) Nusselt number at different temperature levels.

The heat transfer coefficients of TiO_2, CuO, and MWCNT mono nanofluids are approximately 33.83%, 31.10%, and 37.59% greater than pure base fluid, respectively.

According to the Eq. (12.22) used for heat transfer analysis, HTC is directly related to the Nusselt number. Therefore, an increase in one parameter also causes the other to rise. The Nusselt number of pure Syltherm oil is the lowest in comparison to the other HTFs. MWCNTs-TiO_2 HNF has almost 75.7% and MWCNT mono nanofluid has 25.9% higher Nusselt number than respective pure Syltherm oil. Moreover, both performance parameters increase almost linearly with an increase in the temperature. The Nusselt number of MWCNTs-CuO, MWCNTs-CeO_2, TiO_2, and CuO are nearly 1.59%, 2.84%, 75.6%, and 79.85% less than MWCNTs-TiO_2 HNF, accordingly.

PD by HTFs during their circulation into solar collector is a vital parameter in the analysis of heat transfer. Therefore, particular consideration is needed to select the HTFs with less PD to avoid any additional cost. The addition of nanoparticles concentration into base fluid also increases the PD inside the absorber tube. A PD in the present study is more significant for nanofluids as compared to the base fluid; however, its effect will reduce at elevated temperature due to the viscosity of nanofluids decreased. Therefore, a rise in the PD is a constraint buildup due to the utilization of nanofluids, as nanoparticle concentration increases the density and viscosity. This phenomenon causes the stream resistance to enhance with additional pumping power. Fig. 12.9A shows the effect of a rise in the temperature on the PD by the investigated HTFs. It decreases at elevated temperatures due to the reduction in viscosity. MWCNT/Syltherm oil-based mono NF has the highest amount of PD, whereas pure Syltherm oil has the lowest. PD of pure Syltherm oil is approximately 59% less than MWCNT mono NF, while it is 53% and 46% lower than MWCNTs-TiO_2 and MWCNTs-CeO_2 HNFs, respectively.

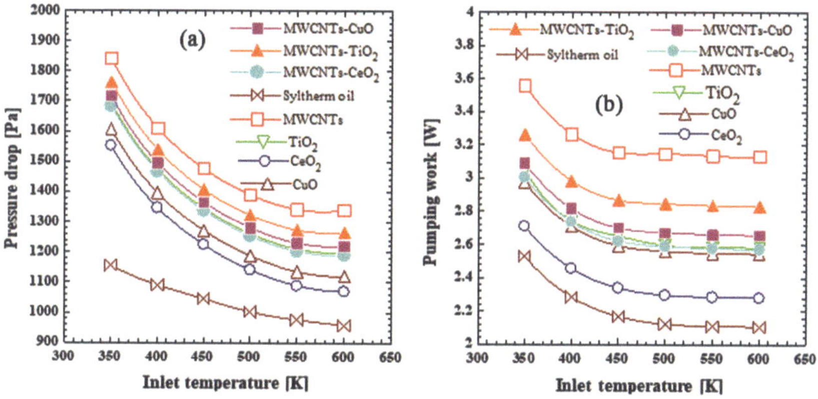

FIGURE 12.9 Temperature impact on (a) pressure drop; (b) pumping work at different temperature levels.

Fig. 12.9B shows the influence of temperature on the pumping work required to move the HTFs in a solar collector loop. It follows the similar footsteps as PD and HTFs with a higher-PD needing more pumping work. Pure Syltherm oil has the lowest pumping work required, which is nearly 29% and 41% less than MWCNTs-TiO_2/Syltherm oil HNF and MWCNT mono NF, respectively. Pumping work for CeO_2, CuO, and TiO_2 is almost 7.2%, 17.8%, and 20.7% higher than pure Syltherm oil, respectively.

The aim of the study is to investigate the possible enhancement in the performance of solar PTC utilizing mono and HNFs. To accomplish this particular goal, a PTC absorber tube operating with pure Syltherm oil is considered as the base or reference case and NFs are compared with this reference case. Figs. 12.10–12.12 show the possible enhancement in thermal efficiency, Nusselt number, and PD of the receiver tube with NFs at three different temperature levels. It can be seen from Fig. 12.10 that a maximum enhancement in efficiency is 2% for MWCNTs-TiO_2 HNF at 350K followed by MWCNTs-CuO and MWCNTs-CeO_2 HNFs, while the lowest is—0.07% for CeO_2 mono NF at 550K. This figure portrays an actual thermal enhancement of the investigated systems utilizing NFs. Mean enhancement at 350 k is 0.013%, whereas at 550K it is 0.007%. When evaluating the HTFs at three temperatures (350K, 450K, and 550K), the CeO_2/Syltherm-800 has 0.003% and MWCNTs-CeO_2/Syltherm-800 has 0.016% mean enhancement in thermal efficiency.

The highest enhancement in Nusselt number is almost 116% for MWCNTs-TiO_2 HNF at 350K, while it is −0.45% for CeO_2 mono NF at 550K (Fig. 12.11). All three HNFs have substantial higher enhancement percentages as compared to the mono NFs. At 450K, enhancement is 7.02%, 15.12%, and 11.84% for CeO_2, MWCNT, and TiO_2 mono nanofluids, respectively. Furthermore, mean enhancement at 450K is 0.434% and it is

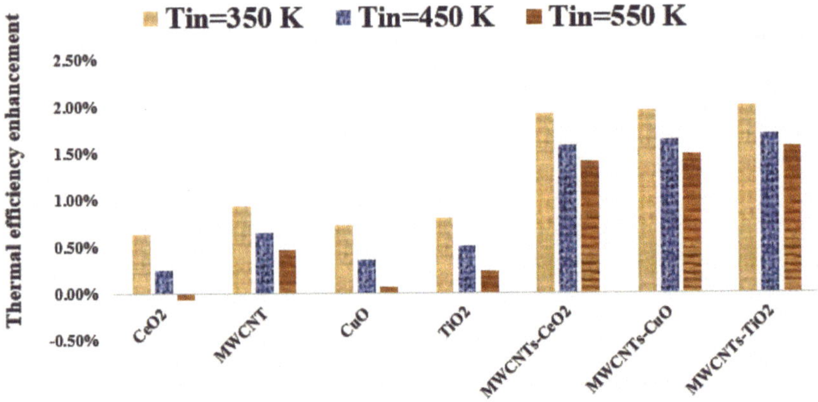

FIGURE 12.10 Thermal efficiency enhancement of examined HTFs at different temperature levels.

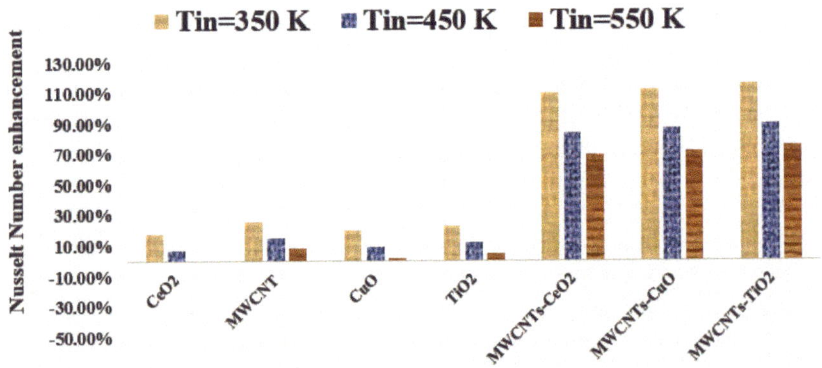

FIGURE 12.11 Nusselt number enhancement of examined HTFs at different temperature levels.

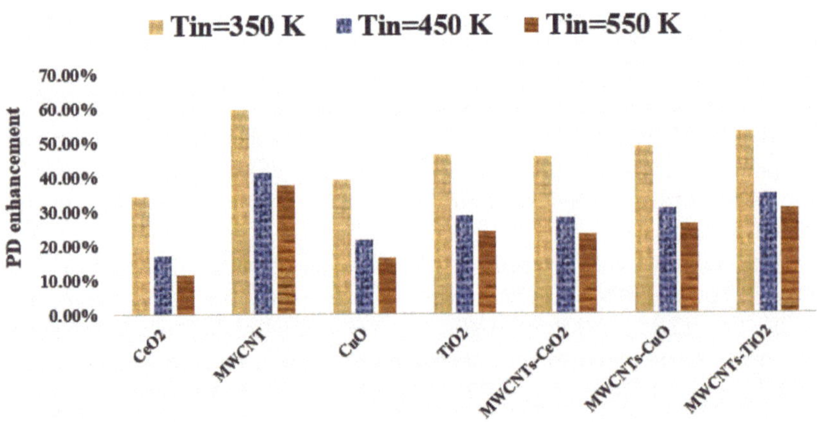

FIGURE 12.12 Pressure drop enhancement of examined HTFs at different temperature levels.

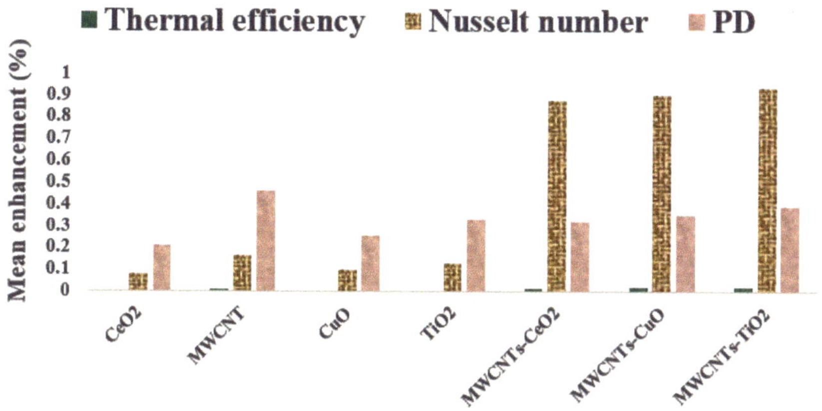

FIGURE 12.13 Mean enhancement comparison of examined HTFs.

0.607% at 350K. Higher values of heat transfer coefficients lead to higher Nusselt numbers. For HTFs. MWCNTs-CeO_2/Syltherm-800 has 109.8% and 69.15% Nusselt number enhancement at 350K and 550K, respectively.

In addition to thermal efficiency and Nusselt number enhancements, Fig. 12.12 explains the PD enhancement for all investigated mono and HNFs at three different temperatures. PD enhancement is higher (almost 59.39%) at 350K for MWCNT mono nanofluid, whereas it is 34.37% for CeO_2 mono NF. For HNFs, MWCNTs-CeO_2/Syltherm-800 and MWCNTs-TiO_2/ Syltherm-800 have almost 45.8% and 53% enhancement at 350K. It is evident that higher thermal efficiency and Nusselt number will lead to higher PD and consequently the additional costs associated with it. Mean enhancement at 350K is approximately 0.466%, while it is 0.241% at 550K.

Fig. 12.13 describes the mean enhancement in thermal efficiency, Nusselt number, and PD of the considered HTFs, using therminol-VP1 as a reference case. This enhancement is taken as the mean at three particular temperatures (350K, 450K, and 550K). The maximum mean enhancement of thermal efficiency is found for MWCNTs-TiO_2 HNF (0.018%) and the minimum is for CeO_2 mono nanofluid (0.003%). The Nusselt number mean enhancement for CeO_2 mono nanofluid is 0.082%, while MWCNTs-TiO_2 HNF has 0.936%. In the case of a PD, the maximum enhancement calculated for MWCNTs mono nanofluid is 0.46%.

Eq. (12.40) is used to calculate the thermohydraulic performance of the PTC using different HTFs in Fig. 12.14. If the value of this criterion is more than 1, flow is treated as enhanced (higher values mean more enhancement). MWCNTs-TiO_2, MWCNTs-CuO, and MWCNTs-CEO_2 HNFs have PEC values of 2.097, 2.073, and 2.064, respectively, at 350K. These are reduced to 27.86%, 28.12%, and 27.88, respectively, as temperature varies from 350K to 600K for the abovementioned HNFs but remain above unity, which indicates the justification of their utilization. In contrast to this, mono

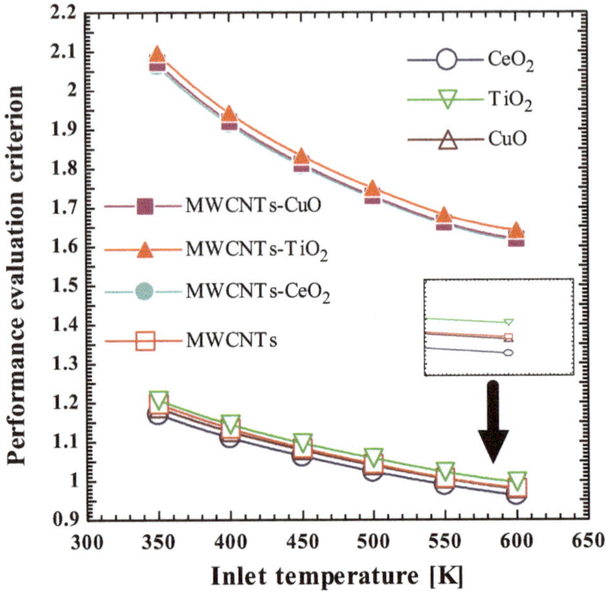

FIGURE 12.14 Performance evaluation criteria of examined HTFs at different temperature levels.

nanofluids have PEC values greater than unity between 350K and 500K, however, PEC values for all investigated mono nanofluids become less than unity when temperature rised from 550K to 600K.

12.4 Conclusion

The present chapter focuses on the utilization, comparison, and evaluation of Syltherm-800-based mono and hybrid nanofluids in PTC at a wide range of operating temperature. PTC absorber tubes with mono nanofluids (MWCNT, TiO_2, CuO, CeO_2) and hybrid nanofluids (MWCNTs-TiO_2, MWCNTs-CuO and MWCNTs-CeO_2) are examined to calculate and compare thermal and exergy efficiency, heat transfer coefficient (h_c), Nusselt number (N_u), PD, and pumping work. Moreover, thermal efficiency, PD, and Nusselt number enhancement is determined by taking pure Syltherm-800 as a reference case. At the end, performance evaluation criteria are evaluated for all the investigated HTFs. The important results of the present work can be summarized as:

- Thermal efficiency of the PTC absorber tube with MWCNTs-TiO_2/Syltherm-800 HNF is the highest (72.61%) as compared to the investigated mono and hybrid nanofluids. It is almost 2%, 1.26%, and 0.08% higher than pure Syltherm-800, CuO/Syltherm-800, and MWCNTs-CeO_2/Syltherm-800, respectively. Mean enhancement in thermal efficiency at

350K and 550K is almost 0.013% and 0.007%, respectively. Thermal efficiency enhancement is approximately 1.96% and 0.81% for MWCNTs-CuO/Syltherm-800 and TiO_2/Syltherm-800, accordingly.

- The highest exergetic efficiency is noted for MWCNTs-CeO_2/Syltherm-800 HNF at 350K, which is approximately 13.26%, while it is nearly 0.10% less than MWCNTs-TiO_2/Syltherm-800 HNF at 600K. The lowest exergy efficiency belongs to pure Syltherm-800, which is 3.6% less than MWCNTs-CuO/Syltherm-800 HNF. In addition, the efficiency of mono nanofluids is in-between pure base fluid and HNFs.
- The highest heat transfer coefficient also belongs to the HNFs; MWCNTs-TiO_2/Syltherm-800 ranges from 821 to 1371, whereas pure base fluid has the least. Heat transfer enhancement at 350K is almost 136% and 31% for MWCNTs-TiO_2/Syltherm-800 HNF and CuO/Syltherm-800 mono nanofluid, respectively. In addition, mean enhancement in Nusselt number at 350K and 450K is 0.607% and 0.434%, accordingly. CeO_2/Syltherm-800 mono nanofluid has the lowest Nusselt number enhancement (18%), whereas MWCNTs-TiO_2/Syltherm-800 HNF has the greatest enhancement (116%).
- MWCNT/Syltherm-800 mono nanofluid has the highest amount of PD (1841 Pa), while CeO_2/Syltherm-800 has the least apart from pure base fluid. Maximum enhancement in the PD is 59.39% for MWCNT/Syltherm-800, whereas MWCNTs-CuO/Syltherm-800 HNF has 48.57% PD enhancement. Mean enhancement at 350K is almost 0.466%, while it is 0.241% at 550K. Moreover, a higher pumping power is associated with the HTFs having maximum PD conditions. Therefore, MWCNT/Syltherm-800 requires greater pumping power, which is 3.55 W, whereas pure base fluid requires the least pumping power. Pumping power enhancement for MWCNTs-TiO_2/Syltherm-800, CuO/Syltherm-800, and TiO_2/Syltherm-800 is approximately 29.12%, 17.76%, and 20.69%, respectively.
- Performance evaluation criteria are evaluated to judge the feasibility of nanofluids. The utilization of all investigated HNFs are suitable as PEC values are more than unity at all investigated temperatures. The value for MWCNTs-TiO_2/Syltherm-800 is almost 2.097, while for MWCNTs-CeO_2/Syltherm-800 it is 2.064. Also, mono nanofluids have PEC values higher than unity at temperatures from 350 to 550K but above that the values become less than unity.

Nomenclature

A_{ap} Aperture area (m^2)
A_r Receiver area (m^2)
C_p Specific heat capacity (J/kg-K)
C_{rat} Concentration ratio

$D_{c,i}$ Inner cover diameter (m)
$D_{c,o}$ Diameter of outer cover (m)
$D_{r,o}$ Receiver outer diameter (m)
D_p Diameter of nanoparticles (nm)
$\dot{E}_X$ Exergy rate (W)
f Friction factor
F_R Heat removal factor
G_b Solar irradiation (W/m^2)
$\boldsymbol{h}$ Coefficient of heat transfer (W/m^2 K)
k_f Thermal conductivity working fluid (W/m-K)
L Length of collector (m)
$\dot{m}_r$ Mass flow rate (kg/s)
N_U Nusselt number
ΔP Pressure drop [Pa]
$\dot{Q}_u$ Useful heat (W)
R_e Reynolds number
$T_{r,av}$ Receiver average temperature (K)
T_c Cover temperature (K)
T_a Environmental temperature (K)
U_L Heat loss coefficient (W/m^2 K)
U_0 Overall heat transfer coefficient (W/m^2 K)
$\dot{W}_p$ Pumping power (W)

Greek Letters

ϵ_{cv} Emissivity of glass cover
ρ_{nf} Density of nanofluid (kg/m^3)
ρ_{np} Density of nanoparticle (kg/m^3)
μ_{nf} Dynamic viscosity of nanofluid [Pas]
η_X Exergy efficiency
η_{th} Thermal efficiency
η_O Optical efficiency
η_{mean} Mean efficiency
η_{pet} Pattela efficiency
σ Stefan Boltzmann constant
φ Total volumetric concentration
φ_1 Concentration of MWCNTs
φ_2 Concentration of TiO_2, CuO, and Ce_2O

Subscripts and superscripts

av/avg Average
amb Ambient
bf Base fluid
co Outer cover
col Collector
conv Convective
cv Cover

h	Hydraulic
nf	Nanofluid
NS	Nanosilica
np-1	MWCNTs nanoparticles
np-2	TiO_2, CuO, and Ce_2O nanoparticles
prod	Produced
PTSC	Parabolic trough solar collector
U	Useful

Abbreviations

CFD	Computational fluid dynamics
CSP	Concentrated solar power
GO	Graphene oxide
HTF	Heat transfer fluid
HNFs	Hybrid nanofluids
MWCNT	Multiwall carbon nanotube
PD	Pressure drop
PTC	parabolic trough collector
PEC	Performance evaluation criteria
PVD	Physical vapor deposition

Acknowledgment

The authors wish to express their great appreciation for the financial support from National International Cooperation Project, China (2016YFE0202000 and 2017YFE0107600), National Natural Science Foundation, China (51976196), and International Cooperation Project of Zhejiang Province (2019C04026).

References

[1] Z. Said, A. Mehmood, Standalone photovoltaic system assessment for major cities of United Arab Emirates based on simulated results, J. Clean. Prod. 142 (2017) 2722–2729.

[2] M.S. Khan, M. Abid, M.A. Bashir, K.P. Amber, S. Khanmohammadi, M. Yan, Thermodynamic and exergoeconomic analysis of a novel solar-assisted multigenerational system utilizing high temperature phase change material and hybrid nanofluid, Energy Convers. Manag. 236 (2021) 113948.

[3] M. Abid, M.S. Khan, T.A.H. Ratlamwala, Comparative energy, exergy and exergo-economic analysis of solar driven supercritical carbon dioxide power and hydrogen generation cycle, Int. J. Hydrog. Energy 45 (9) (2020) 5653–5667.

[4] O.A. Hussein, K. Habib, A.S. Muhsan, R. Saidur, O.A. Alawi, T.K. Ibrahim, Thermal performance enhancement of a flat plate solar collector using hybrid nanofluid, Sol. Energy 204 (2020) 208–222.

[5] M.S. Khan, M. Abid, H.M. Ali, K.P. Amber, M.A. Bashir, S. Javed, Comparative performance assessment of solar dish assisted s-CO2 Brayton cycle using nanofluids, Appl. Therm. Eng. 148 (2019) 295–306.

[6] A.K. Tiwar, V. Kumar, Z. Said, H.K. Paliwal, A review on the application of hybrid nanofluids for parabolic trough collector: recent progress and outlook, J. Clean. Prod. (2021) 126031.

[7] M.S. Khan, M. Abid, M. Yan, T.A. Ratlamwala, I. Mubeen, Thermal and thermodynamic comparison of smooth and convergent-divergent parabolic trough absorber tubes with the application of mono and hybrid nanofluids, Int. J. Energy Res. 45 (3) (2021) 4543–4564.
[8] E. Bellos, C. Tzivanidis, K.A. Antonopoulos, A detailed working fluid investigation for solar parabolic trough collectors, Appl. Therm. Eng. 114 (2017) 374–386.
[9] E. Bellos, C. Tzivanidis, D. Tsimpoukis, Thermal, hydraulic and exergetic evaluation of a parabolic trough collector operating with thermal oil and molten salt based nanofluids, Energy Convers. Manag. 156 (2018) 388–402.
[10] P.D. Myers, D.Y. Goswami, Thermal energy storage using chloride salts and their eutectics, Appl. Therm. Eng. 109 (2016) 889–900.
[11] E. Bellos, C. Tzivanidis, I. Daniil, K.A. Antonopoulos, The impact of internal longitudinal fins in parabolic trough collectors operating with gases, Energy Convers. Manage 135 (2017) 35–54.
[12] M.S. Khan, M. Yan, H.M. Ali, K.P. Amber, M.A. Bashir, B. Akbar, et al., Comparative performance assessment of different absorber tube geometries for parabolic trough solar collector using nanofluid, J. Therm. Anal. Calorim. 142 (6) (2020) 2227–2241.
[13] O.A. Jaramillo, B. Mónica, K.M. Velazquez-Lucho, M. Robles, Parabolic trough solar collector for low enthalpy processes: an analysis of the efficiency enhancement by using twisted tape inserts, Renew. Energy 93 (2016) 125–141.
[14] S.U. Choi, J. Eastman, Enhancing Thermal Conductivity of Fluids with Nanoparticles, Argonne National Lab, IL, 1995.
[15] R. Loni, E. Askari Asli-ardeh, B. Ghobadian, A.B. Kasaeian, Sh Gorjian, Thermodynamic analysis of a solar dish receiver using different nanofluids, Energy 133 (2017) 749–760.
[16] E.F. El-Gazar, W.K. Zahra, H. Hassan, S.I. Rabia, Fractional modeling for enhancing the thermal performance of conventional solar still using hybrid nanofluid: energy and exergy analysis, Desalination 503 (2021) 114847.
[17] S.E. Ghasemi, A.A. Ranjbar, Thermal performance analysis of solar parabolic trough collector using nanofluid as working fluid: a CFD modelling study, J. Mol. Liq. 222 (2016) 159–166.
[18] G. Coccia, G. Di Nicola, L. Colla, L. Fedele, M. Scattolini, Adoption of nanofluids in low enthalpy parabolic trough solar collectors: numerical simulation of the yearly yield, Energy Convers. Manage 118 (2016) 306–319.
[19] A. Mwesigye, Z. Huan, J.P. Meyer, Thermodynamic optimisation of the performance of a parabolic trough receiver using synthetic oil–Al_2O_3 nanofluid, Appl. Energy 156 (2015) 398–412.
[20] E. Bellos, C. Tzivanidis, K.A. Antonopoulos, G. Gkinis, Thermal enhancement of solar parabolic trough collectors by using nanofluids and converging-diverging absorber tube, Renew. Energy 94 (2016) 213–222.
[21] A. Mwesigye, J.P. Meyer, Optimal thermal and thermodynamic performance of a solar parabolic trough receiver with different nanofluids and at different concentration ratios, Appl. Energy 193 (2017) 393–413.
[22] T.R. Shah, H.M. Ali, Applications of hybrid nanofluids in solar energy, practical limitations and challenges: a critical review, Sol. Energy 183 (2019) 173–203.
[23] L. Yang, W. Ji, M. Mao, J.N. Huang, An updated review on the properties, fabrication and application of hybrid-nanofluids along with their environmental effects, J. Clean. Prod. 257 (2020) 120408.
[24] A. Ebrahimi-Moghadam, S. Kowsari, F. Farhadi, M. Deymi-Dashtebayaz, Thermohydraulic sensitivity analysis and multi-objective optimization of Fe3O4/H2O 58

nanofluid flow inside U bend heat exchangers with longitudinal strip inserts, Appl. Therm. Eng. 164 (2020) 114518.

[25] L. Yang, J.-n Huang, W. Ji, M. Mao, Investigations of a new combined application of nanofluids in heat recovery and air purification, Powder Technol. 360 (2020) 956–966.

[26] V. Kumar, J. Sarkar, Particle ratio optimization of Al2O3-MWCNT hybrid nanofluid in minichannel heat sink for best hydrothermal performance, Appl. Therm. Eng. 165 (2020) 114546.

[27] V. Kumar, J. Sarkar, Experimental hydrothermal behavior of hybrid nanofluid for various particle ratios and comparison with other fluids in minichannel heat sink, Int. Commun. Heat. Mass. Transf. 110 (2020) 104397.

[28] M. Beigzadeh, F. Pourfayaz, M.H. Ahmadi, Modeling and improvement of solid oxide fuel cell-single effect absorption chiller hybrid system by using nanofluids as heat transporters, Appl. Therm. Eng. 166 (2020) 114707.

[29] N. Goel, R.A. Taylor, T. Otanicar, A review of nanofluid-based direct absorption solar collectors: design considerations and experiments with hybrid PV/Thermal and direct steam generation collectors, Renew. Energy 145 (2020) 903–913.

[30] M. Heyhat, M. Valizade, S. Abdolahzade, M. Maerefat, Thermal efficiency enhancement of direct absorption parabolic trough solar collector (DAPTSC) by using nanofluid and metal foam, Energy 192 (2020) 116662.

[31] H. Babar, H.M. Ali, Towards hybrid nanofluids: preparation, thermophysical properties, applications, and challenges, J. Mol. Liq. 281 (2019) 598–633.

[32] M. Gupta, V. Singh, S. Kumar, S. Kumar, N. Dilbaghi, Z. Said, Up to date review on the synthesis and thermophysical properties of hybrid nanofluids, J. Clean. Prod. 190 (2018) 169–192.

[33] M.U. Sajid, H.M. Ali, Thermal conductivity of hybrid nanofluids: a critical review, Int. J. Heat. Mass. Transf. 126 (2018) 211–234.

[34] A. Menbari, A.A. Alemrajabi, A. Rezaei, Experimental investigation of thermal performance for direct absorption solar parabolic trough collector (DASPTC) based on binary nanofluids, Exp. Therm. Fluid Sci. 80 (2017) 218–227.

[35] E. Bellos, C. Tzivanidis. Thermal analysis of parabolic trough collector operating with mono and hybrid nanofluids, Sustain Energy Technol. Assess. 2017, ISSN 2213-1388.

[36] A.A. Minea, W.M. El-Maghlany, Influence of hybrid nanofluids on the performance of parabolic trough collectors in solar thermal systems: recent findings and numerical comparison, Renew. Energy 120 (2018) 350–364.

[37] O. Al-Oran, F. Lezsovits, A. Aljawabrah, Exergy and energy amelioration for parabolic trough collector using mono and hybrid nanofluids, J. Therm. Anal. Calorim. (2020) 1–18.

[38] R. Ekiciler, K. Arslan, O. Turgut, B. Kurşun, Effect of hybrid nanofluid on heat transfer performance of parabolic trough solar collector receiver, J. Therm. Anal. Calorim. (2020) 1–18.

[39] X. Li, G. Zeng, X. Lei, The stability, optical properties and solar-thermal conversion performance of SiC-MWCNTs hybrid nanofluids for the direct absorption solar collector (DASC) application, Sol. Energy Mater. Sol. Cell 206 (2020) 110323.

[40] M. Vaka, R. Walvekar, M. Khalid, P. Jagadish, N.M. Mubarak, H. Panchal, Synthesis of hybrid graphene/TiO2 nanoparticles based high-temperature quinary salt mixture for energy storage application, J. Energy Storage 31 (2020) 101540. Available from: https://doi.org/10.1016/j.est.2020.101540.

[41] E.C. Okonkwo, I. Wole-Osho, D. Kavaz, M. Abid, T. Al-Ansari, Thermodynamic evaluation and optimization of a flat plate collector operating with alumina and iron mono and hybrid nanofluids, Sustain. Energy Technol. Assess. 37 (2020) 100636.

[42] E. Kaloudis, E. Papanicolaou, V. Belessiotis, Numerical simulations of a parabolic trough solar collector with nanofluid using a two-phase model, Renew. Energy 97 (2016) 218–229.

[43] A. Alashkar, M. Gadalla, Thermo-economic analysis of an integrated solar power generation system using nanofluids, Appl. Energy 191 (2017) 469–491.

[44] S. Toghyani, E. Baniasadi, E. Afshari, Thermodynamic analysis and optimization of an integrated Rankine power cycle and nano-fluid based parabolic trough solar collector, Energy Convers. Manag. 121 (2016) 93–104.

[45] M.A. Rehan, M. Ali, N.A. Sheikh, M.S. Khalil, G.Q. Chaudhary, T. ur Rashid, et al., Experimental performance analysis of low concentration ratio solar parabolic trough collectors with nanofluids in winter conditions, Renew. Energy 118 (2018) 742–751.

[46] A. Mwesigye, Z. Huan, J.P. Meyer, Thermal performance and entropy generation analysis of a high concentration ratio parabolic trough solar collector with Cu-Therminol® VP-1 nanofluid, Energy Convers. Manag. 120 (2016) 449–465.

[47] A. Khosravi, M. Malekan, M.E. Assad, Numerical analysis of magnetic field effects on the heat transfer enhancement in ferrofluids for a parabolic trough solar collector, Renew. Energy 134 (2019) 54–63.

[48] F-Chart Software, Engineering Equation Solver (EES) <http://www.fchart.com/ees>, 2015.

[49] S.A. Kalogirou, S. Karellas, V. Badescu, K. Braimakis, Exergy analysis on solar thermal systems: a better understanding of their sustainability, Renew. Energy 85 (2016) 1328–1333.

[50] V. Gnielinski, New equations for heat and mass transfer in turbulent pipe and channel flow, Int. Chem. Eng. 16 (2) (1976) 359–368.

[51] T.L. Bergman, F.P. Incropera, D.P. DeWitt, A.S. Lavine, Fundamentals of Heat and Mass Transfer, John Wiley & Sons, 2011.

[52] E.C. Okonkwo, I. Wole-Osho, D. Kavaz, M. Abid, Comparison of experimental and theoretical methods of obtaining the thermal properties of alumina/iron mono and hybrid nanofluids, J. Mol. Liq. 292 (2019) 111377.

[53] G.K. Batchelor, The effect of Brownian motion on the bulk stress in a suspension of spherical particles, J. Fluid Mech. 83 (1) (1977) 97–117.

[54] H.C. Brinkman, The viscosity of concentrated suspensions and solution, J. Chem. Phys. 20 (1952) 571–581.

[55] B.C. Pak, Y.I. Cho, Hydrodynamic and heat transfer study of dispersed fluids with submicron metallic oxide particles, Exp. Heat. Transf. Int. J. 11 (2) (1998) 151–170.

[56] A.A. Minea, Hybrid nanofluids based on Al_2O_3, TiO_2 and SiO_2: numerical evaluation of different approaches, Int. J. Heat. Mass. Transf. 104 (2017) 852–860.

[57] Z. Said, M.E.H. Assad, A.A. Hachicha, E. Bellos, M.A. Abdelkareem, D.Z. Alazaizeh, et al., Enhancing the performance of automotive radiators using nanofluids, Renew. Sustain. Energy Rev. 112 (2019) 183–194.

[58] A.A. Hachicha, Z. Said, S. Rahman, E. Al-Sarairah, On the thermal and thermodynamic analysis of parabolic trough collector technology using industrial-grade MWCNT based nanofluid, Renew. Energy (2020).

Chapter 13

Electronics thermal management applying heat pipes and pulsating heat pipes

Roger R. Riehl[1], Jesús Esarte San Martin[2] and Juncal Estella[2]
[1]GamaTech Thermal Solutions, São José dos Campos, Brazil, [2]NAITEC—Centro Tecnológico en Automoción y Mecatrónica, Pamplona, Spain

13.1 Introduction

The use of thermal control devices with a passive action, such as that found in heat pipes (HPs) and pulsating heat pipes (PHPs), has increasingly gained more attention and importance in electronics cooling projects, due to the reliability and high performance in the heat transfer process. Since HPs have been extensively used as thermal control elements in aerospace missions, such as satellites and probes, their use in domestic and commercial electronics cooling projects has resulted in high heat dissipation capabilities. As a consequence, higher processing speeds have become available in laptop computers and other electronic devices for daily use, as well as contributing to improving radar and telecommunication technologies available for military applications.

The demand for high processing speeds in electronic components is continuously increasing, along with their heat densities, as a result of the more compact architectures found in current printed circuit boards (PCBs). Therefore, it has become clear that projects involving thermal management need to be part of the entire design from the beginning, as a way to avoid the system's shutoff due to temperature overshooting. Applied research in this subject has already identified this pattern, as it can be observed in several designs where thermal management has focused on electronics cooling using HPs [1–5]. Even though HPs can be supplied by several manufacturers, it is a common practice to have customized designs that will present the requirements for that specific application. In parallel, the constant development and refinement of this technology are necessary to obtain devices with thermal capabilities that will meet a specific application [6–11].

Advances in Nanofluid Heat Transfer. DOI: https://doi.org/10.1016/B978-0-323-88656-7.00001-5

When decreasing the size of an electronic component as a resulting evolution of this technology, the processing speeds continuously increase, and a resultant effect is that the heat fluxes that need to be rejected are also observed. Heat flux densities levels have increased considerably as new projects are conceived, resulting in a need to use thermal control devices that will present better capabilities. A common practice is considering devices with low thermal inertia, as current projects require this approach because heat concentration that can cause temperature overshooting needs to be avoided in new designs. Otherwise, electronics operating with temperatures above their design limits will be shut off, thus stopping the PCB operation. Low inertia designs apply HPs and materials to ensure that the thermal control concept will remove the heat once it is generated by any electronic component, efficiently transporting it to a heat sink. The low inertia design approach is specially applied in hermetically sealed systems operating in hostile environments that are close to their temperature limits. One must be aware that the entire thermal management system design considers the cold junction temperature limits of the electronic component, rather than its surface temperature. As a common practice, the material applied to encapsulate the electronic component must be known to give the designer the required information to correctly calculate the component's cold junction temperature during operation, which will then be verified during tests.

Thermal design approaches have successfully implemented HPs and PHPs devices to better control the electronics temperature, ranging from laptop computers to military and aerospace systems. The reliability of all devices will be also dependent on their materials and working fluid chemical compatibility to ensure a lifetime operation. The devices are designed to operate as part of the system's structure, as they are fully integrated to minimize the thermal contact resistance. This method has shown important improvements in the overall thermal system design during their operation, as the most important temperatures were controlled and kept below their limits [12,13].

Since HPs and PHPs rely on their working fluid to efficiently transfer the heat from one end to another there has been a constant search for better options to enhance the device's performance. Working fluids applied to HPs and PHPs must present important characteristics, such as chemical compatibility with all materials used, capability to operate within the temperature range, good thermal capacity, etc., and sometimes they are overlooked and the proper attention is not given. However, important developments have been made to better understand the working fluid importance during the operation of an HP or PHP [6–12], and new options have begun to be considered to enhance these device's thermal capabilities.

In this sense, nanofluids have been considered as important working fluids to improve HPs' and PHPs' thermal performances. A nanofluid is formed by a base fluid (usually a regular working fluid) with the addition of mixed solid nanoparticles. Its application in an HP can be a straightforward procedure that can potentially bring improvement to its performance when compared to the base fluid itself. The capability of obtaining thermal enhancements when using a nanofluid has been considered a great advance to obtain more efficient devices, with great potential to develop more reliable systems [13,14]. Since it was first presented [13], nanofluids have attracted great attention from designers and researchers, resulting in considering this new working fluid in several applications where regular working fluids have reached their physical limits [15,16].

In this work, the features of HPs and PHPs will be presented, depicting their design parameters and construction features, as well as results obtained from real applications. As an important contribution to this type of passive thermal control device, nanofluids will be presented and the information concerning their manufacturing, stability, chemical compatibility, and application will be described.

13.2 Design parameters

The design parameters applied to both HPs and PHPs are related to the understanding of how these devices operate, representing their thermohydraulic behavior in equations with physical meanings, resulting in a representative mathematical model. Mathematical modeling is an important tool to design and predict how HPs and PHPs will operate under specific conditions, related to their geometric characteristics and selected working fluid. A brief description of the applied design parameters used in HPs and PHPs is presented as follows.

13.3 Heat pipes

HPs are high efficient heat transport devices; transferring heat from the source and delivering it to the sink with the minimum temperature drop, without using moving parts. The working fluid inside the HP, which is in its saturated condition, promptly absorbs the heat from the source. Vapor is generated in the porous wick and flows in the direction of the condenser through the center of the tube (vapor diameter). In the condenser, the vapor gives up its latent heat and returns to liquid, being absorbed by the porous wick. By capillary action generated in the wick, liquid flows toward the evaporator to complete its cycle, as represented by Fig. 13.1.

An HP is usually divided into three parts: evaporator, adiabatic, and condenser. However, in some designs, the adiabatic section does not exist. The wick can be formed by a screen mesh, porous sintered powder, or axial grooves, which will be discussed later. The working fluid must be chemically compatible with all materials applied in the HP's construction as a requirement to avoid noncondensable gases (NCGs) generation that compromises its

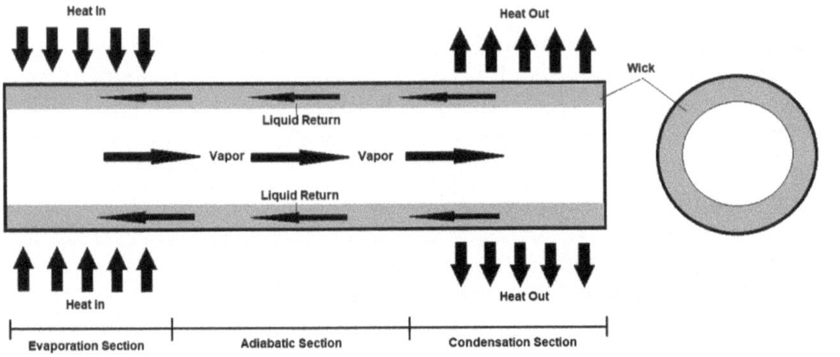

FIGURE 13.1 Operation characteristics of an HP (Schematics by GamaTech).

operation over time. Also, HPs are very sensitive to their orientation against gravity and are often operated horizontally. When the evaporator is below the condenser, it will operate as a thermosyphon and, in most cases, it will cease operation when the evaporator is above the condenser. Table 13.1 presents the influence of each component on the overall performance of an HP.

As already mentioned, chemical compatibility between all parts of an HP plays an important role in its overall performance, especially when dealing with NCGs that can be generated along the device's lifetime. Examples of compatibilities are presented in Table 13.2.

For applications like in aerospace, where an HP needs to operate for up to 20 years without maintenance, this is a very serious issue and proper consideration must be made by designers and manufacturers. There are innumerous publications showing test results from HP using well-known incompatible combinations, such as aluminum alloys and water, as well as the application of other substances (i.e., surfactants) that will cause chemical reactions in the device that are often overlooked by researchers.

Therefore, it becomes clear that when designing an HP, several variables must be considered related to materials (for housing and wick), working fluid, proper welding technique, as well as processes that need to be carefully considered during the fabrication. Such processes involve an adequate and thorough cleaning of all parts, welding under inert atmosphere, high vacuum level to guarantee that the HP is empty and clean, working fluid outgassing to remove any dissolved gases that might result in NCGs generation, among others. The use of incompatible combinations can severely compromise the overall HP performance over time, and it should be avoided at all costs even if the sole intention of researchers is to collect some data.

As a very important subject that is often forgotten or not even considered by many researchers, the NCG generation is related to the working fluid decomposition in the presence of other materials utilized to construct the HP. The decomposition rate can be properly verified according to the

TABLE 13.1 Heat pipe components and their influence on design requirements [17].

Design requirements	Working fluid	Wick material	Case material
Thermal performance			
Transport capacity	S	S	W
Operating temperature range	S	W	W
Temperature drop	M	W	W
Mechanical			
Physical requirements	W	W	M
Wall thickness-internal pressure	W	N	S
Sink–source interface	N	N	S
Dynamic/static loads	W	S	M
Reliability and safety			
Material compatibility	S	S	S
External corrosion	N	N	
Fabrication	M	M	M
Pressure containment/leakage	W	M	S
Toxicity	S	W	W
Gravitational environment			
>1 g	M	M	W
1 g	W	M	W
<1 g			

S, Strong factor; *M*, moderate factor; *W*, weak factor; *N*, negligible.

combination of material and working fluid, related to the temperature to which the HP has been subjected.

The decomposition rate can be verified applying the Arrhenius model, which can be calculated as follows:

$$K = Ae^{\frac{-E_{act}}{RT}} \tag{13.1}$$

The importance of this process shall not be disregarded during the design of an HP. Several investigations have been carried out, especially in the 1970s and 1980s, resulting in important contributions to better understand

TABLE 13.2 Results of experimental compatibility tests [17].

Working fluid	Compatible material	Incompatible material
Water	Stainless steel, copper, silica, nickel, titanium	Aluminum, inconel
Ammonia	Aluminum, stainless steel, cold rolled steel, iron, nickel	Copper
Methanol	Stainless steel, iron, copper, brass, silica, nickel	Aluminum
Acetone	Aluminum, stainless steel, copper, brass, silica	
Freon-11	Aluminum	
Freon-21	Aluminum, iron	
Freon-123	Aluminum	
Heptane	Aluminum	
Dowtherm	Stainless steel, copper, silica	
Lithium	Tungsten, tantalum, molybdenum, niobium	Stainless steel, nickel, inconel, titanium
Sodium	Stainless steel, nickel, inconel, niobium	Titanium
Cesium	Titanium, niobium	
Mercury	Stainless steel	Molybdenum, nickel, tantalum, inconel, titanium, niobium
Lead	Tungsten, tantalum	Stainless steel, nickel, inconel, titanium, niobium
Silver	Tungsten, tantalum	Rhenium

how an HP can show degradation over time regarding the chemical reactions occurring during its operation [18].

Along with the proper consideration regarding chemical compatibility in an HP, the device presents its operation within certain limits that can be determined according to its design parameters. The HPs limitations are summarized and presented in Table 13.3 and adequate explanations about them can be found in Refs. [7,8].

Fig. 13.2 presents the limitations of an HP according to the applied heat flux and operation temperature. The HP is designed to operate within the area below the lines, even though one should be aware that a limitation for a given setup of material and working fluid might not be applied to another. The limitations definition is a requirement to identify the parameters at

TABLE 13.3 Heat pipes limitations [7,8].

Parameter	Equation
Maximum capillary pressure	$\Delta P_{c,\max} = \frac{2\sigma}{r_p}\cos\theta$ (13.2)
Capillary limitation	$\Delta P_{c,\max} \geq \int_{L_{eff}} \frac{dP_l}{dx}dx + \int_{L_{eff}} \frac{dP_v}{dx}dx + \Delta P_{nh} + \Delta P_{ah}$ (13.3)
	$\int_{L_{eff}} \frac{dP_l}{dx}dx = \Delta P_l = \left(\frac{\mu_l}{KA_w\rho_l\lambda}\right)L_{eff}Q_{c,\max}$ (13.4)
	$\int_{L_{eff}} \frac{dP_v}{dx}dx = \Delta P_v = \left(\frac{C(f_v Re_v)\mu_v}{2\left(r_{h,v}\right)^2 A_v\rho_v\lambda}\right)L_{eff}Q_{c,\max}$ (13.5)
	$L_{eff} = 0.5L_e + L_a + 0.5L_c$ (13.6)
	$Q_{c,\max} = \frac{\frac{2\sigma}{r_p} - \Delta P_n - \Delta P_{ah}}{L_{eff}\left[\left(\frac{\mu_l}{KA_w\rho_l\lambda}\right) + \frac{(f_v Re_v)\mu_v}{2r_{h,v}^2 A_v\rho_v\lambda}\right]}$ (13.7)
	$\Delta P_{nh} = \rho_l g d_v \cos\psi$ (13.8)
	$\Delta P_{ah} = \rho_l g L \sin\psi$ (13.9)

(Continued)

TABLE 13.3 (Continued)

Parameter	Equation
Viscous limitation	$Q_v = \frac{A_v r_o^2 \lambda \rho_v P_v}{16 \mu_v L_{eff}}$ (13.10)
Sonic limitation	$Q_s = A_v \rho_v \lambda \left[\frac{\gamma_v R_v T_o}{2(\gamma_{v+1})} \right]^{1/2}$ (13.11)
Entrainment limitation	$Q_e = A_v \lambda \left[\frac{\sigma \rho_v}{2 r_{h,w}} \right]^{1/2}$ (13.12)
Boiling limitation	$Q_b = \left(\frac{2\pi L_e k_{eff} T_v}{\lambda \rho_v ln(r_i/r_v)} \right) \left(\frac{2\sigma}{r_n} - \Delta P_{c,\max} \right)$ (13.13)
Working fluid inventory	$m = A_v L_t \rho_v + A_w L_t \varepsilon \rho_l$ (13.14)
Figure of merit	$N = \frac{\rho_l \lambda \sigma}{\mu_l}$ (13.15)

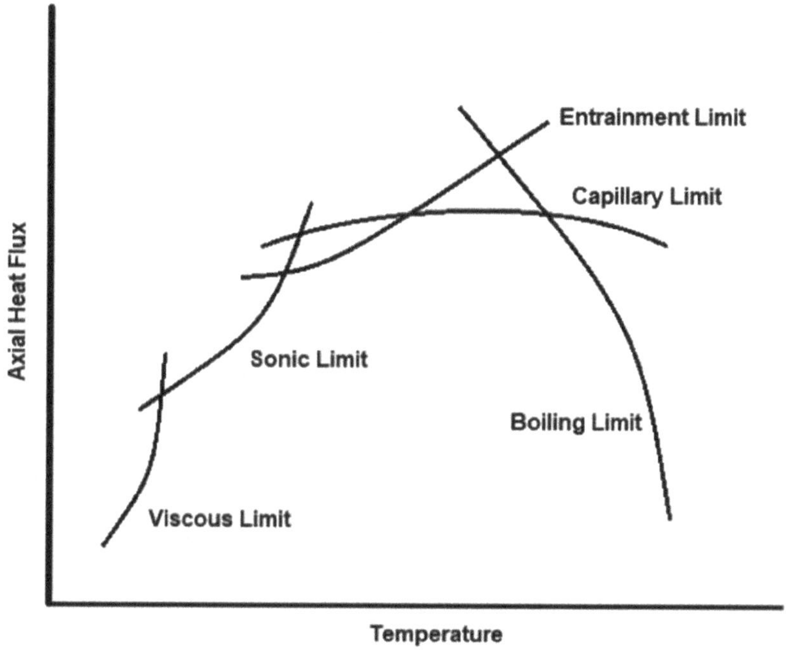

FIGURE 13.2 Limits to heat transport in an HP (Schematics by GamaTech).

which the HP will be operational, also indicating if some change needs to be made in the overall design [9].

Another important parameter that requires full attention from designers is the correct definition of the right amount of working fluid that needs to be present in the HP, so it can present a correct performance. Despite some recent efforts that have shown that authors are interested in defining the most adequate fluid charge and/or fluid inventory for a heat pipe, this parameter has been already defined and well investigated in the past, as presented by Eq. (13.14). This is an important parameter to be considered during the design of a heat pipe, and shall not be simply imposed by a certain percentage of its internal volume as the best option. Physical parameters must be accounted for during the calculation of the fluid's inventory, which considers both vapor and liquid phases, as well as flow through the wick structure and operating temperature range. Eq. (13.14) depicts the required calculation for the fluid's inventory, which does not consider the meniscus recession during the calculation, therefore, the fluid's inventory is larger than necessary. The proper calculation requires the definition of a range of fluid inventory, which is related to the heat pipe's cold and hot operation modes. The cold mode is the lowest temperature at which the heat pipe must operate, and the hot mode is the highest one.

The Figure of Merit is an important parameter for selecting the most indicated working fluid, which will be defined by their capability to perform within the temperature range imposed by the project. For this purpose, Eq. (13.15) can be applied to perform the required calculation to present a comparison between different working fluids as shown by Fig. 13.3, which corresponds to the temperature range between 250K and 380K.

It is important to observe that each working fluid operates in a different temperature range, therefore, the correct selection must rely on the project's requirements considering the capabilities of each substance. More working fluids can be added to this calculation to present a more thorough comparison between different working fluids.

13.3.1 Heat pipe designs

Along with the geometric characteristics and working fluid defined for a given HP design, as well as the temperature range that needs to operate and heat loads, the wick material is an important parameter that requires considerable attention as well. The wick is responsible for generating capillary pressure that will drive the liquid flow from the condenser to the evaporator, ensuring its supply to support the evaporation occurring in this section that serves as the media that transports the heat from the source to where it will be rejected. A wick that is incapable of generating the required amount of

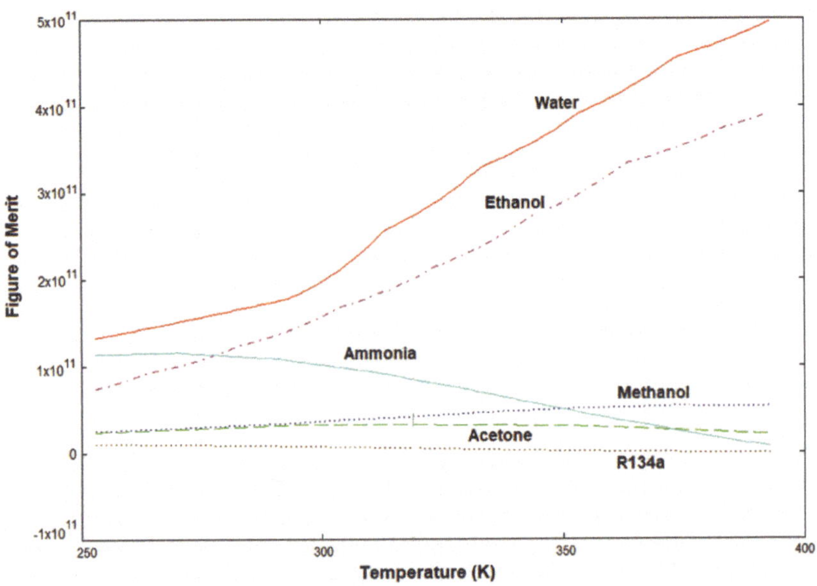

FIGURE 13.3 A figure of merit of various fluids (calculations performed by GamaTech).

capillary pressure will cause the HP to fail, and that can compromise the entire thermal management system.

Several forms of wick can be applied to an HP, being wrapped screen mesh (Fig. 13.4), sintered powder (Fig. 13.5), and axial grooves (Fig. 13.6) that are usually found in aluminum alloys extrusions. Each wick form can be applied at the designer's discretion, however, the axial grooves are the most common ones applied in aerospace projects, since these eliminate the thermal contact resistance between the wick and the HP's housing, whereas the others still can present significant issues regarding this property.

Given the importance of this parameter, it is highly required to correctly calculate the effective thermal conductivity for the selected wick structure, as presented in Table 13.4.

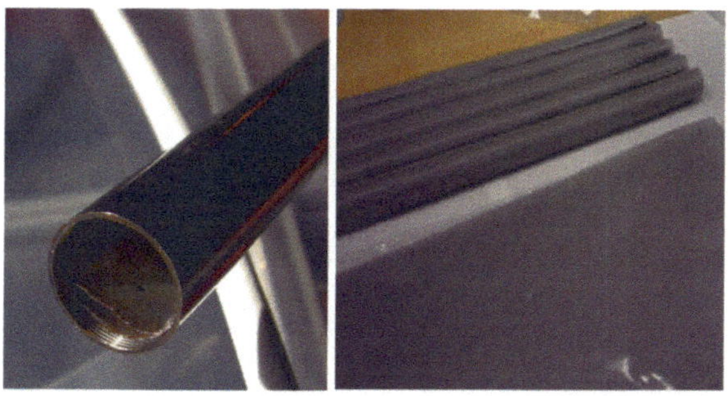

FIGURE 13.4 Wrapped screen mesh (316L stainless steel) (image by GamaTech).

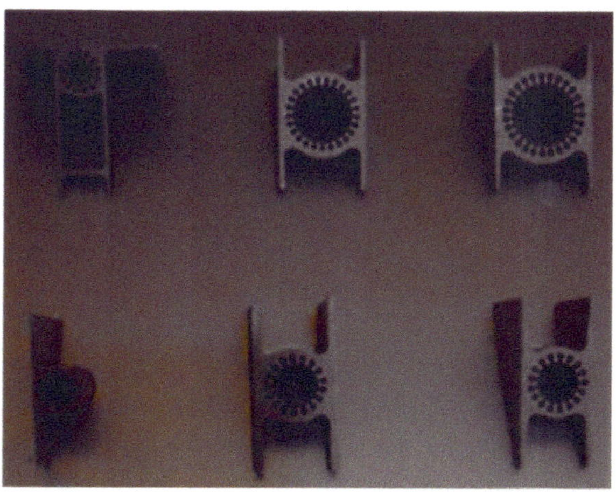

FIGURE 13.5 Axial grooves extrusions in aluminum alloy (image by GamaTech).

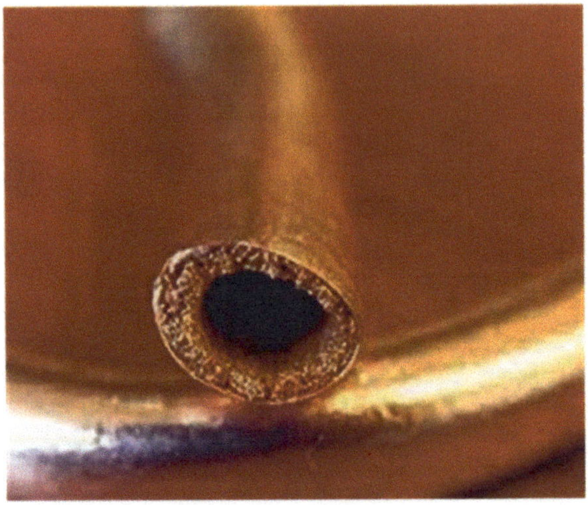

FIGURE 13.6 Sintered powder (copper) (image by GamaTech).

TABLE 13.4 Effective thermal conductivities in heat pipes [7].

Wick structures	Effective thermal conductivity
Wick and liquid in series	$$k_{eff} = \frac{k_1 k_2}{\varepsilon k_w + k_l(1-\varepsilon)} \quad (13.16)$$
Wick and liquid in parallel	$$k_{eff} = \varepsilon k_l + k_w(1-\varepsilon) \quad (13.17)$$
Wrapped screen mesh	$$k_{eff} = \frac{k_l[(k_l+k_w)-(1-\varepsilon)(k_l-k_w)]}{(k_l+k_w)+(1-\varepsilon)(k_l-k_w)} \quad (13.18)$$
Packed spheres	$$k_{eff} = \frac{k_l[(2k_l+k_w)-2(1-\varepsilon)(k_l-k_w)]}{(2k_l+k_w)+(1-\varepsilon)(k_l-k_w)} \quad (13.19)$$
Rectangular grooves	$$k_{eff} = \frac{(w_f k_l k_w \delta) + w k_l (0.185 w_f k_w + \delta k_l)}{(w+w_f)(0.185 w_f k_w + \delta k_l)} \quad (13.20)$$

13.3.2 Application of heat pipes

The applications of HPs are present in several different projects, ranging from aerospace to industrial and defense purposes. They can present different shapes and sizes, given the fact that they need to promote heat transport according to the project's requirements.

FIGURE 13.7 Heat pipes undergoing performance tests (Image by GamaTech).

Fig. 13.7 shows a set of HPs designed for application in oil and gas refineries, utilized to keep the processes' temperatures within defined limits with precision. On the back are a set of 316L stainless steel HPs (19 mm outer diameter; 1.5 m length) and on the front are copper (oxygen-free) HPs with the same geometry. Both sets are performance tested before their integration to the process, designed to operate with water as working fluid.

Fig. 13.8 presents miniature HPs (3 mm outer diameter, 170 mm length) applied to electronics cooling of a telecommunication system, made of copper (oxygen-free) operating with methanol as working fluid. This selection of working fluid was made as the HPs need to operate in an environmental temperature range between −20°C and +60°C. These HPs were later integrated into an aluminum case as presented by Fig. 13.9, where they were directly in contact with the most important electronic components of a PCB, for efficient heat removal and dissipation during their operation.

The configuration presented by Fig. 13.9 was integrated into its designated PCB and tested in an environmental chamber, during a thermal cycling test procedure. The results of these thermal cycling tests are presented in Figs. 13.10 and 13.11, where the PCB was fully operational and the behavior of both HPs was observed. The tests show the HPs operating at a chamber's temperature initiating at −20°C and presenting a successful startup, being tested again for the chamber set at 20°C and 60°C, with the chamber's highest temperature being considered the most critical step of this test. This is because the HP must be able to remove the heat from the components and keep the cold junction temperature under the limit imposed by the component's manufacturer, which is 90°C. The thermal conductance in the presented results is calculated as:

$$G = \frac{Q}{T_{\text{evap}} - T_{\text{cond}}}. \tag{13.21}$$

Fig. 13.11 presents the results for the lower HP, showing that the temperatures were kept below the maximum limit imposed by the electronic components manufacturer. The results show the capability of both HPs to

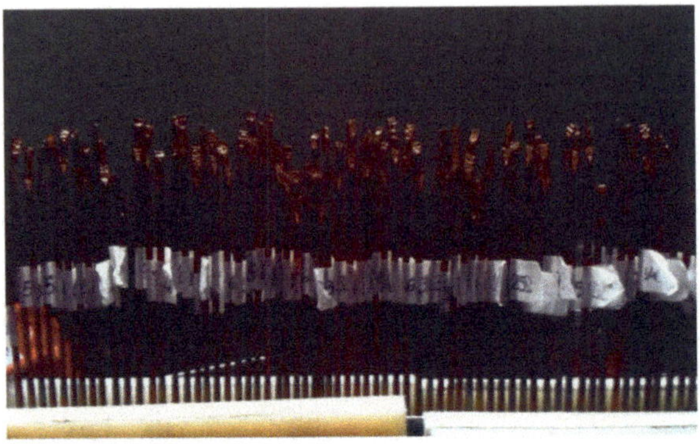

FIGURE 13.8 Miniature HPs (image by GamaTech).

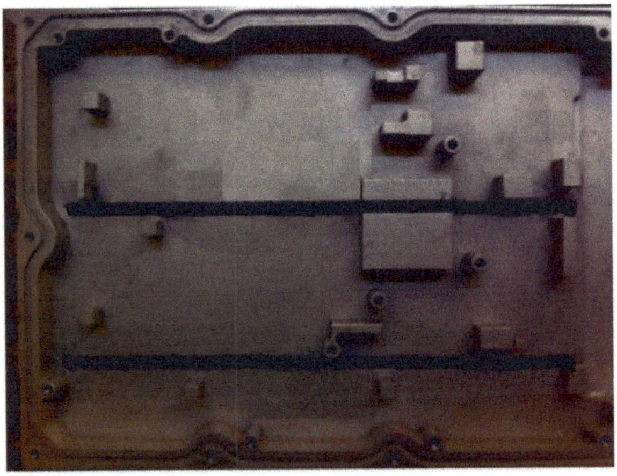

FIGURE 13.9 Miniature HPs integrated into an aluminum case of a PCB (image by GamaTech).

perform the thermal control efficiently while keeping the temperatures of the electronic components below the limits at which they must operate.

The operation of an HP can vary depending on its geometric characteristics, working fluid, heat load, and temperature, but all of them shall present a close to isothermal condition throughout its length. HPs need to present a fast startup, showing temperature rise in all thermocouples (or temperature sensors) with stabilization after a certain period. They must also present a quick response to the variation of the heat load, without indication of temperature overshooting due to a dry-out condition in the evaporator. Fig. 13.12 presents the operation of a 6061 aluminum alloy heat pipe with axial grooves with its cross-section as indicated, charged with ammonia, with a total length of 800 mm.

The test presented by Fig. 13.12 was performed with the HP in horizontal orientation. A skin heater attached to the evaporator was controlled by

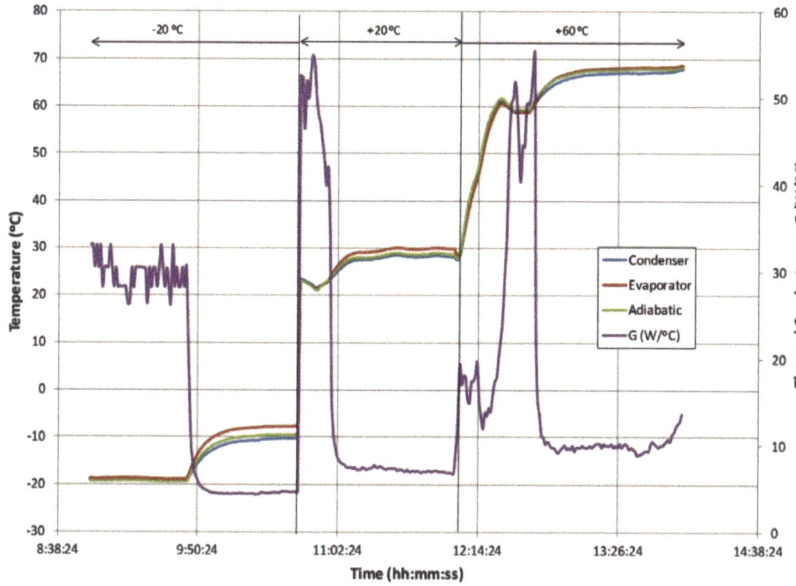

FIGURE 13.10 Integrated environmental test with PCB—upper HP (image by GamaTech).

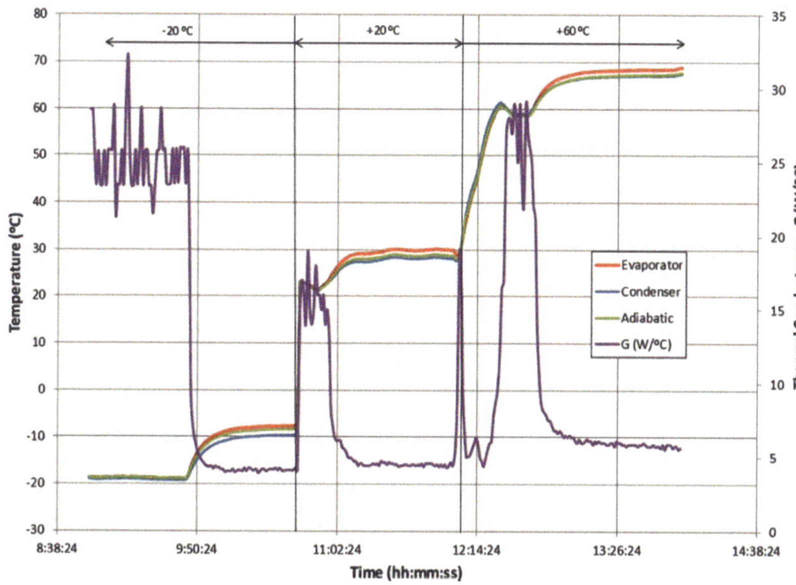

FIGURE 13.11 Integrated environmental test with PCB—lower HP (image by GamaTech).

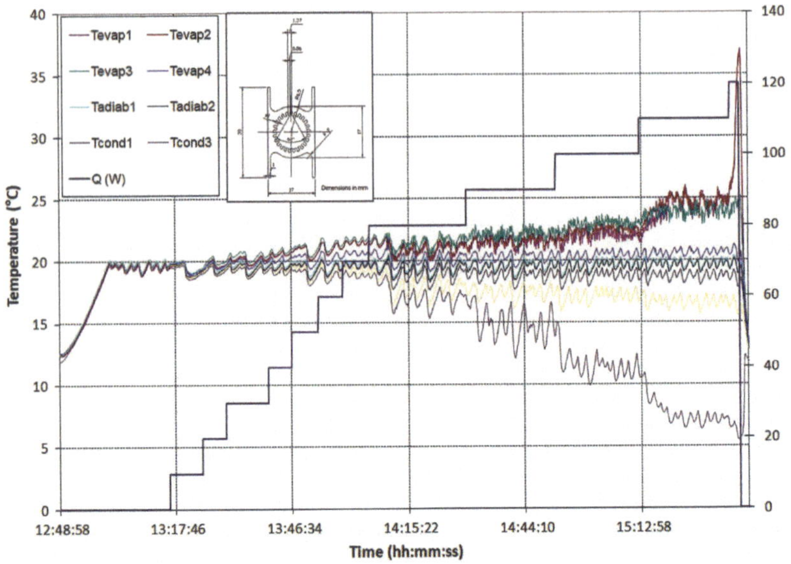

FIGURE 13.12 Axial groove HP test (image by GamaTech).

a digital power supply, whereas the condenser was a cold plate circulating liquid nitrogen (LN_2), also controlled by an automated system that actuates in a flow control valve. The entire system is controlled by a computer that executes the testing procedure, setting the power supply to the programmed heat load while adjusting the LN_2 flow to the cold plate to keep the HP within the desired operating temperature.

This is a common practice while checking the overall heat transport capacity of an HP, which is set to perform with an increase in the heat load administrated to the evaporator. In this case, small heat loads are added to the evaporator until the HP reaches the steady-state condition when a new increase in the heat load is done. This procedure continues until the HP presents a dry-out condition that is characterized by a sudden increase in the evaporator's temperature, which then defines the maximum heat transport that this HP can present. Usually, this test is repeated several times to check the heat pipe capability in performing according to the project's requirements.

13.4 Pulsating heat pipes

Since it was first presented as an alternative passive thermal control device [19], PHPs have been of great interest in the research community, which has greatly contributed to better understanding of how this special type of heat pipe operates. Differently than regular HPs that rely on capillary action to drive the working fluid, PHPs operate by utilizing vapor plugs and liquid slugs motion to transfer the heat from a source to a sink, through a meandering tube bent in several parallel channels, without presenting a wick structure. The main feature of a PHP is the pulsating (or oscillating) flow of the liquid slug being pushed by the vapor plug. Such pulsating flow can be observed from the temperature measurements, as they show a sudden increase followed by a decrease in their values, indicating that the PHP is operating. Fig. 13.13 presents the configurations that a PHP can have.

There have been several contributions to better understand the thermo-fluid dynamics of PHPs configured as both closed and open loops, to establish the physics involved in their operation and how to maximize their thermal performance. Several empirical and analytical models have become available, showing the potential of this special type of heat pipe to be used as a thermal control device in complex thermal management systems. As a consequence, new designing tools have become available and PHPs have been considered as reliable devices for complex applications [20–27], including devices destined for microgravity use [28]. Numerical models and correlations have been conceived to better predict the thermal and hydraulic behavior of PHPs [29–36], which made it possible to design them as thermal control devices for applications that vary from aerospace [22,27] to radars focused on defense purposes [32].

As a general approach, the main equations applied to design a PHP are summarized in Table 13.5, where it is required to check the critical bubble diameter (Eq. 13.22) against the maximum tube diameter (Eq. 13.23) that will be used to manufacture the PHP. The following equations are important to define the maximum heat transport capability of the PHP, but one must have in mind that the verification parameters defined for HPs also apply for this device. Every consideration regarding materials and working fluid capability, as well as manufacturing processes, etc., needs to be observed as well.

The value of Bo is equal to 2 for the critical internal tube diameter. The value of D_{max} must be smaller than D_c to have a fully operational PHP. Usually, PHPs are manufactured with internal diameters that correspond to a fraction of D_c to ensure that the device will perform as needed. Eq. (13.24)

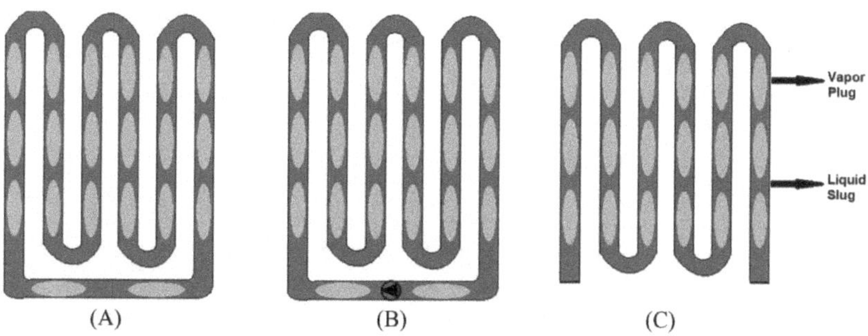

FIGURE 13.13 (A) closed-loop; (B) closed-loop with flow direction valve; (C) open-loop (schematics by GamaTech).

TABLE 13.5 Design parameters for pulsating heat pipes.

Parameter	Equation
Critical diameter	$D_c = Bo\left[\frac{\sigma}{g(\rho_l - \rho_v)}\right]^{0.5}$ (13.22)
Maximum tube diameter	$D_{\max} = 1.836\left[\frac{\sigma}{g\rho_l}\right]^{0.5}$ (13.23)
Minimum tube wall temperature	$T_{\min} = T_{sat} + \frac{4\sigma}{D_c}\frac{T_{sat}}{\lambda\rho_v}$ (13.24)
Maximum heat transfer capacity	$Q_{\max} = \left[Q_{f,\max}\left(\frac{1}{n_b}\right)\left(A_i\frac{L_{eff}}{L_{seg}}\right)\right]$ (13.25)
Maximum heat transport capacity	$Q_{f,\max} = \frac{\lambda\sigma\rho_v}{\mu_v}$ (13.26)
Maximum heat flux to the evaporator	$q'' = \frac{k_l R T_v^2}{r_{in} ln\left(\frac{r_{in}}{r_{in} - \tau_l}\right)\lambda} \ln\left[1 + \frac{2\sigma}{P_v}\left(\frac{1}{r_n} - \frac{1}{2(r_{in} - \tau_l)}\right)\right]$ (13.27)

presents the lowest temperature that one must observe at the PHP's tube wall, so it can initiate its operation. Eq. (13.27) can be used to calculate the liquid film thickness (τ_l), which is important to evaluate whether the PHP is operating under the liquid film evaporation or convective nucleate boiling [3,4]. The maximum heat transport capacity, represented by Eqs. (13.25) and (13.26), serves as a guide to estimate the maximum heat that the PHP can transport from the source to the sink [31,37].

Several publications make available correlations to predict the operation of PHPs according to the flow regimes and superficial velocities of the liquid and vapor. Configurations, where PHPs are set as a so-called 3D [33], should also be especially verified as the working fluid might operate in different gravity orientations. Configurations where the PHP is set as shown in Fig. 13.13B with a flow control valve also require close attention as the working fluid will flow in one direction and the liquid and vapor pressure drop inside the meandering tubes must be determined, so the device can present a stable operation.

Differently than HPs, the PHPs have their working fluid inventory calculated as a fraction or percentage of their internal volume. Authors always related the operation of PHP according to their working fluid inventory, as it can reach an early dry-out condition with a low percentage of their internal volume filled with liquid, whereas presenting unstable operation with higher percentages [22–28]. Therefore, the designer must rely on experience and testing to set the best amount of working fluid to have the PHP performing according to what is required from it.

13.4.1 Application of pulsating heat pipes

The operation of PHPs is quite different than regular HPs since the device needs to have the required conditions to generate the liquid slugs and vapor plugs, as well as the necessary mechanical equilibrium to continuously present a flow of the working fluid. Since pulsations are expected to happen during its operation, high amplitudes in the temperature readings are common as well as higher differences between the evaporator and condenser temperatures when compared to HPs.

Testing a PHP is similar to what is done with regular HP, where the device is assembled in a test rig connected to thermocouples to read the temperatures throughout the device. Fig. 13.14 presents the test performed in an open-loop PHP (Fig. 13.13C), where both ends are clamped separately. The PHP was built with a copper tube (oxygen-free) with 3 mm OD and 1.5 mm ID, presenting 14 parallel channels and 13 curves, presenting 300 mm of length in each section. The working fluid applied to this PHP was methanol with an inventory of 80% inserted in the PHP. The evaporator, adiabatic, and condenser sections presented a length of 100 mm each and the PHP was tested in a horizontal orientation. The heat was applied to the evaporator

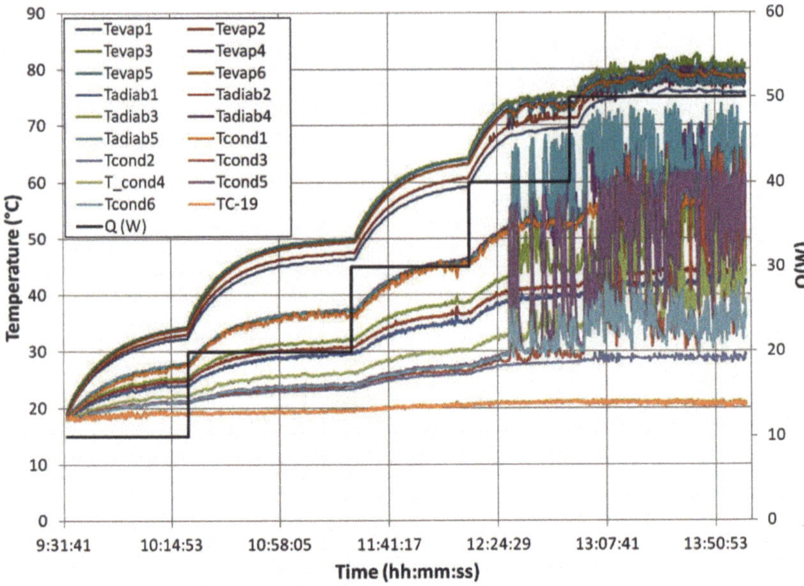

FIGURE 13.14 Open-loop PHP operation (image by GamaTech).

using a skin heater connected to a digital power supply and the condensation was done by air with a velocity of 3.0 m/s. Fig. 13.14 presents the results for this test, where intense pulsations can be observed from the beginning of the test at low power. As the heat load is increased, so are the pulsations observed in all three sections of the PHP, showing an intense flow of vapor plugs pushing the liquid slugs, promoting the heat transport from the source to the sink.

Even though PHPs present intense pulsations/oscillation in their temperatures during operation, this device has been applied to several projects where conventional HPs would present limitations, especially those where the heat source needs to be above the heat sink. In this case, regular HP cannot operate due to the lack of enough capillary pressure to move the working fluid. In this case, PHPs can be applied as they show little sensitivity to the gravity force [29,31,37]. The same PHP, when tested with a different working fluid, presents the capability of each fluid in keeping the evaporator temperature as low as possible, as this is one of the requisites imposed on a thermal control device. Keeping the heat source temperature as low as possible indicates the device's capability in transferring the heat efficiently, as this is linked to the working fluid selection. Verification of the variables that present a direct effect on the thermal control device's performance proves to be an important task during their design, as shown in Table 13.1. Fig. 13.15 presents the results for the evaporator section

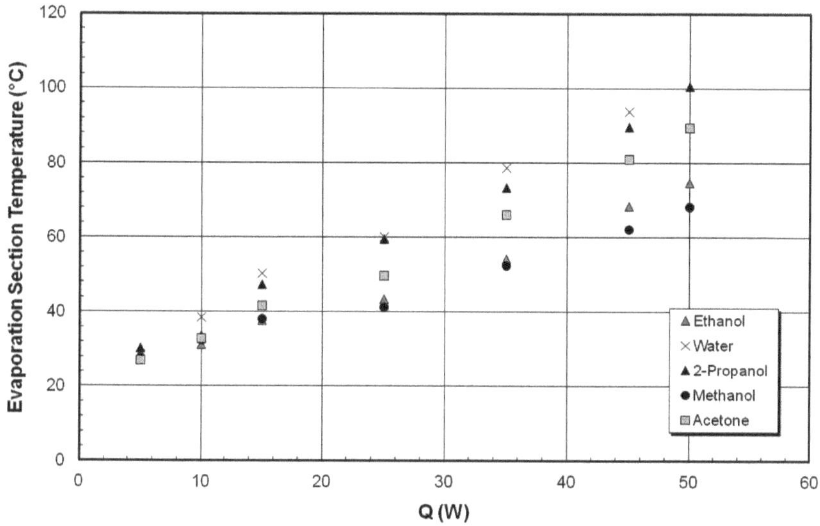

FIGURE 13.15 Evaporation section temperature comparison with different working fluids (image by GamaTech).

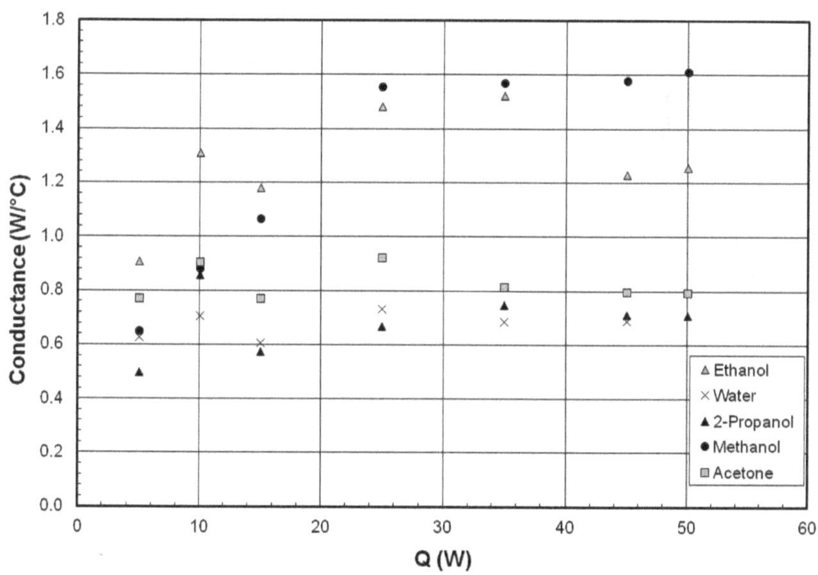

FIGURE 13.16 Thermal conductances for the PHP operating with different working fluids (image by GamaTech).

temperatures (after reaching steady-state) considering different working fluids and Fig. 13.16 presents the thermal conductance observed during the PHP's operation.

As depicted in Figs. 13.15 and 13.16, the PHP can operate with different temperature conditions depending on the applied working fluid, as shown for the specific configuration of the PHP. In this specific case, methanol presents lower evaporator section temperatures and higher thermal conductances. The selection of the correct working fluid can result in a better thermal control capability by the PHP and the HP, which must be chosen by the designer.

Due to their capability in performing in adverse conditions, along with their reliability during the heat transport from the source to the sink, PHPs have been successfully integrated into several projects during the last years presenting different configurations, as shown by Fig. 13.17. For the presented configurations, the thermal control of PCBs was done by PHPs, built with oxygen-free copper tubing with water, for an inventory of 50% of their internal volume. Due to the nature of their application, their condenser was connected to a liquid cooling system working as the heat sink, which was able to dissipate the heat load to the ambient in a remote section.

Pulsating HPs present a very complex thermal control device, even though their construction is really simple, and they present a low cost when compared to regular HPs. They can be built without major technological needs, being able to perform with high heat loads at any orientation. However, the same procedures applied to manufacture an HP must also be applied to a PHP, to bring them up to a level of reliability that is needed for a complex thermal management system.

13.5 Nanofluids capabilities and models

Nanofluids have the potential of improving HPs and PHPs thermal performances without changes on the devices' designs, by simply substituting the working fluid by them. However, proper considerations must be performed as the nanofluid may impose different limitations to the thermal control

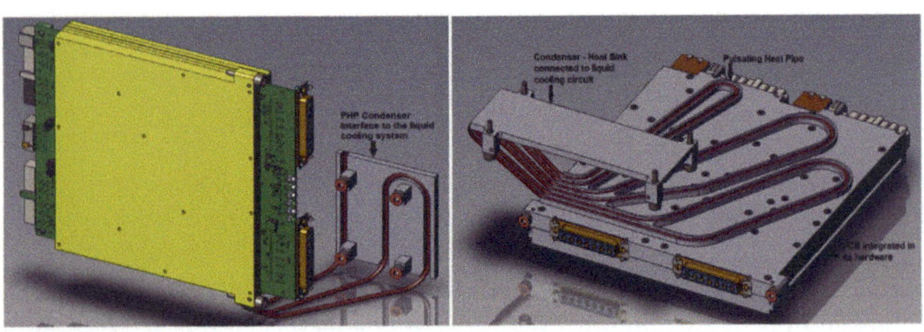

FIGURE 13.17 Integration of PHPs for electronics cooling (image by GamaTech).

device, as the addition of solid nanoparticles to the base fluid will cause changes in the transport properties, such as density and viscosity.

As an important subject to be investigated, several publications have become available describing the potentials of using nanofluids to enhance the thermal management capabilities of many devices [12,13,16,18,21,38]. However, some issues need to be properly addressed before applying nanofluids in any thermal control device, which are related to their stability, chemical compatibility, sedimentation along time, etc., and can potentially impact their final performance [18,38–44]. Perhaps, the most important contribution that could be obtained nowadays is a reliable model to predict the nanofluid's stability and expected transport properties values, as different authors will come up with different results for the same combination of solid nanoparticle size and concentration and base fluid [16].

Different correlations to predict the nanofluid thermal properties end up with confusing results, which makes it difficult to select which one should be used. As shown by Table 13.6, each model to predict the nanofluid thermal conductivity is advised to be used for a given condition, neither one considering important properties of the solid nanoparticles such as purity, size, and shape (with the percentage of particles with such characteristics).

As an undesirable consequence, different results are obtained leading to additional uncertainties in the use of nanofluids that compromise their potential application in the enhancement of thermal control devices, because imprecision in the results is observed, as presented by Table 13.7 for CuO and water nanofluid. There are important variations in the results, considering the same range of solid nanoparticles size, volume fraction added to the base fluid, temperature, and sonication time. Since the current models do not consider other parameters as mentioned above, better approaches are required.

Such a variation in the models' results for thermal conductivity is also observed for other transport properties. As presented by Table 13.8, there are even other issues in published results that do not consider chemical incompatibilities between solid nanoparticles and base fluid, as previously presented by Table 13.2.

The fact that a combination of Al_2O_3 and water has been used to obtain results for a viscosity correlation should not be considered at all for application in HPs and PHPs, knowing that this is one of the most incompatible pairs and NCGs will be generated at a very high rate. Given the uncertainties in such correlations, the model defined by Einstein [54] is still the best tool to be used when considering nanofluids for HPs and PHPs applications. Therefore, when considering a nanofluid to be applied in a thermal control device, one should be aware of choosing the best combination of solid nanoparticles and base fluid, focusing on chemical compatibility and stability, to guarantee that the device will operate as expected.

TABLE 13.6 Thermal conductivity models applied to nanofluids.

Author	Model	Comments
Maxwell [45]	$k_{eff} = k_p + 2k_f + 2\phi_p(k_p - k_f)$ $k_f = k_p + 2k_f - 2\phi_p(k_p - k_f)$	Based on spherical particles, random suspensions which must be under conduction solution theory through stationary conditions
Hamilton [46]	$k_{eff} = \frac{k_p + (n-1)k_f + (n-1)\phi_p(k_p - k_f)}{k_p + (n-1)k_f - \phi(k_p - k_f)}$	For high concentrations of spherical particles under conditions of different effective medium theory
Prasher et al. [47]	$k_{eff} = \left(1 + ARe^m Pr^{0.333}\phi_p\right)\left[\frac{k_p + 2k_f + 2\phi_p(k_p - k_f)}{k_p + 2k_f - \phi_p(k_p - k_f)}\right]k_f$	Obtained from Maxwell model and included the effects of correction generated by the Brownian motion
Koo and Kleinstreuer [48]	$k_{eff} = k_{static} + k_{Brownian} k_{static} = \frac{k_p + 2k_f + 2\phi_p(k_p - k_f)}{k_p + 2k_f - \phi_p(k_p - k_f)}$ $k_{Brownian} = 5 \times 10^4 \beta\phi_p\rho_p c_p \sqrt{\frac{k_B T}{\rho_p D}}\left(T, \phi_p\right)$	Considers the effects of surrounding liquid motion with random nanoparticles movement. Based on static Maxwell theory and dynamic effect of Brownian motion
Yu and Choi [21]	$k_{eff} = \frac{k_p + 2k_f + 2\phi(k_{pe} - k_f)(1+\beta)^3}{k_{pe} + 2k_f - \phi(k_{pe} - k_f)(1+\beta)^3} k_f$ $k_{pe} = \frac{2(1-\gamma) + (1+\beta)^3(1+2\gamma)\gamma}{-(1-\gamma) + (1+\beta)^3 + (1+2\gamma)} k_p$	It was based on Maxwell model but additionally taking into account the effects of nanolayer thickness and thermal conductivity parameters

TABLE 13.7 Thermal enhancement ratio for CuO and water nanofluids.

Author	CuO nanoparticle size (nm)	Thermal enhancement ratio	Volume fraction (vol.%)	Temperature (°C)	Sonication time (h)
Karthikeyan et al. [49]	8.0	1.020	0.020	20	0.5
	8.0	1.080	0.090	20	0.5
	8.0	1.130	0.100	20	0.5
	8.0	1.190	0.300	20	0.5
	8.0	1.250	0.800	20	0.5
	8.0	1.316	1.000	20	0.5
Nemade et al. [50]	33.0	1.197	0.500	55	1
	42.0	1.134	0.500	55	0.75
	46.0	1.124	0.500	55	0.5
	53.5	1.087	0.500	55	0.25
Khedkar et al. [51]	25.0	1.050	0.010	26	1.5
	25.0	1.120	0.020	26	1.5
	25.0	1.130	0.030	26	1.5
	25.0	1.160	0.040	26	1.5
	25.0	1.170	0.050	26	1.5
	25.0	1.320	0.075	26	1.5

(*Continued*)

TABLE 13.7 (Continued)

Author	CuO nanoparticle size (nm)	Thermal enhancement ratio	Volume fraction (vol.%)	Temperature (°C)	Sonication time (h)
Wang et al. [52]	42.5	1.080	0.020	25	Not informed
	42.5	1.100	0.040	25	
	42.5	1.110	0.100	25	
	42.5	1.125	0.150	25	
	42.5	1.160	0.400	25	
Pryia et al. [53]	50.0	1.020	0.004	28	6
	50.0	1.060	0.008	28	6
	50.0	1.100	0.012	28	6
	50.0	1.130	0.016	28	6
	50.0	1.050	0.004	50	6
	50.0	1.160	0.008	50	6
	50.0	1.250	0.012	50	6
	50.0	1.320	0.016	50	6
	50.0	0.950	0.004	55	6
	50.0	1.240	0.008	55	6
	50.0	1.330	0.012	55	6
	50.0	1.430	0.016	55	6

TABLE 13.8 Viscosity models for nanofluids.

Author	Model	Comments
Einstein [54]	$\mu_{eff} = \left(1 + 2.5\phi_p\right)\mu_f$	Based on phenomenological hydrodynamic equation for infinitely diluted suspensions of spheres with no interaction between spheres. Work well for maximum volume concentration of 2%
Brinkman [55]	$\mu_{eff} = \frac{1}{(1+2.5\phi_p)^{2.5}}\mu_f = \left(1 + 2.5\phi + 4.375\phi_p^2 + \ldots\right)\mu_f$	Extended Einstein's model by considering the effect of addition of one solute molecule to an existing solution.
Buongiorno [56]	$\mu_{eff} = \left(1 + 39.11\phi_p + 533.9\phi_p^2\right)\mu_f$ $\mu_{eff} = \left(1 + 5.45\phi_p + 108.2\phi_p^2\right)\mu_f$	Curve fitting from experimental data of Al_2O_3-water nanofluid
Nguyen et al. [57]	$\mu_{eff} = \mu_f 0.904e^{0.148\phi_p} \mu_{eff} = \left(1 + 0.025\phi_p + 0.015\phi_p^2\right)\mu_f$	Curve fitting from experimental data of Al_2O_3-water nanofluid
Chen et al. [58]	$\mu_{nf} = \mu_{bf}\left[1 + 10.6\phi_p + \left(10.6\phi_p\right)^2\right]$	Adjusted model for experimental vs theoretical data by considering the rheological effects of shear-rate
Kulkami et al. [59]	$ln\left(\mu_{eff}\right) = A\left(\frac{1}{T}\right) - BA = 20587\phi_p^2 + 15857\phi_p + 1078.3$ $B = -107.12\phi_p^2 + 53.54\phi_p + 2.8715$	Curve fiting from experimental CuO-water: $5\% < \phi_p < 15\%$ dp = 29 nm; 278 < T(K) < 323; shear rate = 100 l/s

13.6 Nanofluids in heat transfer systems: pros and cons

Nanofluid (NF) is a fluid in which nanoparticles (particles of size 1–100 nm) are in suspension. Doping the base fluid results in an increase in thermal conductivity which leads to an improvement in heat transfer. Nanofluids are currently being investigated for their great potential in HPs to increase heat transfer compared to conventional working fluids.

The heat transfer capacity of a nanofluid depends on several factors such as the concentration of the dispersed nanoparticles, the dispersion methodology, size, shape, and thermal conductivity of the nanoparticles, the base fluid type, nanofluid operating temperature, surfactant addition, and pH values, among others.

The use of nanofluids in heat transfer systems has several pros and cons that should be considered. The main positive aspect of using nanofluids is that the addition of certain nanoparticles has the potential to enhance the thermal conductivity so they can increase the energy efficiency in different devices such as HPs [60–62]. Nanoparticles have a larger surface area compared to micrometer-sized particles so, since heat transfer is a surface area-related property, it is possible to achieve better effects with a lower concentration. Furthermore, since boiling is the predominant heat transfer mechanism in HP, it is to be expected that by improving it the HP performance will increase. As shown in several studies, nanofluids improve boiling, which makes them a key element in improving HP performance. On the other hand, one of the main drawbacks of nanofluids is their stability. When a stable nanofluid is not achieved in the long term, problems of agglomeration and sedimentation of nanoparticles appear, and nanofluid thermal performance changes, decreasing device efficiency. Nanofluid stability can be improved by adding surfactants. However, these surfactants can decrease nanofluid thermal conductivity and increase its viscosity, negatively impacting the heat transfer. Therefore, stability is one of the main challenges that currently is focused on in many research efforts, so it is very important to know which factors are important to achieve a stable nanofluid.

However, contradictory results are often found in the literature related to nanofluids and more research is still needed to understand the mechanisms and real effect of nanoparticles on HP thermal performance.

13.6.1 Types of nanoparticles used in nanofluids

Depending on the type of nanoparticle used in nanofluid, several groups can be distinguished [60,63–70], which can also be better visualized in Figs. 13.18 and 13.19:

- *Metal-based nanofluids*. Metallic nanoparticles are appropriate to enhance the thermal conductivity of the fluid which makes them suitable for heat transfer applications. Nanoparticles such as Ag, Au, and Cu are used to prepare metal-based nanofluids. One of the main problems is the

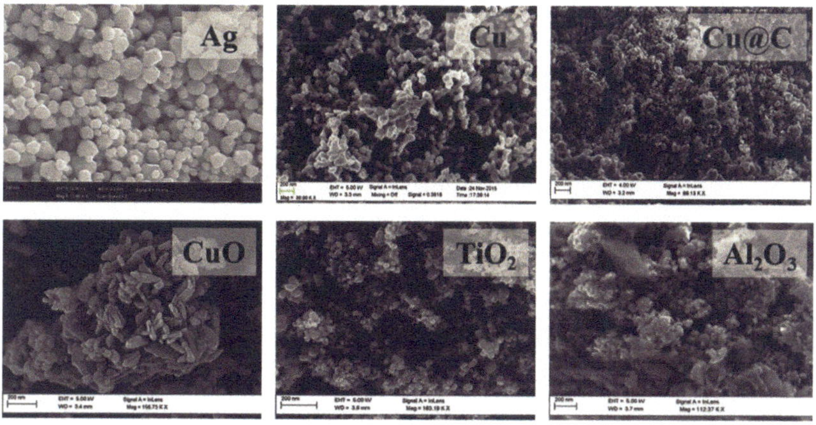

FIGURE 13.18 Examples of nanoparticles used for nanofluid formulation: metallic such as Ag, Cu, and Cu coated with C to avoid oxidation and metal oxides such as CuO, TiO_2, and Al_2O_3 (SEM images by NAITEC).

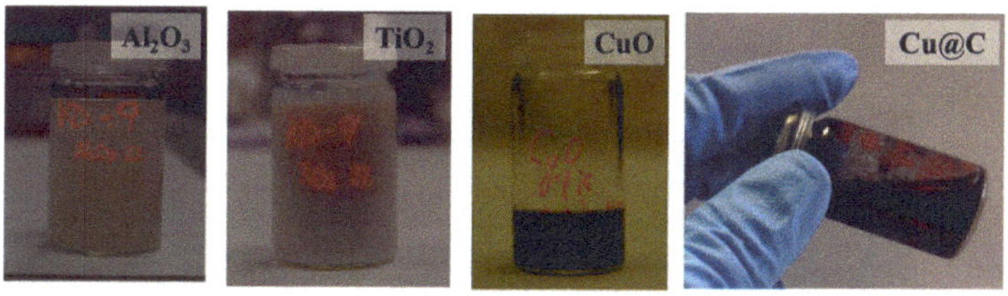

FIGURE 13.19 Examples of nanofluids based on different nanoparticles: Al_2O_3, TiO_2, CuO, and Cu coated with C (nanofluids formulated by NAITEC).

oxidation of these types of nanoparticles, which could be minimized by protecting nanoparticles with a coating layer of other materials such as polyvinylpyrrolidone (PVP) or carbon.

- *Metal oxide-based nanofluids*. Metal oxide nanoparticles are among the most conventional ones due to their stable dispersion in base fluid and their lower cost than metallic ones. Nanoparticles such as Al_2O_3, CuO, TiO_2, ZnO, Fe_3O_4, SiO_2, or CeO_2 are used to prepare metal oxide-based nanofluids.
- *Carbon-based nanofluids*. Carbon-based nanoparticles or nanosheets can significantly enhance the thermal properties of the base fluids. Nanoparticles or nanosheets such as carbon blacks, carbon nanotubes (SWCNTs, MWCNTs), graphene, or graphite are used to prepare carbon-based nanofluids. These materials have a very high cost, which limits their use.
- *Hybrid-based nanofluids*. Recently, many research works are focus on hybrid nanofluids, i.e., a nanofluid prepared with a combination of

different nanoparticles dispersed in a base fluid. According to several authors, using hybrid nanofluid can lead to higher improvement in thermal performance. Some of these combinations are MWCNTs-TiO_2, Graphene oxide-Al_2O_3, SiC-CuO/C, CNT-Fe_3O_4, TiO_2-Ag, Mg-TiO_2, or ZnO-TiO_2.

Depending on application several working fluids, pure or combined, can be used as base fluid to prepare the nanofluid such as water, ethylene glycol, glycerol, methanol, acetone, paraffin wax, or oils.

13.6.2 Nanofluid preparation

There are two main methods of nanofluid preparation: one-step and two-step method [68,70–73], which can be defined as follows:

1. One-step method
   In nanofluid formulation using the one-step method, the nanoparticles are simultaneously synthesized and dispersed in the working fluid; no intermediate steps (nanoparticle drying and storage and their dispersion in the working fluid) to minimize the nanoparticle aggregation probability are needed. This method has several advantages, such as preventing oxidation, which is a key issue when using metallic nanoparticles such as Cu or Ag. Also, it presents better nanoparticles dispersion, preventing the formation of aggregates and benefits long-term stability. In addition, this synthesis route allows for greater control over the size and shape of the nanoparticles. However, this method also has its drawbacks. The most important is its high production cost in comparison with the one-step method, which generally makes it difficult to use in industrial applications. In addition, unreacted reactants remaining in the end product can cause problems for implementation and, since this method is only valid for fluids with low vapor pressure, the field of applications in which it can be used is very limited.
2. Two-step method
   In this method, the nanofluid is prepared in two steps. First, the NPs are generated and in a second step they are dispersed in the base fluid using several techniques such as high-speed mechanical stirring or high-power ultrasound. This method is the most widely used due to its lower cost, which makes it feasible to scale up and use in industrial applications. One of the disadvantages of this method is the lower stability of the nanofluids obtained due to the agglomeration and sedimentation of nanoparticles. Due to its nanometric size, strong van der Waals forces appear among nanoparticles forming aggregates, even before they are dispersed in the fluid, which sediment. There are several strategies to avoid aggregates formation and improve nanofluid stability such as the addition of dispersants or surfactants, pH control, dispersion optimized parameters,

or surface modification. Nanofluids used in HPs are generally prepared by this method.

13.6.3 Nanofluid stability strategies

Nanofluid stability is one of the main challenges to overcome to achieve the implementation of nanofluids in heat pipe applications on an industrial scale. When a nanofluid is not stable, nanoparticles tend to agglomerate due to Van der Waals attraction forces and precipitate, so the nanofluid loses its heat transfer potential. In addition, these precipitated nanoparticles can lead to deposit formation in the HPs during and after cycles [69,74].

However, this deposition phenomenon has different interpretations in the literature. On the one hand, some studies claim that the formation of these layers has a negative effect on nanofluid thermal conductivity because there are fewer nanoparticles in suspension, reducing the heat transfer of the system. Other studies indicate that the decrease in heat transfer is due to the reduction of nucleation sites as a consequence of nanoparticles entrapment in the cavities of the HP container during deposition [75,76]. On the other hand, other research works claim that precisely the formation of this porous layer on the pipe's surface during the HP operation increases the number of nucleation sites and in turn the HP thermal performance [39,77–79]. Therefore, again, contradictory results are found in the literature, and further and deeper research are needed to understand the mechanisms of nanofluids.

There are several strategies aiming to obtain stable nanofluids [71,72,74,80,81]:

- *Dispersion method.* Various dispersion methods can be used, alone or in combination to establish a dispersion methodology, such as magnetic stirring, milling, high-speed mechanical stirring, ultrasonic bath, or high-power ultrasound (Fig. 13.20). In addition, depending on the chosen dispersion method, various parameters can be optimized: time (Fig. 13.21), frequency, power amplitude, speed, stop/start cycles, etc. The ideal dispersion conditions for a particular nanofluid must be established taking into account the specific nanoparticles (chemical nature, percentage, size) and base fluid (water, alcohol, glycol, acetone). For example, the application of long periods of high-power ultrasound generates a lot of heat which in ductile nanoparticles such as Cu, causes sintering of the particles, losing their initial shape and size related to heat transfer performance.
- *Surfactants.* Surfactants are materials that minimize the surface tension between two liquids or between a liquid and a solid. The suitable type of surfactant and its percentage should be selected according to the type of nanoparticle and base fluid. The surfactants can be cationic (positively charged), anionic (negatively charged), amphoteric (negatively and

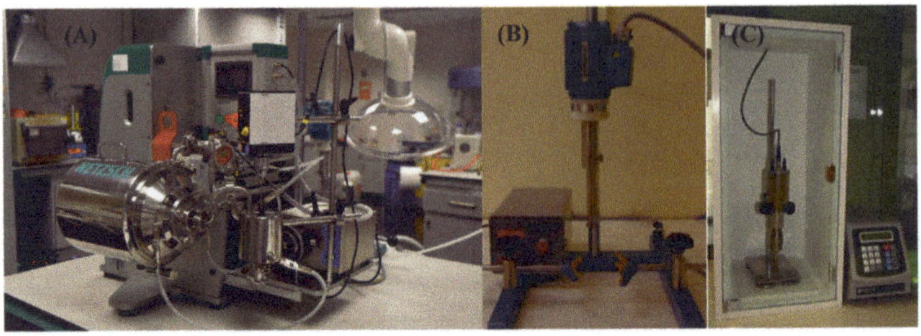

FIGURE 13.20 Examples of dispersion methods: (A) ball mill, (B) high-speed mechanical stirrer, and (C) ultrasonic probe (dispersion equipment property of NAITEC).

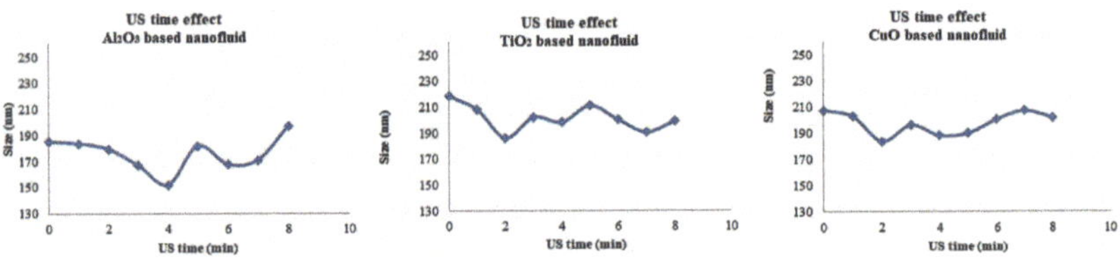

FIGURE 13.21 Example of ultrasonic time (US time) effect on particle size in different nanofluids (US time study and nanofluids developed by NAITEC).

positively charged), and nonionic (noncharged). Among the different surfactants on the market, it is possible to find sodium dodecyl sulfate (SDS), dodecyltrimethylammonium bromide (DTAB), polyvinylpyrrolidone (PVP, oleic acid, gum Arabic (GA), cetyltrimethylammonium bromide (CTAB), polyvinyl alcohol (PVA), Tween 80, or Triton X-100.

It is very important to keep in mind that the addition of dispersants favors nanofluid stability (Fig. 13.22), but it can also negatively affect other properties such as surface tension, viscosity, and thermal conductivity. Therefore, although they are generally necessary to achieve stable nanofluids, it is very important to minimize their use to reduce their adverse impact on heat transfer mechanisms.

- *Surface modification of nanoparticles.* This strategy involves functionalizing the original nanoparticle with chemical groups that favor the stabilization of the suspension, avoiding the use of other additives such as dispersants.
- *pH control.* The stability of nanofluids depends to a large extent on the pH of the colloidal system. As the electrokinetic properties affect the nanofluid stability the pH control can increase stability due to strong repulsive forces.

FIGURE 13.22 Example of carbon-based nanofluid (A) with and (B) without surfactant (nanofluid developed by NAITEC).

13.6.4 Nanofluid stability mechanisms

To achieve stable nanofluids, it is crucial to overcome the attractive van der Waals forces. There are two main dispersion mechanisms: electrostatic stabilization and steric stabilization [72,82,83].

- *Electrostatic stabilization*. Nanoparticles dispersed in fluids present electrical charges on their surface, when they are of the same sign (positive or negative) repulsive forces appear among nanoparticles. With the use of additives, such as several surfactants, it is possible to increase these charges and make them all of the same sign.

  DLVO theory explains the electrostatic stabilization of colloidal particles in their Brownian movement. This theory states that the interaction between electrostatically stabilized particles results from a balance between the attractive van der Waals forces and the electrical double-layer repulsive ones. A charge is generated on the surface of the particles and a second, more diffuse layer of oppositely charged ions is created around them. As the particles approach each other the charges provide a very effective barrier against contact. As the thickness of this second layer increases the stabilization increases with it, i.e., the wider the second layer, the more stable the fluid.
- *Steric stabilization*. Another approach to stabilize nanoparticles is by steric hindrance. Steric stabilization can be achieved by binding macromolecules, usually polymer chains, on the nanoparticle surface as a steric barrier between the nanoparticles and the van der Waals attractive forces. When two particles try to join, their polymeric chains interlock, leaving the particles with no freedom of movement. Therefore, the entropy decreases and allows the particles to remain isolated and avoid contact with each other.

Both stability mechanisms, electrostatic and steric, can be combined to stabilize nanoparticle dispersions. This combined mechanism is referred to as electrosteric stabilization.

13.6.5 Nanofluid stability evaluation

There are several methods and techniques used to assess the nanofluid's stability, and some of the equipment used is shown in Fig. 13.23. The most common methods are [70,71,80]:

- *Visual inspection.* With this method, images are taken over time to identify phase separation and/or sedimentation of particles. It has the disadvantage that if the sedimentation rate or phase separation is not clear, it is difficult to establish comparisons and requires long times to obtain results.
- *Zeta potential.* Zeta potential is a measure of the electrostatic (or charge) repulsion or attraction between particles and is one of the fundamental parameters that affect stability. Zeta potential measurement provides detailed information on the causes of flocculation, dispersion, and aggregation, and is used to improve the formulation of nanofluids. The zeta potential is not only a function of the particles, it also depends on the liquid they are dispersed. Any variation in the liquid's chemical–ionic composition modifies the zeta potential value. For zeta potential measurements it is necessary to dilute the sample and a common discussion is whether such dilution replicates the behavior of the original fluid. In general, nanofluids that have a zeta potential above $+30\,mV$ or below $-30\,mV$ are considered stable.
- *Dynamic light scattering (DLS) method.* Dynamic light scattering (DLS) allows measuring not only the particles' size but also their size distribution (Fig. 13.24). The particle Brownian motion makes the laser light scatter at different intensities. By analyzing these light intensity fluctuations and taking into consideration the Stokes–Einstein relation,

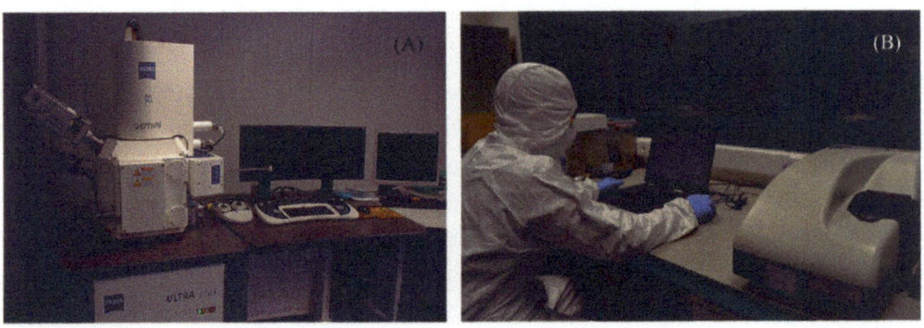

FIGURE 13.23 Examples of equipment used to evaluate nanofluid stability: (A) scanning electron microscope (SEM); and (B) zeta-sizer to measure size and distribution size of nanoparticles and zeta potential values (equipment property of NAITEC).

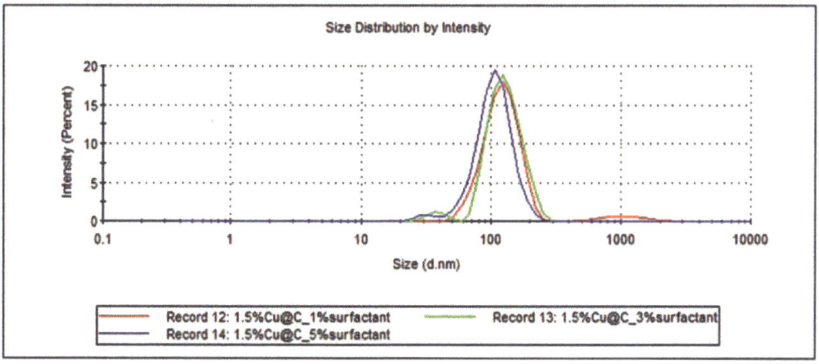

FIGURE 13.24 Example of differences in particle size distribution depending on surfactant percentage (Cu coated with C-based nanofluids formulated by NAITEC).

the Brownian speed is determined and so is the particle size. As in the case of zeta potential, DLS measurements require diluted samples.

- *Electron microscopy*. Electron microscopy techniques (scanning electron microscopy "SEM" and transmission electron microscopy "TEM") allow observation of the morphology and size of nanoparticles. They are commonly used to measure nanoparticle aggregates and compare them with original nanoparticle' sizes to assess whether their formation could be related to particle sedimentation problems.
- *Absorbance spectrometry*. This technique is used to quantitatively analyze the stability of the dispersion by measuring the absorption spectrum as a function of the nanoparticle concentration according to Beer–Lamberts law. This method is applicable when the concentration of nanoparticles is low. If the concentration is high, dilution of the samples is necessary which may not replicate the stability of the original nanofluid.
- *Three omega method*. The 3ω method measures the nanofluid thermal conductivity variance as a consequence of particles agglomeration and/or sedimentation. This technique is based on the principle of the linear relation between nanofluid thermal conductivity and nanoparticles concentration (volume, mass concentration). It is not a widely used technique probably because of the frequent contradictory results obtained in the variations of thermal conductivity with the volume concentration of the nanoparticles.

13.6.6 Nanofluids after operation cycles

During their lifetime, nanofluids undergo different phase changes, bubble formation, movement, thermal changes, etc. As a consequence of these phenomena, the properties and stability of the nanofluid can be significantly altered after operating cycles.

There are not many studies related to the characteristics of nanofluids before and after operation cycles, probably because of the difficulty of discerning the phenomena that occur and their consequences. Cacua et al. [66] review some of these investigations and find very contradictory results in the literature. For example, Grab et al. [84] found a significant increase in the nanoparticle size of TiO_2 and Au nanofluids as a consequence of their agglomeration after operation cycles. They also observed that after a prolonged shutdown, the nanoparticles sedimented and formed a porous layer on the evaporator which affected the thermal performance. However, Liu et al. [85] studied nanofluids of CuO nanoparticles in water as a base fluid and found no difference in performance between freshly prepared nanofluids and nanofluids that had been in the thermosyphon for 2 weeks without operation cycles. Other authors such as Chen et al. [75] who worked with functionalized SiO_2 nanoparticles found no porous layer deposited in the evaporator after evaluating their performance 1 week after the first test. Do and Jang [86] contributed with numerical studies using water-based Al_2O_3 nanofluids and concluded that nanoparticles deposition on the evaporator surface was responsible for the enhancement in HP thermal performance. Other authors observed an increase in heat transfer area in the evaporator by forming a thin porous layer [87–89]. Therefore, it seems that the performance of the nanofluid after operation cycles is not always the same and depends on many factors (type of nanoparticle, concentration, size, base fluid, etc.) so, again, more research is needed to better understand the mechanisms involved and obtain clearer conclusions about the performance of the nanofluid under operating conditions.

13.6.7 Nanofluids applications

Important developments have been achieved using nanofluids in HPs and PHPs, being applied to the thermal management of electronic components in PCBs [4,5], which has resulted in devices with reliable performances and able to operate during their lifetime expectancy with minor NCG generation. In this case, the nanofluid has been obtained with CuO nanoparticles with 95% purity and 29 nm in size (90% of the nanoparticles present such size) and deionized water and the base fluid. A total amount of 3.5%w.w (by weight) of nanoparticles was added to the base fluid, being mixed in an ultrasonic bath for 2 hours without indication of sedimentation after 48 hours.

The nanofluid was then inserted in the PHP with the same geometric characteristics used to obtain the results of Fig. 13.14, undergoing the same testing pattern as described then, to verify the nanofluid's potential in operating in this thermal device. Fig. 13.25 shows the results for this test, where the intensity of pulsations increased by the addition of the nanoparticles.

The presence of the nanoparticles in the working fluid is believed to generate additional nucleation sites which enhance the pulsations that will be

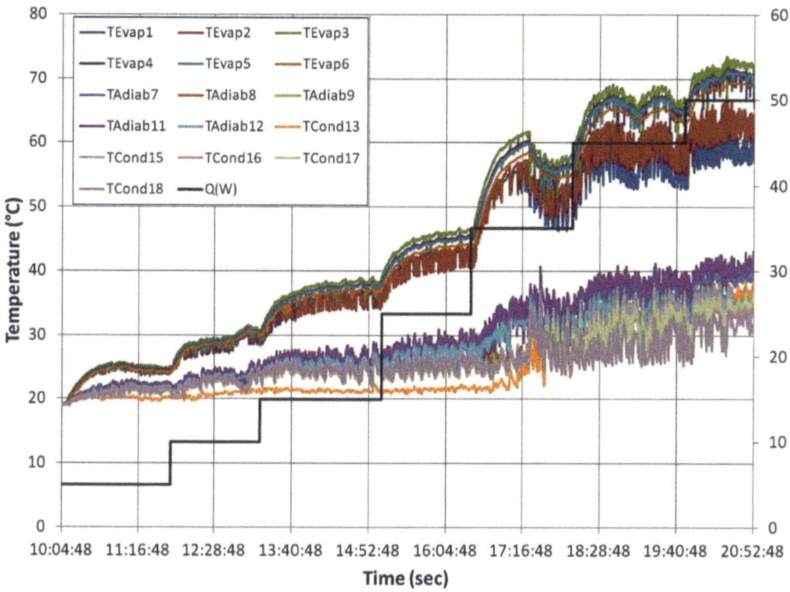

FIGURE 13.25 PHP operation with CuO-water nanofluid (image by GamaTech).

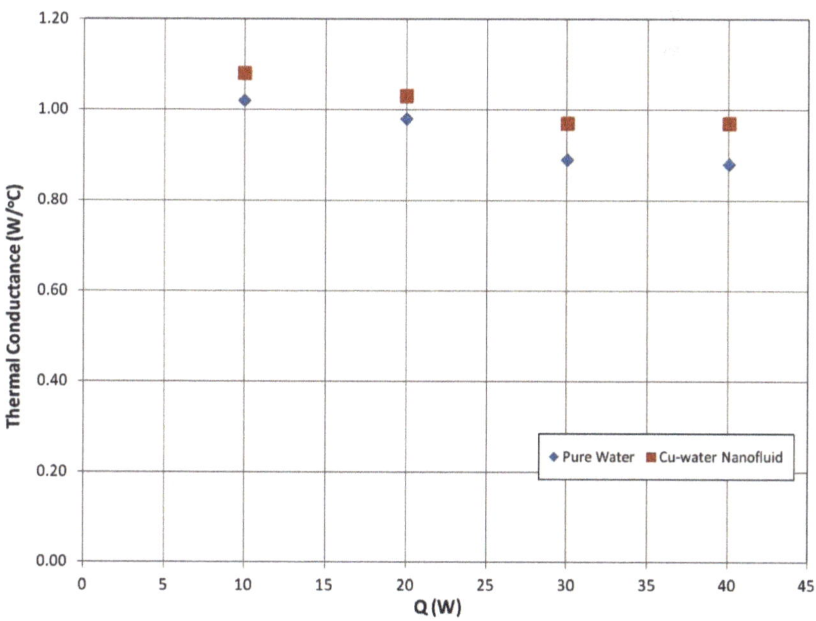

FIGURE 13.26 Thermal conductances for the nanofluid operation with the PHP (image by GamaTech).

visible during the entire PHP operation [18,31,38]. The nanofluid directly influences the operation to obtain lower evaporator section temperatures, which is a desirable condition in thermal control devices, also causing an impact on the thermal conductances. Fig. 13.26 summarizes the results for the thermal conductances obtained for the PHP.

13.7 Concluding remarks

There are, indeed, potential areas that can benefit from using nanofluids to enhance the device's thermal dissipation capabilities. In this matter, PHPs have been shown to be potential candidates to use nanofluids in a wide range of applications, because they operate without the presence of a wick structure that can result in localized pressure drops.

The technological advances in obtaining nanoparticles have resulted in more stable nanofluids, allowing the determination of thermophysical properties that are important to better evaluate their overall performance in the thermal management system.

In general, the following can be concluded:

- Heat pipes and pulsating HPs are important devices in the designs of modern thermal management systems, especially where concentrated heat fluxes demand more reliable thermal devices.
- The technological level of development presented by HPs and pulsating HPs, from their design to their final integration, indicates them to be mature enough to be applied to any area that demands reliable passive thermal management, such as electronics cooling.
- Nanofluids is a promising field that requires further investigation for a better understanding of the physical mechanisms that allow improving the heat transfer capacities of two-phase heat transfer devices.
- Nanofluids can be applied to a wide range of equipment that requires enhancement of thermal capacities.
- As a major issue, nanofluid stability continues to be a challenge and requires a more thorough investigation.
- The vast amount of experimental research conducted to date positions nanofluids as a key enabling technology to advance cooling and thermal management for the foreseeable future.

Nomenclature

ΔP_n	normal hydrostatic pressure drop
ΔP_{ah}	axial hydrostatic pressure drop
ε	wick's porosity
δ	groove depth
γ	specific heats ratio
λ	latent heat of vaporization

θ	working fluid's wettability angle
ψ	heat pipe's angle with horizontal
σ	working fluid's surface tension
μ_l	liquid dynamic viscosity
μ_v	vapor dynamic viscosity
ρ_l	liquid density
ρ_v	vapor density
τ_l	liquid film thickness
ϕ	mass fraction of solid nanoparticles
A	empirical relationship between temperature and rate coefficient (s^{-1})
A_i	internal cross-section area
A_v	vapor's flow area
A_w	wick's area
Bo	Bond number
D_c	critical diameter
D_{max}	maximum tube diameter
d_v	vapor diameter
E_{act}	activation energy of the reaction, based on the combination of the working fluid and the materials involved (J/mol)
ϕ	mass fraction of solid nanoparticles
f_v	vapor's friction factor
g	gravity's acceleration
Q	heat load
Q_{max}	maximum heat transfer capacity
$Q_{f,max}$	maximum heat transport capacity
q''	maximum heat flux to the evaporator
R	universal gas constant
Re_v	vapor's Reynolds number
L	overall heat pipe's length
L_a	adiabatic length
L_e	evaporation length
L_c	condensation length
L_{eff}	effective heat pipe's length
L_{seg}	segment length of the PHP
m	working fluid inventory (kg)
P_v	vapor pressure
r_{in}	internal tube radius
r_o	tube's outside radius
r_p	mean pore radius
r_n	critical nucleation site radius
$r_{h,w}$	wick structure hydraulic radius
$r_{h,v}$	vapor flow hydraulic radius
w_f	width of groove fin
w	wire spacing
K	wick's permeability
K	decomposition rate (Eq. 13.1)
k_{eff}	effective thermal conductivity

k_l liquid thermal conductivity
k_w wick's thermal conductivity
T absolute temperature
T_{sat} saturation temperature
T_{min} minimum tube wall temperature
Tevap evaporator temperature
Tadiab adiabatic temperature
Tcond condenser temperature

References

[1] R.R. Riehl, Surveillance system heat transfer enhancement using single-phase flow with nanofluid, in: Heat Powered Cycles Conference, Alkmaar, The Netherlands, 9–12 September 2012.

[2] R.R. Riehl, L. Cachuté,. Thermal control of surveillance systems using heat pipes and pulsating heat pipes, in: 49th AIAA/ASME/SAE/ASEE Joint Propulsion Conference & Exhibit 11th International Energy Conversion Engineering Conference, San Jose, CA, 14–17 July 2013.

[3] R.R. Riehl, L. Cachuté, Thermal management of surveillance equipments electronic components using pulsating heat pipes, in: IEEE-ITherm Conference—Intersociety Conference on Thermal and Thermomechanical Phenomena in Electronics Systems, Orlando, FL, 27–30 May 2014.

[4] D. Silva, E. Marcelino, R.R. Riehl, Thermal performance comparison between water-copper and water-stainless steel heat pipes, in: 50th AIAA/ASME/SAE/ASEE Joint Propulsion Conference & Exhibit 12th International Energy Conversion Engineering Conference, Orlando, FL, July 27–29 2015.

[5] R.R. Riehl, Passive thermal management systems using pulsating heat pipes, in IX Minsk International Seminar Heat Pipes, Heat Pumps, Refrigerators, Power Sources, Minsk, Belarus, 07–10 September 2015.

[6] A. Faghri, Heat Pipe Science and Technology, Taylor and Francis, Washington, DC, 1995.

[7] S.W. Chi, Heat Pipe Theory and Practice, Hemisphere Publishing Corporation, Washington, DC, 1976.

[8] G. Peterson, An Introduction to Heat Pipe Modeling, Testing and Applications, John Wiley and Sons, New York, 1994.

[9] D.A. Reay, P.A. Kew, Heat Pipes-theory, Design and Applications, fifth ed., Elsevier's Science & Technology, Oxford, 2006.

[10] J.H. Rosenfeld, N.J. Gernert, Life test results for water heat pipes operating at 200°C to 300°C, in: Proceedings of the STAIF-Space Technology and Applications International Forum, 10–14 February, Albuquerque, NM, 2008.

[11] W.G. Anderson, S. Tamanna, C. Tarau, J.R. Hartenstine, D. Ellis, Intermediate temperature heat pipe life tests, in: Proceedings of the 16th International Heat Pipe Conference, Lyon France, 20–24 May 2012.

[12] K.N. Shukla, Heat pipe for aerospace applications—an overview, Journal of Electronics Cooling and Thermal Control 5 (2015) 1.

[13] S.U.S. Choi, Enhancing thermal conductivity of fluids with nanoparticles, ASME FED 231 (1995) 99–105.

[14] S.K. Das, S.U.S. Choi, W. Yu, T. Pradeep, Nanofluids Science and Technology, first ed., John Wiley & Sons Inc, New Jersey, 2008.

[15] L. Colla, L. Fedele, S. Mancin, S. Bobbo, D. Ercole, O. Manca, Nano-PCMs for electronics cooling applications, in: Proceedings of the 5th International Conference on Micro/Nanoscale Heat and Mass Transfer ASME-MNHMT, 4–6 January, Biopolis, Singapore, 2016.

[16] E. Marcelino, R.R. Riehl, A review on thermal performance of CuO-water nanofluids applied to heat pipes and their characteristics, in: Proceedings of the 15th IEEE Intersociety Conference on Thermal and Thermomechanical Phenomena in Electronic Systems (ITherm), 31 May–3June 2016, Las Vegas NV, USA.

[17] D. de O. Silva, Experimental Investigation of the Thermal Performances of Screen Mesh Wick Heat Pipes Operating in Mid-Level Temperatures, Doctoral Thesis, National Institute for Space Research (INPE), 173p, 2018.

[18] R.R. Riehl, S.H.S. Murshed, Life time expectancy prediction and ageing process of heat pipes using nanofluids, Heat Transf. Eng. (2020). Available from: https://doi.org/10.1080/01457632.2020.1818423.

[19] H. Akachi, F. Polasek, P. Stulc, Pulsating heat pipes, in: Proceedings of the 5th International Heat Pipe Symposium, Melbourne, Australia, 1996, pp. 208–217.

[20] B.Y. Tong, T.N. Wong, K.T. Ooi, Closed-loop pulsating heat pipe, Appl. Therm. Eng. 21 (2001) 1845–1862.

[21] W. Yu, S.U.S. Choi, The role of interfacial layers in the enhanced thermal conductivity of nanofluids: a renovated Maxwell model, J. Nanopar. Res. 5 (2003) 167–171.

[22] A.A.M. Delil, Pulsating and Oscillating Heat Transfer Devices in Acceleration Environments from Microgravity to Supergravity, SAE Paper # 2001-01-2240, 2001.

[23] L. Lin, R. Ponnapan, J. Leland, Experimental investigation of oscillating heat pipes, AIAA J. Thermo. Heat Transf 15 (2001) 395–400.

[24] H. Yang, S. Khandekar, M. Groll, Operational limit of closed loop pulsating heat pipes, Appl. Therm. Eng. 28 (2008) 49–59.

[25] S. Khandekar, P. Charoesawan, M. Groll, P. Terdtoon, Closed-loop pulsating heat pipes part B: visualization and semi-empirical modeling, Appl. Therm. Eng. 23 (2003) 2021–2033.

[26] P. Srikrishna. A.R. Hari Kumar, S.U.M. Reddy, G.S.V.L. Narasimham, Experimental investigation of a single closed loop pulsating heat pipe, in: Proceedings of the 24th National and 2nd International ISHMT-ASTFE Heat and Mass Transfer Conference (IHMTC-2017), 27–30 December 2017, BITS Pilani, Hyderabad, India.

[27] L. Pagliarini, L. Cattani, F. Bozzoli, M. Mameli, S. Filippeschi, S. Rainieri, et al., Thermal characterization of a multi-turn pulsating heat pipe in microgravity conditions: statistical approach to the local wall-to-fluid heat flux, Int. J. Heat Mass Transfer 169 (2021) 120930.

[28] P. Supirattanakul, S. Rittidech, B. Bubprachot, Application of a closed-loop oscillating heat pipe with check valves (CLOHP) on performance enhancement in air conditioning system, Energy Build. 43 (2011) 1531–1535.

[29] M. Mameli, M. Marengo, S. Zinna, Numerical model of a multi-turn closed loop pulsating heat pipe: effects of the local pressure losses due to meanderings, Int. J. Heat Mass Transf 55 (2012) 1036–1047.

[30] Y. Miyazaki, H. Kawai, N. Iwata, H. Ogawa, Operating limit of oscillating heat pipe, in: Proceedings of the 16th International Heat Pipe Conference, Lyon, France, 20–24 May 2012.

[31] R.R. Riehl, Evaluation of the thermal-hydro-dynamics behavior of an open loop pulsating heat pipe, https://doi.org/10.13140/RG.2.1.1024.9443.

[32] D. Bastakoti, H. Zhanga, D. Lia, W. Caia, F. Li, An overview on the developing trend of pulsating heat pipe and its performance, Appl. Therm. Eng. 141 (2018) 305–332.

[33] J. Qu, J. Zhao, Z. Rao, Experimental investigation on the thermal performance of three-dimensional oscillating heat pipe, Int. J. Heat Mass Transf 109 (2017) 589–600.

[34] A. Takawale, S. Abraham, A. Sielaff, P.S. Mahapatra, A. Pattamatta, P. Stephan, A comparative study of flow regimes and thermal performance between flat plate pulsating heat pipe and capillary tube pulsating heat pipe, Appl. Therm. Eng. 149 (2019) 613–624. Available from: https://doi.org/10.1016/j.applthermaleng.2018.11.119.

[35] P. Charoensawan, P. Terdtoon, P. Tantakom, P. Ingsuwan, Effect of evaporator section lengths, number of turns, and working fluid on internal flow patterns of a vertical closed-loop oscillating heat pipe, in: Proceeding of the 7th International Heat Pipe Symposium. Jeju Korea, pp. 360–367, 2003.

[36] N. SanthiSree, N.V.V.S. Sudheer, P. Bhramara, Experimental evaluation of closed-loop pulsating heat pipe with water-based working fluids, Int. J. Mech. Prod. Eng. Resear. Develop. 9 (2019) 11–20.

[37] R.R. Riehl, Passive thermal management of surveillance systems using pulsating heat pipes, in: IX Minsk International Seminar "Heat Pipes, Heat Pumps, Refrigerators, Power Sources," Minsk, Belarus, 07–10 September 2015.

[38] R.R. Riehl, N. Santos, Water-copper nanofluid application in an open loop pulsating heat pipe, Appl. Therm. Eng. 42 (2012) 6–10.

[39] Z.-H. Liu, Y.-Y. Li, A new frontier of nanofluid research – application of nanofluids in heat pipes, Int. J. Heat Mass Transf., IJHMT-6786–6797, Elsevier, October 2011.

[40] N.R. Karthikeyan, J. Philip, B. Raj, Effect of clustering on the thermal conductivity of nanofluids, Mater. Chem. Phys. 109 (1) (2008) 50–55.

[41] K. Nemade, S. Waghuley, A novel approach for enhancement of thermal conductivity of CuO/H2O based nanofluids, Appl. Thermal Eng 95 (2016) 271–274.

[42] B.R.R. Bapu, S. Kaylvizhi, S. Murugavalli, Two-phase cooling with nanofluid for highly dense electronic systems-on-chip – a pilot study, Microelectron. Reliab. 108 (May 2020) 113640.

[43] X.J. Wang, X.F. Li, Influence of pH on nanofluids' viscosity and thermal conductivity, Chin. Phys. Lett. 26 (5) (2009) 056–061.

[44] K.R. Priya, K.S. Suganthi, K.S. Rajan, Transport properties of ultra-low concentration CuO–water nanofluids containing non-spherical nanoparticles, Int. J. Heat Mass Transf. 55 (2012) 4734–4743.

[45] J.C. Maxwell, A Treatise on Electricity and Magnetism, second ed., Claredon Press, Oxford, 1881.

[46] R.L. Hamilton, Thermal conductivity of heterogeneous two-component systems, Ind. Eng. Chem. Fundam. 1 (1962) 182–191.

[47] R. Prasher, P. Bhattacharya, P.E. Phelan, Thermal conductivity of nanoscale colloidal solutions (nanofluids), Phys. Rev. Lett. 94 (2005) 1–4.

[48] J. Koo, C. Kleinstreuer, Laminar nanofluid flow in microheat-sinks, Int. J. Heat Mass Trans. 48 (2005) 2652–2661.

[49] N.R. Karthikeyan, J. Philip, B. Raj, Effect of clustering on the thermal conductivity of nanofluids, Mater. Chem. Phys. 109 (1) (2008) 50–55.

[50] K. Nemade, S. Waghuley, A novel approach for enhancement of thermal conductivity of CuO/H_2O based nanofluids, Appl. Thermal Eng. 95 (2016) 271–274.

[51] R.S. Khedkar, S.S. Sonawane, K.L. Wasewar, Influence of CuO nanoparticles in enhancing the thermal conductivity of water and monoethylene glycol-based nanofluids, Int. Comm. Heat Mass Transf. 39 (2012) 665–669.

[52] X.J. Wang, X.F. Li, Influence of pH on nanofluids' viscosity and thermal conductivity, Chin. Phys. Lett. 26 (5) (2009) 056–061.

[53] K.R. Priya, K.S. Suganthi, K.S. Rajan, Transport properties of ultra-low concentration CuO–water nanofluids containing non-spherical nanoparticles, Int. J. Heat Mass Transf. 55 (2012) 4734–4743.

[54] A. Einstein, Eine neue bestimmung der moleküldimensionen, Ann. Phys. 324 (1906) 289–306.

[55] H. Brinkman, The viscosity of concentrated suspensions and solutions, J. Chem. Phys. 20 (1952) 571.

[56] J. Buongiorno, Convective transport in nanofluids, J. Heat Transf. 128 (2006) 240–250.

[57] C.T. Nguyen, F. Desgranges, G. Roy, N. Galanis, T. Mafe, S. Boucher, et al., Temperature, and particle-size dependent viscosity data for water-based nanofluids—hysteresis phenomenon, Int. J. Heat Fluid Flow 28 (2007) 1492–1506.

[58] H. Chen, Y. Ding, C. Tan, Rheological behaviour of nanofluids, New J. Phys. 9 (2007) 367.

[59] D.P. Kulkarni, D.K. Das, S.L. Patil, Effect of temperature on rheological properties of copper oxide nanoparticles dispersed in propylene glycol and water mixture, J Nanosci. Nanotechnol. 7 (2007) 2318–2322.

[60] M.A. Nazari, M.H. Ahmadi, M. Sadeghzadeh, M.B. Shafii, M. Goodarzi, A review on application of nanofluid in various types of heat pipes, J. Cent. South Univ. 26 (5) (2019) 1021–1041.

[61] M.M. Tawfik, Experimental studies of nanofluid thermal conductivity enhancement and applications: a review, Renew. Sustain. Energy Rev. 75 (2017) 1239–1253.

[62] J. Esarte, R. Riehl, J. Estella, Nanofluid as advanced cooling technology. Success stories, in: Heat Transfer-Design, Experimentation and Applications, IntechOpen, 2020.

[63] R.K. Sahu, S.H. Somashekhar, P.V. Manivannan, Investigation on copper nanofluid obtained through micro electrical discharge machining for dispersion stability and thermal conductivity, Procedia Eng. 64 (2013) 946–955.

[64] N.A. Che Sidik, H.A. Mohammed, O.A. Alawi, S. Samion, A review on preparation methods and challenges of nanofluids, Int. Commun. Heat Mass Transf. 54 (2014) 115–125.

[65] H. Ghorabaee, M.R.S. Emami, M. Shafahi, Effect of nanofluid and surfactant on thermosyphon heat pipe performance, Heat Transf. Eng. 41 (2020) 1829–1842.

[66] K. Cacua, R. Buitrago-Sierra, E. Pabón, A. Gallego, C. Zapata, B. Herrera, Nanofluids stability effect on a thermosyphon thermal performance, Int. J. Therm. Sci. 153 (2020) 106347.

[67] M.A. Nazari, R. Ghasempour, M.H. Ahmadi, A review on using nanofluids in heat pipes, J. Therm. Anal. Calorim. 137 (2019) 1847–1855.

[68] E.C. Okonkwo, I. Wole-Osho, I.W. Almanassra, Y.M. Abdullatif, T. Al-Ansari, An updated review of nanofluids in various heat transfer devices, J. Therm. Anal. Calorim. (2020).

[69] A.L.S. Sudhan, K. Ramachandran, A.B. Solomon, C.P. Jawahar, Research progress on performance enhancement of heat pipes: a review, J. Therm. Anal. Calorim. (2021).

[70] S. Chakraborty, P.K. Panigrahi, Stability of nanofluid: a review, Appl. Therm. Eng. 174 (2020) 115259.

[71] A. Qamar, Z. Anwar, H. Ali, R. Shaukat, S. Imran, A. Arshad, et al., Preparation and dispersion stability of aqueous metal oxide nanofluids for potential heat transfer applications: a review of experimental studies, J. Therm. Anal. Calorim. (2020).

[72] V. Çorumlu, A. Ozsoy, M. Ozturk, Evaluation of heat transfer mechanisms in heat pipe charged with nanofluid, Arab. J. Sci. Eng. 44 (2019) 5195–5213.

[73] R. Sureshkumar, S. Tharves Mohideen, N. Nethaji, Heat transfer characteristics of nanofluids in heat pipes: a review, Renew. Sustain. Energy Rev. 20 (2013) 397–410.

[74] K. Cacua, R. Buitrago-Sierra, B. Herrera, E. Pabón, S.M. Sohel Murshed, Nanofluids' stability effects on the thermal performance of heat pipes. a critical review, J. Therm. Anal. Calorim. 136 (2019) 1597–1614.

[75] Y.J. Chen, P.Y. Wang, Z.H. Liu, Application of water-based SiO2 functionalized nanofluid in a loop thermosyphon, Int. J. Heat Mass. Transf 56 (2013) 59–68.

[76] S. Khandekar, Y.M. Joshi, B. Mehta, Thermal performance of closed two-phase thermosyphon using nanofluids, Int. J. Therm. Sci 47 (2008) 659–667.

[77] N.K. Gupta, A.K. Tiwari, S.K. Ghosh, Heat transfer mechanisms in heat pipes using nanofluids – a review, Exp. Therm. Fluid Sci. 90 (2018) 84–100.

[78] R. Tanshen, B. Munkhbayar, J. Nine, H. Chung, H. Jeong, Effect of functionalized MWCNTs/water nanofluids on thermal resistance and pressure fluctuation characteristics in oscillating heat pipe, Int. Commun. Heat Mass Transf. 48 (2013) 93–98.

[79] Z. Wan, J. Deng, B. Li, Y. Xu, X. Wang, Y. Tang, Thermal performance of a miniature loop heat pipe using water—copper nanofluid, Appl. Therm. Eng. 78 (2015) 712–719.

[80] S. Mukherjee, S. Paria, Preparation and stability of nanofluids-a review, IOSR J. Mech. Civil Eng. 9 (2013) 63–69.

[81] K.A. Jehhef, M.A. Al Abas Siba, Effect of surfactant addition on the nanofluids properties: a review, Acta Mech. Malaysia (AMM) 2 (2) (2019) 01–19. Zibeline International Publishing.

[82] F. Yu, Y. Chen, X. Liang, J. Xu, C. Lee, Q. Liang, et al., Dispersion stability of thermal nanofluids, Prog. Nat. Sci.: Mater. Int. 27 (5) (2017) 531–542.

[83] W. Yu, H. Xie, A review on nanofluids: preparation, stability mechanisms, and applications, J. Nanomater. 1–17 (2012) ID435873.

[84] T. Grab, U. Gross, U. Franzke, M.H. Buschmann, Operation performance of thermosyphons employing titania and gold nanofluids, Int. J. Therm. Sci 86 (2014) 352–364.

[85] Z.H. Liu, X.F. Yang, G.L. Guo, Effect of nanoparticles in nanofluid on thermal performance in a miniature thermosyphon, J. Appl. Phys 102 (2007) 013526.

[86] K.H. Do, S.P. Jang, K. Hyung, S. Pil, K.H. Do, S.P. Jang, Effect of nanofluids on the thermal performance of a flat micro heat pipe with a rectangular grooved wick, Int. J. Heat Mass Transf. 53 (2010) 2183–2192.

[87] R. Hajian, M. Layeghi, K. Abbaspour Sani, Experimental study of nanofluid effects on the thermal performance with response time of heat pipe, Energy Convers. Manage. 56 (2012) 63–68.

[88] T. Yousefi, S.A. Mousavi, B. Farahbakhsh, M.Z. Saghir, Experimental investigation on the performance of CPU coolers: effect of heat pipe inclination angle and the use of nanofluids, Microelectron. Reliab. 53 (2013) 1954–1961.

[89] G. Huminic, A. Huminic, Numerical study on heat transfer characteristics of thermosyphon heat pipes using nanofluids, Energy Convers. Manage. 76 (2013) 393–399.

Chapter 14

Role of nanofluids in microchannel heat sinks

Ali Turab Jafry[1], Sohail Malik[1], Naseem Abbas[2], Uzair Sajjad[3], Hafiz Muhammad Ali[4,5] and Chi-Chuan Wang[6]

[1]*Faculty of Mechanical Engineering, Ghulam Ishaq Khan Institute of Engineering Sciences and Technology, Topi, Pakistan,* [2]*Department of Mechanical Engineering, Sejong University, Seoul, South Korea,* [3]*Department of Energy and Refrigerating Air-Conditioning Engineering, National Taipei University of Technology, Taipei, Taiwan,* [4]*Mechanical Engineering Department, King Fahd University of Petroleum and Minerals, Dhahran, Saudi Arabia,* [5]*Interdisciplinary Research Center for Renewable Energy and Power Systems (IRC-REPS), King Fahd University of Petroleum and Minerals, Dhahran, Saudi Arabia,* [6]*Department of Mechanical Engineering, National Yang Ming Chiao Tung University, Hsinchu, Taiwan*

Nomenclature

a	Particle radius (m)
C	Constant
C_p	Specific heat capacity (J/kg · K)
d_p	Particle diameter (m)
D	Characteristic dimension (m)
D_h	Hydraulic diameter (m)
f	Friction factor
h	Convective heat transfer coefficient (W/m^2 · K)
k	Thermal conductivity (W/m · K)
L	Length of channel (m)
Nu	Nusselt number
Pe	Peclet number
Pr	Prandtl number
Re	Reynolds number
T	Temperature (K)
$\overline{V}$	Average fluid velocity (m/s)
v, V	Flow velocity (m/s)
x^+	Nondimensional flow distance
x	Location in the axial flow
ΔP	Pressure drop (Pa)

Advances in Nanofluid Heat Transfer. DOI: https://doi.org/10.1016/B978-0-323-88656-7.00003-9

Greek Letters

ρ	Density of fluid (kg/m^3),
μ	Dynamic viscosity of fluid (Pa · s).
α	Thermal diffusivity (m^2/s)
ϕ	Volume concentration of nanoparticles (%)
δ	Distance between two spherical nanoparticles (nm)
β	Ratio of nanolayer thickness to particle radius
τ_w	Wall shear stress (Pa)
α_c	Aspect ratio of channel
κ_H	Hagenbach factor
κ_i	Inlet loss coefficients
κ_o	Outlet loss coefficients

Subscripts

app	Apparent
B	Brownian motion
c, ch	Channel
d	Dynamic
EG	Ethylene glycol
f*, *BF	Base fluid
NF	Nanofluid
p*, *NP	Nanoparticles
pl	Nanoparticle with interface layer
Reg	Regression
s	Static
w	Water

14.1 Introduction

With the further miniaturization of electronic devices, the heat density also rises enormously. Hence, it requires a sufficient heat dissipation means, and nanofluids have been proved to be quite effective especially in further augmentation in liquid cooling. Yet, the nanofluids can be further integrated with microchannel heat exchangers to boost even higher heat dissipation. This is because the fluid flow in a microchannel yields a better thermal performance due to its higher surface area to volume ratio, and the heat transfer coefficient is dramatically increased. This can be made clear from Nu $(h \times D/k) =$ constant for the fully developed laminar flow where h is proportional to $1/D$.

Nanofluids, composed of a mixture of nanoparticles in suspension within a base fluid, can further improve the cooling efficacy of microchannels in addition to their geometric advantage. This technology has led to the development of various types of nanofluids, and microfluidic channel geometries to achieve better hydrodynamic and thermal performance. The nanoparticles enhance the overall thermal conductivity of the liquid with marginal changes in its viscosity. It is important that the viscosity of the nanofluid must not increase drastically as it will require higher pumping power to flow the fluid in microchannels. This increase in thermal conductivity leads to a smaller and lighter cooling system with comparable performance to conventional ones. However, the suspension of nanofluids is prone to sedimentation and agglomeration issues, which need to be resolved for long-term use. Moreover, there are various factors affecting the thermal conductivity and flow performance of nanofluids in microchannel heat sinks (MCHS).

Hence, in this chapter, firstly the major thermophysical characteristics are described in detail with their mathematical models. Afterwards, the flow performance of nanofluids in microchannels is described in association with various geometries. Lastly, various channel geometries are described which can be used for improved thermal performance. The reader of this chapter should have basic knowledge of heat transfer and fluid mechanics to completely understand its contents.

14.2 Key characteristics of nanofluids

Nanofluids have shown significant potential among many applications. They are composed of suspensions containing nanometer-sized particles, tubes or fibers. Metallic, carbon-based, ceramic, or nonmetallic particles can be employed. Moreover, hybrid nanofluids made from a mixture of different suspensions also reveal interesting flow behavior. For instance, a binary mixture of iron and copper oxide revealed distinct Newtonian and non-Newtonian behavior at low and high concentrations, respectively [1]. The common base fluid includes water, oil, or sometimes ethylene glycol (EG). Many researchers have reported improvement in thermal conductivity with very small concentrations (normally less than 1% volume fraction of nanoparticles), especially in microchannel application. Such improvements are related to the particle characteristics, size and shape, volume concentration, and the type of base fluid. Researchers are now trying to stabilize the nanoparticle suspensions for prolonged periods, increase the thermal conductivity, reduce the pressure drop, and reduce erosion problems in MCHS. Therefore, before using nanofluids in MCHS, the knowledge of their thermophysical properties is essential for characterizing performance. In this section, the authors will discuss the basic properties with examples.

14.2.1 Thermophysical properties

Before listing the theoretical and experimental models for various thermophysical properties subject to laminar or turbulent flow conditions, it is important to understand some basic dimensionless numbers in the analysis [2]. A list of dimensionless numbers typically used for heat transfer applications is in the following:

Reynolds number, which is a ratio of inertial to viscous forces:

$$Re = \frac{\rho v D}{\mu}$$

where ρ is density of fluid (kg/m^3), v is flow velocity (m/s), D is the characteristic dimension (m), and μ is dynamic viscosity of fluid (Pa · s).

Prandtl number is defined as the ratio of momentum diffusivity to thermal diffusivity:

$$Pr = \frac{C_p \mu}{k}$$

where C_p is the specific heat (J/kg · K), and k is the thermal conductivity (W/m · K).

Peclet number is the ratio of advection to diffusion:

$$Pe = \frac{v d_p}{\alpha} = RePr$$

where d_p is the particle diameter [characteristic length (m)], α is the thermal diffusivity given by:

$$\alpha = \frac{k}{\rho C_p}$$

The heat transfer correlations developed for pure fluid could also be used to describe the thermal behavior of nanofluids. However, in order to use these correlations, one must use the appropriate thermal and transport properties developed for nanofluids. Since the aim is to increase heat transfer rate for MCHS, the main parameters upon analysis are density, viscosity, heat capacity, and finally the thermal conductivity of the nanofluid. The two-phase mixture equations are used to find the thermophysical properties of nanofluids. The **density** of a nanofluid (ρ_{NF}) can be found from mixing rule [3,4]:

$$\rho_{NF} = (1 - \phi)\rho_{BF} + \phi \rho_{NP}$$

where ϕ represents volume concentration of nanoparticles, ρ_{BF} is the density of base fluid, and ρ_{NP} is the density of nanoparticles.

The **specific heat capacity** (C_{pNF}(J/kgK)) of nanofluid, and can be computed using [2]:

$$C_{pNF} = \frac{(1 - \phi)(\rho C_p)_{BF} + \phi(\rho C_p)_{NP}}{(1 - \phi)\rho_{BF} + \phi \rho_{NP}}$$

The **viscosity** of the base fluid increases with the presence of nanoparticles, and hence more pumping power is required to allow flow. However, the viscosity is not only dependent on nanoparticles volume concentration, but is also a function of temperature, particle geometry, mixture combination, surfactants, etc. Thus, to date, experimental studies have been conducted, and

TABLE 14.1 Classical and recent models (theoretical and experimental) for estimating viscosity of nanofluids.

Authors	Equation	Comments
Theoretical models		
Einstein model [6]	$\frac{\mu_{NF}}{\mu_{BF}} = 1 + 2.5\phi$	Valid for low concentrations $\phi < 1\%$, under predicts at higher concentration [4]
Brinkman model [7]	$\frac{\mu_{NF}}{\mu_{BF}} = \frac{1}{(1-\phi)^{2.5}}$	Spherical particles and also valid for moderate concentrations
Batchelor model [8]	$\frac{\mu_{NF}}{\mu_{BF}} = 1 + 2.5\phi + 6.2\phi^2$	Spherical particles
Vajjha [9]	$\frac{\mu_{NF}}{\mu_{BF}} = Ae^{C\phi}$ $A = 0.9197, C = 22.8539$ $\mu_{BF} = Ae^{\frac{B}{T}}$ $A = 0.555 \times 10^{-3}, B = 2664$	CuO nanofluid in 60:40 (EG/water) with $0.01 \leq \phi \leq 0.06$ $293K \leq T \leq 363K$
Yang et al. [10]	$\mu_{NF} = \mu_s + \mu_d$ $= \frac{\mu_{BF}}{(1-\phi_{pl})^{2.5}} + \frac{\rho_{pl} v_B (d_p + 2h)^2}{72C\delta}$	Water-based Newtonian nanofluids. $\mu_s + \mu_d$ are the static and dynamic components of viscosity h is the interface layer thickness, C is the correction factor
Experimental models		
Drew and Passman [11]	$\frac{\mu_{NF}}{\mu_{BF}} = 1 + 2.5\phi$	$\phi < 5vol\%$
Tseng and Lin [12]	$\frac{\mu_{NF}}{\mu_{BF}} = 13.47\exp(35.98\phi)$	TiO_2 nanoparticles in water ($\phi = 0.05 - 0.12$)
Maiga et al. [13]	$\frac{\mu_{NF}}{\mu_{BF}} = 1 + 7.3\phi + 123\phi^2$	Al_2O_3 nanoparticles in water
Maiga et al. [13]	$\frac{\mu_{NF}}{\mu_{BF}} = 1 - 0.19\phi + 306\phi^2$	Al_2O_3 nanoparticles in ethylene glycol
Song et al. [14]	$\frac{\mu_{NF}}{\mu_{BF}} = 1 + 56.5\phi$	SiO_2 nanoparticles in water
Grag et al. [15]	$\frac{\mu_{NF}}{\mu_{BF}} = 1 + 11\phi$	Cu nanoparticles in ethylene glycol
Williams et al. [5]	$\mu_{NF} = \mu_{BF} \times \exp\left[\frac{4.91\phi}{(0.2092 - \phi)}\right]$	Al_2O_3 nanoparticles in water
Masoumi et al. [16]	$\mu_{NF} = \mu_{BF} + \frac{\rho_p V_B d_p^2}{72C\delta}$ $\delta = \sqrt[3]{\frac{\pi}{6\phi}} d_p$	d_p: particle diameter, V_B: Brownian velocity, ρ_p: density of particleAl$_2$O$_3$ nanoparticles in water

the appropriate models have been developed with a set of limitations. Table 14.1 summarizes some of the theoretical and experimental models used to determine the effective viscosity of nanofluids. A comparison of viscosity data of various Al_2O_3 nanofluids with classical and recent empirical models is presented in Fig. 14.1. It is observed that with an increase in nanoparticle concentration, the viscosity of nanofluids is increased. Moreover, several models underpredict the viscosity with the exception of the model proposed by Williams et al. [5].

The individual thermal conductivities of the base liquid k_f and nanoparticles k_p determine the value of the nanofluid k_{NF}. Another factor is the volume fraction ϕ of nanoparticles used. Many researchers have proposed different models to estimate the thermal conductivity of nanofluids. Different classical and recent models for estimating the effective value of a nanofluid's thermal conductivity are summarized in Table 14.2.

Fig. 14.2 below shows the results for a dimensionless model for estimating the k_{NF}.

Based on the previous literature, the thermal conductivity is determined for different nanofluids. The goal is to enhance this factor as much as possible using a variety of factors summarized in Fig. 14.3.

The effect of these parameters on thermal conductivity of nanofluids is discussed in forthcoming sections.

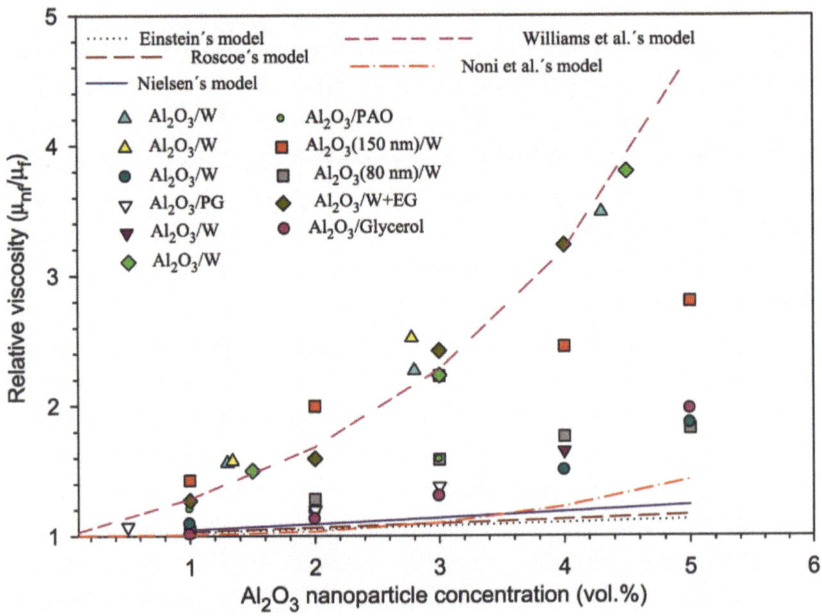

FIGURE 14.1 A comparison of experimental and model predictions for determining the viscosity of Al_2O_3 nanofluids as a function of nanoparticle concentration [17].

TABLE 14.2 Mathematical models for estimating thermal conductivities of different types of nanofluids using theoretical and experimental approaches.

Authors	Equation	Comments
Theoretical Models		
Hamilton and Crosser [18]	$\frac{k_{NF}}{k_f} = \frac{k_p + (n-1)k_f + (n-1)\phi(k_p - k_f)}{k_p + (n-1)k_f - \phi(k_p - k_f)}$	$\frac{k_p}{k_f} > 100$
Davis [19]	$\frac{k_{NF}}{k_f} = 1 + \frac{3(\alpha - 1)}{(\alpha + 2) - \phi(\alpha - 1)}\left[\phi + f(\alpha)\phi^2 + 0(\phi^3)\right]$	$\alpha = \frac{k_p}{k_f}$
Lu and Lin [20]	$\frac{k_{NF}}{k_f} = 1 + a\phi + b\phi^2$	Spherical and nonspherical particles
Prasher et al. [21]	$\frac{k_{NF}}{k_f} = \left(1 + A\mathrm{Re}^m \mathrm{Pr}^{0.333}\phi\right)\left[\frac{k_p + 2k_f + 2\phi(k_p - k_f)}{k_p + 2k_f - \phi(k_p - k_f)}\right]$	$n = 3$, A and m are constants.
Xue [22]	$\frac{k_{NF}}{k_f} = \frac{1 - \phi + 2\phi\frac{k_p}{k_p - k_f}\ln\frac{k_p + k_f}{2k_f}}{1 - \phi + 2\phi\frac{k_f}{k_p - k_f}\ln\frac{k_p + k_f}{2k_f}}$	Nanospheres with interfacial shell
Experimental Models		
Li and Peterson [23]	$\frac{k_{NF} - k_f}{k_f} = 0.764\phi + 0.0187(T - 273.15) - 0.462$ $\frac{k_{NF} - k_f}{k_f} = 3.761\phi + 0.0179(T - 273.15) - 0.307$	Al_2O_3 nanoparticles in water CuO nanoparticles in water
Chandrsekar et al. [24]	$\frac{k_{NF}}{k_f} = \frac{k_p + (n-1)k_f + (n-1)(1+\beta)^3\phi(k_p - k_f)}{k_p + (n-1)k_f - (1-\beta)^3\phi(k_p - k_f)} + \frac{C\phi(T - T_0)}{\mu_f k_f a^4}$	Al_2O_3, CuO and TiO_2 nanoparticles in water, TiO_2 nanoparticles in EG β: ratio of nanolayer thickness to particle radius a. C is constant
Patel et al. [25]	$\frac{k_{NF}}{k_f} = \left(1 + 0.135\left(\frac{k_p}{k_f}\right)^{0.273}\phi^{0.467}\left(\frac{T}{20}\right)^{0.547}\left(\frac{100}{d_p}\right)^{0.234}\right)$	Metallic and oxide nanofluids
Godson et al. [26]	$\frac{k_{NF}}{k_f} = 0.9692\phi + 0.9508$	Ag nanoparticles in water

(Continued)

TABLE 14.2 (Continued)

Authors	Equation	Comments
Corcione [27]	$\frac{k_{NF}}{k_f} = 1 + 4.4\mathrm{Re}^{0.4}\mathrm{Pr}^{0.66}\left(\frac{T}{T_{fr}}\right)^{10}\left(\frac{k_p}{k_f}\right)^{0.03}\phi^{0.66}$	Al_2O_3 nanoparticles in water
Hosseini et al. [28]	$\frac{k_{NF}}{k_f} = 1 + J\left(\frac{\phi^a}{d_p^{\,b}}\right)\left(\frac{T}{T_0}\right)^c$	J, a, b, and c are constants.

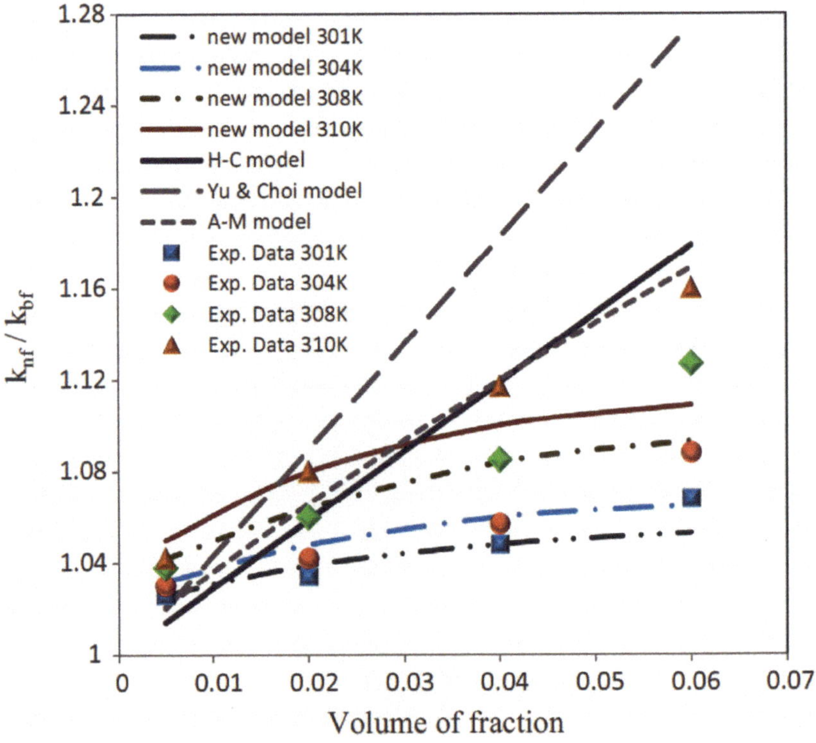

FIGURE 14.2 Comparative investigation by Hosseini et al. [28] referred to as a new model with experimental data [29] for 47 nm nanoparticles and various volume fractions for k_{NF} value.

14.2.2 Effect of nanoparticle concentration

Generally, the addition of well-dispersed, minute quantity of nanoparticles in base fluids results in an increase of thermal conductivity of nanofluids. Upon the effect of particle loading, it is found that higher nanoparticle concentration leads to improved thermal conductivity of nanofluids. For example,

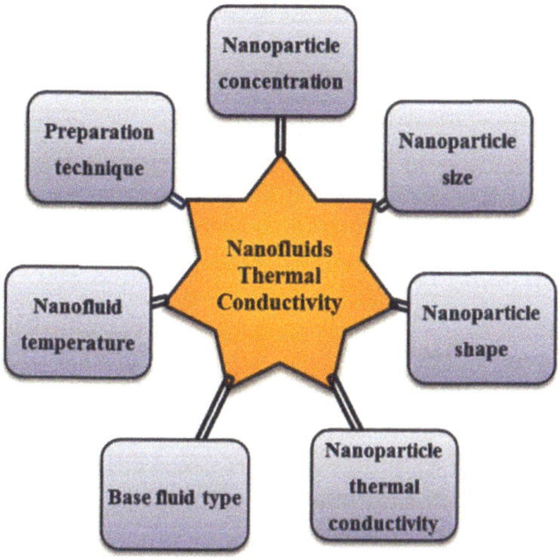

FIGURE 14.3 Various factors that influence the thermal conductivity of nanofluids [30].

Kang et al. [31] reported that by increasing particle concentration from 0.1% to 0.39%, the thermal conductivity of silver/water nanofluid is improved from 3% to 11%. The rising trend may behave linearly or nonlinearly depending on the type of suspension and base fluid. A study by Xie et al. on SiC nanoparticles revealed a linear increase of thermal conductivity from 1%–4% volume fraction range [32]. Another study by Liu et al. on MWCNTs at less than 1% volume fraction also shows an almost linear trend of increasing thermal conductivity [33]. Using TiO_2 nanoparticles, Murshed et al. discovered a nonlinear relationship at volumetric loading of 0.005–0.02 and a linear relationship from 0.02–0.05 for the rise in thermal conductivity [34]. Some studies reporting a nonlinear trend were by Hong et al. using Fe nanofluids from 0.2% to 0.6% volume fraction [35], Wen and Ding using CNT nanofluid from 0% to 1% volume concentration [36], and TiO_2 from 1% to 3% weight percentage [37]. It is also important to note that the increasing rate of thermal conductivity with higher volume fraction is different for various studies on the same type of nanoparticles. Among the existing literature, the majority of works conclude a linear increase with added particle concentration while some reveal an increase to a plateau at certain threshold concentrations, followed by a decline [38].

14.2.3 Effect of nanoparticle size

One of the critical factors is the overall dimension of a nanoparticle used for achieving better thermal characteristics. Nanoparticle size contributes not only to stability but also to the effective thermal conductivity. In most of the studies, thermal conductivity is inversely proportional to the nanoparticle size. Yet this is applicable for metal nanoparticles, as well as oxides/nonmetal nanoparticles. For a quantitative evaluation about the size effect, to offer the same thermal conductivity for a 35 nm particle nanofluid having 0.052% concentration, a 100 nm particle size nanofluid would need to increase its concentration to 7.5%. This effect could be generally attributed to higher Brownian motion and surface effect in smaller particles. However, it is not conclusive at this point. The inverse relationship was confirmed in many studies, such as by Mintsa et al. using alumina and copper oxide nanofluids [39], Chon et al. using alumina nanofluid with size ranging from 11 to 150 nm [40], Prasher et al. with particle radius ranging from 7 to 20 nm [41], and Chopkar et al. using $Al_{70}Cu_{30}$ from 0 to 80 nm crystalline size [42]. However, other factors such as particle shape tend to deviate from this inverse trend. For instance, multiwall CNTs and cylindrical shapes with large size tend to have a positive influence on thermal conductivity compared with that of smaller spherical shapes [32,43]. Similarly, CuO (33 nm) showed higher thermal conductivity compared to SiO_2 (12 nm) with the same base fluid [43].

14.2.4 Effect of nanoparticle shape

Analysis of particle shape is difficult since it is infeasible to compare different shapes based on the sizes/geometry. Therefore, few approximations based on experimental data are summarized here for comparison of similar sizes. The influence of particle shape on the thermal conductivity was investigated by Jeong et al. [44], achieving an additional increase of 6% thermal conductivity by varying the ZnO nanoparticles from spherical to nearly rectangular at 5% volume fraction. Similar results were observed where the thermal conductivity of cylindrical shapes outperformed that of spherical ones. For instance, TiO_2 nanoparticles in rod shape (10 nm $\times$ 40 nm, diameter by length) have higher thermal conductivity compared to 15 nm (diameter) spherical ones [34]. Moreover, thermal conductivity also increased by increasing the length-to-diameter ratio for nanotubes in suspension [45]. In another study, using multiwall CNTs showed a nonlinear increase in thermal conductivity when compared to the available model predictions [46]. However, a few results also showed negligible influence upon different shapes.

14.2.5 Effect of nanoparticle thermal conductivity

On average, the thermal conductivity of the nanoparticles also impacts the thermal conductivity of the nanofluid to some extent. However, it should be noted that the actual values of both conductivities are completely different. Based on the thermal conductivity enhancement level, three different categories are identified. The first category shows the highest thermal conductivity in association with the *advanced structural materials* such as carbon nanotubes, graphene, and the like. Followed by the second category with the *metallic nanoparticles* such as Ag, Au, Cu, Fe, and many more, and high thermal conductivity prevails in nanofluids. Finally, the third category of nanofluid contains the *oxides nanoparticles* such as TiO_2, CuO, Al_2O_3, ZnO, SiC, SiO_2, etc. [47]. Normally their attained thermal conductivities are inferior to the prior categories.

14.2.6 Effect of base fluid

The thermal conductivity of the nanofluid is not proportional to the thermal conductivity of the base fluid irrespective of shape of nanoparticles as reported by most of the published literature. Meaning that higher overall *enhancement* prevails for the greater *disparity* among the nanoparticle and base fluid thermal conductivity. This was confirmed by various studies among base fluids for EG and water. Hwang et al. compared the thermal conductivity of MWCNTs in water ($k_{f\_w} = 0.613$ W/mK) and oil ($k_{f\_oil} = 0.107$ W/mK) [43]. The results showed an increase of 7% in water and 8.7% in oil at volume fractions of 1% and 0.5%, respectively, indicating that lower thermal conductivity in base fluid, such as oil, led to a greater rise even at low volume fraction. Similar results were observed by Xie et al. when comparing thermal conductivity values of alumina nanoparticles in water ($k_{f\_w} = 0.604$ W/mK), EG ($k_{f\_EG} = 0.258$ W/mK), and pump oil ($k_{f\_oil} = 0.141$ W/mK) [48]. The results showed higher thermal conductivities for all nanofluids, however, in increasing order of water, then EG, and greatest enhancement for pump oil, respectively. It is important to note that the enhancement is a comparison of the nanofluid with its base fluid value, and not its absolute value. The absolute value of these nanofluids will be closer to their base fluid values, meaning water as a base fluid would still have higher absolute value of thermal conductivity compared to oil. Some results have also shown the opposite trend [33]. The recent trend of mixing two base fluids also reveals different results. For instance, EG–water mixture with 50:50% mixing ratio showed improved thermal conductivity compared to using only conventional base fluids.

14.2.7 Effect of nanofluid temperature

Many studies have investigated the effect of temperature on improving the thermal conductivity of nanofluids. Temperature has a direct influence on the Brownian motion of nanoparticles. Moreover, it also affects the clustering of nanofluids. An increase in Brownian motion would result in enhancement in thermal conductivity. By contrast, the clustering imposes a negative impact on Brownian motion. Hence, an increase in temperature may not always result in a positive outcome. Abareshi et al. [49] reported an increase in thermal conductivity from 2.8% to 8.9% with a temperature rise from 10°C to 40°C for Fe_3O_4 (1%) in water. Patel et al. [50] also showed that the temperature and thermal conductivity of nanofluids are directly proportional to each other. This was later confirmed by many other reports. Temperature rise revealed a nonlinear enhancement in thermal conductivity by using nanotubes [51,52]. Moreover, the thermal conductivity follows similar trend to that of the base fluid [53]. This was concluded by other studies in water, EG, and EG–water mixtures [54,55]. However, some studies showed an opposite trend despite having small dependence on temperature [56]. To summarize, the Brownian motion, clustering, and dispersion stability of nanofluids are affected by temperature which makes the prediction complex for determining the role of temperature to enhance thermal conductivity.

14.2.8 Effect of preparation technique

As discussed in the previous section, the nanofluid dispersion technique plays a vital role in not only its stability but also the thermal conductivity enhancement since it affects the morphological properties of the nanoparticles [57,58]. Two additional important factors will be discussed here, that is, the pH value of the nanofluid and the sonication time.

Zhu et al. [59] found that the thermal conductivity is improved by increasing pH value and peaks around 8–9, followed by a steady decline for further raising the pH value. This tendency is also confirmed by Xie et al. [48]. Moreover, it is also seen that the addition of a chemical dispersant along with pH control of suspension results in improvement of the thermal conductivity. Dispersant types such as Nanosperse AQ, potassium oleate surfactant (OAK^+), or oleic acid resulted in improved working properties and heat transfer rates [30,60].

Sonication is the application of sound energy to mix particles in a sample, such as producing nanofluids using nanoparticles. It results in even dispersion of nanoparticles in fluids and can also break down microsized aggregates. The amount of time for sonication (homogenization) plays a critical role for its long-term stability as a suspension. Khedkar et al. [61] noticed a direct relation of improvement in thermal conductivity with an increase in sonication time up to a certain limit. This was attributed to higher Brownian motion of nanoparticles and agglomeration. Hence, a moderate sonication time is preferred since a smaller time can lead to untangled nanoparticle bundles, whereas a longer time can cause the breakdown of agglomerates. For instance, an increase in thermal conductivity occurs up to 50 min of sonication, after which a saturation state is observed [35]. Buonomo et al. [62] reported a Nano-Flash method for measuring thermal conductivity and also reported a minimum sonication time for obtaining a stable nanofluid suspension.

14.2.9 Summary

In conclusion, the thermal conductivity improves if volume fraction of nanoparticles is increased to a certain limit, keeping in mind the viscosity of the nanofluid. The enhancement is also linked to an increase in temperature. However, the particle size shows an opposite trend with higher thermal conductivities at smaller particle size at the nanometer scale. The dynamic viscosity of nanofluids decreases with an increase in temperature but increases with the volume concentration of nanoparticles. Nanofluids also exhibit non-Newtonian behavior and in some cases viscoelasticity. However, each of these studies have limited application range, as described in previous sections. Therefore, the readers are recommended to carefully choose the correct correlation based on the various factors of temperature, volume concentration, and the size and type of nanoparticles chosen for achieving the desired thermal conductivity range.

14.3 Flow of nanofluids in microchannels

Two important thermofluid characteristics need to be considered in a MCHS; namely the required pumping power to overcome viscous effects, and high heat transfer coefficient to facilitate high heat dissipation. Therefore, the thermal performance can be improved by in-depth understanding of these two processes. The heat transfer rate is related to the surface area to volume ratio, meaning a smaller channel size would result in an improved heat transfer coefficient. Additionally, improving the thermal conductivity of the fluid (by manipulating nanoparticles) also results in a higher heat transfer coefficient.

Morini summarized the experimental investigations on convective heat transfer for single-phase flow in microchannels [63]. It was concluded from many studies that both the friction factor and the Nusselt number in microchannels showed inconsistent behavior compared to conventional theory. The Fanning friction factor f is defined as the ratio of the wall shear stress τ_w to the flow kinetic energy per unit volume:

$$f = \frac{\tau_w}{1/2\rho\overline{V}^2}$$

Here $\overline{V}$ is the average fluid velocity. This Fanning friction factor is linked to the pressure drop by:

$$f = \frac{\Delta P}{2\rho\overline{V}^2}\frac{D_h}{L}$$

Here, ΔP is the pressure drop, L is the length of channel, and D_h is the hydraulic diameter. For fully developed laminar flow in a round tube $f = \frac{16}{\mathrm{Re}}$. For turbulent regime, several different models can be used.

Another important aspect is the developing flow in microchannels. Since only short lengths are investigated in microchannels, it is quite possible that the developing region is dominant rather than the fully developed region. A nondimensional flow distance to characterize the entrance region is used:

$$x^+ = \frac{x}{D_h \mathrm{Re}}$$

where x^+ represent the nondimensional flow distance and x is the location in the axial flow direction. The flow is considered fully developed for a value of $x^+ = 0.05$. In order to determine the effect of the entire channel, the developing region adopts the so-called apparent friction factor f_{app}.

$$\Delta P = \frac{2(f_{app}\mathrm{Re})\mu \overline{V} x}{D_h^2}$$

The two components for friction factor are the fully developed region of flow and the pressure defect region given by the Hagenbach factor κ_H in the following equation:

$$\Delta P = \frac{2(f\mathrm{Re})\mu VL}{D_h^2} + \frac{\kappa(x)\rho_f V^2}{2}$$

The Hagenbach factor starts from 0 to $\kappa_H(\infty)$ for fully developed flow and depends on the aspect ratio of channel α_c. For a rectangular channel, the following equation determines the Hagenbach factor with 0.04% accuracy:

$$\kappa(\infty) = 0.6796 + 1.2197\alpha_c + 3.3089\alpha_c^2 - 9.5921\alpha_c^3 + 8.9089\alpha_c^4 - 2.9959\alpha_c^5$$

Moreover, the entrance and exit losses must also be considered when developing the overall pressure loss equation. Hence, the pressure drop consists of the entrance, exit, developing, and fully developed regions by the respective friction factor equations.

$$\Delta P = \frac{\rho \overline{V}^2}{2}\left[\kappa_i + \kappa_o + \frac{f_{app}L}{D_h}\right]$$

Here, κ_i and κ_o are the inlet and outlet loss coefficients.

Since the liquid flow is in the continuum regime, the Navier–Stokes equations, along with continuity, momentum, and energy equations are valid for microchannels. Moreover, it is also important to note that the much higher pumping power prevails at turbulent flow. Hence, the laminar flow regime is more preferred in typical applications.

A number of studies have been conducted on determining the flow performance of nanofluids on MCHS. Kuppusamy et al. [64] reported the flow characteristics for triangular grooved MCHS. Various types of nanoparticles (alumina, silica, CuO, ZnO), subject to volume fraction (0.01–0.04), particle diameter (25–80 nm), and types of base fluids (water, EG, and engine oil) were investigated. Their results indicated an enhancement of heat transfer at the expense of increasing the pressure drop. Table 14.3 tabulates the hydraulic characteristics such as friction factors for various channel geometries used. Note that Pourmehran et al. [65] used a fin-shaped MCHS with Al_2O_3–water and CuO–water as nanofluid.

As seen from Fig. 14.4, the pressure difference increases with increasing Reynolds number. Hence, more pumping power is required by raising the flow velocity [70]. Moreover, the increase in particle concentration also leads to a minor increase in pressure drop since it is directly related to the viscosity of the nanofluid. This result indicates a significant benefit of using nanofluid where the particle concentration increases with a very minor penalty of pressure drop. Similarly, when compared to water, the pumping power is only moderately higher, indicating its superior performance in MCHS applications.

14.4 Thermal performance of nanofluids in microchannels

As mentioned earlier, the friction coefficient f is the main flow parameter to characterize pressure drop penalty. For heat transfer performance, the Nusselt number Nu is used to evaluate the convective heat transfer performance. In practice, Nu is a function of Re, Pr, x^* (spatial coordinate), and geometry configuration.

$$Nu = f(x^*, Re, Pr)$$

Radiation is normally neglected in the modeling of nanofluid. However, Puttaswmay et al. [71] considered the nonlinear thermal radiation for heat transfer rate in a hybrid nanofluid in a vertical microchannel. Their obtained numerical results demonstrated that radiation parameter and nanoparticle volume fraction reduce the thermal energy of the hybrid nanofluid. For characterizing the performance of the heat and mass transfer behaviors, some commonly used dimensionless parameters are listed in Table 14.4.

Different researchers use different correlations for their experimental data based on their test configurations. However, Nusselt number is the most commonly used performance parameter to represent the heat transfer performance in the literature. Different Nusselt number correlations developed by researchers for thermal performance enhancement along with the quantified improvements in heat transfer are tabulated in Table 14.5.

After investigating variations in Nu for any given configuration the overall thermal performance can be evaluated. There is no linear or direct relationship of thermal performance that can be generalized for any specific type of nanofluid or microchannel configuration. The thermal parameters are indirectly connected to the flow parameters that are well explained above and it can be generalized that the decrease in flow parameters of the working fluid will enhance the thermal performance of the microchannel-based

TABLE 14.3 Hydraulic characteristics of nanofluids in microchannel heat sinks (MCHS).

Authors	Channel geometry	Model
Kuppusamy et al. [64]	Triangular grooved MCHS	$\Delta P = \frac{2(fRe)\mu VL}{D_h} + \frac{\kappa(x)\rho_f V^2}{2}$ $\kappa(\infty) = 0.6796 + 1.2197\alpha_c + 3.3089\alpha_c^2 - 9.5921\alpha_c^3 + 8.9089\alpha_c^4 - 2.9959{\alpha_c}^5$ $fRe = 96(1 - 1.3553\alpha_c + 1.9467\alpha_c^2 - 1.7012\alpha_c^3 + 0.9564\alpha_c^4 - 0.2537\alpha_c^5)$
Pourmehran et al. [65]	Fin shaped MCHS	$f = \frac{2\Delta P}{\rho \overline{V}^2}\frac{D_h}{L} = \frac{96}{\text{Re}}\left(\frac{\alpha_c}{1+\alpha_c}\right)$ *where* $\alpha_c = \frac{h_{ch}}{w_{ch}}$ *(Aspect ratio)*
Hatami and Ganji [66]	Fin-shaped MCHS	$f = h_a\left[\frac{T_s - T_{nf}}{\frac{qwL}{1-\varepsilon k_s}}\right]^p$, *where* $h_a = \frac{\overline{\text{Nu}}k_{nf}}{D_h}$ h_a convection heat transfer coefficient, ε is porosity, subscript s means solid, p power law index
Sakanova et al. [67]	Rectangular double layer and double side MCHS	$f = \frac{\Delta P}{2\rho \overline{V}^2}\frac{D_h}{L}$
Li et al. [68]	Triangular cavities and rectangular ribs	$f = \frac{2\Delta P}{\rho \overline{V}^2}\frac{D_h}{L}$
Ho et al. [69]	Rectangular microchannels	$f = \frac{\Delta P}{2\rho \overline{V}^2}\frac{D_h}{L}$

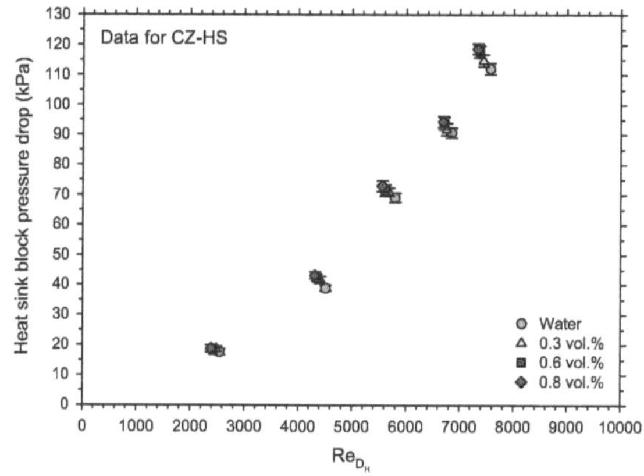

(A) data for CZ-HS

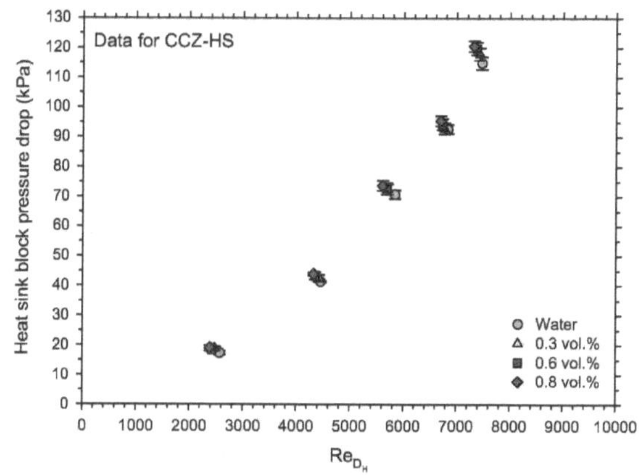

(B) data for CCZ-HS

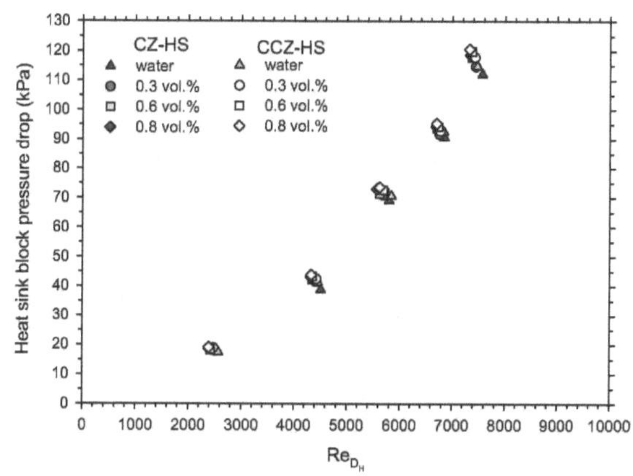

(C) effect of single cross-cut flow channel

FIGURE 14.4 Comparison of heat sink pressure drop for (A) continuous zigzag flow channel (CZ-HS) (B) single crosscutting zigzag flow channel (CCZ-HS) and (C) combined CZ-HS and CCZ-HS for effect of single cross-cut flow channel, required for SiO_2-water nanofluid against Reynolds number and varying particle volume fraction [70].

TABLE 14.4 Important dimensionless numbers for analysis in heat and mass transfer.

Name	Correlation
Biot number	$Bi = \frac{hL}{k_s}$
Mass transfer Biot number	$Bi_m = \frac{h_m L}{D_{AB}}$
Bond number	$Bo = \frac{g(\rho_l - \rho_v)L^2}{\sigma}$
Coefficient of friction	$C_f = \frac{\tau_s}{\rho V^2}$
Eckert number	$Ec = \frac{V^2}{C_p(T_s - T_\infty)}$
Fourier number	$Ec = \frac{\alpha l}{L^2}$
Mass transfer Fourier number	$Fo_m = \frac{D_{AB}t}{L^2}$
Grashof number	$Gr_L = \frac{g\beta(T_s - T_\infty)L^3}{v^2}$
Colburn *j* factor	$j_H = StPr^{2/3}$
Colburn *j* factor	$j_m = St_m Sc^{2/3}$
Jacob number	$\frac{c_p(T_s - T_{sat})}{h_{fg}}$
Lewis number	$L_e = \frac{\alpha}{D_{AB}}$
Nusselt number	$Nu_L = \frac{hL}{k_f}$
Schmidt number	$Sc = \frac{v}{D_{AB}}$
Sherwood number	$Sh_L = \frac{h_m L}{D_{AB}}$
Stanton number	$St = \frac{h}{\rho V c_p} = \frac{Nu_L}{Re_L Pr}$
Mass transfer Stanton number	$St_m = \frac{h_m}{V} = \frac{Sh_L}{Re_L Sc}$
Weber number	$We = \frac{\rho V^2 L}{\sigma}$

device. The question to explore in this case will be to quantify the enhancement in thermal performance at the expense of a given flow degradation. However, entropy generation is another parameter being analyzed in parallel to evaluate the sustainability of a given cooling device.

14.5 Entropy analysis of nanofluid-based microchannel heat sinks

As the nanofluid-based heat exchangers work in a closed cycle, the sustainability and usability of the enclosed working fluid must be considered.

TABLE 14.5 Nusselt number correlations and enhancement for various nanofluid types.

Authors	Correlations formed	*Nu* enhancement	Configurations and nano fluid types
Kumar et al. [72]	$Nu_{Reg} = 0.02327 Re^{0.8} Pr^{0.4} (1+\varphi)^{1.969}$ $16,000 \leq Re \leq 30,000$	14.7% increment at 0.06% volume fraction	Experimental Fe_3O_4/ distilled water
Arani and Amani [73]	$Nu_{NF} = 0.006 Re^{0.86} Pr \varphi^{0.35} (d_{p,c}/d_f)^{0.1}$ where $d_{p,c} = -90 + 18.667 d_p - 0.75 d_p^2 + 0.00833 d_p^3$ $d_f = 0.396 nm, d_p = 10, 20, 30, 50\ (nm)$ $10,000 \leq Re \leq 50,000$	6% deviation	Experimental TiO_2/ distilled water
Arani and Amani [74]	$Nu_{Re} = 0.0041 Re^{0.83} Pr^{1.35} (1 + \varphi^{9.836})$ $8000 \leq Re \leq 51,000$	–	Experimental TiO_2/ distilled water
Jafarimoghaddam and Aberoumand [75]	$Nu = 1.81 \mathrm{Re}^{0.141} Pr^{0.7835} (0.003\varphi + 0.5)$ $Re < 160$	Heat transfer coefficient from 15.3%–36.7% enhancement	Experimental Cu/EG-nanofluid

Along with thermal and hydraulic performance, the entropy generation of the device is also analyzed closely for assessing effective life cycle. The degradation of the nanofluid and fouling of surface channels from external contamination will result in additional entropy generation of the device, and accordingly degrade the overall efficiency. One cause of entropy generation is fouling which is the sedimentation of unwanted material along with nanoparticles agglomeration on the walls, therefore resulting in decreased thermal and hydraulic performance over time. This sedimentation may be removed by changing operating conditions to restore the MCHS back to its initial conditions, but its long-time effect may still deteriorate and offset the system performance during cyclic operation. The life cycle of fouling is shown in Fig. 14.5.

A lot of work has been recently carried out to quantify the degradation of the recharging microchannels using nanofluids. Samal et al. [76] used water-based graphene–silver hybrid nanofluid and studied the entropy generation behavior as the thermal and hydraulic performance enhances. Monaledi and Makinde [77] found that nanoparticles near the walls can increase the entropy generation while those near the center line have minimal effect on entropy generation. Mosavi et al. [78] studied ferrofluid flow in the microchannel with twisted porous ribs and found that triple layered channels performed better in terms of total entropy generation. Karimipour et al. [79] analyzed the injection of nanofluids in a magnetic field and observed an 82.8% increase in entropy generation. All prior studies show that the entropy generation for cyclic nanofluid plays a key role to define the estimated life

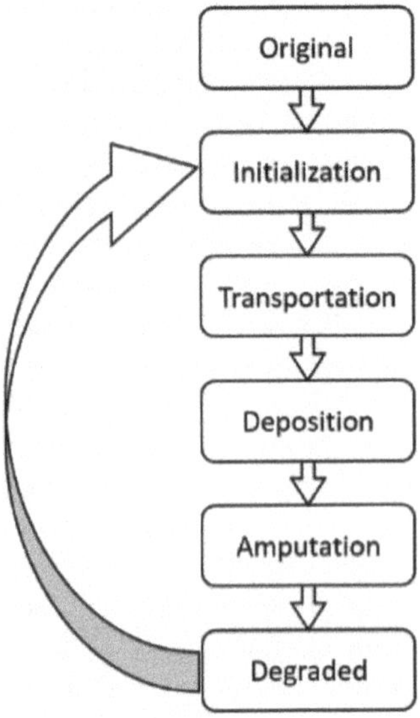

FIGURE 14.5 Entropy generation due to fouling in microchannels.

of a microchannel device. It also has added a new parameter to consider along with hydrothermal performance and pumping power.

14.6 Geometry effect of microchannels

After selection of the working fluid, nanoparticles, and other parameters, the performance enhancement can easily be attainable by changing the fluid flow environment. This is facilitated through changing the geometry of microchannels. This geometry variation can be broadly divided into three categories:

1. Changing cross-sectional shape of channels
2. Adding geometrical inclusions/exclusions in the cross-section
3. Varying the path of the fluid

The basic objective of doing any of the above methods is to increase the opportunity for the nanofluid to interact with the microchannel walls resulting in increased heat transfer, as explained earlier. This can cause flow disruption, obstruction, or surface roughness. This performance enhancement is achieved passively but an increase in pressure drop eventually causes more pumping power.

14.6.1 Changing shape of channels

The basic shape of MCHS is circular, and some modified geometries are tabulated in Table 14.6.

14.6.2 Adding geometrical inclusions/exclusions in the cross section

Similarly, adopting flow obstruction in the MCHS for some examples is listed in Table 14.7.

14.6.3 Varying path of the fluid

In this type of flow disruption, the overall path of the microchannel is varied to gain longer contact surface area and contact time for the nanofluids. Some path shapes and their examples are shown in Table 14.8.

These shapes are provided as available options for the reader while further variations, combinations, and addition of complexities in these shapes can provide increments/decrements in the overall behavior of the MCHS.

By selecting a specific geometry and pathway of the microchannel, an optimized dimension is sought for the best performance of the heat sink for the designed parameters. The optimization can give an increase up to 50% or sometimes even more in the performance indicators. Generally the overall

TABLE 14.6 Various microchannel cross-sectional shapes investigated by different researchers for use in MCHS applications using nanofluids.

Authors	Shape name	Example sketch
Memon et al. [80]	Rectangular	
Memon et al. [80]	Triangular	
Mohammed et al. [81]	Trapezoidal	
Ghani et al. [82]	Slanted	
Pourmehran et al. [65]	Fin shaped	
Roshani et al. [83]	Miniature plate	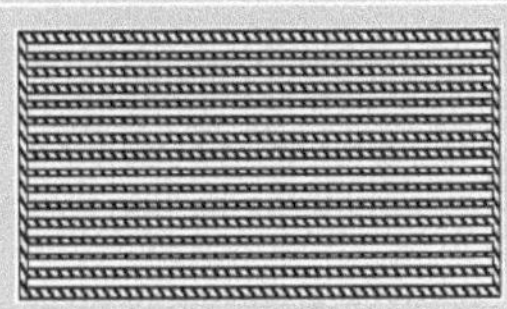

performance of MCHS is enhanced by decreasing the aspect ratio and increasing the channel width ratio for any given geometry. It can be concluded that varying the cross section and path shapes of microchannels and introducing inclusions/exclusions in the fluid path cast a positive effect on the heat transfer enhancement of MCHS.

14.7 Future advances and challenges

Some of the difficulties and possibilities for future research in the field of heat transfer are mentioned in this section for possible improvements in the use of

TABLE 14.7 Inclusion/exclusion geometries used for heat transfer enhancement applications.

Author(s)	Inclusion/exclusion	Geometry
Li et al. [68]	Rib	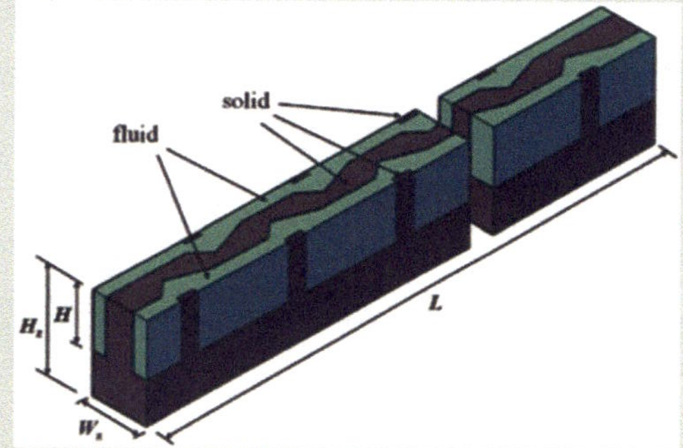
Khodabandeh et al. [84]	Cavity	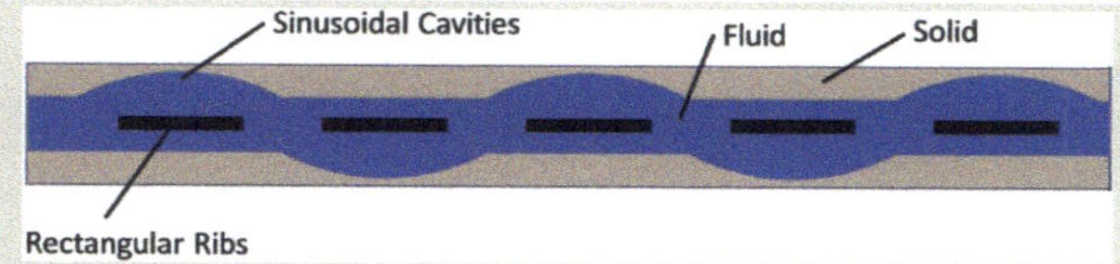
Rehman et al. [85]	Dimple/protrusions	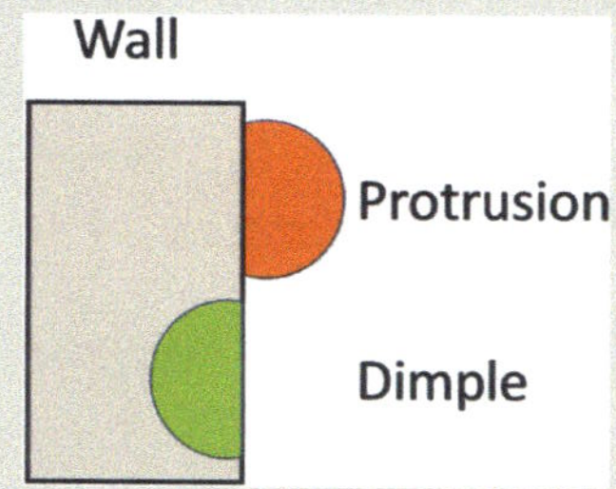

(*Continued*)

TABLE 14.7 (Continued)

Author(s)	Inclusion/exclusion	Geometry
Abouali and Baghernezhad [86]	Grooved	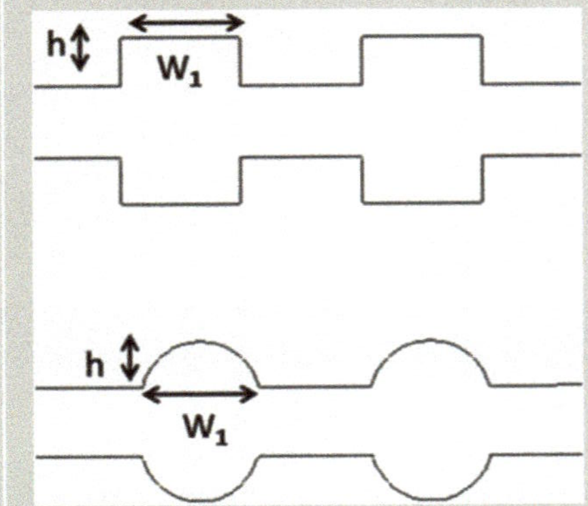
Hosseinirad and Hormozi [87]	Finned	
Nguyen et al. [88]	Corrugated	

TABLE 14.8 Various fluid path of microchannels for heat transfer enhancement.

Authors	Shape Name	Schematic
Lyu et al. [89]	Fractal	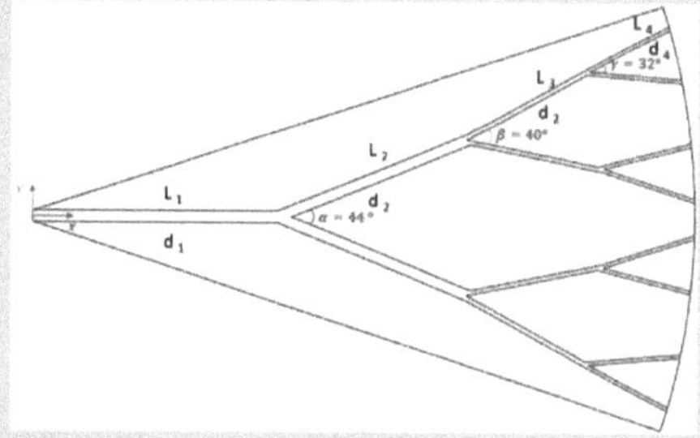
Srivastava et al. [90]	Layered	
Bazdar et al. [91]	Wavy	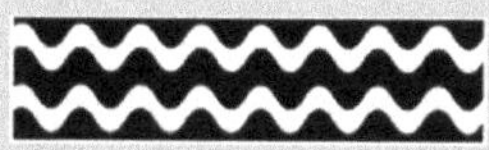
Schmid et al. [92]	Free hanging with base closed	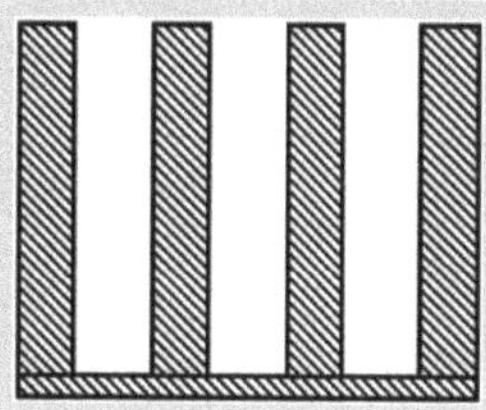
Wangikar et al. [93]	Serpentine	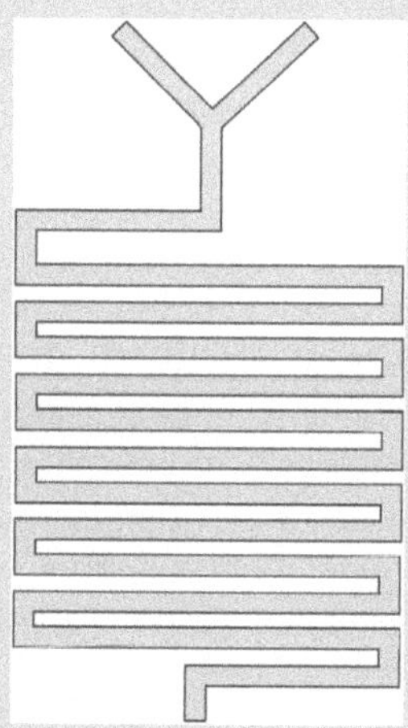

nanofluids. First, the preparation of nanofluids is costly and complicated, thus making its mass production a difficult task. Also, during its preparation, additional research is required to determine long-term stability of nanoparticles with influencing factors, such as sonication time, the type of nanofluid, and its volume fraction, to reduce agglomeration and sedimentation in the microchannels. Additionally, more work is required to improve the classical models for the prediction of thermal and flow properties for quicker analysis. It is also important to note that the protocol for analysis must be the same for all the investigations for better comparison among the literature. The modern statistical models of machine learning and algorithms must be carefully chosen to analyze thermal and flow prediction in a much faster and more accurate way.

Future studies can be related to many factors affecting the performance of nanofluids in microchannels. For instance, the impact of surface tension needs to be investigated in detail as it might lead to a stable version of a nanofluid. Similarly, the high polar solvents should be studied in detail to determine the influence on thermal conductivity of nanofluids. The effect of additives with an aim to improve thermal performance and decrease the viscosity of nanofluids is another important area of research. The operational life of MCHS is also an important factor to be considered, which is related to the total number of cycles completed by the working fluid. The performance may deteriorate with time along with entropy generation which is another important factor that needs to be addressed. Complex geometries with extremely small size are hard to manufacture along with added cost. In service thermodynamic fatigue is an important factor deciding the future of MCHS. Eventually, cost-effective and accurate manufacturing processes are necessary in making nanofluids a successful commercial product in MCHS. Moreover, a simple and affordable way of making microchannels must be developed with the capability to operate with less pumping power and maintenance cost.

14.8 Conclusions

This chapter summarized the thermophysical properties of nanofluids, the measurement of viscosity and thermal conductivity, and comparative analysis of the flow and thermal performance of nanofluids in MCHS. This should provide the reader with enough knowledge to determine the type of nanofluid, and accurately predict the thermal performance based on the theoretical modeling provided for various channel geometries and types of nanofluids.

The thermal conductivity increases with volume fraction of nanoparticles. The rise in temperature also results in increase in thermal conductivity. However, the particle size shows an opposite trend with higher thermal conductivities at smaller particle size at the nanometer scale. The dynamic viscosity of nanofluids decreases with an increase in temperature but increases with the volume concentration of nanoparticles. Nanofluids also exhibit non-Newtonian behavior and in some cases viscoelasticity. It can be concluded that varying cross section and path shapes of microchannels and introducing inclusions/exclusions in the fluid path cast a positive effect on the heat transfer enhancement of MCHS.

The concluding remark for nanofluids is that nanofluids possess a promising future in electronic cooling devices as microchannels improve the surface area to volume ratio. Therefore, it would result in the reduction of cooling device size and improve the thermal performance at the same time. It is also shown that advanced nanoparticles and metal nanoparticles with higher thermal conductivities greatly improve the thermal performance of nanofluids. Lastly, the nanoparticle size and volume fraction, the channel shape, and dimensions play key roles in determining the friction factor for flow analysis and Nusselt number for thermal characteristics of the nanofluid in microchannels.

References

[1] M. Bahrami, M. Akbari, A. Karimipour, M. Afrand, An experimental study on rheological behavior of hybrid nanofluids made of iron and copper oxide in a binary mixture of water and ethylene glycol: non-Newtonian behavior, Exp. Therm. Fluid Sci. 79 (2016) 231–237.

[2] G. Huminic, A. Huminic, Application of nanofluids in heat exchangers: a review, Renew. Sustain. Energy Rev. 16 (2012) 5625–5638.

[3] Y. Xuan, W. Roetzel, Conceptions for heat transfer correlation of nanofluids, Int. J. Heat. Mass. Transf. 43 (2000) 3701–3707.

[4] S. Nabati Shoghl, J. Jamali, M. Keshavarz Moraveji, Electrical conductivity, viscosity, and density of different nanofluids: an experimental study, Exp. Therm. Fluid Sci. 74 (2016) 339–346.

[5] W. Williams, J. Buongiorno, L.-W. Hu, Experimental investigation of turbulent convective heat transfer and pressure loss of alumina/water and zirconia/water nanoparticle colloids (nanofluids) in horizontal tubes, J. Heat. Transf. 130 (2008).

[6] A. Einstein, Eine neue Bestimmung der Moleküldimensionen, Ann. Phys. 324 (1906) 289–306.

[7] H.C. Brinkman, The viscosity of concentrated suspensions and solutions, J. Chem. Phys. 20 (1952) 571.

[8] G.K. Batchelor, The effect of Brownian motion on the bulk stress in a suspension of spherical particles, J. Fluid Mech. 83 (2006) 97–117.

[9] R.S. Vajjha, D.K. Das, P.K. Namburu, Numerical study of fluid dynamic and heat transfer performance of Al_2O_3 and CuO nanofluids in the flat tubes of a radiator, Int. J. Heat. Fluid Flow. 31 (2010) 613–621.

[10] J. Yang, N. Zhao, Z. Li, C. Sun, A combined theory model for predicting the viscosity of water-based Newtonian nanofluids containing spherical oxide nanoparticles, J. Therm. Anal. Calorim. 135 (2019) 1311–1321.

[11] S.L.P. Donald, A. Drew, Theory of Multicomponent Fluids, 135, Springer-Verlag, New York, 1999.

[12] W.J. Tseng, K.-C. Lin, Rheology and colloidal structure of aqueous TiO2 nanoparticle suspensions, Mater. Sci. Eng. A 355 (2003) 186–192.

[13] S.E.B. Maïga, S.J. Palm, C.T. Nguyen, G. Roy, N. Galanis, Heat transfer enhancement by using nanofluids in forced convection flows, Int. J. Heat. Fluid Flow. 26 (2005) 530–546.

[14] S. Song, C. Peng, M.A. Gonzalez-Olivares, A. Lopez-Valdivieso, T. Fort, Study on hydration layers near nanoscale silica dispersed in aqueous solutions through viscosity measurement, J. Colloid Interface Sci. 287 (2005) 114–120.

[15] J. Garg, B. Poudel, M. Chiesa, J.B. Gordon, J.J. Ma, J.B. Wang, et al., Enhanced thermal conductivity and viscosity of copper nanoparticles in ethylene glycol nanofluid, J. Appl. Phys. 103 (2008) 074301.

[16] N. Masoumi, N. Sohrabi, A. Behzadmehr, A new model for calculating the effective viscosity of nanofluids, J. Phys. D: Appl. Phys. 42 (2009) 055501.

[17] S.M.S. Murshed, P. Estellé, A state of the art review on viscosity of nanofluids, Renew. Sustain. Energy Rev. 76 (2017) 1134–1152.

[18] R.L. Hamilton, O.K. Crosser, Thermal conductivity of heterogeneous two-component systems, Ind. Eng. Chem. Fundam. 1 (1962) 187–191.

[19] R.H. Davis, The effective thermal conductivity of a composite material with spherical inclusions, Int. J. Thermophys. 7 (1986) 609–620.

[20] S.Y. Lu, H.C. Lin, Effective conductivity of composites containing aligned spheroidal inclusions of finite conductivity, J. Appl. Phys. 79 (1996) 6761–6769.

[21] R. Prasher, P. Bhattacharya, P.E. Phelan, Thermal conductivity of nanoscale colloidal solutions (nanofluids), Phys. Rev. Lett. 94 (2005) 025901.

[22] Q.Z. Xue, Model for thermal conductivity of carbon nanotube-based composites, Phys. B: Condens. Matter 368 (2005) 302–307.

[23] C.H. Li, G.P. Peterson, Experimental investigation of temperature and volume fraction variations on the effective thermal conductivity of nanoparticle suspensions (nanofluids), J. Appl. Phys. 99 (2006) 084314.

[24] M. Chandrasekar, S. Suresh, R. Srinivasan, A.C. Bose, New analytical models to investigate thermal conductivity of nanofluids, J. Nanosci. Nanotechnol. 9 (2009) 533–538.

[25] H.E. Patel, T. Sundararajan, S.K. Das, An experimental investigation into the thermal conductivity enhancement in oxide and metallic nanofluids, J. Nanopart. Res. 12 (2010) 1015–1031.

[26] L. Godson, B. Raja, D.M. Lal, S. Wongwises, Experimental investigation on the thermal conductivity and viscosity of silver-deionized water nanofluid, Exp. Heat. Transf. 23 (2010) 317–332.

[27] M. Corcione, Rayleigh-Bénard convection heat transfer in nanoparticle suspensions, Int. J. Heat. Fluid Flow. 32 (2011) 65–77.

[28] S.M. Hosseini, M.R. Safaei, M. Goodarzi, A.A.A.A. Alrashed, T.K. Nguyen, New temperature, interfacial shell dependent dimensionless model for thermal conductivity of nanofluids, Int. J. Heat. Mass. Transf. 114 (2017) 207–210.

[29] C.H. Li, G.P. Peterson, The effect of particle size on the effective thermal conductivity of Al_2O_3-water nanofluids, J. Appl. Phys. 101 (2007) 044312.

[30] M.M. Tawfik, Experimental studies of nanofluid thermal conductivity enhancement and applications: a review, Renew. Sustain. Energy Rev. 75 (2017) 1239–1253.

[31] H.U. Kang, S.H. Kim, J.M. Oh, Estimation of thermal conductivity of nanofluid using experimental effective particle volume, Exp. Heat. Transf. 19 (2006) 181–191.

[32] H. Xie, J. Wang, T. Xi, Y. Liu, Thermal conductivity of suspensions containing nanosized SiC particles, Int. J. Thermophys. 23 (2002) 571–580.

[33] M.-S. Liu, M. Ching-Cheng Lin, I.T. Huang, C.-C. Wang, Enhancement of thermal conductivity with carbon nanotube for nanofluids, Int. Commun. Heat. Mass. Transf. 32 (2005) 1202–1210.

[34] S.M.S. Murshed, K.C. Leong, C. Yang, Enhanced thermal conductivity of TiO_2—water based nanofluids, Int. J. Therm. Sci. 44 (2005) 367–373.

[35] T.-K. Hong, H.-S. Yang, C.J. Choi, Study of the enhanced thermal conductivity of Fe nanofluids, J. Appl. Phys. 97 (2005) 064311.

[36] D. Wen, Y. Ding, Effective thermal conductivity of aqueous suspensions of carbon nanotubes (carbon nanotube nanofluids), J. Thermophys. Heat. Transf. 18 (2004) 481–485.

[37] W. Dongsheng, D. Yulong, Natural convective heat transfer of suspensions of titanium dioxide nanoparticles (nanofluids), IEEE Trans. Nanotechnol. 5 (2006) 220–227.

[38] A. Ceylan, K. Jastrzembski, S.I. Shah, Enhanced solubility Ag-Cu nanoparticles and their thermal transport properties, Metall. Mater. Trans. A 37 (2006) 2033–2038.

[39] H.A. Mintsa, G. Roy, C.T. Nguyen, D. Doucet, New temperature dependent thermal conductivity data for water-based nanofluids, Int. J. Therm. Sci. 48 (2009) 363–371.

[40] C.H. Chon, K.D. Kihm, S.P. Lee, S.U.S. Choi, Empirical correlation finding the role of temperature and particle size for nanofluid (Al_2O_3) thermal conductivity enhancement, Appl. Phys. Lett. 87 (2005) 153107.

[41] R. Prasher, P.E. Phelan, P. Bhattacharya, Effect of aggregation kinetics on the thermal conductivity of nanoscale colloidal solutions (nanofluid), Nano Lett. 6 (2006) 1529–1534.

[42] M. Chopkar, P.K. Das, I. Manna, Synthesis and characterization of nanofluid for advanced heat transfer applications, Scr. Mater. 55 (2006) 549–552.

[43] Y. Hwang, J.K. Lee, C.H. Lee, Y.M. Jung, S.I. Cheong, C.G. Lee, et al., Stability and thermal conductivity characteristics of nanofluids, Thermochim. Acta 455 (2007) 70–74.

[44] J. Jeong, C. Li, Y. Kwon, J. Lee, S.H. Kim, R. Yun, Particle shape effect on the viscosity and thermal conductivity of ZnO nanofluids, Int. J. Refrig. 36 (2013) 2233–2241.

[45] X. Zhang, H. Gu, M. Fujii, Effective thermal conductivity and thermal diffusivity of nanofluids containing spherical and cylindrical nanoparticles, Exp. Therm. Fluid Sci. 31 (2007) 593–599.

[46] S.U.S. Choi, Z.G. Zhang, W. Yu, F.E. Lockwood, E.A. Grulke, Anomalous thermal conductivity enhancement in nanotube suspensions, Appl. Phys. Lett. 79 (2001) 2252–2254.

[47] L. Yang, J. Xu, K. Du, X. Zhang, Recent developments on viscosity and thermal conductivity of nanofluids, Powder Technol. 317 (2017) 348–369.

[48] H. Xie, J. Wang, T. Xi, Y. Liu, F. Ai, Q. Wu, Thermal conductivity enhancement of suspensions containing nanosized alumina particles, J. Appl. Phys. 91 (2002) 4568–4572.

[49] M. Abareshi, E.K. Goharshadi, S. Mojtaba Zebarjad, H. Khandan Fadafan, A. Youssefi, Fabrication, characterization and measurement of thermal conductivity of Fe_3O_4 nanofluids, J. Magn. Magn. Mater. 322 (2010) 3895–3901.

[50] H.E. Patel, S.K. Das, T. Sundararajan, A.S. Nair, B. George, T. Pradeep, Thermal conductivities of naked and monolayer protected metal nanoparticle based nanofluids: manifestation of anomalous enhancement and chemical effects, Appl. Phys. Lett. 83 (2003) 2931–2933.

[51] M. Xing, J. Yu, R. Wang, Experimental investigation and modelling on the thermal conductivity of CNTs based nanofluids, Int. J. Therm. Sci. 104 (2016) 404–411.

[52] A. Shahsavar, M.R. Salimpour, M. Saghafian, M.B. Shafii, An experimental study on the effect of ultrasonication on thermal conductivity of ferrofluid loaded with carbon nanotubes, Thermochim. Acta 617 (2015) 102–110.

[53] M.P. Beck, Y. Yuan, P. Warrier, A.S. Teja, The thermal conductivity of alumina nanofluids in water, ethylene glycol, and ethylene glycol + water mixtures, J. Nanopart. Res. 12 (2010) 1469–1477.
[54] W. Yu, H. Xie, L. Chen, Y. Li, Enhancement of thermal conductivity of kerosene-based Fe3O4 nanofluids prepared via phase-transfer method, Colloids Surf. A: Physicochem. Eng. Asp. 355 (2010) 109–113.
[55] H. Li, L. Wang, Y. He, Y. Hu, J. Zhu, B. Jiang, Experimental investigation of thermal conductivity and viscosity of ethylene glycol based ZnO nanofluids, Appl. Therm. Eng. 88 (2015) 363–368.
[56] W. Duangthongsuk, S. Wongwises, Measurement of temperature-dependent thermal conductivity and viscosity of TiO2-water nanofluids, Exp. Therm. Fluid Sci. 33 (2009) 706–714.
[57] X. Wang, X. Xu, S.U.S. Choi, Thermal conductivity of nanoparticle—fluid mixture, J. Thermophys. Heat. Transf. 13 (1999) 474–480.
[58] Y. Xuan, Q. Li, Heat transfer enhancement of nanofluids, Int. J. Heat. Fluid Flow. 21 (2000) 58–64.
[59] D. Zhu, X. Li, N. Wang, X. Wang, J. Gao, H. Li, Dispersion behavior and thermal conductivity characteristics of Al2O3–H2O nanofluids, Curr. Appl. Phys. 9 (2009) 131–139.
[60] T. Parametthanuwat, N. Bhuwakietkumjohn, S. Rittidech, Y. Ding, Experimental investigation on thermal properties of silver nanofluids, Int. J. Heat. Fluid Flow. 56 (2015) 80–90.
[61] R.S. Khedkar, S.S. Sonawane, K.L. Wasewar, Influence of CuO nanoparticles in enhancing the thermal conductivity of water and monoethylene glycol based nanofluids, Int. Commun. Heat. Mass. Transf. 39 (2012) 665–669.
[62] B. Buonomo, O. Manca, L. Marinelli, S. Nardini, Effect of temperature and sonication time on nanofluid thermal conductivity measurements by nano-flash method, Appl. Therm. Eng. 91 (2015) 181–190.
[63] G.L. Morini, Single-phase convective heat transfer in microchannels: a review of experimental results, Int. J. Therm. Sci. 43 (2004) 631–651.
[64] N.R. Kuppusamy, H.A. Mohammed, C.W. Lim, Thermal and hydraulic characteristics of nanofluid in a triangular grooved microchannel heat sink (TGMCHS), Appl. Math. Comput. 246 (2014) 168–183.
[65] O. Pourmehran, M. Rahimi-Gorji, M. Hatami, S.A.R. Sahebi, G. Domairry, Numerical optimization of microchannel heat sink (MCHS) performance cooled by KKL based nanofluids in saturated porous medium, J. Taiwan. Inst. Chem. Eng. 55 (2015) 49–68.
[66] M. Hatami, D.D. Ganji, Thermal and flow analysis of microchannel heat sink (MCHS) cooled by Cu–water nanofluid using porous media approach and least square method, Energy Convers. Manag. 78 (2014) 347–358.
[67] A. Sakanova, S. Yin, J. Zhao, J.M. Wu, K.C. Leong, Optimization and comparison of double-layer and double-side micro-channel heat sinks with nanofluid for power electronics cooling, Appl. Therm. Eng. 65 (2014) 124–134.
[68] Y.F. Li, G.D. Xia, D.D. Ma, Y.T. Jia, J. Wang, Characteristics of laminar flow and heat transfer in microchannel heat sink with triangular cavities and rectangular ribs, Int. J. Heat. Mass. Transf. 98 (2016) 17–28.
[69] C.J. Ho, L.C. Wei, Z.W. Li, An experimental investigation of forced convective cooling performance of a microchannel heat sink with Al_2O_3/water nanofluid, Appl. Therm. Eng. 30 (2010) 96–103.

[70] W. Duangthongsuk, S. Wongwises, An experimental investigation on the heat transfer and pressure drop characteristics of nanofluid flowing in microchannel heat sink with multiple zigzag flow channel structures, Exp. Therm. Fluid Sci. 87 (2017) 30–39.

[71] V. Puttaswmay, G. Bijjanal Jayanna, S. Doranalu Onkarappa, Heat transfer and irreversibility rate in MHD flow of a hybrid nanofluid with Newton boundary condition, slip flow, and nonlinear thermal radiation, Heat. Transf. 50 (2021) 3342–3365.

[72] N.T. Ravi Kumar, P. Bhramara, B.M. Addis, L.S. Sundar, M.K. Singh, A.C.M. Sousa, Heat transfer, friction factor and effectiveness analysis of Fe_3O_4/water nanofluid flow in a double pipe heat exchanger with return bend, Int. Commun. Heat. Mass. Transf. 81 (2017) 155–163.

[73] A.A. Abbasian Arani, J. Amani, Experimental investigation of diameter effect on heat transfer performance and pressure drop of TiO_2–water nanofluid, Exp. Therm. Fluid Sci. 44 (2013) 520–533.

[74] A.A. Abbasian Arani, J. Amani, Experimental study on the effect of TiO_2–water nanofluid on heat transfer and pressure drop, Exp. Therm. Fluid Sci. 42 (2012) 107–115.

[75] A. Jafarimoghaddam, S. Aberoumand, On the implementation of Cu/ethylene glycol nanofluids inside an annular pipe under a constant wall temperature boundary condition, Heat Transf.—Asian Res. 46 (2017) 647–655.

[76] S.K. Samal, M.K. Moharana, Thermo-hydraulic and entropy generation analysis of recharging microchannel using water-based graphene–silver hybrid nanofluid, J. Therm. Anal. Calorim. 143 (2021) 4131–4148.

[77] R.L. Monaledi, O.D. Makinde, Entropy generation analysis in a microchannel Poiseuille flows of nanofluid with nanoparticles injection and variable properties, J. Therm. Anal. Calorim. 143 (2021) 1855–1865.

[78] A. Mosavi, S.N. Sedeh, D. Toghraie, A. Karimipour, Analysis of entropy generation of ferrofluid flow in the microchannel with twisted porous ribs: the two-phase investigation with various porous layers, Powder Technol. 380 (2021) 349–357.

[79] A. Karimipour, D. Bahrami, R. Kalbasi, A. Marjani, Diminishing vortex intensity and improving heat transfer by applying magnetic field on an injectable slip microchannel containing FMWNT/water nanofluid, J. Therm. Anal. Calorim. 144 (2021) 2235–2246.

[80] S.A. Memon, M.B. Sajid, M.S. Malik, A.B.S. Alquaity, M.M.U. Rehman, T.A. Cheema, et al., Investigation of the thermal performance of salt hydrate phase change of nanoparticle slurry flow in a microchannel, J. Chem. 2019 (2019) 5271923.

[81] H.A. Mohammed, P. Gunnasegaran, N.H. Shuaib, Influence of various base nanofluids and substrate materials on heat transfer in trapezoidal microchannel heat sinks, Int. Commun. Heat. Mass. Transf. 38 (2011) 194–201.

[82] I.A. Ghani, N.A.C. Sidik, N. Kamaruzaman, Hydrothermal performance of microchannel heat sink: the effect of channel design, Int. J. Heat. Mass. Transf. 107 (2017) 21–44.

[83] M. Roshani, S. Ziaeddin Miry, P. Hanafizadeh, M. Ashjaee, Hydrodynamics and heat transfer characteristics of a miniature plate pin-fin heat sink utilizing Al_2O_3–water and TiO_2–water nanofluids, J. Therm. Sci. Eng. Appl. 7 (2015).

[84] E. Khodabandeh, S.A. Rozati, M. Joshaghani, O.A. Akbari, S. Akbari, D. Toghraie, Thermal performance improvement in water nanofluid/GNP–SDBS in novel design of double-layer microchannel heat sink with sinusoidal cavities and rectangular ribs, J. Therm. Anal. Calorim. 136 (2019) 1333–1345.

[85] M.M.U. Rehman, T.A. Cheema, F. Ahmad, A. Abbas, M.S. Malik, Numerical investigation of heat transfer enhancement and fluid flow characteristics in a microchannel heat

sink with different wall/design configurations of protrusions/dimples, Heat. Mass. Transf. 56 (2020) 239–255.

[86] O. Abouali, N. Baghernezhad, Numerical investigation of heat transfer enhancement in a microchannel with grooved surfaces, J. Heat. Transf. 132 (2010).

[87] E. Hosseinirad, F. Hormozi, Influence of shape, number, and position of horizontal mini-fins on thermal-hydraulic performance of minichannel heat sink using nanofluid, Heat. Transf. Eng. 38 (2017) 892–903.

[88] Q. Nguyen, D. Bahrami, R. Kalbasi, Q.-V. Bach, Nanofluid flow through microchannel with a triangular corrugated wall: heat transfer enhancement against entropy generation intensification, Math. Methods Appl. Sci. (2020). n/a.

[89] Z. Lyu, F. Pourfattah, A.A.A. Arani, A. Asadi, L.K. Foong, On the thermal performance of a fractal microchannel subjected to water and kerosene carbon nanotube nanofluid, Sci. Rep. 10 (2020) 7243.

[90] P. Srivastava, R.I. Patel, A. Dewan, A study on thermal characteristics of double-layered microchannel heat sink: effects of bifurcation and flow configuration, Int. J. Therm. Sci. 162 (2021) 106791.

[91] H. Bazdar, D. Toghraie, F. Pourfattah, O.A. Akbari, H.M. Nguyen, A. Asadi, Numerical investigation of turbulent flow and heat transfer of nanofluid inside a wavy microchannel with different wavelengths, J. Therm. Anal. Calorim. 139 (2020) 2365–2380.

[92] G. Schmid, L.-H. Yang, T.-H. Yang, S.-L. Chen, Influence of inter-fin base length on thermal performance of free hanging horizontal base heat sinks, Appl. Therm. Eng. 108 (2016) 1226–1236.

[93] S.S. Wangikar, P.K. Patowari, R.D. Misra, Numerical and experimental investigations on the performance of a serpentine microchannel with semicircular obstacles, Microsyst. Technol. 24 (2018) 3307–3320.

Chapter 15

Nanofluids for enhanced performance of building thermal energy systems

Naseem Abbas[1], Muhammad Bilal Awan[2], Mohsin Ali Badshah[3], Uzair Sajjad[4] and Waseem Raza[5]

[1]Department of Mechanical Engineering, Sejong University, Seoul, South Korea, [2]Department of Mechanical Engineering, University of Central Punjab, Lahore, Pakistan, [3]Department of Chemical Engineering and Materials Science, The Henry Samueli School of Engineering, University of California, Irvine, CA, United States, [4]Department of Energy and Refrigerating Air-Conditioning Engineering, National Taipei University of Technology, Taipei, Taiwan, [5]Department of Mechanical Engineering, Jeju National University, Jeju, South Korea

15.1 Introduction

The building sector emits 40% of the total carbon emissions by consuming one-third of the world's primary energy [1]. Building thermal energy systems are considered as the major contributor in increasing the building energy demand [2]. Enhancing the performance of these building energy systems is key to reducing carbon emissions and ensuring sustainability in the building energy sector. Building thermal energy systems can be potentially divided into three categories, that is, solar thermal-based systems, nonsolar systems, and thermal energy storage (TES) systems. Solar thermal energy systems generally include photovoltaic thermal (PVT) systems, absorption chillers, and fluid heating systems such as concentrated tube, evacuated tube panels, etc. Nonsolar energy systems include refrigerators, heating, ventilation, and air conditioning (HVAC) systems for cooling and heating applications and boilers for hot water supply. TES devices include sensible heat storage devices, such as hot water tanks, and latent heat storage (LHS) devices, such as phase change material (PCMs). Available research has shown that the performance of building thermal systems can be substantially raised by enhancing their heat transfer characteristics [3,4]. In building thermal systems, smaller size, low heat transfer coefficients, and low thermal conductivity are considered the main hindrance to enhancing the heat transfer characteristics

Advances in Nanofluid Heat Transfer. DOI: https://doi.org/10.1016/B978-0-323-88656-7.00014-3

of these systems. An innovative way to enhance the heat transfer characteristics of these systems without altering the system configuration or size is by using nanoparticles. Nanoparticles act as nanofluids when suspended in the base or working fluids such as water, oils, refrigerants, methanol, and ethylene glycol [5]. With the estimate of an increase in HVAC energy demand by 70% by 2050 [6], established technology such as nanofluid application in building energy systems will help to save a considerable amount of energy.

Generally, nanofluids are classified based on the type of particles. Widely used nanoparticles are based on carbon, metal, metal oxides, and metal hybrids. These nanoparticles form a uniform mixture when mixed with base fluids. The selection of nanoparticles differs for different applications and the main consideration for the selection of any nanoparticle should focus on fluid stability and potential improvement in the intrinsic properties such as viscosity, density, specific heat, and thermal conductivity of the base fluid [7]. Available research has shown that for effective results, the selection of optimized properties of nanofluids such as base fluid, fluid flow rate, shape, dimension, and volume fraction of nanoparticles is important [8]. Fig. 15.1 shows the thermal conductivities of mostly used nanoparticles and base fluids. It can be observed that the difference between the thermal conductivity of widely used base fluids and nanoparticles is huge and a small fraction of these particles can considerably enhance the thermal conductivity of base fluids.

15.2 Overview of domain knowledge related to nanofluids

Nanofluid is generally prepared by using the single-step method or the two-step method. In the single-step method, the production and

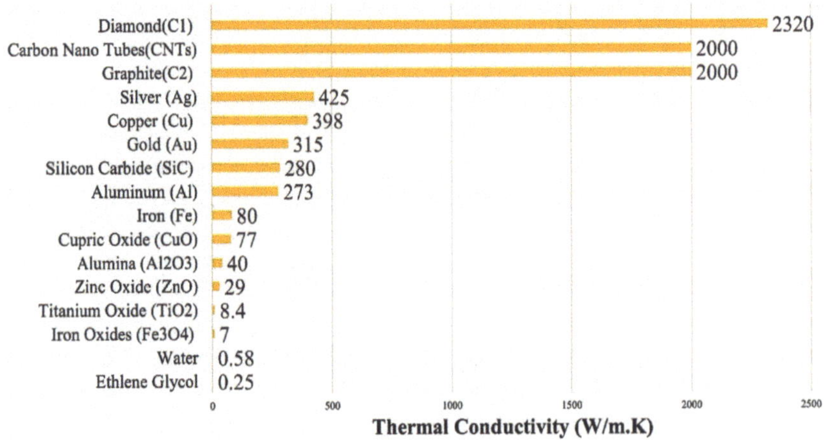

FIGURE 15.1 Thermal conductivities of base fluid and nanoparticles.

suspension of nanoparticles into the base fluid are done simultaneously. This method is considered costly but advantageous in reducing the agglomeration of nanoparticles. Conversely, in the two-step method, nanoparticles are produced and dispersed separately. This method is preferred over the single-step method because of being cost-effective and less labor-intensive. For homogeneity of the mixture stabilizing agents, homogenizer, stirrers, or ultrasonicates can be used. Both methods are presented in Fig. 15.2 [9].

The heat transfer capacity of nanofluids is considered dependent on several parameters that need to be optimized for maximum effectiveness. The mathematical formulation of the important parameters is presented below [10].

The volumetric concentration of nanoparticles can be calculated using Eq. (15.1).

$$\varnothing = \frac{\rho_f \varnothing_m}{\rho_f \varnothing_m + \rho_s \quad (1 - \varnothing_m)} \tag{15.1}$$

where $\varnothing_m$ is the mass fraction of nanoparticles.

The density of nanofluid can be calculated using Eq. (15.2).

$$\rho_{nf} = \varnothing \rho_s + (1 - \varnothing)\rho_f \tag{15.2}$$

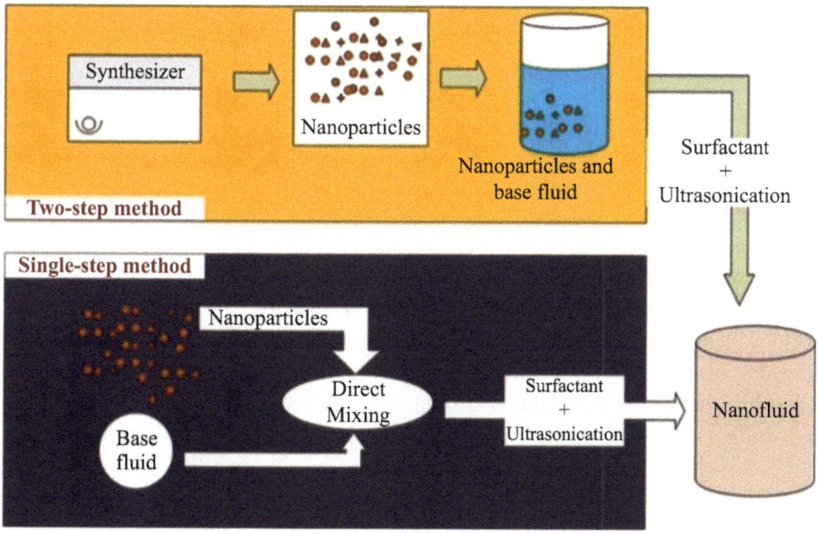

FIGURE 15.2 Preparation method of nanofluids [9].

Reynold's number is considered an important parameter for the control of heat transfer of fluids flow inside pipes and can be calculated using Eq. (15.3).

$$R_e = \frac{\rho_{nf} v D}{\mu_{nf}} \tag{15.3}$$

The Hamilton–Crosser model is used for calculating the thermal conductivity of the nanofluid, as presented in Eq. (15.4).

$$K_{nf} = k_f \left[\frac{k_s + (n-1)k_f - (n-1)\varnothing(k_f - k_s)}{k_s + (n-1)k_f + \varnothing(k_f - k_s)} \right] \tag{15.4}$$

Specific heat of nanofluids can be calculated using Eq. (15.5) and Eq. (15.6). In Eq. (15.6), it is assumed that nanoparticles and base fluids are in thermal equilibrium.

$$C_{pnf} = (1 - \varnothing)C_{pf} + \varnothing C_{ps} \tag{15.5}$$

$$C_{pnf} = \frac{(1 - \varnothing)\rho_f C_{pf} + \varnothing \rho_s C_{ps}}{\rho_{nf}} \tag{15.6}$$

The volumetric flow rate of nanofluid can be calculated by using Eq. (15.7).

$$V_{nf} = v A_{p,i} \tag{15.7}$$

Pumping power to circulate nanofluids can be calculated using Eq. (15.8).

$$V_{nf} W = \frac{V_{nf} \Delta P}{\eta_P} \tag{15.8}$$

where ΔP is the pressure loss and η_P is the pump efficiency.

It is worthwhile to note that the above formulations are not standardized. Several studies have further modified these formulations for achieving more accuracy.

Although the above parameters are helpful in parametric calculations, still accurate models are needed to accurately predict the heat transfer enhancement in nanofluid applications. Few theoretical studies have highlighted some parameters to aid in the development of an accurate heat transfer enhancement model by considering several internal forces, for instance in Ref. [11], it was concluded that Brownian motion significantly influences the thermal conductivity of the nanofluids. In Ref. [12], it was reported that nanofluids not only experience Brownian motions but also experience electrical double layer, drag, van der Waals, and thermophoresis forces. In Ref. [13], it was deduced that during evaporation, molecular layer formation could be the cause of better heat transfer.

Although the lack of an accurate predictive heat transfer model is one of the hurdles in sensitive applications, current research has shown that the use of nanofluid is advantageous in enhancing the effectiveness of building thermal systems. Based on the available studies, a review of nanofluids in enhancing the effectiveness of building thermal energy systems is provided below.

15.3 Role of nanofluids in efficiency enhancement of building energy systems

15.3.1 Nanofluids for performance enhancement of photovoltaic thermal systems

Nanofluids have been proven effective in enhancing the performance of the PVT system [14,15]. Fig. 15.3 indicates the schematic of a generalized way to use nanofluids in PVT applications.

Nanofluids have played a vital role in enhancing the electrical, thermal, and overall performance of PVT systems. Table 15.1 has summarized the performance enhancement in terms of the thermal, electrical, and overall efficiency of the PVT system by using various nanofluids. It was found that SiC nanofluid showed the leading performance compared to other nanofluids, listed in Table 15.1. Based on the available studies the following key

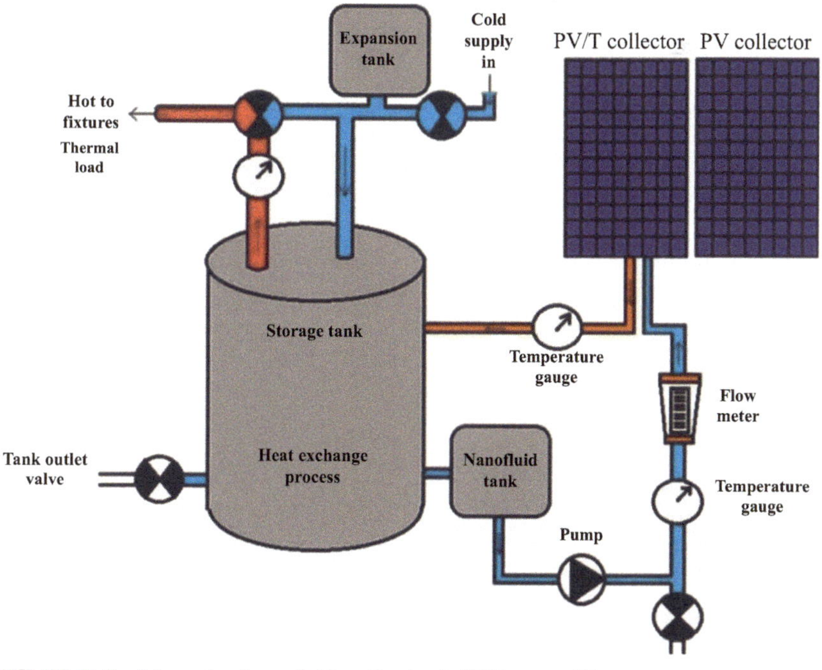

FIGURE 15.3 Schematic of nanofluid application in PVT system [9].

TABLE 15.1 Summary of PVT system performance improvement with the introduction of nanofluids.

Authors	Type of nanoparticle	Concentration and size	Base fluid	Usage of nanofluid	Remarks
Cui and Zhu [16]	MgO	0.02 wt.%, 0.06 wt.%, 0.10 wt.%	Water	Coolant	The electrical performance of the system was increased to 15.9%
Taylor et al. [17]	SiO_2	Therminol VP-1: 0–0.01 wt.%; 20–50 nm	Water and Therminol	Spectral filter	The overall performance of the system was enhanced by 4%–8% compared to the water-based system
Chandrasekar et al. [18]	Al_2O_3 and CuO	0.1 vol.%	Water	Coolant	The electrical performance was increased to 9.4%, and 9.8%, when CuO nanofluid and Al_2O_3 nanofluid was used as cooling fluid, respectively
Faizal et al. [19]	CuO, SiO_2, TiO_2, and Al_2O_3	0.02 wt.% to 0.032 wt.% for all	Water	Coolant	The nanofluids-based system enhanced the system's overall efficiency compared to the conventional systems
Karami and Rahimi [20]	Boehmite ($AlOOH. \times H_2O$)	0.01, 0.1, 0.3 wt.%; 5–10 nm	Water	Coolant	The electrical performance was enhanced to 37.67%
Sardarabadi et al. [21]	SiO_2	1, 3 wt%, 11–14 nm	Water	Coolant	The thermal performance was enhanced by 7.6% and 12.8% for nanofluid with 1 wt.% and 3 wt.%, respectively, and the overall performance was also enhanced by 3.6% and 7.9% for nanofluid with 1 wt.% and 3 wt.%, respectively
Xu and Kleinstreuer [22]	Al_2O_3	5 vol.%, 38.4 nm	Water	Coolant	The overall system performance was enhanced by 20%

Siddharth et al. [17]	Ag, Au	0.00019 vol.% & 0.00025 vol.%; 5 nm each	Water	Coolant	Both silver (Ag) and Gold (Au) nanofluid-based systems showed enhanced overall system performance
Said et al. [23]	TiO_2	0.1, 0.3 vol.%	Water	Coolant	The energy performance of the system was increased by 76.6% at 0.1 wt.%
Jing et al. [24]	SiO_2	0.5, 1, 2 vol.%; 5, 10, 25, 50 nm	Water	Coolant	The exergetic performance of the system was increased by 9.5%
Hussien et al. [25]	Al_2O_3	0.1%–0.5%; 30 nm	Water	Coolant	The thermal performance of the system was enhanced to 34.4% and the electrical performance was enhanced to 12.1%
Rejeb et al. [26]	Al_2O_3 and Cu	0.001%–1.5%; 10 nm	Water and ethylene glycol	Coolant	The overall system performance was enhanced and the best performance was obtained using the Cu-based nanofluids
Sharaf et al. [27]	Al_2O_3	15–40 nm	Water and synthetic oil	Coolant	Al_2O_3-synthetic oil-based system showed better performance compared to Al_2O_3/water-based system
Radwan et al. [28]	Al_2O_3, SiC	0–4 vol.%; 20 nm	Water	Coolant	SiC showed better performance than Al_2O_3, and up to 19% improvement was achieved in electrical efficiency
Shamani et al. [22]	SiO_2, TiO_2 and SiC	0.5, 1, 2 wt.%	Water	Coolant	SiC nanofluid system showed better performance with the highest electrical performance of 13.52%
Sardarabadi et al. [29]	Al_2O_3, TiO_2 and ZnO	10–30 nm ZnO: 2 wt.%, 10–25 nm	Water	Coolant	The electrical performance for Al_2O_3, ZnO, and TiO_2 was enhanced by 6.36%, 6.46%, and 6.54%, respectively
Hassani et al. [30]	CNTs and Ag	0.1 vol.% and 0.001 vol.%; 15 nm and 10 nm	Water	Coolant	CNTs-based nanofluids showed better performance compared to the conventional system

(Continued)

TABLE 15.1 (Continued)

Authors	Type of nanoparticle	Concentration and size	Base fluid	Usage of nanofluid	Remarks
Noghrehabadi et al. [31]	SiO_2	1 wt.%	Water	Coolant	The overall system performance was increased by 62%
Al-Waeli et al. [32]	SiC	3 wt.%	Water	Coolant	The overall system performance was enhanced by 88.9%
Das and Kalita [33]	CuO	N/A	Water	Coolant	The exergy was enhanced by 57.78%, whereas the electrical performance was enhanced by 0.5%
Lari et al. [34]	Ag	0.5 vol.%; 10 nm	Water	Coolant	The thermal and electrical performances were enhanced by 13% and 8.5% respectively
Ebaid et al. [35]	TiO_2, Al_2O_3	(0.01, 0.05 and 0.1) wt.%	water–polyethylene glycol mixture	Coolant	The overall system performance was enhanced using a nanofluids-based system
Lee et al. [36]	CuO, Al_2O_3	0.05 wt.%	Water	Coolant	Thermal performance was increased by 21.30% and 15.40% with CuO and Al_2O_3, respectively
Ebaid et al. [37]	Al_2O_3, TiO_2	0.01%, 0.05%, and 0.1%	Water	Coolant	The overall performance was enhanced from 2.89% to 11.08%

factors are identified for performance enhancement of PVT system by using nanofluids:

- Materials and size of nanoparticles: performance of the nanofluid alters with nanoparticles sizes. Different nanoparticles have different behavior at different temperatures and rates because of the variations in size and type of particles. For each application, base fluid, flow channel, and flow rate, different sizes of nanoparticles are required that must be selected after running parametric analysis to avoid agglomeration and to achieve maximum performance of any system.
- Type of fluidic flow: type of fluidic flow has a critical role in enhancing the performance of the PVT system. Selecting the high-velocity laminar flow tends to have a positive impact on the performance of the system as it provides better suspension of particles which ultimately enhance the performance of the system.
- Flow rate: nanofluids' flow rate is an important factor to enhance the performance of the PVT system. PVT system performance can increase till a threshold point, increasing the flow rate beyond the threshold point can negatively impact the efficiency of the system. Selecting the optimum point flow rate increases the performance and the life of the PVT system.
- The concentration of nanofluid: nanofluids concentration has a significant role in enhancing the performance of the PVT system. Selecting the optimum concentration increases the performance and stability of the system. For instance, 3 wt.% of SiC in water was proved highly effective.

15.3.2 Nanofluids for performance enhancement of heating, ventilation, and air conditioning systems

Since building sector energy consumption is rapidly increasing, it is essential to develop new heat transfer fluids to reduce costs and meet the industrial demand. There are several ways to improve the efficiency of HVAC system: (1) increasing the condenser heat rejection rate, (2) reducing the power consumption of the compressor by employing intelligent control, (3) increasing evaporator heat absorption rate, and (4) by decreasing the volumetric flow rate of the refrigerant inside the system by reducing its temperature. Advances in the field of nanotechnology have opened several ways to enhance the effectiveness of refrigerant-based systems.

Nanofluid usage in PVT systems and TES systems is simpler compared to HVAC and refrigeration systems. Both HVAC and refrigeration systems involve phase change and compression processes, and these processes enhance the complexity of using nanofluids in these systems. Nanoparticles blend with refrigerants is termed as a nanorefrigerant. Current research has shown an inconsistent relationship between data of boiling heat transfer coefficient among nanorefrigerants and nanofluids but an overall increase in boiling heat transfer is observed by using the nanorefrigerants [38].

Nanoparticles are used in refrigerant-based systems in two ways, that is, either by suspending nanoparticles in the refrigerant or using them as refrigerant–nanolubricant pair. A significant decrease in friction coefficient has been observed by replacing the conventional lubrication oil with nanolubricants [39]. Furthermore, nanolubricants have also been proved effective in reducing the compressor shell temperature; in Ref. [40], 3°C–5°C reduction in compressor shell temperature was observed when using the nanolubricant.

Condensing heat transfer and boiling heat transfer phenomenon are considered critical when using nanorefrigerants, as both of these processes involve phase change. Heat transfer during these processes greatly depends on the thermophysical properties, including viscosity and thermal conductivity of the nanorefrigerants. Though all thermophysical properties like density, latent and specific heat, etc. are considered important, surface tension is of the utmost importance. Surface tension is directly related to the liquid superheat needed for the bubble existence [38]. Other properties such as specific heat are also reported as a critical factor for performing the energy and exergy analysis of the nanorefrigerant-based system [41].

Several studies have highlighted the importance of using nanoparticles in the refrigerant-based systems for performance enhancement. Peng et al. [42], for instance, performed an experiment to predict the heat transfer performance of refrigerant-based nanofluids: R-113 and copper oxide (CuO) mixture. The nanorefrigerant showed better performance as compared to the solely refrigerant-based system, and the maximum enhancement was found to be around 29.7%.

An example of heat pump heating cycle power consumption and coefficient of performance (COP) is shown in Fig. 15.4. The power consumption of R-22 and TiO_2 is lower than R-22 after a flow rate of 0.275 L/s.

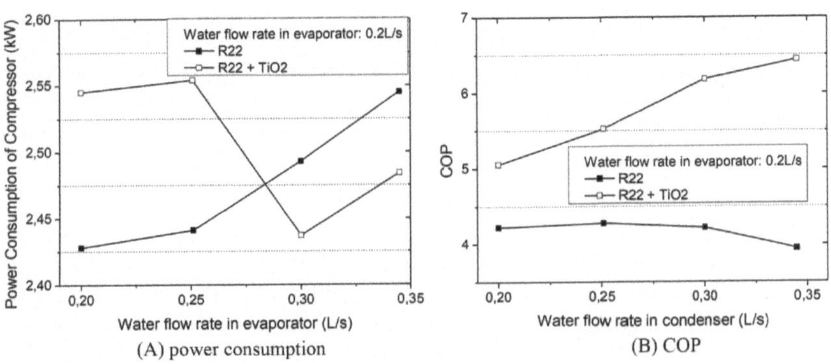

FIGURE 15.4 Heat pump heating cycle power consumption and COP [43].

Fig. 15.4B shows the superiority of R-22 + TiO_2 in terms of COP, as the use of nanofluid increases the COP and decreases the power consumption.

In another study [44], R-600a and TiO_2 mixture was used as a nanorefrigerant. The study showed that utilizing TiO_2 with the R-600a refrigerant reduced the energy consumption by 10%. In Ref. [43], R-134a and Al_2O_3 mixture was used. The COP was enhanced by 25.7% with a mass fraction of 0.01%. In Ref. [45], titanium oxide (TiO_2) nanoparticles were mixed with R-600a to prepare the nanorefrigerant. Results showed a 78% increase in the system COP. Sanukrishna et al. [13] prepared a mixture of copper oxide (CuO) nanoparticles in R-134a and polyalkylene glycol. An increase of 12.67% was observed in the thermal conductivity and 37% enhancement in the flow boiling was also observed. Overall an increase of 7.5% was observed in the COP. Many other studies have been done on the HVAC systems using different nanofluids, as summarized in Table 15.2.

These studies have shown that advances in the field of nanofluids have enabled the preparation of stable nanorefrigerants that can provide better heat flux, which results in high evaporator performance because of better pool boiling. Moreover, nanorefrigerants have been proved effective in reducing the overall energy consumption of HVAC systems by providing better COP, heat transfer coefficient, and freezing capacity, and by reducing the discharge pressure, and controlling the irreversibility. Though overall COP of the refrigerants-based system is observed to be improved with the use of nanorefrigerants, the pressure drop is also observable in these systems [53]. Pressure drop is because of the increase in viscosity of the refrigerants. This pressure drop increases the lift between compressor inlet and outlet that demands more compressor power to reach the desired pressure but this power increase is overcome by the increase in heat transfer due to the introduction of nanoparticles in the nanorefrigerants. Moreover, the requirement of low discharge pressure also negates the negative effect of low suction pressure.

Other important factors in refrigerant-based systems are pool boiling, flow boiling, and condensation that are described in details in Ref. [38].

15.3.3 Nanofluids for performance enhancement of thermal storage systems

Nanofluids usage has also been reported extensively in the literature to enhance the performance of thermal storage systems. Thermal energy in the building can be potentially stored by utilizing sensible and LHS systems. LHS systems such as PCMs in the form of paraffin and salt hydrates have been proven to be more effective as compared to sensible heat storage systems. In LHS, phase change can be in different transitional forms such as solid–solid, solid–liquid, solid–gas, and liquid–gas forms. In solid–solid transition, material stores heat by transforming its crystalline structure. Such

TABLE 15.2 Performance evaluation of HVAC systems using nanofluid.

Authors	Configuration	Study method	Nanofluid	Remarks
Abbas et al. [46]	Air-conditioning	Experimental	R134a + Carbon Nanotubes	4.2% increase in COP of the system was observed
Subareesh et al. [47]	Refrigeration	Experimental	R12 + TiO_2	COP of the system increased by 17%
Kumar et al. [48]	Refrigeration	Experimental	TiO_2 + R600a	11% reduction in energy consumption of the system was observed
Fedele et al. [49]	Heat pump	Experimental	R-134a + R-22	Little improvement in the performance of the system was observed
Bi et al. [50]	Refrigeration	Experimental	R-600a + TiO_2	9.6% savings of energy was achieved
Xing et al. [40]	Refrigeration	Experimental	R-600a + C60	5%~6% improvement in COP was achieved
Kumar and Elensezhian [51]	Refrigeration	Experimental	R-134a + Al_2O_3	10.3% reduction in energy consumption was achieved
Subramani and Prakash [52]	Refrigeration	Experimental	R-134a + Al_2O_3	25% reduction in power consumption of the system was observed

materials offer better design flexibility. Examples of solid–solid transition materials are KHF_2 and Penta-glycerine. Solid and liquid transition to gas offers maximum heat storage capacity but high volume changes during phase change makes these systems complex. The most practical and widely used transition is solid–liquid transition. This transition offers high latent heat of fusion with small volume changes and small temperature swings. These materials have better energy density as compared to sensible heat storage systems, which enhances the effectiveness of PCMs in storing more energy with similar size compared to the sensible heat storage systems. The main three types of PCMs are summarized in Table 15.3.

For effective utilization of PCMs in buildings, Hawes et al. [54] considered encapsulation, direct incorporation, and immersion as the most promising methods. These techniques have been proven to be effective in reducing the peak heat flux [55]. Encapsulation can be done at the micro- and macro-level [56]. In the microencapsulation technique, small PCM particles are enclosed in high-strength thin polymeric films that must be compatible with the construction materials. This technique offers better heat transfer with easy to use application but has the disadvantage of affecting the mechanical strength of the structure. In macroencapsulation PCM is enclosed in pouches, spheres, panels, or tubes. This method offers more versatility with effective heat storage capability but needs protection against destruction and the requirement of much more work for integration into the building structure [57]. Direct incorporation is one of the cheapest methods, where liquid or powdered PCM is directly added to the building material. Energy storage

TABLE 15.3 Summary of available phase change materials.

Material	Advantages	Disadvantages	Example
Organic	Chemically stable, high latent heat of fusion, nonreactive and nonsegregating properties	Low thermal conductivity, costly, and flammable	Paraffins such as Acetic acid, Phenol, etc., and nonparaffins such as fatty acids
In-organic	Low cost, high availability, high conductivity, high density, high specific heat, and low volume changes	Corrosive, segregation, and supercooling	Salt hydrates such as $Na_2SO_40.10H_2O$, $LiNO_30.3H_2O$, etc
Eutectics	Nonsegregating, congruent properties. These have advantages of both organic and inorganic materials that mainly depend upon the combination of materials used	As these materials are mainly mixtures, they will have slightly less effective properties as compared to organic and inorganic materials	$Na_2SO_4 + NaCl + KCl + H_2O$, $Na_2SO_4 + NaCl + H_2O$ etc

wallboard is a prime example of this technique [58]. In the immersion technique, the porous building material is dipped into the melted PCMs, where PCM enters the pores, and once the building material cools, PCM remains in the pores and provides effective heat storage applications. However, leakage can occur after several years of usage [59]. Marani and Nehdi [60] have provided a detailed review of the effective utilization of PCM incorporation techniques into building materials.

It is reported that the homogeneous and stable dispersion of nanoparticles in the base fluids can substantially enhance the physical and thermal performance of both of the TES systems compared to the conventional systems alone [61,62]. Though PCMs have gained a lot of attraction because of the wide temperature range, high latent heat, and congruent melting temperatures, these PCMs are redundant in real applications because of the relatively low thermodynamics (thermal conductivity, specific heat, heat of fusion and density) characteristics. It is observed that mixing the carbon-based materials, metals, and metal oxides, such as graphite sheets, graphene sheets, SWCNT, MWCNT, carbon flakes, active carbon, carbon fibers silver (Ag), copper (Cu), auminum (Al), aluminum oxide (Al_2O_3), magnesium oxide (MgO), zinc oxide (ZnO), and titanium dioxide (TiO_2), with the PCMs enhances thermodynamic properties, that is, melting and solidification of PCMs, heat transfer, and thermal conductivity [63]. Table 15.4 has summarized various studies considering thermal storage systems and nanofluids. It is observed that the CNTs dispersed in the wax have a significant effect on the latent heat of the system [64]. Gavarrell et al. [65] reported the MWCNT/water had rapid heating of the system for only 3% volume fraction. Kumaresan et al. [66] also reported the enhancement in thermal conductivity as the MWCNT concentration increased from 0.15% to 0.60%. Wang et al. [67] reported no change in latent heat using the 2% graphite nanoparticle, however they found a significant reduction in the supercooling characteristics that reduced from 9.9°C to almost 0.0°C. Zeng et al. [68] reported the enhancement in the thermal conductivity of the PCM with Ag nanoparticles dispersed in 1-tetradecanol (TD). They also reported the increase in thermal conductivity by increasing the suspension of nanoparticles. Wu et al. [69] showed a significant reduction of 20% in the cooling time using the Al_2O_3-H_2O nanofluid. Harikrishnan et al. [70] prepared the CuO nanoparticles-based nanofluid with fractions 0.1, 0.3, 0.5, 0.8, and 1.0 wt.%, and achieved 40.24% enhancement in thermal conductivity with 1.0 wt.%. Mo et al. [71] dispersed the TiO_2 nanoparticles in different ethylene glycol solutions (12%, 22%, and 34%) and achieved 1.3°C drops in the phase change temperature with 34% ethylene glycol solution. He et al. [72] used TiO_2 nanoparticles in a saturated $BaCl_2$ solution and observed enhancement in the thermal conductivity by 12.76% and reduction in supercooling by 84.92%. Wu et al. [73] performed the study using the Al, Cu, and C/Cu nanoparticles in a paraffin solution. They observed that the Cu suspended nanoparticle in the paraffin

TABLE 15.4 Summary of various materials application for thermal storage systems.

References	Type of nanoparticle	Concentration and size	Base fluid	Remarks
Kasaeian et al. [74]	MWCNT	0.2% to 0.3% (weight fraction)	Mineral oil	The performance of the system was enhanced up to 5%–7% by using the 0.3%, MWCNT/ mineral oil nanofluid
Liu et al. [75]	CuO	D = 50 nm, 1.2 wt.%	Water	CuO-based nanofluid enhanced the system performance by 12.4%
He et al. [76]	CuO	D = 25–50 nm and 0.01, 0.02, 0.04, 0.1, 0.2 wt.%	Deionized water	CuO-based nanofluid influences the optical performance of the system which enhanced the absorption capacity of the system
Faizal et al. [77]	MWCNT	0.2 wt.% and 0.4 wt.%	N/A	MWCNT nanofluid reduced the surface area and enhanced the overall system performance
Nasrin et al. [78]	Al_2O_3	5.0 vol.%	Water	Al_2O_3-based nanofluid enhanced the heat transfer by 19%. Al_2O_3 had a higher heat transfer due to convection

(Continued)

TABLE 15.4 (Continued)

References	Type of nanoparticle	Concentration and size	Base fluid	Remarks
Yousefi et al. [79]	MWCNT	D = 10–30 nm, 0.2 wt.%	Water	Higher performance was achieved with MWCNT suspension
Yousefi et al. [80]	Al_2O_3	D = 15 nm, 0.2 and 0.4 wt.%	Water	0.2 wt.% Al_2O_3 nanoparticles enhanced the system performance up to 28.3%
Taylor et al. [81]	Graphite, Al, Cu, Ag	20	Therminol VP-1	The thermal energy is enhanced by using all types of nanofluids
Taylor et al. [82]	TiO_2, Ag, Cu, Al, Au, and graphite	N/A	Water	The absorbance of the system was enhanced using the nanofluids, which enhanced the overall performance of the system
Tyagi et al. [83]	Al_2O_3	0–20	Water	Al_2O_3 volume fraction enhanced the system performance up to 2%

solution significantly enhanced the thermal performance. They also observed that the cooling and heating time also reduced by 28.2% and 30.3% using the Cu nanoparticles suspensions in paraffin.

15.4 Barriers

Although nanofluids have been proven to be effective in enhancing the performance of building energy systems, several barriers exist in the practical implementation of nanofluids in building energy systems. As most of the building energy systems have closed-loop small sizes, hence the agglomeration of nanoparticles is one of the leading problems. Other common problems observed are corrosion, pressure drop, and increase in the pumping power [84]. A nontechnical hurdle is the nonstandardization of the nanofluids and lack of accurate theoretical models, which add uncertainty to the available technical data. Nanofluids also observe various kinds of forces during circulation such as van der Waals forces, electric double layer forces, drag forces, Brownian forces, thermophoresis, and interfacial layers that can potentially build bridges between base fluids and nanoparticles [85]. A few studies have also reported sedimentation as a major problem [9]. Although these barriers exist, simple solutions have been proposed for nanofluids' effective usage. For example, stabilizing agents can be added to the base fluid to reduce interfacial forces but it is not considered a permanent solution. Because of being at the experimental stage, the cost of nanofluids is high but once the use of nanofluid starts spreading to all fields, considerable reduction in cost is expected.

Overall, in building energy systems, the use of nanofluids in the TES systems and PVT systems is relatively more stable as compared to their use in refrigeration or HVAC systems. In refrigeration and HVAC systems, working fluid has to pass through the compressor and slight agglomeration of the particles can compromise the safety of the whole system, hence the size of the particles in the HVAC systems should be carefully selected.

Conclusively, though nanofluids' practical application in the building energy system is in its infancy, current research has shown that in the future this technology can prove to be highly effective in reducing the overall electricity cost of building operations.

15.5 Conclusions

This chapter has reviewed the advances of nanofluids in the field of building thermal energy systems including PVT, refrigeration and HVAC systems, and TES systems. A major advancement was found in the field of refrigeration and HVAC systems, where nanoparticles were used to prepare nanorefrigerants. Because of the involvement of the compression process, the preparation of nanorefrigerants demanded better suspension and a relatively smaller size. Nanorefrigerants in refrigeration systems have shown significant improvement in terms of COP, cooling capacity, thermal conductivity, and overall efficiency. A reduction in compressor friction and wear can be achieved as well. It is anticipated that use of nanorefrigerants will result in smaller systems sizes and less electrical power. Though the use of nanorefrigerants decreases the suction pressure that demands more compression power to achieve the desired pressure, better thermodynamic characteristics of nanorefrigerants, and the requirement of low discharge pressure allows smaller compressors to be used that decreases the overall compression power of the system. Similar performance enhancement was also observed in PVT and TES systems. Among the TES systems, PCMs showed a high potential of utilizing nanofluid. MWCNT, silver (Ag), copper oxide (CuO), aluminum (Al), aluminum oxide (Al_2O_3), magnesium oxide (MgO), zinc oxide (ZnO), silicon carbide (SiC), iron oxide (Fe_3O_4), and titanium dioxide (TiO_2) were the more frequently used nanoparticles in building applications. Silicon carbide (SiC)-based nanofluid performance in the PVT system was considerably high. SiC is a widely available and cheap nanoparticle as compared to the others. Hence, its use in the PVT system is recommended. Overall, all nanofluids have proven to be effective in enhancing the performance of the energy systems. But the studies reviewed were either experimental or theoretical and thus this demands a considerable amount of work to allow the practical use of nanofluids in building applications.

References

[1] T. Abergel, C. Delmastro, Tracking Buildings 2020, Paris, 2020 https://www.iea.org/reports/tracking-buildings-2020

[2] HVAC Energy Breakdown, 2013. https://www.energy.gov.au/publications/hvac-factsheet-energy-breakdown

[3] U. Stritih, Heat transfer enhancement in latent heat thermal storage system for buildings, Energy Build. (2003). Available from: https://doi.org/10.1016/j.enbuild.2003.07.001.

[4] Z. Guo, A review on heat transfer enhancement with nanofluids, J. Enhanc. Heat. Transf. (2020). Available from: https://doi.org/10.1615/JEnhHeatTransf.2019031575.

[5] B.A. Bhanvase, D.P. Barai, S.H. Sonawane, N. Kumar, S.S. Sonawane, Intensified heat transfer rate with the use of nanofluids, Handbook of Nanomaterials for Industrial Applications, Elsevier, Chennai, India, 2018. Available from: https://doi.org/10.1016/B978-0-12-813351-4.00042-0.

[6] D. Ürge-Vorsatz, L.F. Cabeza, S. Serrano, C. Barreneche, K. Petrichenko, Heating and cooling energy trends and drivers in buildings, Renew. Sustain. Energy Rev. (2015). Available from: https://doi.org/10.1016/j.rser.2014.08.039.

[7] M. Izadi, M. El Haj Assad, Use of nanofluids in solar energy systems, in: Des. Perform. Optim. Renew. Energy Syst., 2021. https://doi.org/10.1016/b978-0-12-821602-6.00017-1.

[8] N. Abbas, M.B. Awan, M. Amer, S.M. Ammar, U. Sajjad, H.M. Ali, et al., Applications of nanofluids in photovoltaic thermal systems: a review of recent advances, Phys. A Stat. Mech. Its Appl. (2019). Available from: https://doi.org/10.1016/j.physa.2019.122513.

[9] A. Kaggwa, J.K. Carson, Developments and future insights of using nanofluids for heat transfer enhancements in thermal systems: a review of recent literature, Int. Nano Lett. (2019). Available from: https://doi.org/10.1007/s40089-019-00281-x.

[10] D.P. Kulkarni, D.K. Das, R.S. Vajjha, Application of nanofluids in heating buildings and reducing pollution, Appl. Energy. (2009). Available from: https://doi.org/10.1016/j.apenergy.2009.03.021.

[11] K.S. Suganthi, K.S. Rajan, Metal oxide nanofluids: review of formulation, thermo-physical properties, mechanisms, and heat transfer performance, Renew. Sustain. Energy Rev. (2017). Available from: https://doi.org/10.1016/j.rser.2017.03.043.

[12] H. Farzaneh, A. Behzadmehr, M. Yaghoubi, A. Samimi, S.M.H. Sarvari, Stability of nanofluids: molecular dynamic approach and experimental study, Energy Convers. Manage. (2016). Available from: https://doi.org/10.1016/j.enconman.2015.12.044.

[13] S.S. Sanukrishna, A.S. Vishnu, M. Jose, Prakash, Nanorefrigerants for energy efficient refrigeration systems, J. Mech. Sci. Technol. (2017). Available from: https://doi.org/10.1007/s12206-017-0746-4.

[14] S.M.S. Murshed, K.C. Leong, C. Yang, Investigations of thermal conductivity and viscosity of nanofluids, Int. J. Therm. Sci. (2008). Available from: https://doi.org/10.1016/j.ijthermalsci.2007.05.004.

[15] M. Chandrasekar, S. Suresh, A. Chandra Bose, Experimental investigations and theoretical determination of thermal conductivity and viscosity of Al2O3/water nanofluid, Exp. Therm. Fluid Sci. (2010). Available from: https://doi.org/10.1016/j.expthermflusci.2009.10.022.

[16] Y. Cui, Q. Zhu, Study of photovoltaic/thermal systems with MgO-water nanofluids flowing over silicon solar cells, in: Asia-Pacific Power Energy Eng. Conf. APPEEC, 2012. https://doi.org/10.1109/APPEEC.2012.6307203.

[17] R.A.T. and T.P.O.S. Saroha, T. Mittal, P.J. Modi, V. Bhalla, V. Khullar, H. Tyagi, Theoretical analysis and testing of nanofluids-based solar photovoltaic/thermal hybrid collector, J. Heat. Transf. 137 (2015). Available from: https://doi.org/10.1115/1.4030228.

[18] M. Chandrasekar, S. Suresh, T. Senthilkumar, M. Ganesh Karthikeyan, Passive cooling of standalone flat PV module with cotton wick structures, Energy Convers. Manage. (2013). Available from: https://doi.org/10.1016/j.enconman.2013.03.012.

[19] M. Faizal, R. Saidur, S. Mekhilef, M.A. Alim, Energy, economic and environmental analysis of metal oxides nanofluid for flat-plate solar collector, Energy Convers. Manage. 76 (2013) 162–168. Available from: https://doi.org/10.1016/j.enconman.2013.07.038.

[20] N. Karami, M. Rahimi, Heat transfer enhancement in a PV cell using Boehmite nanofluid, Energy Convers. Manage. 86 (2014) 275–285. Available from: https://doi.org/10.1016/j.enconman.2014.05.037.

[21] M. Sardarabadi, M. Passandideh-Fard, S. Zeinali Heris, Experimental investigation of the effects of silica/water nanofluid onPV/T (photovoltaic thermal units), Energy (2014). Available from: https://doi.org/10.1016/j.energy.2014.01.102.

[22] Z. Xu, C. Kleinstreuer, Computational analysis of nanofluid cooling of high concentration photovoltaic cells, J. Therm. Sci. Eng. Appl. 6 (2014) 031009. Available from: https://doi.org/10.1115/1.4026355.

[23] Z. Said, M.A. Sabiha, R. Saidur, A. Hepbasli, N.A. Rahim, S. Mekhilef, et al., Performance enhancement of a flat plate solar collector using titanium dioxide nanofluid and polyethylene glycol dispersant, J. Clean. Prod. 92 (2015) 343–353. Available from: https://doi.org/10.1016/j.jclepro.2015.01.007.

[24] D. Jing, Y. Hu, M. Liu, J. Wei, L. Guo, Preparation of highly dispersed nanofluid and CFD study of its utilization in a concentrating PV/T system, Sol. Energy (2015). Available from: https://doi.org/10.1016/j.solener.2014.11.008.

[25] H.A. Hussien, A.H. Noman, A.R. Abdulmunem, Indoor investigation for improving the hybrid photovoltaic /thermal system performance using nanofluid (AL_2O_3-water), Eng. Tech. J. (2015).

[26] O. Rejeb, M. Sardarabadi, C. Ménézo, M. Passandideh-Fard, M.H. Dhaou, A. Jemni, Numerical and model validation of uncovered nanofluid sheet and tube type photovoltaic thermal solar system, Energy Convers. Manage. 110 (2016) 367–377. Available from: https://doi.org/10.1016/j.enconman.2015.11.063.

[27] O.Z. Sharaf, M.F. Orhan, Thermodynamic analysis and optimization of densely-packed receiver assembly components in high-concentration CPVT solar collectors, Energy Convers. Manage. 121 (2016) 113–144. Available from: https://doi.org/10.1016/j.enconman.2016.05.012.

[28] A. Radwan, M. Ahmed, S. Ookawara, Performance enhancement of concentrated photovoltaic systems using a microchannel heat sink with nanofluids, Energy Convers. Manage. 119 (2016) 289–303. Available from: https://doi.org/10.1016/j.enconman.2016.04.045.

[29] M. Sardarabadi, M. Passandideh-Fard, Experimental and numerical study of metal-oxides/water nanofluids as coolant in photovoltaic thermal systems (PVT), Sol. Energy Mater. Sol. Cells 157 (2016) 533–542. Available from: https://doi.org/10.1016/j.solmat.2016.07.008.

[30] S. Hassani, R. Saidur, S. Mekhilef, R.A. Taylor, Environmental and exergy benefit of nanofluid-based hybrid PV/T systems, Energy Convers. Manage. 123 (2016) 431–444. Available from: https://doi.org/10.1016/j.enconman.2016.06.061.

[31] E.H.A.R. Noghrehabadi, M. Moravej, An experimental investigation on the performance of a symmetric conical solar collector using SiO_2/water nanofluid, Trans. Phenom. Nano Micro Scales 5 (2016) 23–29. Available from: https://doi.org/10.7508/tpnms.2017.01.003.

[32] A.H.A. Al-Waeli, K. Sopian, M.T. Chaichan, H.A. Kazem, H.A. Hasan, A.N. Al-Shamani, An experimental investigation of SiC nanofluid as a base-fluid for a photovoltaic thermal PV/T system, Energy Convers. Manage. 142 (2017) 547–558. Available from: https://doi.org/10.1016/j.enconman.2017.03.076.

[33] D.D.P. Kalita, Performance Improvement of a Novel Flat Plate Photovoltaic Thermal (PV/T) System Using Copper Oxide Nanoparticle—Water as Coolant, in: Int. Conf. Nano Energy Water, 2017, pp. 97–104. https://link.springer.com/chapter/10.1007/978-3-319-63085-4_14

[34] M.O. Lari, A.Z. Sahin, Design, performance and economic analysis of a nanofluid-based photovoltaic/thermal system for residential applications, Energy Convers. Manage. (2017). Available from: https://doi.org/10.1016/j.enconman.2017.07.045.

[35] M.S.Y. Ebaid, A.M. Ghrair, M. Al-Busoul, Experimental investigation of cooling photovoltaic (PV) panels using (TiO_2) nanofluid in water-polyethylene glycol mixture and (Al_2O_3) nanofluid in water-cetyltrimethylammonium bromide mixture, Energy Convers. Manage. (2018). Available from: https://doi.org/10.1016/j.enconman.2017.10.074.

[36] J.H. Lee, S.G. Hwang, G.H. Lee, Efficiency improvement of a photovoltaic thermal (PVT) system using nanofluids, Energies (2019). Available from: https://doi.org/10.3390/en12163063.

[37] M.S.Y. Ebaid, M. Al-busoul, A.M. Ghrair, Performance enhancement of photovoltaic panels using two types of nanofluids, Heat. Transf. (2020). Available from: https://doi.org/10.1002/htj.21745.

[38] V. Nair, P.R. Tailor, A.D. Parekh, Nanorefrigerants: a comprehensive review on its past, present and future, Int. J. Refrig. (2016). Available from: https://doi.org/10.1016/j.ijrefrig.2016.01.011.

[39] K. Lee, Y. Hwang, S. Cheong, L. Kwon, S. Kim, J. Lee, Performance evaluation of nano-lubricants of fullerene nanoparticles in refrigeration mineral oil, Curr. Appl. Phys. (2009). Available from: https://doi.org/10.1016/j.cap.2008.12.054.

[40] M. Xing, R. Wang, J. Yu, Application of fullerene C60 nano-oil for performance enhancement of domestic refrigerator compressors, Int. J. Refrig. (2014). Available from: https://doi.org/10.1016/j.ijrefrig.2013.12.004.

[41] I.M. Shahrul, I.M. Mahbubul, S.S. Khaleduzzaman, R. Saidur, M.F.M. Sabri, A comparative review on the specific heat of nanofluids for energy perspective, Renew. Sustain. Energy Rev. (2014). Available from: https://doi.org/10.1016/j.rser.2014.05.081.

[42] H. Peng, G. Ding, W. Jiang, H. Hu, Y. Gao, Heat transfer characteristics of refrigerant-based nanofluid flow boiling inside a horizontal smooth tube, Int. J. Refrig. (2009). Available from: https://doi.org/10.1016/j.ijrefrig.2009.01.025.

[43] H. Li, W. Yang, Z. Yu, L. Zhao, The performance of a heat pump using nanofluid ($R22 + TiO_2$) as the working fluid-an experimental study, Energy Procedia (2015). https://doi.org/10.1016/j.egypro.2015.07.158.

[44] I. Dincer, M.M. Hussain, I. Al-Zaharnah, Analysis of sectoral energy and exergy use of Saudi Arabia, Int. J. Energy Res. 28 (2004) 205–243. Available from: https://doi.org/10.1002/er.962.

[45] D.S. Adelekan, O.S. Ohunakin, J. Gill, O.E. Atiba, I.P. Okokpujie, A.A. Atayero, Experimental investigation of a vapour compression refrigeration system with 15 nm TiO_2-R600a nano-refrigerant as the working fluid, Procedia Manuf. (2019). https://doi.org/10.1016/j.promfg.2019.06.079

[46] M. Abbas, R.G. Walvekar, M.T. Hajibeigy, S. Farhood, Efficient air-condition unit by using nano-refrigerant, EURECA (2013).

[47] R. Krishna Sabareesh, N. Gobinath, V. Sajith, S. Das, C.B. Sobhan, Application of TiO 2 nanoparticles as a lubricant-additive for vapor compression refrigeration systems—an experimental investigation, Int. J. Refrig. (2012). Available from: https://doi.org/10.1016/j.ijrefrig.2012.07.002.

[48] R.R. Kumar, K. Sridhar, M. Narasimha, Heat transfer enhancement in domestic refrigerator using R600a/mineral oil/nano-Al_2O_3 as working fluid, Int. J. Comput. Eng. Res. (2013).

[49] L. Fedele, L. Colla, S. Minetto, M. Scattolini, F. Bellomare, S. Bobbo, et al., Nanofluids characterization and application as nanolubricants in heat pump systems, Sci. Technol. Built Environ. (2015). Available from: https://doi.org/10.1080/23744731.2015.1023164.

[50] S. Bi, K. Guo, Z. Liu, J. Wu, Performance of a domestic refrigerator using TiO_2-R600a nano-refrigerant as working fluid, Energy Convers. Manage. (2011). Available from: https://doi.org/10.1016/j.enconman.2010.07.052.

[51] D.S. Kumar, R. Elansezhian, Experimental study on Al_2O_3—R134a nano refrigerant in refrigeration system, Int. J. Mod. Eng. Res. (2012).

[52] N. Subramani, M. Prakash, Experimental studies on a vapour compression system using nanorefrigerants, Int. J. Eng. Sci. Technol. (1970). Available from: https://doi.org/10.4314/ijest.v3i9.8.

[53] D. Yang, B. Sun, H. Li, X. Fan, Experimental study on the heat transfer and flow characteristics of nanorefrigerants inside a corrugated tube, Int. J. Refrig. (2015). Available from: https://doi.org/10.1016/j.ijrefrig.2015.04.011.

[54] D.W. Hawes, D. Feldman, D. Banu, Latent heat storage in building materials, Energy Build. (1993). Available from: https://doi.org/10.1016/0378-7788(93)90040-2.

[55] Y. Zhang, G. Zhou, K. Lin, Q. Zhang, H. Di, Application of latent heat thermal energy storage in buildings: state-of-the-art and outlook, Build. Environ. (2007). Available from: https://doi.org/10.1016/j.buildenv.2006.07.023.

[56] A.M. Khudhair, M.M. Farid, A review on energy conservation in building applications with thermal storage by latent heat using phase change materials, Energy Convers. Manage. (2004). Available from: https://doi.org/10.1016/S0196-8904(03)00131-6.

[57] P. Schossig, H.M. Henning, S. Gschwander, T. Haussmann, Micro-encapsulated phase-change materials integrated into construction materials, Sol. Energy Mater. Sol. Cell (2005). Available from: https://doi.org/10.1016/j.solmat.2005.01.017.

[58] D. Feldman, D. Banu, D. Hawes, E. Ghanbari, Obtaining an energy storing building material by direct incorporation of an organic phase change material in gypsum wallboard, Sol. Energy Mater. (1991). Available from: https://doi.org/10.1016/0165-1633(91)90021-C.

[59] H. Kaasinen, The absorption of phase change substances into commonly used building materials, Sol. Energy Mater. Sol. Cells (1992). Available from: https://doi.org/10.1016/0927-0248(92)90118-9.

[60] A. Marani, M.L. Nehdi, Integrating phase change materials in construction materials: critical review, Constr. Build. Mater. (2019). Available from: https://doi.org/10.1016/j.conbuildmat.2019.05.064.

[61] P.K. Nagarajan, J. Subramani, S. Suyambazhahan, R. Sathyamurthy, Nanofluids for solar collector applications: a review, Energy Procedia (2014). https://doi.org/10.1016/j.egypro.2014.12.017.

[62] A.N. Al-Shamani, M.H. Yazdi, M.A. Alghoul, A.M. Abed, M.H. Ruslan, S. Mat, et al., Nanofluids for improved efficiency in cooling solar collectors—a review, Renew. Sustain. Energy Rev. (2014). Available from: https://doi.org/10.1016/j.rser.2014.05.041.

[63] G. Raam Dheep, A. Sreekumar, Influence of nanomaterials on properties of latent heat solar thermal energy storage materials—a review, Energy Convers. Manage. (2014). Available from: https://doi.org/10.1016/j.enconman.2014.03.058.

[64] A. Zabalegui, D. Lokapur, H. Lee, Nanofluid PCMs for thermal energy storage: latent heat reduction mechanisms and a numerical study of effective thermal storage performance, Int. J. Heat. Mass. Transf. (2014). Available from: https://doi.org/10.1016/j.ijheatmasstransfer.2014.07.051.

[65] P. Gimenez-Gavarrell, V.D. Romanin, S. Fereres, Latent heat of fusion and melting temperature of molten salt based carbon nanotube suspensions used as phase change materials, in: ASME 2015 9th Int. Conf. Energy Sustain. ES 2015, Collocated with ASME 2015 Power Conf. ASME 2015 13th Int. Conf. Fuel Cell Sci. Eng. Technol. ASME 2015 Nucl. Forum, 2015. https://doi.org/10.1115/ES2015-49791.

[66] V. Kumaresan, R. Velraj, S.K. Das, The effect of carbon nanotubes in enhancing the thermal transport properties of PCM during solidification, Heat. Mass. Transf. (2012). Available from: https://doi.org/10.1007/s00231-012-0980-3.

[67] F. Wang, C. Zhang, J. Liu, X. Fang, Z. Zhang, Highly stable graphite nanoparticle-dispersed phase change emulsions with little supercooling and high thermal conductivity for cold energy storage, Appl. Energy (2017). Available from: https://doi.org/10.1016/j.apenergy.2016.11.122.

[68] J.L. Zeng, L.X. Sun, F. Xu, Z.C. Tan, Z.H. Zhang, J. Zhang, et al., Study of a PCM based energy storage system containing Ag nanoparticles, J. Therm. Anal. Calorim. (2007). Available from: https://doi.org/10.1007/s10973-006-7783-z.

[69] S. Wu, D. Zhu, X. Li, H. Li, J. Lei, Thermal energy storage behavior of Al_2O_3-H_2O nanofluids, Thermochim. Acta (2009). Available from: https://doi.org/10.1016/j.tca.2008.11.006.

[70] S. Harikrishnan, A.A. Roseline, S. Kalaiselvam, Preparation and thermophysical properties of water-glycerol mixture-based CuO nanofluids as PCM for cooling applications, IEEE Trans. Nanotechnol. (2013). Available from: https://doi.org/10.1109/TNANO.2013.2265753.

[71] S. Mo, K. Zhu, T. Yin, Y. Chen, Z. Cheng, Phase change characteristics of ethylene glycol solution-based nanofluids for subzero thermal energy storage, Int. J. Energy Res. (2017). Available from: https://doi.org/10.1002/er.3599.

[72] Q. He, S. Zeng, S. Yin, S. Wang, Experimental investigation on nucleation supercooling degree of TiO_2-H_2O nanofluids for cool storage, Adv. Mater. Res. (2012). Available from: https://doi.org/10.4028/http://www.scientific.net/AMR.550-553.2723.

[73] S. Wu, D. Zhu, X. Zhang, J. Huang, Preparation and melting/freezing characteristics of Cu/paraffin nanofluid as phase-change material (PCM), Energy Fuels (2010). Available from: https://doi.org/10.1021/ef9013967.

[74] A. Kasaeian, S. Daviran, R.D. Azarian, A. Rashidi, Performance evaluation and nanofluid using capability study of a solar parabolic trough collector, Energy Convers Manage. (2015). Available from: https://doi.org/10.1016/j.enconman.2014.09.056.

[75] Z.H. Liu, R.L. Hu, L. Lu, F. Zhao, H.S. Xiao, Thermal performance of an open thermosyphon using nanofluid for evacuated tubular high temperature air solar collector, Energy Convers Manage. (2013). Available from: https://doi.org/10.1016/j.enconman.2013.04.010.

[76] Q. He, S. Wang, S. Zeng, Z. Zheng, Experimental investigation on photothermal properties of nanofluids for direct absorption solar thermal energy systems, Energy Convers Manage. (2013). Available from: https://doi.org/10.1016/j.enconman.2013.04.019.

[77] M. Faizal, R. Saidur, S. Mekhilef, Potential of size reduction of flat-plate solar collectors when applying MWCNT nanofluid, in: IOP Conf. Ser. Earth Environ. Sci. (2013). https://doi.org/10.1088/1755-1315/16/1/012004

[78] R. Nasrin, M.A. Alim, A.J. Chamkha, Effects of physical parameters on natural convection in a solar collector filled with nanofluid, Heat. Transf.—Asian Res. (2013). Available from: https://doi.org/10.1002/htj.21026.

[79] T. Yousefi, E. Shojaeizadeh, F. Veysi, S. Zinadini, An experimental investigation on the effect of pH variation of MWCNT-H_2O nanofluid on the efficiency of a flat-plate solar collector, Sol. Energy (2012). Available from: https://doi.org/10.1016/j.solener.2011.12.003.

[80] T. Yousefi, F. Veysi, E. Shojaeizadeh, S. Zinadini, An experimental investigation on the effect of Al_2O_3-H_2O nanofluid on the efficiency of flat-plate solar collectors, Renew. Energy (2012). Available from: https://doi.org/10.1016/j.renene.2011.08.056.

[81] R.A. Taylor, P.E. Phelan, T.P. Otanicar, C.A. Walker, M. Nguyen, S. Trimble, et al., Applicability of nanofluids in high flux solar collectors, J. Renew. Sustain. Energy (2011). Available from: https://doi.org/10.1063/1.3571565.

[82] R.A. Taylor, P.E. Phelan, T.P. Otanicar, R. Adrian, R. Prasher, Nanofluid optical property characterization: towards efficient direct absorption solar collectors, Nanoscale Res. Lett. (2011). Available from: https://doi.org/10.1186/1556-276X-6-225.

[83] H. Tyagi, P. Phelan, R. Prasher, Predicted efficiency of a low-temperature nanofluid-based direct absorption solar collector, J. Sol. Energy Eng. Trans. ASME (2009). Available from: https://doi.org/10.1115/1.3197562.

[84] O. Mahian, A. Kianifar, S.A. Kalogirou, I. Pop, S. Wongwises, A review of the applications of nanofluids in solar energy, Int. J. Heat. Mass. Transf. (2013). Available from: https://doi.org/10.1016/j.ijheatmasstransfer.2012.10.037.

[85] N. Sizochenko, M. Syzochenko, A. Gajewicz, J. Leszczynski, T. Puzyn, Predicting physical properties of nanofluids by computational modeling, J. Phys. Chem. C (2017). Available from: https://doi.org/10.1021/acs.jpcc.6b08850.

Chapter 16

Ionic nanofluids: preparation, characteristics, heat transfer mechanism, and thermal applications

Tayyab Raza Shah[1], Chao Zhou[1], Hafiz Muhammad Rizwan[2], Muhammad Abdullah[3], Asad Iqbal[3], Awais Awan[3] and Hafiz Muhammad Ali[4,5]

[1]*College of Engineering, Peking University, Beijing, P.R. China,* [2]*Faculty of Mechanical Engineering, GIK Institute of Engineering Sciences and Technology, Topi, Pakistan,* [3]*Mechanical Engineering Department, Quaid-E-Azam College of Engineering and Technology, Sahiwal, Pakistan,* [4]*Mechanical Engineering Department, King Fahd University of Petroleum & Minerals (KFUPM), Dhahran, Saudi Arabia,* [5]*Interdisciplinary Research Center for Renewable Energy and Power Systems (IRC-REPS), King Fahd University of Petroleum and Minerals, Dhahran, Saudi Arabia*

Abbreviations

C_p	Heat capacity (specific) (J/kg K)
K	Thermal conductivity (W/m K)
L	Width of cavity (m)
N_u	Nusselt number (average)
Nu_L	Nusselt number (local)
P	Pressure (N/m^2)
Pr	Prandtl number
Ra	Rayleigh number
u	Component of velocity (x-direction)
U	Dimensionless component of velocity (x0-direction)
v	Component of velocity (y-direction)
V	Dimensionless component of velocity (y-direction)
w	Length of hot element (m)

Subscripts

f	Base fluid
Ionf	Ionanofluid

Advances in Nanofluid Heat Transfer. DOI: https://doi.org/10.1016/B978-0-323-88656-7.00017-9

nf Water-based nanofluid
r Ratio

16.1 Introduction

Ionanofluids are prepared by dispersing nanoparticles in ionic liquids (also known as modern salt at room temperature). Ionic liquids have tunable thermophysical characteristics and their features could be tailored by varying the anion–cation ratio as per the requirements of the application [1–5]. A list of the most commonly used ionic liquids as the base fluid for making ionanofluids is presented in Table 16.1. Several general characteristics of ionic liquids are as follows:

1. Ionic liquids have high viscosity, and the viscosity of ionic liquids tends to decrease with increasing temperature since the temperature augmentation causes the intermolecular attractive forces to weaken. The high viscosity of ionic liquids makes them suitable for duction heat transfer. The viscosity of ionic liquids can be tailored by introducing some amount of low viscosity fluids. Nanoparticle loading further tends to increase the viscosity of the fluids. High viscosity poses challenges like high frictional losses and high pumping power due to high-pressure drops during the flow.
2. Thermal conductivity of ionic liquids either stays unchanged or tends to decrease when the temperature of the fluid is increased. The thermal conductivity of ionic liquids is also low which is increased by the loading of high thermal conductivity nanoparticles.
3. Ionic liquids carry good specific heat values and the specific heat tends to increase with increasing temperature and nanoparticle loading.
4. Density of ionic liquids tends to decrease with an increase in temperature. However, nanoparticle loading makes the density of the fluid higher due to the nanolayered effects and intermolecular attraction.

Characteristics of ionanofluids are very much dependent on the method of preparation, ionic liquid characteristics, nanoparticle characteristics (like particle size, particle shape, particle rheological and structural characteristics), and operational conditions [7–10].

This chapter comprises an explanation of methods of ionanofluids preparation with a discussion on the positives and negatives of either of the preparation methodologies. Subsequently, a coherent discussion on thermophysical characteristics and the important parameters that affect the performance of ionanofluids is presented. Furthermore, potential thermal applications of ionanofluids have been discussed. Eventually, a brief discussion on some of the challenges that the ionanofluids face has been included.

TABLE 16.1 List of some of the most commonly used ionic liquids as base fluids for ionanofluid preparation [6].

Name		Density (g/cm^3)	Heat capacity (J/g-°C)	Thermal conduction (W/m-k)
Therminol VP-1		1.140 at 60°C	1.774 at 100°C	0.135 at 25°C
Therminol 68		0.860 at 25°C	1.880 at 100°C	0.123 at 25°C
Shell Thermia B		0.940 at 25°C	2.180 at 100°C	0.132 at 25°C
Syltherm 800		1.020 at 25°C	1.750 at 100°C	0.134 at 25°C
[EMIM][BF_4]	1-Ethyl-3-methylimidazolium tetrafluoroborate	1.218 at 60°C	1.281 at 100°C	0.200 at 25°C
[BMIM][BF_4]	1-Butyl-3-methylimidazolium tetrafluoroborate	1.181 at 60°C	1.627 at 100°C	0.186 at 25°C
[BMIM][Tf_2N]	1-Butyl-3-methylimidazolium bis(trifluoromethyl sulfonyl)imide	1.384 at 60°C	1.430 at 100°C	0.127 at 25°C
[HMIM][BF_4]	1-Hexyl-3-methylimidazolium tetrafluoroborate	1.145 at 60°C	2.355 at 100°C	0.162 at 25°C
[C_2mim][CH_3SO_3]	1-Ethyl-3-methylimidazolium methane sulfonate	1.214 at 60°C	1.773 at 60°C	0.188 at 25°C
[C_4mim][NTf_2]	3-Butyl-3-octylimidazolium bis[trifluoromethyl sulfonyl]imide	1.386 at 60°C	1.740 at 60°C	0.126 at 25°C
[C_2mim][$EtSO_4$]	1-Ethyl-3-methyl imidazolium methyl sulfate	1.241 at 25°C	1.615 at 25°C	0.175 at 25°C
[C_4mim][Cl]	1-*n*-butyl-3-methylimidazolium chloride	1.086 at 25°C	1.849 at 25°C	0.176 at 25°C
[C_4mim][Br]	1-Butyl-3-methylimidazolium bromide	1.293 at 25°C	1.447 at 25°C	0.160 at 25°C
[C_4mim][I]	1-Butyl-3-methylimidazolium iodide	1.489 at 25°C	1.142 at 25°C	0.131 at 25°C
[C_4mim][BF_4]	1-Butyl-3-methylimidazolium tetrafluroborate	1.210 at 25°C	1.612 at 25°C	0.163 at 25°C

16.2 Preparation methods

Preparation methodologies of ionanofluids are the same as those of conventional nanofluids. Ionanofluids could be prepared by either the one-step method of preparation in which nanoparticle synthesis [through physical vapor deposition process (PVD)] and nanoparticle dispersion in the base fluid, that is, the ionic liquid, is carried out simultaneously, or the two-step preparation method in which the nanoparticles are first synthesized through some chemical and mechanical processing and then the particles are dispersed in the base fluid. Methods of nanofluid preparation have been schematically presented in Fig. 16.1.

The preparation method of ionanofluids holds vital significance as the performance of the fluid and thermophysical properties are highly influenced by the way the ionanofluid is prepared. The parameters that impact the selection of the method include cost, fluid quantity, type of base fluid, type of nanoparticles, and the desired application of the fluid. Among the aforementioned preparation method selection criteria, the cost is the most crucial one since the single-step method is quite expensive compared to the two-step method. Moreover, the quantity of the fluid is also a decisive factor when selecting the preparation methodology for a fluid. For the production of bulk quantity of fluid, the two-step method is the most suitable one since it is less

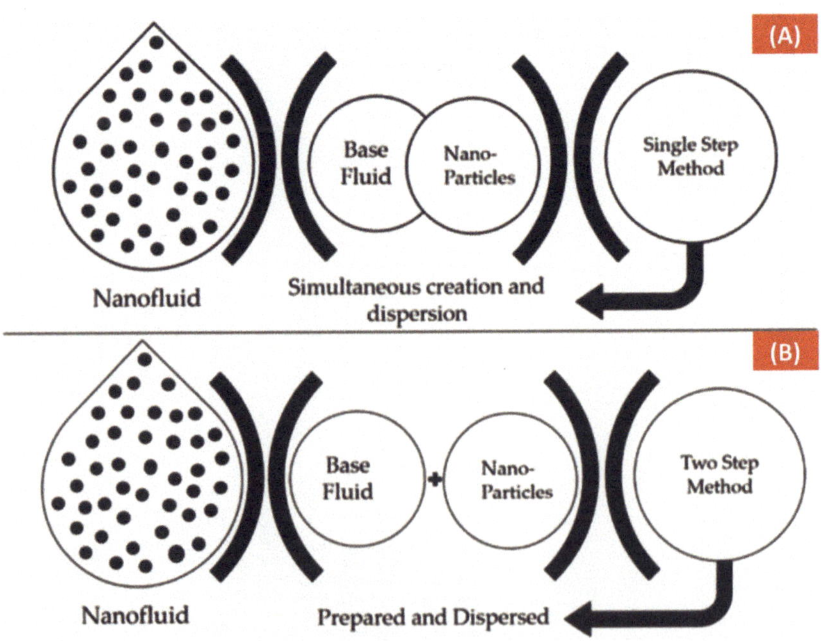

FIGURE 16.1 Nanofluid preparation method: (A) single-step and (B) two-step [11].

expensive and yields a larger quantity of ionanofluids. However, the stability and thermophysical characteristics of the nanofluids prepared through the single-step method are quite superior to those prepared through the two-step preparation method.

Even though the single-step method yields better ionanofluids in terms of thermophysical characteristics and colloidal stability, it is scarcely used in ionanofluid production due to its high cost and complexity, as reported by Minea and Murshed [12]. The two-step method is the most widely used method of preparing the ionanofluids, as reported in the literature. Jóźwiak et al. [13] experimentally evaluated the thermal performance of MWCNT-$[C_2C_1im][SCN]$ ionanofluid. They prepared the nanofluid sample through the two-step method. They prepared three samples of ionanofluids with 0.0531 vol.%, 0.0930 vol.%, and 0.133 vol.% concentration of MWCNT nanoparticles in the base fluid. They sonicated the samples for mere 10 minutes, and they observed complete sedimentation after six months. Wang et al. [14] prepared gold-based ionanofluid through the two-step method and they employed a novel technique of two-phase process, as depicted in Fig. 16.2. Ionanofluid samples prepared through both of the processes were stirred for some time to evenly disperse the nanoparticles in the base fluid. However, the absorption analysis (as shown in Fig. 16.3) showed that nanoparticle clusters tend to form even during the preparation process in the case of the two-phase process.

Amde et al. [15,16] also used a two-step method to prepare ZnO-$[HMIM][PF_6]$ ionanofluid and performed ultrasonication for 90 minutes to

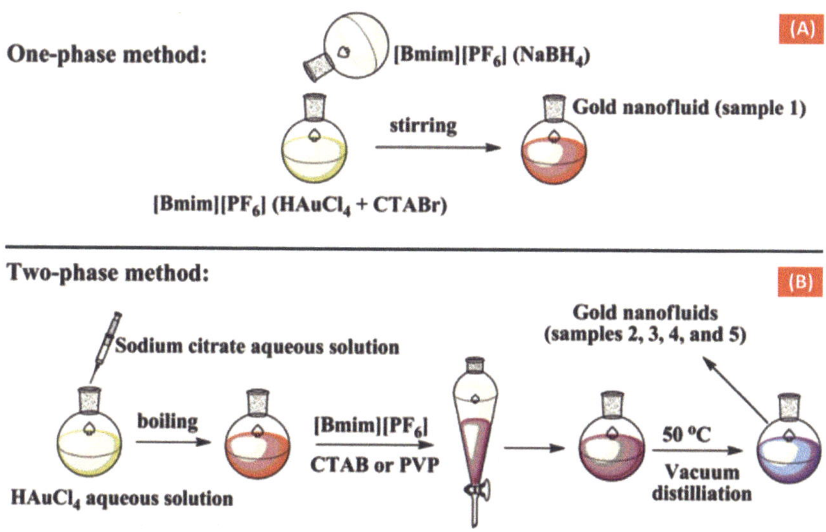

FIGURE 16.2 Preparation methodology of gold-based ionanofluids via one-phase process (A) and two-phase process (B) [14].

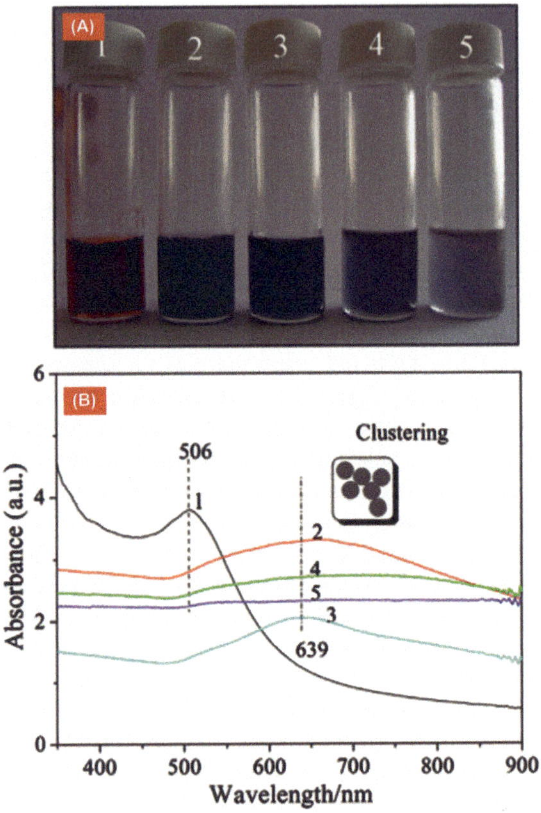

FIGURE 16.3 Gold-based ionanofluid samples (A) and light absorption through the ionanofluid samples [14].

homogenize the colloidal suspension of nanoparticles. The preparation scheme and the step-wise illustration have been presented in Fig. 16.4. Similarly, Wu et al. [17] employed the same preparation method for TiO_2-based ionanofluid. Deb et al. [18] prepared the ZnS-based ionanofluid using the same method. Alizadeh et al. [19] prepared graphene-based ionanofluid (with [HMIM][PF_6] ionic liquid as base fluid) by the two-step method. Their work was intended to test the viscosity, electrical conductivity, and surface tension of the ionanofluid in response to the temperature change. Oster et al. conducted an extensive study to evaluate the heat capacity and density of various ionic liquids and ionanofluids with different base fluid and nanoparticle compositions.

In the literature, the two-step method has been predominantly used for ionanofluid preparation due to process simplicity and lower cost. However, the one-step method is preferred for better results, as there are no processes like nanoparticle drying, storage, transportation, sonication, etc., which reduces the nanoparticle clustering probability. In the two-step

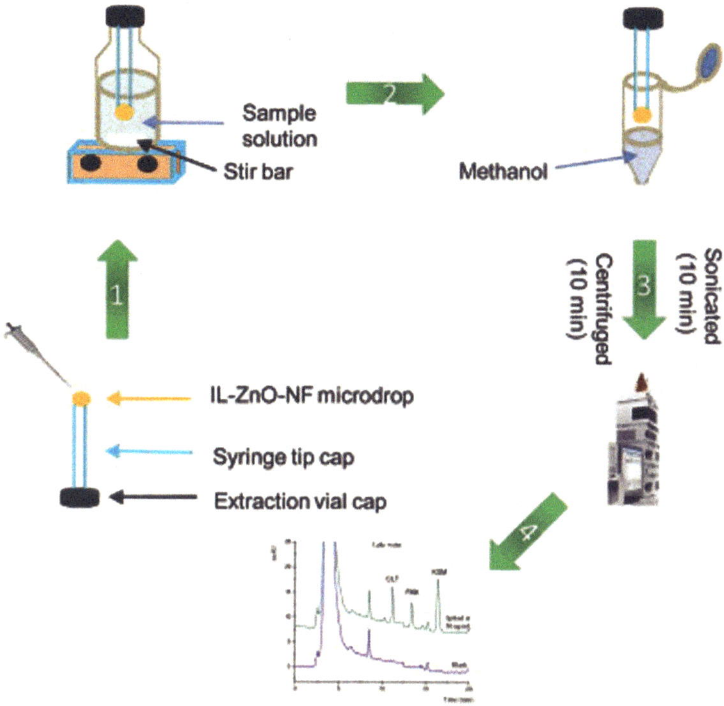

FIGURE 16.4 Preparation scheme of ZnO-[HMIM][PF_6] ionanofluid [15].

method, the following major considerations are taken into account to harness better results.

1. Since the nanoparticles are synthesized separately in the two-step method, the nanoparticle size must be uniform, and the drying process of the nanoparticles must be carefully performed.
2. It must be made sure to keep the nanoparticles away from moisture during the packaging and transportation process.
3. Magnetic stirring and ultrasonication must be carried out at optimum frequency temperature for an appropriate period. Prolonged sonication or high temperature can cause cluster formation, boiling of base fluid that will eventually lead to chemical and physical composition deterioration, and nanoparticle surface/structure breakdown.
4. It is recommended to use a mechanical homogenizer to break the nanoparticle clusters and uniformly distribute the nanoparticles in the base fluid.

Some of the most important steps that help to prepare effective nanofluids are presented in Fig. 16.5.

After the preparation of ionanofluid, some analysis/techniques are employed to evaluate the dispersibility and uniformity of nanoparticles in the base fluid. Some of the postpreparation tests/analyses have been presented in Fig. 16.6.

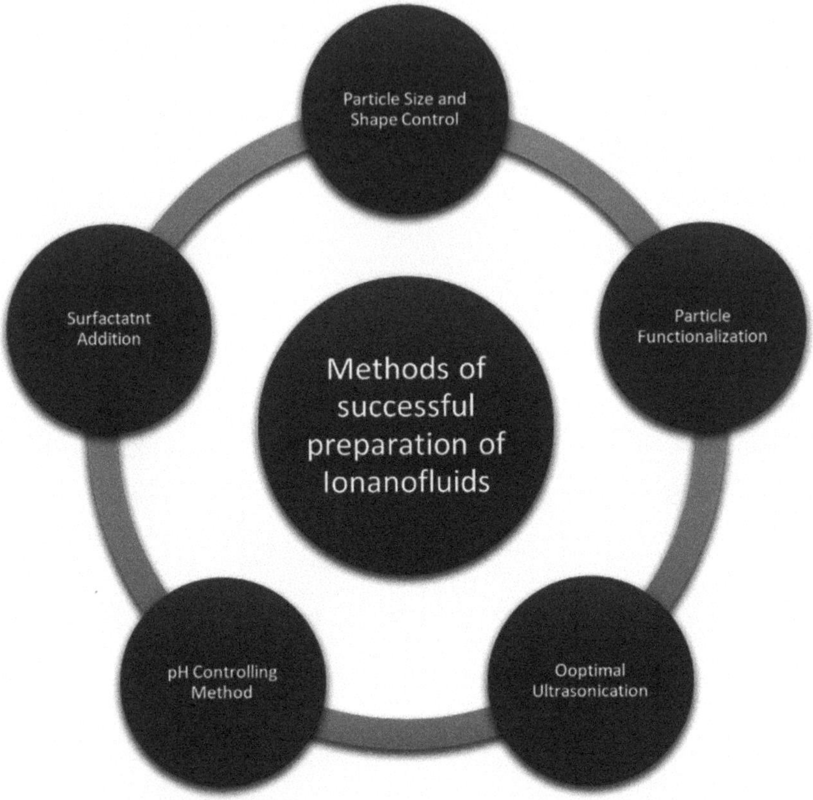

FIGURE 16.5 Nanoparticle stability improvement techniques.

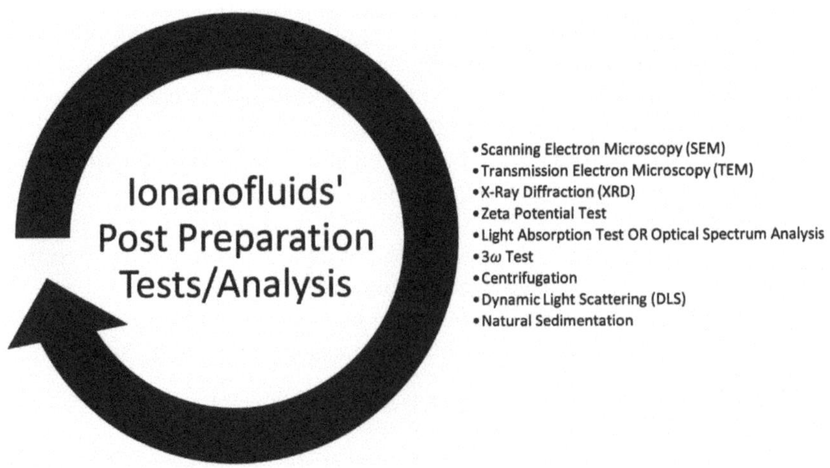

FIGURE 16.6 Postpreparation analysis of ionanofluids.

16.3 Characteristics

Ionanofluids possess very attractive characteristics that are essential for a heat transfer fluid. Some of the most important characteristics of ionanofluids that have brought them into the limelight include viscosity, density, thermal conductivity, and specific heat capacity.

Characters of ionanofluids are influenced by various parameters and the value of these characteristics is very much dependent on the type of response against the change in those parameters.

16.3.1 Thermal conductivity

The thermal conductivity of the ionanofluids has been reported to be much better than the base fluid, that is, ionic liquids. The factors influencing the thermal conductivity of ionanofluids have been presented in Fig. 16.7.

Ionic liquids are reported to possess relatively low thermal conductivity as compared to some of the most commonly used heat transfer fluids. However, ionic liquids offer tunable characteristics by varying the cation–anion ratio in the fluid. Therefore, to increase the thermal conductivity of the ionic liquids, nanoparticles of high thermal conductivities are disseminated, and positive results have been reported in most of the recent studies. Moreover, the thermal conductivity of the ionic liquids has been reported to manifest no significant response to the temperature change (sometimes it tends to decrease with an increase in temperature) [6]. Even though the presence of nanoparticles increases the thermal conductivity of

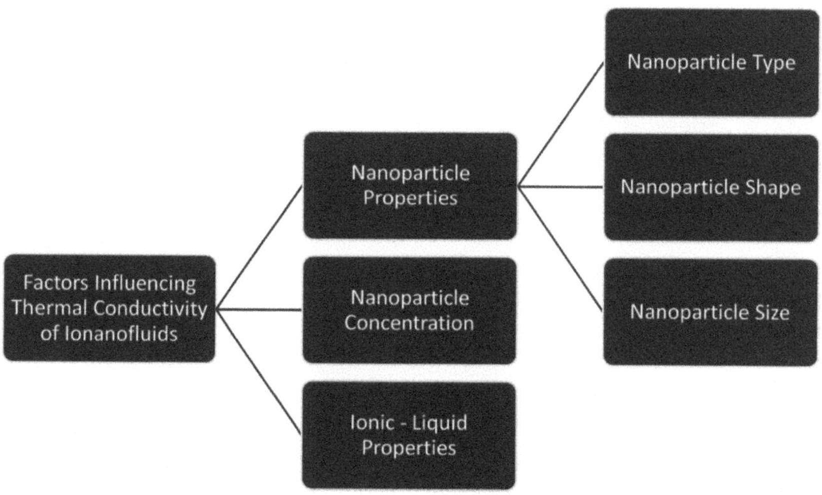

FIGURE 16.7 Important factors affecting thermal conductivity of ionanofluids.

the ionic liquids, the change in thermal conductivity with a temperature change is very minimal. An increase in thermal conductivity has been reported up to 30% [12].

The influence of the nanoparticle concentration on the thermal conductivity of the ionanofluids is the same as that of simple nanofluids. An increase of nanoparticle concentration is observed to increase the thermal conductivity of ionanofluids due to the increased number of nanoparticles resulting in higher particle interaction and greater Brownian motion [20].

The size and shape of nanoparticles have a vital influence on the thermal conductivity of ionanofluids [21]. The greater size of nanoparticles offers a higher surface area for convective heat transfer, whereas smaller nanoparticle size offers greater Brownian motion which leads to the higher thermal conductivity of ionanofluids.

Furthermore, a deeper insight into thermal conductivity and mechanism of heat transfer through the ionanofluids has been presented in section 16.4.

16.3.2 Viscosity

Ionic liquids have been reported to have higher viscosity as compared to conventional thermal fluids [22]. The introduction of nanoparticles in ionic liquids further increases viscosity. However, an increase in temperature reduces the viscosity of ionic liquids and ionanofluids. The nanoparticle concentration increase increases the viscosity of ionanofluids to a great extent and viscosity increases in turn increase the friction factor and increase pumping power. An increase in nanoparticle extent leads to higher interaction between nanoparticles and enables them to form clusters which cause a higher viscosity of the fluid. The greater size of nanoparticles also causes higher viscosity of the fluid. The shape of nanoparticles also has a notable effect on the viscosity of the ionanofluids.

16.3.3 Specific heat

The effect of nanoparticle addition on the specific heat of ionic liquids is quite inconclusive as some of the researchers have reported a positive impact of the addition of nanoparticles on the specific heat, while others have reported either no influence or a negative effect. Similarly, the temperature has been reported to have the same effect as the nanoparticle concentration has on the specific heat of the ionanofluids.

16.3.4 Density

The density of ionanofluids is the least discussed property yet it has a significant effect on heat transfer performance and has been observed to undergo positive change when temperature and nanoparticle concentration are increased.

16.4 Heat transfer mechanism

Ionic liquid-based nanofluids show different kinds of properties that permit characterization of modern heat transfer fluids by adapting the mechanism of cation and anion to achieve the required physical and chemical properties. The ionic nanofluids are considered environmentally friendly because they are not flammable or unstable at atmospheric conditions and they can also be recycled. The heat transfer of ionanofluids as well as of conventional nanoliquids is the most important phenomenon to assess the thermal performance and it can be calculated experimentally easily (nevertheless, there are more studies on numerical evaluation as compared to the number of experimental studies). However, the heat transfer properties of ionanofluids, that is, thermal conductivity and viscosity, are assumed to be the most important, and they should be investigated properly because if their values increase the heat transfer effect will be increased. The overall energy consumption for ionic-nanofluids is reduced due to the increase in the value of the heat transfer effect.

The thermal conductivity of different ionanofluids is the main feature to evaluate the heat transfer effect at different values of temperature. The plot of some ionanofluids with different values of thermal conductivity against different values of temperature is presented in Fig. 16.8.

Minea et al. [24] numerically evaluated the heat transfer performance of the ionanofluids. The heat transfer mechanism for ionanofluids was investigated in two cases with different locations of the heater. The heater is positioned at the lowest surface of the enclosure (case-1). In case-1, the heater is positioned at the left corner of the upright wall. The heater's length was half of the eddy produced in the enclosure. It was concluded from the comparison that the natural convection effect is produced owing to the ionanofluid filled in the cavity. The ionic nanofluids are obtained by adding some nanoparticles in the basic fluids with specific concentrations. In this case, the [C_4MIM] and [NTf_2] were used as base liquids and the nanoparticles of Al_2O_3 with [0.5−1.5 vol.%] concentration were added to form ionic nanofluids [24]. They investigated the influence of Rayleigh number (Ra), nanoparticle concentration, and thermophysical properties on convective heat transfer performance. They observed a very insignificant effect of temperature on Nusselt number (Nu). Nanoparticle concentration increased the viscosity which in turn increased the conduction heat transfer phenomenon. However, the effect of Ra on Nu was quite significant as shown in Fig. 16.9.

Most of the studies on ionanofluids are based on numerical appraisals of thermophysical characteristics and heat transfer performance. The governing equations with nondimensional numbers, energy equations, and momentum equations for two-dimensional steady flow are presented as follows.

$$X = \frac{x}{L}, \quad Y = \frac{y}{L}, \quad U = \frac{uL}{\alpha_f}, \quad V = \frac{vL}{\alpha_f}, \quad P = \frac{PL^2}{\rho_{ionf}\alpha_f^2} \tag{16.1}$$

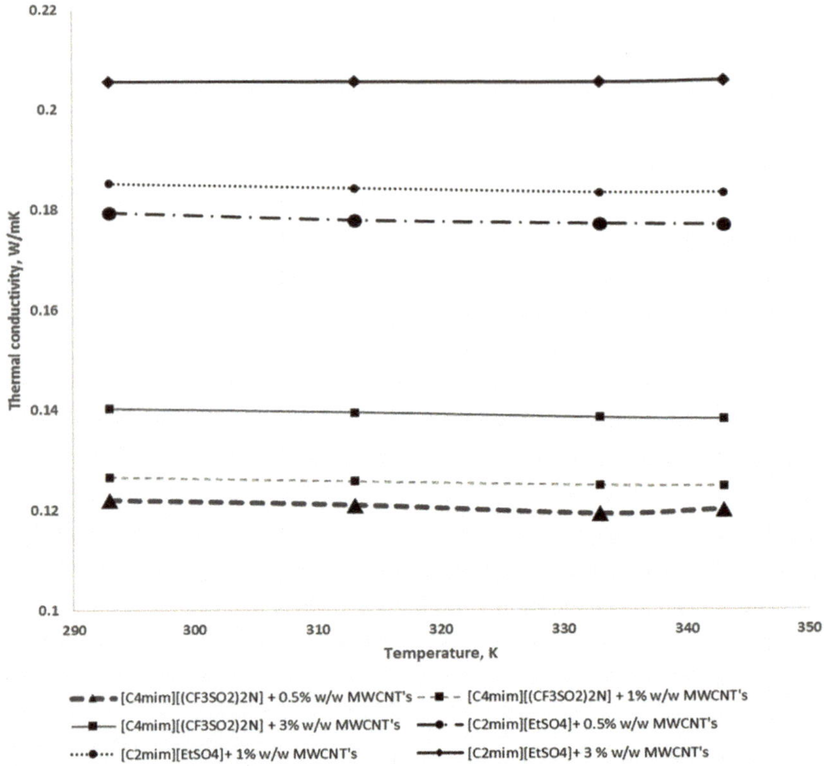

FIGURE 16.8 Thermal conductivity variation of ionanofluid against the temperature [23].

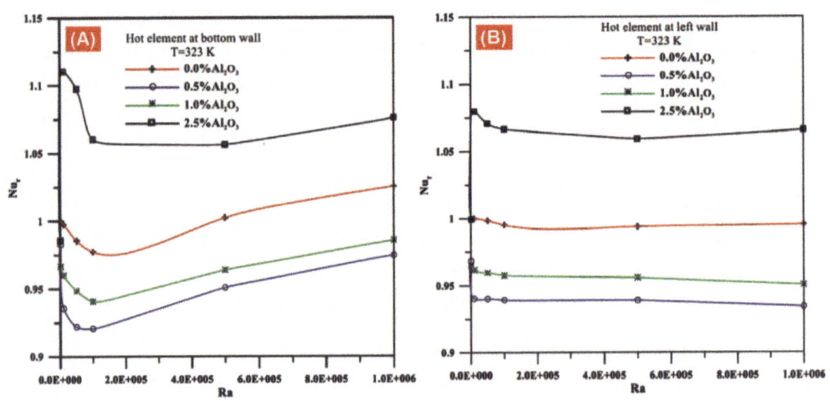

FIGURE 16.9 Effect of Ra on Nu of ionanofluids [24].

$$Pr = \frac{\nu_f}{\alpha_f}, \quad Pr = \frac{\beta_f L^3 (T_h - T_c)}{\nu_f \alpha_f}, \quad W = \frac{w}{L} \tag{16.2}$$

Whereas,

$$\alpha = \frac{k}{\rho c_p}, v = \frac{u}{\rho} \tag{16.3}$$

The dimensionless equations of energy, momentum, and continuity in two-dimensional steady-state flow are presented as follows.

$$\frac{\partial U}{\partial X} + \frac{\partial V}{\partial Y} = 0 \tag{16.4}$$

$$U\frac{\partial U}{\partial X} + V\frac{\partial U}{\partial X} = -\frac{\partial P}{\partial X} + \frac{u_{ionf}}{\rho_{ionf}\alpha_f^2}\left(\frac{\partial^2 V}{\partial X^2} + \frac{\partial^2 U}{\partial Y^2}\right) \tag{16.5}$$

$$U\frac{\partial U}{\partial X} + V\frac{\partial U}{\partial X} = -\frac{\partial P}{\partial X} + \frac{u_{ionf}}{\rho_{ionf}\alpha_f^2}\left(\frac{\partial^2 V}{\partial X^2} + \frac{\partial^2 U}{\partial Y^2}\right) + \frac{\beta_{ionf}}{\beta_f}RaPr\theta \tag{16.6}$$

$$U\frac{\partial U}{\partial X} + V\frac{\partial U}{\partial X} = \frac{\alpha_{ionf}}{\alpha_f}\left(\frac{\partial^2 \theta}{\partial X^2} + \frac{\partial^2 \theta}{\partial Y^2}\right) \tag{16.7}$$

The comparison of results with simulation results shows that minor concentrations of Al_2O_3 in the ionic fluids result in higher Nu as compared to the nanofluids having water as base fluid [24]. On the other hand, at high volume concentrations, the viscosity gets increased, and the heat transfer through conduction heat transfer in the liquid becomes more dominant than the convection.

The enhancement in the healing process generally can be characterized by the theory of ideal gas mixture but this approach is not valid for ionanofluids. In the case of ionanofluids, the C_P of CNT is less than the ionic liquid, therefore the improvement in the heat transfer effect is detected. According to the thermal equilibrium phenomena between the ionanofluids and the ionic liquids, the values obtained are not accurate enough to explain the transfer effect.

There are different ways to explain the heat transfer process in the ionic liquids like (1) the C_p of the value of nanoparticles is much higher than the ionic liquids and is associated with the phonon spectrum with distinct values, and notably the influence of nanoparticle size is insignificant; and (2) due to the collision of nanoparticles and base fluid molecules, the greater surface area in a unit mass of nanoparticles elevates the thermal resistance and eventually results in extra thermal storage [25].

The enhancement in the healing process of ionic nanofluids depends upon semisolid-like liquids. The heat transfer mechanism for ionanofluids in the case of a semisolid can be analyzed on the surface of nanoparticles instead of interacting layers of fluid. The simulation modeling of nanolayers including thermal analysis shows that due to the liquid nanolayers on the surface of nanoparticles the heat transfer takes place.

The heat transfer phenomenon of incompressible ionanofluid between two infinite parallel circular chambers was investigated by Justyna et al. [26] To observe the mechanism of heat transfer, three nanoparticle types (Cu, Al_2O_2, and TiO_2) with concentrations ranging from 0.1 to 0.7 vol.%, were dispersed in the ionic base fluid. The effects of a squeeze number (S), nanoparticle concentration, heat source (Hs), and Prandtl number on Nu were mathematically analyzed. It was concluded from the analysis that Cu is the finest choice for the increase of the heat transfer via optimized thermal parameter [26].

Ionic liquids, in the form of salts, have a complicated structure and ionic liquids depend upon ionic interaction. Taking into account the ionic structures, the ionic fluids can interact with compounds, for example, 1-alkyl-3-methylimidazolium-based mixtures. Ionic liquids contain delocalized electron configuration past the N_1-C_2-N_3 atoms. Between C_4 and C_5 there is a double bond at the opposite side of the ring. Moreover, there is a weak delocalization in the middle region as shown in Fig. 16.10.

The ionic fluids are examined from the perspective of a system of interacting cations and anions due to the attraction forces as well as a network hydrogen bond. The size of an anion is a more critical feature in this networking because if the size of anion is bigger the ionic nature is increased and thus the delocalization process is increased. Therefore, the position and size of the ionic liquids are different.

The ionanofluids can be prepared by adding some nanosized particles to the heat transfer fluids and it is an easy way to improve the heat transfer effect. The thermophysical characteristics like density thermal conductivity and heating capacity make the ionic liquids more preferable when used as base fluids in nanofluids. The ionanofluids were investigated at a temperature above 100°C and it was observed that ionanofluids used for the average to the large-temperature thermal system have distinct characteristics as

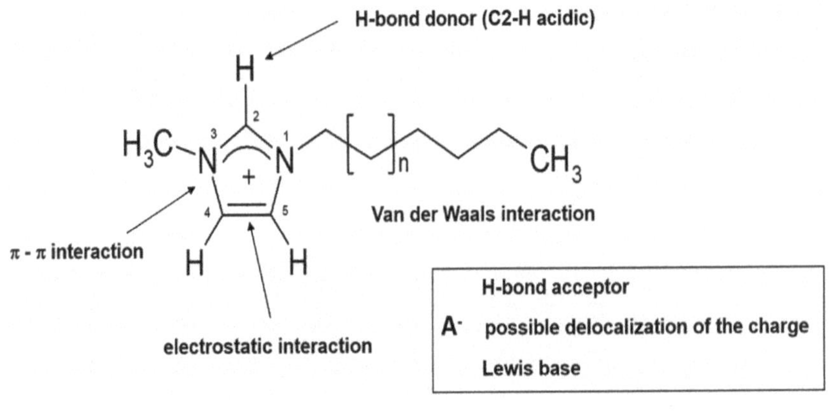

FIGURE 16.10 Ionic liquid atomic structure [26].

compared to the nanofluids having water and EG as base fluid. Hence, a research gap is present to find the wide range of temperatures for the ionanofluids system for heat transfer.

The ionanofluids are prepared by adding some nanoadditives to the ionic liquids. The graphene is more prominent for the preparation of ionanofluids by adding it into the base fluid ([HMIM]BF_4) at a temperature below 100°C. The experimental analysis for evaluating the thermophysical properties above 100°C has been performed as well by Liu et al. [27]. The thermophysical properties for the [HMIM][BF_4] ionic liquid and ionanofluids containing graphene nanoadditives are considered in a temperature range of 20°C–215°C and the results of thermophysical properties (thermal conductivity, viscosity, and density) are presented in Figs. 16.11–16.14. Moreover, the thermal stability of ionic base fluid and graphene nanoadditives-based ionanofluids was analyzed, and the potential of the graphene nanoadditives-based ionanofluids for high thermal systems has been studied [27].

The [HMIM][BF_4] ionic liquid-based nanofluid containing graphene nanoadditives displayed greater thermal stability owing to a higher disintegration temperature of 440°C. Results showed that as the temperature varies from 25°C to 200°C, the value of thermal conductivity of ionanofluids and the base fluid changes in a linear manner with the temperature, while their viscosities manifest a significant drop, specifically in the temperature range

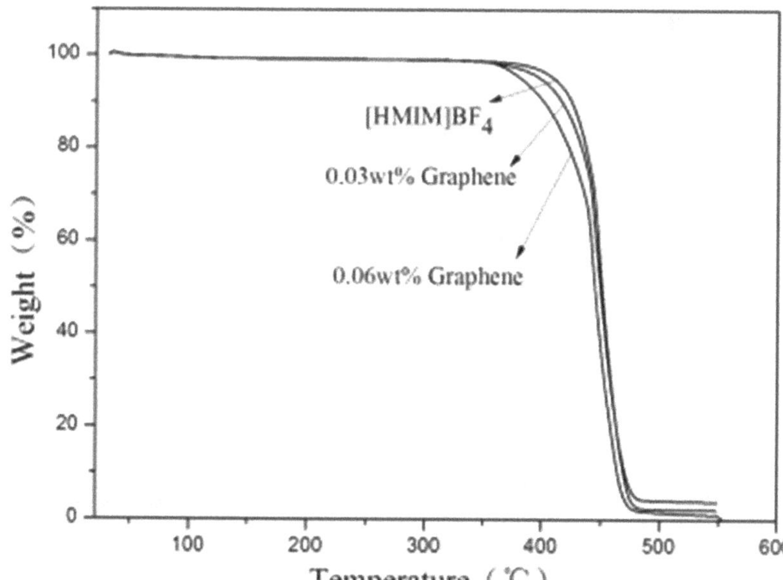

FIGURE 16.11 The concentration of [HMIM][BF_4] and the graphene-based ionanofluids against temperature [27].

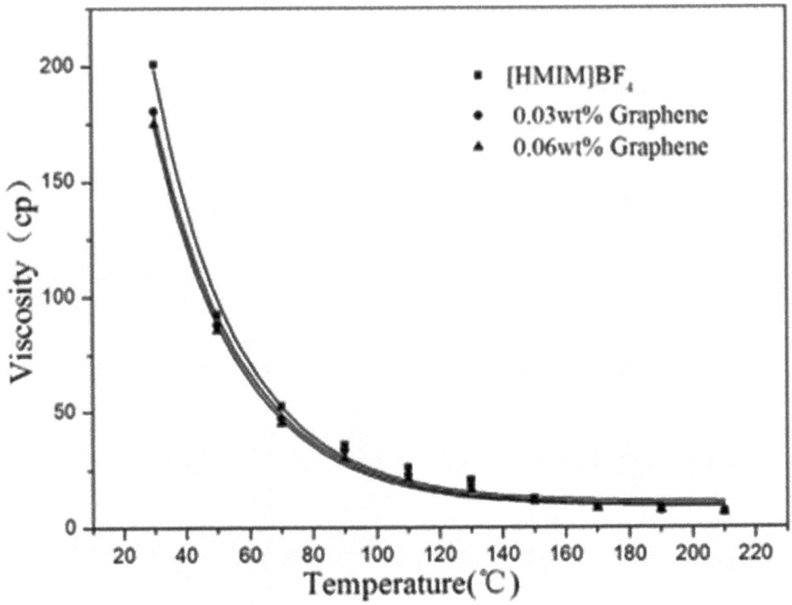

FIGURE 16.12 Change in viscosity of ionanofluids against the temperature [27].

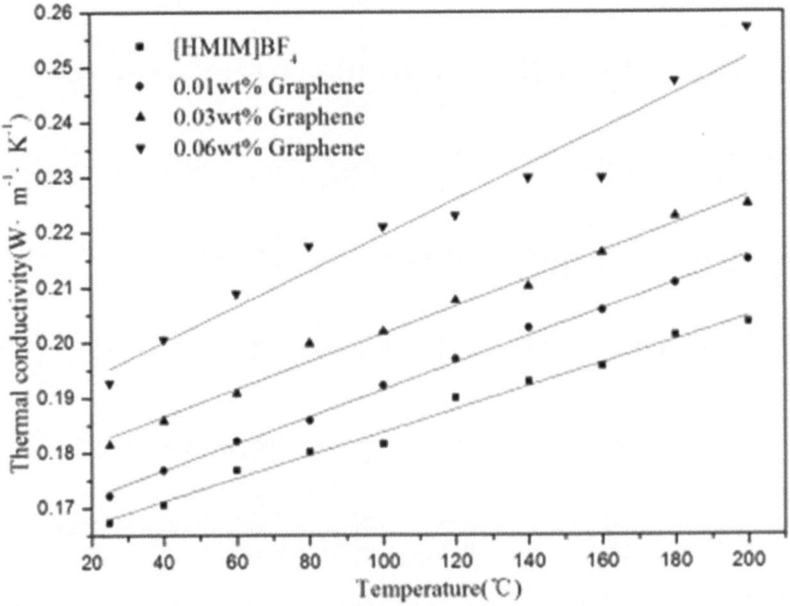

FIGURE 16.13 Thermal conductivity variation of ionanofluid against the temperature [27].

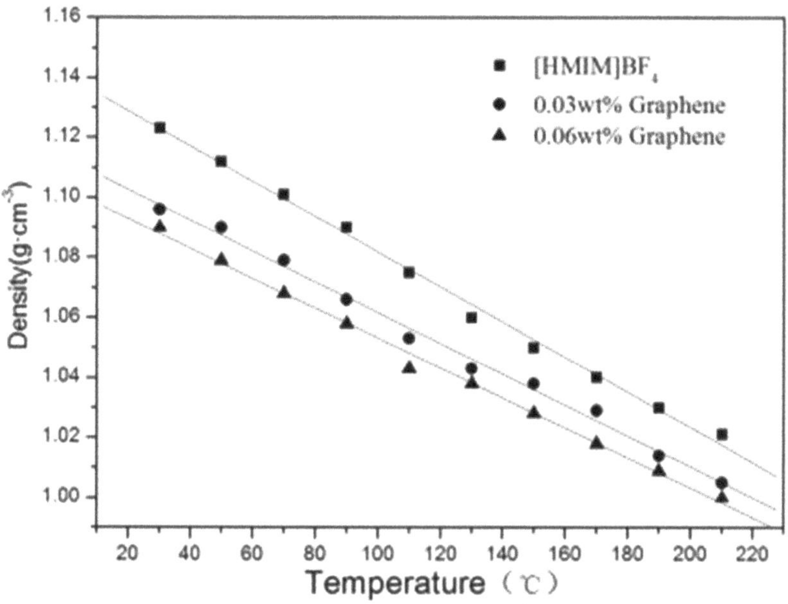

FIGURE 16.14 Ionanofluid density changes against the temperature variation [27].

of 30°C–90°C. As the temperature increases from 20°C to 200°C, the 0.06 wt.% graphene-based ionanofluid displays a prominent improvement in thermal conductivity of 15.2% to 22.9%, a significant decrease in viscosity 13.1%, a small decline in C_p, and a density of 3% and 2.8% was reported, respectively, compared to the base fluid. The thermodynamic characteristics of the base ionic fluid and the graphene-based ionanofluids are far superior to the existing conventional thermofluids and VP − 1. These features make them most suitable for the state-of-the-art thermal systems in the medium- to high-temperature range [27].

The ionic fluids with suspended nanoparticles are considered for solar application in the future due to their specific properties. This phenomenon has a greater impact on the heat transfer mechanism of ionic fluids because when the nanoparticles are added to the ionic fluids the heat transfer effect is increased due to convection. The behavior or change in the value of viscosity of ionanofluids is characterized by rheological tests. The viscosity of ionanofluids $[C_2MIM][CH_3SO_3]$ with the addition of alumina nanoparticles was observed and the value of viscosity was evaluated using a rheometer.

A comparison of all the other fluids with ionanofluids was made to check the Newtonian behavior. Fig. 16.15 shows that the shear rate for all the fluids is high compared to the ionic fluids. It can be seen that the value of viscosity is 0.113 Pa-s at 298.15K. However, the magnitude of viscosity is increased with the addition of nanoparticles. The highest rate of increase in viscosity is

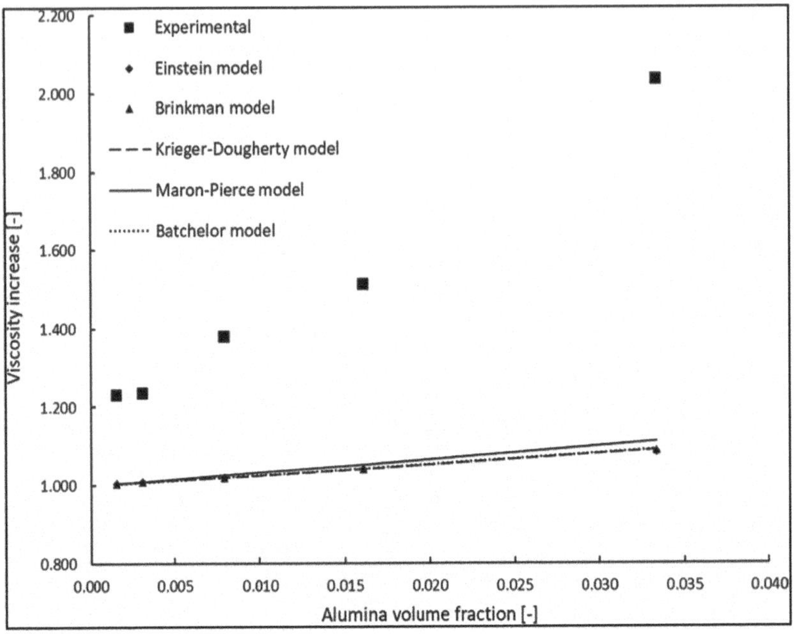

FIGURE 16.15 Experimental and numerical result comparison of ionanofluids viscosity variation against nanoparticle concentration [28].

attained by adding the 10 wt.% of Al_2O_3, that is up to 251.3%. Furthermore, in Fig. 16.15, the experimental results of previous literature and different models are presented which consist of both theoretical equations and experimental results. It shows the volume fraction of Al_2O_3 nanoparticles added in ionic fluids. The investigational data renovation from mass to volume fraction was made to have a better comparison of results [28].

The ionic liquid 1-butyl-3-methylimidazolium tetrafluoroborate [BMIM][BF_4] was studied with the addition of nanodiamond particles [29]. The viscosity of ionanofluids is decreased with the increase in temperature but it is gradually increased with the increase in the concentration of nanodiamond. This is because when the temperature of the ionic fluids is increased the random motion between molecules and the collision between molecules is increased as well. This behavior of ionic fluid helps to explain the viscosity difference between solids and liquids. According to the theory of cohesion force, the attraction forces between the molecules of liquids are stronger than the gases. Therefore, when the temperature has increased the collision between the molecules is increased but on the other hand, the cohesive forces between molecules become weaker. This phenomenon also leads to the decrease in surface tension and shear stress of ionanofluids. It was concluded that cohesion tends to dominate over the force of interaction, which results in a viscosity decrease

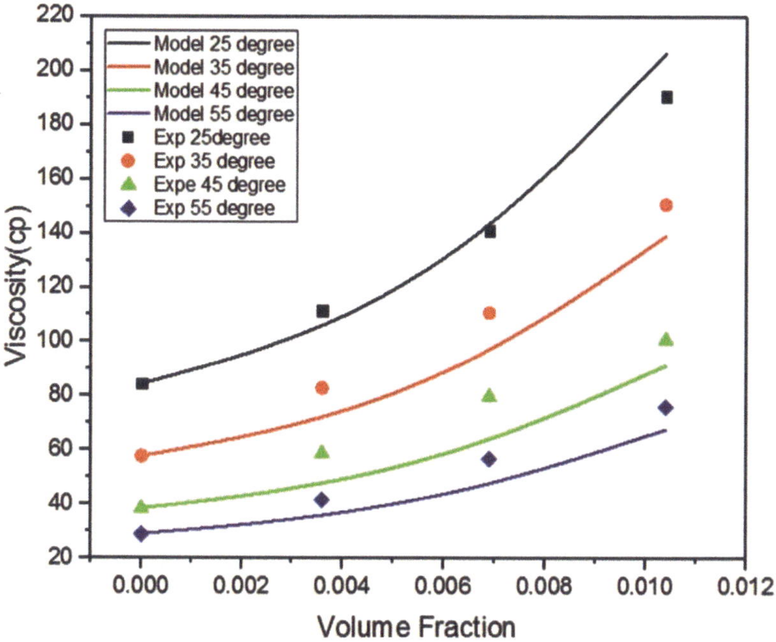

FIGURE 16.16 Viscosity variation of ionanofluids for varying nanoparticle loading and temperature [29].

owing to the temperature escalation. It can be concluded that the viscosity of the ionanofluids is reduced when the temperature is increased, while the value of viscosity becomes larger when the concentration of the nanoparticles is increased. The graph of viscosity of ND: [BMIM][BF_4] against various temperatures and nanoparticle loadings is shown in Fig. 16.16 [29].

The rheological trend of ionanofluids was analyzed by the steady shear rate experiments and oscillatory tests. The steady shear rate experiments were made to find the value of viscosity and shear rate as against time and the temperature. However, the dynamic or oscillatory tests were performed to obtain the value of storage and waste modulus's a function of time and temperature. Rheology is a technique that is used in nanotechnology to characterize the different factors of ionic liquids or ionanofluids. The rheological parameters like viscosity, temperature, pressure drop, and heat transfer are observed during the rheological analysis.

The density of the ionanofluids also reduces when the temperature of the ionic liquid is increased. However, the value of density can be increased by enhancing the concentration of nanoparticles. The density of the ionanofluids can be measured using different volume fractions of nanoparticles at variable ranges of temperature. A thermodynamic model can be used to calculate the value of density with respect to time, temperature, and volumetric fraction.

The thermodynamic model is given in the following equations.

$$V^E = V - V^{id} \tag{16.8}$$

In Eq. (16.8), V^E, V^{id}, and V are excess, ideal, and real volumes of ionanofluids, respectively.

While the real ionanofluid volume can be calculated from Eq. (16.9).

$$V^{id} = X_1 V^0 + X \tag{16.9}$$

The density is a significant feature that is used to explain the heat transfer phenomena in ionanofluids. The density of the solid nanoparticles has a higher value of density as compared to the nanofluids, which is the reason that when nanoadditives were introduced in the ionic liquids the density of the ionanofluids then increases [30]. The dimensionless numbers are very important to observe the flow conditions that mostly depend upon the density. The values Re and Pr are very significant to identify whether the flow is turbulent or laminar if the fluid is dependent upon the density. The density is one of the parameters of the ionanofluids which is studied less than the others because it can be estimated easily.

The density of some ionic fluids is decreased when the fixed concentration of nanoparticles is added to the mixture. The study of ionic fluid with constant concentration (0.003 M) of Ru nanoadditives at variable temperatures ranging from 20°C to 70°C results in a slight reduction in the value of density, as shown in Fig. 16.17. However, it is due to the very small number of nanoparticles added in the ionic fluids. Otherwise if the concentration of nanoparticles is increased then the density of ionic fluids is improved [23].

Fig. 16.17 shows that when the value of temperature gradually increases the value of density is reduced. This trend indicates that the temperature carries a greater influence on the density of ionanofluids.

Heat transfer is the most significant factor in many industrial products; however, the heat transfer rate of conventional fluids is very low due to the low thermal conductivity. Furthermore, to improve the heat transfer rate of liquids, thermal characteristics, specifically thermal conductivity, need to be investigated, and the influential factors are required to be examined keenly. A minute augmentation in the value of thermal conductivity of the thermofluids can lead to significant improvement in thermal efficiency. Moreover, the energy intake demand of thermal appliances can also be reduced in this way. The value of thermal conductivity can be improved by adding the nanoparticles to ionic or conventional fluids.

The researchers have been trying to optimize the thermal conductivity of the thermal fluids for a long time. Several important factors have a decisive effect on the thermal conductivity of the nanofluids like viscosity, density, velocity, the geometry of the system, and the frictional effect. The thermal conductivity of ionanofluids can be calculated by Eq. (16.10).

$$\lambda = \alpha.\rho.C_p \tag{16.10}$$

where α represents thermal diffusivity [31].

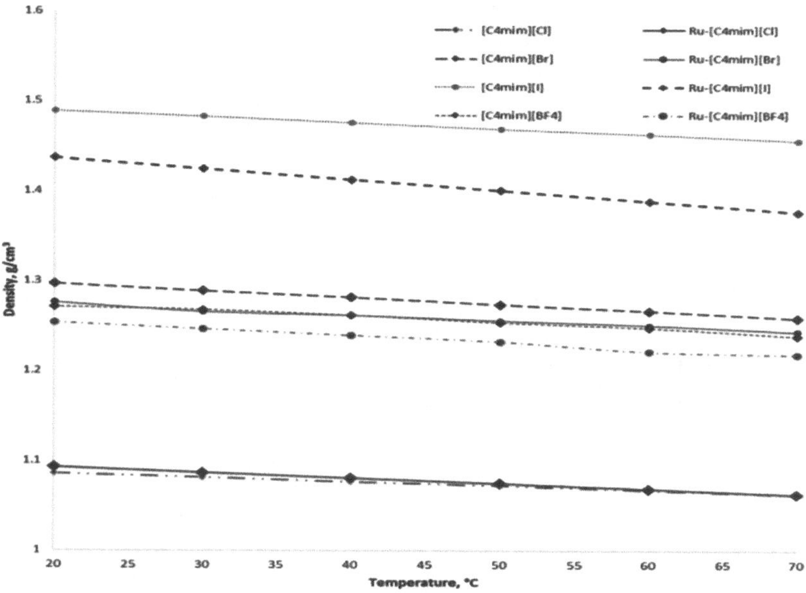

FIGURE 16.17 Temperature effect on Ru-ionanofluids density [23].

The thermal conductivity of ionanofluids is in general studied as a function of temperature, nanoparticle loading extent, that is, the concentration and dispersion, and material characteristics like shape and size of nanoparticles. The key mechanism credited for the thermal conductivity augmentation is the Brownian motion of the molecules of ionanofluids, which results from the suspension of nanoparticles. The result of temperature changes on the thermal conductivity of ionic liquids has been extensively studied in the past research owing to the significance of temperature in thermal applications. Furthermore, the improvement in the value of thermal conductivity is accomplished by the variation of temperature to check the impact of the temperature on the thermal performance of ionanofluids.

The thermal conductivity of ionanofluids is investigated against the change in temperature keeping nanoparticles' concentration to be the same. The results revealed a minor increment in the value of thermal conductivity when nanoparticles were loaded in the base fluid. The concentration of nanoparticles added into the ionic fluids has a very doubtful trend in the case of thermal conductivity.

The topmost enhancement of 4% in the value of thermal conductivity of Ru-ionanofluid was reported at 20°C. However, in some cases, a minute decrease in the thermal conductivity has also been reported, but in general, the thermal conductivity of ionanofluids tends to increase with the increment of the concentration of nanoparticles. It was also reported that the thermal conductivity of the ionanofluids tends to mount as the size of the anions is smaller [32].

The thermal conductivity of nanofluids containing CNT depends upon the temperature and concentration of the nanoparticles. In the literature, the comparison of CNT-based nanofluids and CNT–ionanofluids has extensively been done and the comparative analysis reveals that the heat transfer mechanism in both types of fluids can be evaluated based on thermal conductivity. It has been reported that the thermal conductivity of the CNT–ionanofluids tends to be higher, thus resulting in greater heat transfer phenomena [33].

Ionanofluids containing 1 wt.% of MWCNT nanoparticles in $[C_2mim][NTf_2]$ and $[C_2mim][EtSO_4]$ were studied to evaluate the change in thermal conductivity against temperature by Castro et al. [33]. They also examined the thermal conductivity of water-based nanofluids containing MWCNT and SWCNT at 1 wt.% concentration of nanoparticles. The final results revealed that at the normal temperature, there was an increase of 35.5% in thermal conductivity of $[C_4\ mim][NTf_2]$-based ionanofluids as compared to the base ionic fluid. However, $[C_2mim][EtSO_4]$-based ionanofluids displayed a mere 8.5% enhancement in thermal conductivity as compared to the base fluid. Nevertheless, the temperature was found to have an insignificant effect on thermal conductivity.

The thermal conductivities of $[C_6mim][BF_4]$, $[C_4mim][PF_6]$, $[C_6mim][PF_6]$, $[C_4mim][CF_3SO_3]$, and $[C_4mpyrr][(CF_3SO_2)_2N]$ were investigated for various temperature values (293K to 353K) a pressure of at 0.1 MPa. It was found that the change in the value of thermal conductivity is very linear, and the trend of linearity is the same until $\pm 0.6\%$, as shown in Fig. 16.18. Some other researchers also investigated the thermal conductivity of $[C_4mim][PF_6]$ and $[C_6mim][PF_6]$ for a temperature range of 294–335K and at pressure up to 20 MPa [34]. Vacuum processing, that is, heating and drying, was employed to synthesize the samples. The absorption purification process was employed for purifying the samples using charcoal as absorbent.

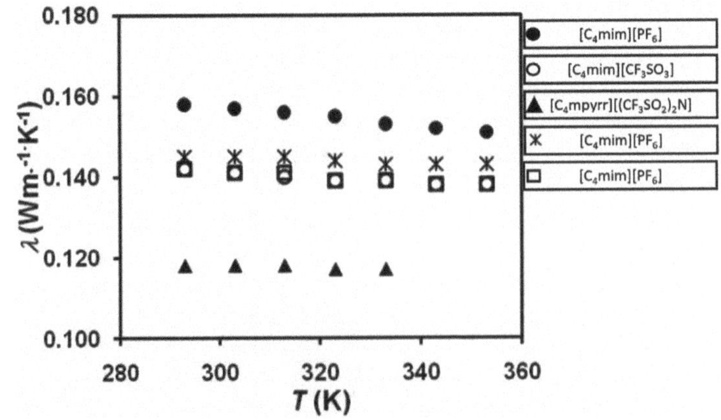

FIGURE 16.18 Thermal conductivity of ionic liquids against the temperature [34].

C_p is the magnitude of heat energy required to raise the temperature of a material by a unit degree, which plays a vital role for evaluation of the thermal performance of nanofluids. C_p tends to increase with the increase of temperature. However, the value of C_p tends to drop with the increase in density [35]. The specific heat is studied for both the ionic liquids and the ionanofluids with nanoparticles of alumina. The range of temperature for the nanoparticles' addition process of ionic liquids was 283.15K to 333.15K. The observed results, presented in Fig. 16.19, show that the value of specific

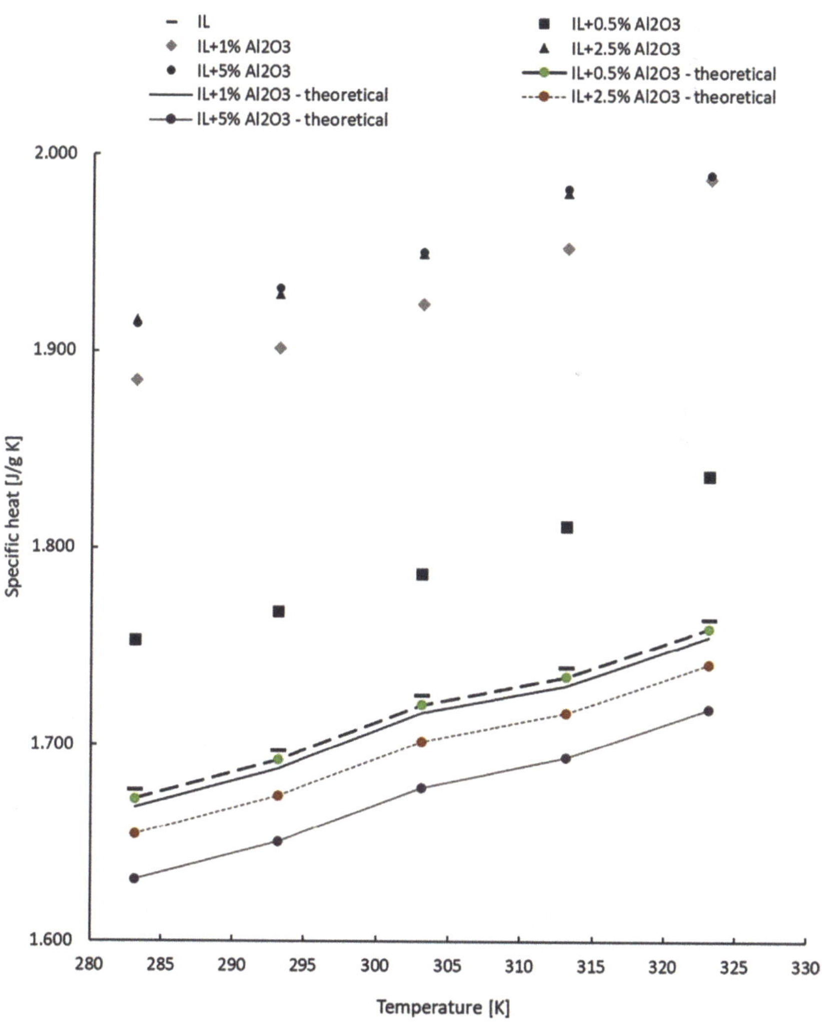

FIGURE 16.19 Comparison of experimental and numerical values of specific heat against the temperature [28].

heat is increased when the 2.5 wt.% of nanoparticles were dispersed in the ionic liquid. The magnitude of specific heat of ionanofluids was compared with previous studies with different concentrations of nanoparticles at the same range of temperature, and the same trend was obtained.

The C_p of ionanofluids has a greater impact on the practical use of the ionanofluids in thermal systems. The major use of ionic liquids as heat transfer fluids is found mainly in the heat exchangers in chemical plants and solar thermal power generation, and their performance is subject to the magnitude of the specific heat capacity.

16.5 Thermal applications

Ionanofluids carry very suitable characteristics for thermal energy applications—as a thermal energy carrier fluid—since the conventional fluids with higher thermal conductivity have been reported to have lower thermal energy storage capability [11,36–43]. However, ionanofluids show substantial thermal energy storage characteristics. For the aforementioned reason, ionanofluids have been used for applications in solar energy systems, material lubrication, surface treatment, biotechnology/biomedical, electronic component cooling, etc. Brief details of some of the studies that have appraised the potential of various ionanofluids for a variety of applications have been given in Table 16.2.

16.6 Future prospects and challenges

Ionanofluids can replace conventional thermal fluids owing to their much superior characteristics and performance in thermal applications, particularly in solar energy systems. Some of the challenges are linked with the ionanofluids.

1. Ionanofluids have high viscosity compared to conventional fluids and nanofluids, which requires a high pumping power to displace the fluid and results in high frictional losses.
2. High viscosity of ionic liquids results in highly viscous ionanofluids and makes them suitable for conduction heat transfer, whereas high fluid viscosity reduces the rate of convection heat transfer. Therefore, conventional fluids like water and ethylene-glycol are used to make binary liquids with the ionic liquids.
3. Very insignificant change in thermal conductivity of ionanofluids is observed under the temperature change. Whereas the conventional nanofluids display an increase in thermal conductivity and heat transfer rate at high temperatures, making them suitable for thermal applications. However, ionanofluids do not offer this specific characteristic due to there being no change in thermal conductivity.
4. Suspension stability of nanoparticles in ionic liquids remains a challenge.

TABLE 16.2 Applications of ionanofluids.

Reference	Nanoparticle	Base fluid	Application	Major findings
[44]	Fe_3O_4	Ionic liquid	Solar energy	An increase in ionic radius was reported. Greater anionic size results in greater solar absorption due to reduced transmittance. Higher solar absorption leads to superior thermal conversion. Therefore, ionic base fluids with nanoadditives have been regarded to carry the noteworthy potential for solar-thermal applications
[45]	MoS_2	Solvent-free ionic nanofluid	Surface protection	Nanofluids of ionic base fluids were found to have much lower a stable coefficient of friction. Moreover, the Ionanofluid was observed to carry self-heating lubricant features. Ionanofluid could be employed to protect the surfaces from scratches and wear effects as it is capable to form a protective layer over the surface
[46]	Al_2O_3	([C_4mim][NTf_2]) ionic liquids	Thermal application	An anomalous augmentation of 11% in thermal conductivity and 49% in the heat capacity was observed at 0.9 vol.% nanoadditive loading. ionanofluids depicted shear thinning under a high shearing rate and a drastic decline in viscosity was reported as well
[47]	2D MXenes (Ti_3C_2)	Ionic Liquid	Hybrid solar photovoltaic/ thermal systems	They analyzed the performance of Ionanofluid in a photovoltaic thermal system and reported an overwhelming performance enhancement was reported. An improvement of 78.5% in thermal efficiency, 18.7% in electrical efficiency, and 6% in coefficient of heat transfer was reported. The addition of MXene nanoparticles results in improved thermophysical characteristics of the ionic base fluid

(Continued)

TABLE 16.2 (Continued)

Reference	Nanoparticle	Base fluid	Application	Major findings
[48]	TiO_2	Pyr_{14} TFSI ionic liquid	Power sources	They employed Ionanofluid as an electrolyte in Li/$Limn_2O_4$ cells and obtained a 131 mAh/g discharge rate at 25°C which is about 96%. Whereas conventional electrolytes can deliver as much as an 88% discharge rate
[49]	Graphene Oxide	[Bmim][NTf_2]	Solar power plants	A massive boost of 6.5% in thermal conductivity and 27% in heat capacity was reported. Extensive analysis of results revealed that the density and the viscosity tend to reduce with the mounting temperature
[50]	Carbon-coated Ni	[HMIM][NTf_2] ionic liquid	Solar energy	They reported that the combined use of various types of nanoparticles (i.e., hybrid ionanofluids) results in a greater reduction in spectral transmittance due to the synergistic effect of different nanoparticles as compared to the mono ionanofluids. Moreover, at higher nanoparticle loading (concentration) greater augmentation in solar spectral absorption was reported
[51]	Graphene	[HMIM]BF_4 ionic liquid	Solar energy	Considering the results reported in the study, the efficiency of the solar system improved with increasing solar irradiance as well as the receiver's height. However, a reduction in efficiency was observed when the nanoadditive concentration was increased. Solar receiver's efficiency was observed to be 70% at 0.005 wt.% nanoparticle loading at 600K and 20×1000 W/m^2 for 5 cm receiver

[52]	SiC	[HMIM]BF_4 ionic liquid	Solar energy	The prepared silicon-carbide-based Ionanofluid exhibited anomalous nanosuspension stability (for 20 days). Thermal characteristics were also found to improve with the dispersion of nanoadditives in the ionic base fluid. A considerable improvement in thermal conductivity and viscosity was reported
[27]	Graphene	[HMIM]BF_4 ionic liquid	Renewable energy	Ionic base fluid, that is, [HMIM][BF_4] and graphene-based ionanofluids display greater thermal stability owing to a high decomposition temperature of 400 °C. Ionanofluid of 0.006 wt.% nanoadditive loading depicted thermal conductivity augmentation from 15.2% to 22.9% when the temperature increased from room temperature to 200 °C. Moreover, viscosity was found to reduce from 4.6% to 13.3%, and specific heat was reduced by 3% during the temperature change. Besides, a 2.8% reduction in density was also reported
[53]	Nanoparticle enhanced ionic liquid	[C_4mim][NTf_2], [C_4mpyrr][NTf_2], 1% Al_2O_3	Solar energy	Dispersion of nanoadditives in the ionic base fluid enhanced the thermal conductivity by as much as 6% and the heat capacity got improved by 23%. Moreover, a 20% improvement in the coefficient of heat transfer was also reported in the tested range of temperature
[54]	SiC, Graphene	[H_{mim}]SF_4	Thermal application	Ionanofluids of silicon-carbide and graphene were studied for potential application in thermal systems as a thermal transporting fluid. The study was conducted in a laminar flow regime and it was reported that in laminar flow thermal parameters tend to improve quite impressively. The ratio of pumping power was observed to be lesser than 1 for all of the studied cases. Therefore, the proposed Ionanofluid was recommended to be the best alternative to conventional thermal fluids

(*Continued*)

TABLE 16.2 (Continued)

Reference	Nanoparticle	Base fluid	Application	Major findings
[55]	Gel polymer electrolyte	Pyrrolidinium-based ionanofluid	Lithium batteries	Ionanofluid of 50 wt.% nanoadditive loading was tested for Li-ion batteries. Exceptional ionic conductivity of 2.3X 10^{-3} S cm^{-1} at 25°C as well as sufficient Li-ion transmittance and better electrochemical stability was reported. Ionanofluid electrolyte basted battery depicted better cycling stability and capacity of discharge due to optimized features. Moreover, comprehensive quantitative data has been presented in the research article for better insights
[56]	MXene	Aqueous ionic liquid	PV/T system	As per the results, the topmost enhancement of 47% in thermal conductivity of ionanofluid was observed at 0.2 wt.% nanoparticle loading of titania. Titania insertion however results in minute viscosity uplift. Moreover, an improvement of 13.95% in electrical efficiency and an 81.15% hike in thermal efficiency of the photovoltaic thermal system was observed. A notable augmentation of 12.6% in thermal coefficients was achieved as compared to the aqua-silica nanofluid and 2% as compared to MXene-palm oil nanofluid
[57]	CNT	Solvent-free ionic nanofluid	Polyurethane thermoplastic elastomer	Nanotubes of carbon dispersed in ionic base fluid result in the formation of highly stable ionanofluid and the CNT carries decreased breaking strength and modulus of elasticity however, 50% greater toughness characteristics are obtained in this way
[58]	Silica	Eutectic molten salt-based nanofluid	Thermal applications	An insignificant effect of nanoparticle loading was observed on the melting point and the latent of the ionic solar fluid. Besides, they proposed an accurate correlation for predicting the heat capacity of the solar salt-based ionanofluid as a function of temperature, and nanoadditive diameter and concentration

[59]	Graphene	([EMIm]Ac) ionic liquid	Absorption refrigeration	Insertion of graphene nanoadditives enhanced the thermal conductivity as well as reduced the specific heat of the nanofluid even at small nanoparticle concentrations. An increase of 27.7% in nanofluid viscosity and a 43.3% increment in thermal conductivity were reported at a temperature of 373.15K. About 3.62% reduction in specific heat was observed at 5 m.% nanoparticle loading and 363.15K temperature
[60]	Silica (SiO_2)	Solvent-free ionic nanofluid	Smart lubrication	Silica-based ionanofluid was found to carry excellent lubrication potential due to its highly thick double electric layer. An electric potential of 1.5 V yielded a very low friction coefficient and thus better surface lubrication
[61]	Graphene oxide-silica hybrid nanocore	Solvent-free ionic nanofluid	High-performance lubricating addictive	Graphene-silica-based hybrid nanocomposite depicts liquid characteristics and carries tremendous potential for lubrication applications due to inordinate dispersion stability

16.7 Conclusions

In this chapter, ionanofluids—ionic liquids containing nanoparticles dispersion—have been studied in detail. Insights into preparation method, characteristics, heat transfer mechanism, thermal applications, and challenges of ionanofluids have been presented. A coherent and logical discussion on the thermophysical and physiochemical phenomenon of ionanofluids has been included in this chapter. Ionic liquids have tunable characteristics, which make them very suitable to replace the conventional base fluids of nanofluids. Therefore, several researchers have tested ionanofluids for thermal applications. Due to several complications like high viscosity and the slight change in thermal conductivity at varying temperatures, researchers have tested binary solutions of ionic liquids with conventional base fluids like water and ethylene glycol.

Overall, ionanofluids have very strong prospects for a broad range of applications and yet there are some challenges still to be addressed, such as the stability of nanoparticle dispersion and the high viscosity of ionanofluids.

References

[1] E.I. Chereches, J.I. Prado, M. Chereches, A.A. Minea, L. Lugo, Experimental study on thermophysical properties of alumina nanoparticle enhanced ionic liquids, J. Mol. Liq. (2019) 291. Available from: https://doi.org/10.1016/j.molliq.2019.111332.

[2] D.P. Soman, S. Karthika, P. Kalaichelvi, T.K. Radhakrishnan, Impact of viscosity of nanofluid and ionic liquid on heat transfer, J. Mol. Liq. 291 (2019) 111349. Available from: https://doi.org/10.1016/j.molliq.2019.111349.

[3] D. Deb, S. Bhattacharya, Ion transport in surface functionalized SnO2 nanoparticles dispersed imidazolium ionanofluids: Decoupling from structural relaxation, J. Mol. Liq. 285 (2019) 697–706. Available from: https://doi.org/10.1016/j.molliq.2019.04.101.

[4] E.I. Chereches, A.A. MINEA, K.V. SHARMA, A complex evaluation of [C2mim] [CH3SO3]—alumina nanoparticle enhanced ionic liquids internal laminar flow, Int. J. Heat. Mass. Transf. 154 (2020) 119674. Available from: https://doi.org/10.1016/j.ijheatmasstransfer.2020.119674.

[5] X. Zhang, L. Pan, L. Wang, J.J. Zou, Review on synthesis and properties of high-energy-density liquid fuels: hydrocarbons, nanofluids and energetic ionic liquids, Chem. Eng. Sci. 180 (2018) 95–125. Available from: https://doi.org/10.1016/j.ces.2017.11.044.

[6] A.A. Minea, Overview of ionic liquids as candidates for new heat transfer fluids, Int. J. Thermophys. (2020) 41. Available from: https://doi.org/10.1007/s10765-020-02727-3.

[7] S.Y. Gu, X.F. Gao, Y.H. Zhang, Synthesis and characterization of solvent-free ionic molybdenum disulphide (MoS2) nanofluids, Mater. Chem. Phys. 149 (2015) 587–593. Available from: https://doi.org/10.1016/j.matchemphys.2014.11.012.

[8] W.M. El-Maghlany, A.A. Minea, Ionanofluids natural convection heat transfer and entropy generation in a rectangular cavity: viscosity influence, J. Mol. Liq. 338 (2021) 116651. Available from: https://doi.org/10.1016/j.molliq.2021.116651.

[9] J.P. Vallejo, J.M. Liñeira del Río, J. Fernández, L. Lugo, Tribological performance of silicon nitride and carbon black Ionanofluids based on 1-ethyl-3-methylimidazolium methanesulfonate, J. Mol. Liq. 319 (2020) 114335. Available from: https://doi.org/10.1016/j.molliq.2020.114335.

[10] A.H. Sharifi, I. Zahmatkesh, A.M. Mozhdehi, A. Morsali, F.F. Bamoharram, Stability appraisement of the alumina-brine nanofluid in the presence of ionic and non-ionic disparents on the alumina nanoparticles surface as heat transfer fluids: quantum mechanical study and Taguchi-optimized experimental analysis, J. Mol. Liq. (2020) 319. Available from: https://doi.org/10.1016/j.molliq.2020.113898.

[11] H.M. Ali, T.R. Shah, H. Babar, Z.A. Khan, Application of nanofluids for thermal management of photovoltaic modules: a review, Microfluid. Nanofluidics, InTech, 2018. Available from: http://doi.org/10.5772/intechopen.74967.

[12] A.A. Minea, S.M. Sohel Murshed, Ionic liquids-based nanocolloids—a review of progress and prospects in convective heat transfer applications, Nanomaterials (2021) 11. Available from: https://doi.org/10.3390/nano11041039.

[13] B. Jóźwiak, G. Dzido, A. Kolanowska, R.G. Jędrysiak, E. Zorębski, H.F. Greer, et al., From lab and up: superior and economic heat transfer performance of ionanofluids containing long carbon nanotubes and 1-ethyl-3-methylimidazolium thiocyanate, Int. J. Heat. Mass. Transf. 172 (2021) 121161. Available from: https://doi.org/10.1016/j.ijheatmasstransfer.2021.121161.

[14] B. Wang, X. Wang, W. Lou, J. Hao, Ionic liquid-based stable nanofluids containing gold nanoparticles, J. Colloid Interface Sci. 362 (2011) 5–14. Available from: https://doi.org/10.1016/j.jcis.2011.06.023.

[15] M. Amde, Z.Q. Tan, R. Liu, J.F. Liu, Nanofluid of zinc oxide nanoparticles in ionic liquid for single drop liquid microextraction of fungicides in environmental waters prior to high performance liquid chromatographic analysis, J. Chromatogr. A 1395 (2015) 7–15. Available from: https://doi.org/10.1016/j.chroma.2015.03.049.

[16] M. Amde, J.F. Liu, Z.Q. Tan, D. Bekana, Ionic liquid-based zinc oxide nanofluid for vortex assisted liquid liquid microextraction of inorganic mercury in environmental waters prior to cold vapor atomic fluorescence spectroscopic detection, Talanta 149 (2016) 341–346. Available from: https://doi.org/10.1016/j.talanta.2015.12.004.

[17] X. Wu, X. Li, M. Yang, H. Zeng, S. Zhang, R. Lu, et al., An ionic liquid-based nanofluid of titanium dioxide nanoparticles for effervescence-assisted dispersive liquid–liquid extraction for acaricide detection, J. Chromatogr. A 1497 (2017) 1–8. Available from: https://doi.org/10.1016/j.chroma.2017.03.005.

[18] D. Deb, B. Dutta, S. Bhattacharya, Viscosity decoupled charge transport in surface functionalized ZnS nanoparticle dispersed imidazolium ionanofluids, Mater. Res. Bull. 116 (2019) 22–31. Available from: https://doi.org/10.1016/j.materresbull.2019.03.028.

[19] J. Alizadeh, M. Keshavarz Moraveji, An experimental evaluation on thermophysical properties of functionalized graphene nanoplatelets ionanofluids, Int. Commun. Heat. Mass. Transf. 98 (2018) 31–40. Available from: https://doi.org/10.1016/j.icheatmasstransfer.2018.08.008.

[20] T.C. Paul, R. Mahamud, J.A. Khan, Multiphase modeling approach for ionic liquids (ILs) based nanofluids: Improving the performance of heat transfer fluids (HTFs), Appl. Therm. Eng. 149 (2019) 165–172. Available from: https://doi.org/10.1016/j.applthermaleng.2018.12.039.

[21] L. Zhang, L. Chen, J. Liu, X. Fang, Z. Zhang, Effect of morphology of carbon nanomaterials on thermo-physical characteristics, optical properties and photo-thermal conversion performance of nanofluids, Renew. Energy 99 (2016) 888–897. Available from: https://doi.org/10.1016/j.renene.2016.07.073.

[22] X. Wang, Y. Chi, T. Mu, A review on the transport properties of ionic liquids, J. Mol. Liq. 193 (2014) 262–266. Available from: https://doi.org/10.1016/j.molliq.2014.03.011.

[23] A. Adriana, S.M.S. Murshed, A review on development of ionic liquid based nano fluids and their heat transfer behavior, Renew. Sustain. Energy Rev. 91 (2018) 584–599. Available from: https://doi.org/10.1016/j.rser.2018.04.021.

[24] A. Minea, W.M. El-maghlany, Natural convection heat transfer utilizing ionic nanofluids with temperature-dependent thermophysical properties, Chem. Eng. Sci. 174 (2017) 13–24. Available from: https://doi.org/10.1016/j.ces.2017.08.028.

[25] T.C. Paul, A.K.M.M. Morshed, E.B. Fox, J.A. Khan, International Journal of Heat and Mass Transfer Experimental investigation of natural convection heat transfer of Al 2 O 3 Nanoparticle Enhanced Ionic Liquids (NEILs), Heat. Mass. Transf. 83 (2015) 753–761. Available from: https://doi.org/10.1016/j.ijheatmasstransfer.2014.12.067.

[26] Ł. Justyna, M. Paszkiewicz, A. Krukowska, A. Malankowska, A. Zaleska-medynska, Ionic liquids for nano- and microstructures preparation. Part 1: properties multifunctional role, Adv. Colloid Interface Sci. 230 (2016) 13–28. Available from: https://doi.org/10.1016/j.cis.2015.08.006.

[27] J. Liu, F. Wang, L. Zhang, X. Fang, Z. Zhang, Thermodynamic properties and thermal stability of ionic liquid-based nanofluids containing graphene as advanced heat transfer fluids for medium-to-high-temperature applications, Renew. Energy 63 (2014) 519–523. Available from: https://doi.org/10.1016/j.renene.2013.10.002.

[28] E.I. Chereches, D. Bejan, C. Ibanescu, M. Danu, A. Adriana, Ionanofluids with [C2 mim][CH3 SO3] ionic liquid and alumina nanoparticles: an experimental study on viscosity, specific heat and electrical conductivity, Chem. Eng. Sci. 229 (2021) 116140. Available from: https://doi.org/10.1016/j.ces.2020.116140.

[29] S. Jorjani, M. Mozaffarian, G. Pazuki, A novel nanodiamond based IoNano fluid: experimental and mathematical study of thermal properties, J. Mol. Liq. 271 (2018) 211–219. Available from: https://doi.org/10.1016/j.molliq.2018.08.116.

[30] B. Bakthavatchalam, K. Habib, R. Saidur, B. Baran, K. Irshad, Comprehensive study on nano fluid and ionano fluid for heat transfer enhancement: a review on current and future perspective distilled water, J. Mol. Liq. 305 (2020) 112787. Available from: https://doi.org/10.1016/j.molliq.2020.112787.

[31] J.C. Thermodynamics, J.M.P. França, F. Reis, S.I.C. Vieira, M.J.V. Lourenço, F.J.V. Santos, et al., Thermophysical properties of ionic liquid dicyanamide (DCA), J. Chem. Thermodyn. 79 (2014) 248–257. Available from: https://doi.org/10.1016/j.jct.2014.05.008.

[32] A.A. Minea, M.G. Moldoveanu, O. Dodun, Thermal conductivity enhancement by adding nanoparticles to ionic liquids, in: Solid State Phenomena, 261, Trans Tech Publications, Ltd., 2017, pp. 121–126. Available from: https://doi.org/10.4028/www.scientific.net/ssp.261.121.

[33] C.A. Nieto De Castro, S.M.S. Murshed, M.J.V. Lourenço, F.J.V. Santos, M.L.M. Lopes, J. M.P. França, Enhanced thermal conductivity and specific heat capacity of carbon nanotubes ionanofluids, Int. J. Therm. Sci. 62 (2012) 34–39. Available from: https://doi.org/10.1016/j.ijthermalsci.2012.03.010.

[34] C.A.N. De Castro, M.J.V. Lourenço, A.P.C. Ribeiro, E. Langa, S.I.C. Vieira, P. Goodrich, et al., Thermal properties of ionic liquids and IoNanoFluids of imidazolium and pyrrolidinium liquids, J. Chem. Eng. Data 55 (2010) 653–661. Available from: https://doi.org/10.1021/je900648p.

[35] C. Hermida-Merino, A.B. Pereiro, J.M.M. Araújo, C. Gracia-Fernández, J.P. Vallejo, L. Lugo, M.M. Piñeiro, Graphene ionanofluids, thermal and structural characterization, Nanomaterials 9 (2019) 1549. Available from: https://doi.org/10.3390/nano9111549.

[36] H.M. Ali, H. Babar, T.R. Shah, M.U. Sajid, M.A. Qasim, S. Javed, Preparation techniques of TiO2 nanofluids and challenges: a review, Appl. Sci. 8 (2018) 587. Available from: https://doi.org/10.3390/app8040587.

[37] T.R. Shah, H.M. Ali, M.M. Janjua, On aqua-based silica (Sio2−water) nanocoolant: convective thermal potential and experimental precision evaluation in aluminum tube radiator, Nanomaterials 10 (2020) 1−23. Available from: https://doi.org/10.3390/nano10091736.

[38] F. Abbas, H.M. Ali, T.R. Shah, H. Babar, M.M. Janjua, U. Sajjad, et al., Nanofluid: potential evaluation in automotive radiator, J. Mol. Liq. 297 (2020) 112014. Available from: https://doi.org/10.1016/j.molliq.2019.112014.

[39] T.R. Shah, H. Babar, H.M. Ali, Energy harvesting: role of hybrid nanofluids, emerging nanotechnologies for renewable energy, Micro Nano Technol (2021) 173−211. Available from: https://doi.org/10.1016/b978-0-12-821346-9.00011-0.

[40] F. Abbas, H.M. Ali, M. Shaban, M.M. Janjua, T.R. Shah, M.H. Doranehgard, et al., Towards convective heat transfer optimization in aluminum tube automotive radiators: Potential assessment of novel Fe2O3-TiO2/water hybrid nanofluid, J. Taiwan. Inst. Chem. Eng. 124 (2021) 424−436. Available from: https://doi.org/10.1016/j.jtice.2021.02.002.

[41] T.R. Shah, H.M. Ali, Applications of hybrid nanofluids in solar energy, practical limitations and challenges: a critical review, Sol. Energy 183 (2019) 173−203. Available from: https://doi.org/10.1016/j.solener.2019.03.012.

[42] H. Ali, T.R. Shah, H. Babar, A.M. Ali, Hybrid nanofluids: techniques and challenges of stability enhancement. 4th Int. Conf. Adv. Mech. Eng., Istambul, 2018, pp. 60−76.

[43] T.R. Shah, H. Koten, H.M. Ali, Performance effecting parameters of hybrid nanofluids, Hybrid Nanofluids for Convection Heat Transfer, Elsevier, 2020, pp. 179−213. Available from: http://doi.org/10.1016/B978-0-12-819280-1.00005-7.

[44] P. Cao, Y. Li, Y. Wu, H. Chen, J. Zhang, L. Cheng, et al., Role of base fluid on enhancement absorption properties of Fe3O4/ionic liquid nanofluids for direct absorption solar collector, Sol. Energy 194 (2019) 923−931. Available from: https://doi.org/10.1016/j.solener.2019.11.039.

[45] S. Gu, Y. Zhang, B. Yan, Solvent-free ionic molybdenum disulfide (MoS2) nanofluids with self-healing lubricating behaviors, Mater. Lett. 97 (2013) 169−172. Available from: https://doi.org/10.1016/j.matlet.2013.01.058.

[46] T.C. Paul, A.K.M.M. Morshed, E.B. Fox, J.A. Khan, Thermal performance of Al2O3 nanoparticle enhanced ionic liquids (NEILs) for concentrated solar power (CSP) applications, Int. J. Heat. Mass. Transf. 85 (2015) 585−594. Available from: https://doi.org/10.1016/j.ijheatmasstransfer.2015.01.071.

[47] B. Bakthavatchalam, K. Habib, R. Saidur, N. Aslfattahi, S.M. Yahya, A. Rashedi, et al., Optimization of thermophysical and rheological properties of mxene ionanofluids for hybrid solar photovoltaic/thermal systems, Nanomaterials 11 (2021) 1−28. Available from: https://doi.org/10.3390/nano11020320.

[48] P. Bose, D. Deb, S. Bhattacharya, Ionic liquid based nanofluid electrolytes with higher lithium salt concentration for high-efficiency, safer, lithium metal batteries, J. Power Sources 406 (2018) 176−184. Available from: https://doi.org/10.1016/j.jpowsour.2018.10.050.

[49] A. Hosseinghorbani, M. Mozaffarian, G. Pazuki, Application of graphene oxide IoNanofluid as a superior heat transfer fluid in concentrated solar power plants, Int. Commun. Heat. Mass. Transf. (2020) 111. Available from: https://doi.org/10.1016/j.icheatmasstransfer.2019.104450.

[50] L. Zhang, J. Liu, G. He, Z. Ye, X. Fang, Z. Zhang, Radiative properties of ionic liquid-based nanofluids for medium-to-high-temperature direct absorption solar collectors, Sol. Energy Mater. Sol Cell 130 (2014) 521–528. Available from: https://doi.org/10.1016/j.solmat.2014.07.040.

[51] J. Liu, Z. Ye, L. Zhang, X. Fang, Z. Zhang, A combined numerical and experimental study on graphene/ionic liquid nanofluid based direct absorption solar collector, Sol. Energy Mater. Sol Cell 136 (2015) 177–186. Available from: https://doi.org/10.1016/j.solmat.2015.01.013.

[52] W. Chen, C. Zou, X. Li, An investigation into the thermophysical and optical properties of SiC/ionic liquid nanofluid for direct absorption solar collector, Sol. Energy Mater. Sol Cell 163 (2017) 157–163. Available from: https://doi.org/10.1016/j.solmat.2017.01.029.

[53] T.C. Paul, A.K.M.M. Morshed, J.A. Khan, Nanoparticle Enhanced Ionic Liquids (NEILS) as working fluid for the next generation solar collector, Procedia Eng. 56 (2013) 631–636. Available from: https://doi.org/10.1016/j.proeng.2013.03.170.

[54] G. Huminic, A. Huminic, Heat transfer capability of ionanofluids for heat transfer applications, Int. J. Thermophys. 42 (2021) 1–17. Available from: https://doi.org/10.1007/s10765-020-02757-x.

[55] D. Deb, P. Bose, S. Bhattacharya, Gel-polymer electrolytes plasticized with pyrrolidinium-based ionanofluid for lithium battery applications, Ion. (Kiel.) 27 (2021) 123–136. Available from: https://doi.org/10.1007/s11581-020-03807-y.

[56] L. Das, K. Habib, R. Saidur, N. Aslfattahi, S.M. Yahya, F. Rubbi, Improved thermophysical properties and energy efficiency of aqueous ionic liquid/mxene nanofluid in a hybrid pv/t solar system, Nanomaterials 10 (2020) 1–26. Available from: https://doi.org/10.3390/nano10071372.

[57] S.Y. Gu, L.L. Liu, B.B. Yan, Effects of ionic solvent-free carbon nanotube nanofluid on the properties of polyurethane thermoplastic elastomer, J. Polym. Res. 21 (2014) 356. Available from: https://doi.org/10.1007/s10965-014-0356-0.

[58] D. Ercole, O. Manca, K. Vafai, An investigation of thermal characteristics of eutectic molten salt-based nanofluids, Int. Commun. Heat. Mass. Transf. 87 (2017) 98–104. Available from: https://doi.org/10.1016/j.icheatmasstransfer.2017.06.022.

[59] F.F. Zhang, F.F. Zheng, X.H. Wu, Y.L. Yin, G. Chen, Variations of thermophysical properties and heat transfer performance of nanoparticleenhanced ionic liquids, R. Soc. Open. Sci. (2019) 6. Available from: https://doi.org/10.1098/rsos.182040.

[60] Y. Guo, G. Liu, G. Li, F. Zhao, L. Zhang, F. Guo, et al., Solvent-free ionic silica nanofluids: Smart lubrication materials exhibiting remarkable responsiveness to weak electrical stimuli, Chem. Eng. J. 383 (2020) 123202. Available from: https://doi.org/10.1016/j.cej.2019.123202.

[61] Y. Guo, L. Guo, G. Li, L. Zhang, F. Zhao, C. Wang, et al., Solvent-free ionic nanofluids based on graphene oxide-silica hybrid as high-performance lubricating additive, Appl. Surf. Sci. 471 (2019) 482–493. Available from: https://doi.org/10.1016/j.apsusc.2018.12.003.

Chapter 17

Hybrid nanofluids towards advancement in nanofluids for heat sink

Furqan Jamil[1], Taha Baig[1,2], Hafiz Muhammad Ali[3,4], Muhammad Shehryar[2], Shahab Khushnood[1] and Faisal Hassan[2]
[1]Mechanical Engineering Department, University of Wah, Wah Cantonment, Pakistan, [2]Mechanical Engineering Department, University of Engineering and Technology, Taxila, Pakistan, [3]Mechanical Engineering Department, King Fahd University of Petroleum and Minerals, Dhahran, Saudi Arabia, [4]Interdisciplinary Research Center for Renewable Energy and Power Systems (IRC-REPS), King Fahd University of Petroleum and Minerals, Dhahran, Saudi Arabia

Nomenclature

Abbreviation

Al	Aluminum
Al_2O_3	Aluminum oxide
CuO	Copper oxide
EG	Expanded graphite
GNPs	Graphene nanoplatelets
ME PCM	Microencapsulated phase change material
MWCNTs	Multiwall carbon nanotubes
NEPCM	Nanoenhanced phase change material
NPs	Nanoparticles
PCM	Phase change material
SiO_2	Silicon oxide
TiO_2	Titanium oxide

Symbols

°C	Temperature
k	Thermal conductivity
V	Volt
W	Watt

Advances in Nanofluid Heat Transfer. DOI: https://doi.org/10.1016/B978-0-323-88656-7.00006-4

17.1 Introduction

With the rapid advancement in electronic technology, electronic appliances are continuously being utilized in our day-to-day life. Currently electronic products generate an enormous amount of heat which affects their performance and decrease their reliability. The heat flux in electronic components cooled by air has almost reached its limit of 100 W/cm^2. Thermal performance of an electronic device decreases with an increase in temperature. Therefore, it is necessary to maintain the temperature of electronic components at the threshold. Heat sinks are used for heat dissipation to a fluid medium in order to maintain the temperature of electronic components at an optimum value. The heat transfer rate depends on the geometry of the heat sink and the type of fluid being used as the medium.

Mukherjee et al. [1] reviewed the stability and preparation of nanofluids. Nanofluid, a basic result of nanotechnology, has become a subject of fascination because of its exceptional heat transfer properties in different fields, including cooling, power, atomic, space, microelectronics, and biomedical machines. However, the planning and adjustment of such liquids are surely an item of concern for better arrangements. For the last decade various exploration and advancement works have been done with regard to the amalgamation and solidness of such materials. In this commitment, a brief survey has been introduced to give an update about the synthesis and steadiness strategies for nanofluids.

Devendiran et al. [2] presented a review about the preparation, applications, and characterization of different nanofluids. The collection of appropriate fluid for heat transfer is significant for the design of heat transfer systems. Nanofluid, a colloidal blend made of a base liquid and a nanoparticle, is another form of heat transfer liquid that is becoming a liquid with high potential in heat transfer applications because of improved thermal conductivity. This paper surveys the planning of metal and metal oxides nanofluids and the different methods used to examine the physical and synthetic qualities of nanofluids. Thermophysical and heat transfer properties of nanofluids including the improved conductivity value are introduced.

Ba et al. [3] presented a study of the preparation and stability of nanofluids. Graphene has drawn much consideration from the science world due to its mechanical and thermal properties. Graphene nanofluid is very important for its simple combination, longer suspension dependability, higher conductivity value, lower disintegration, consumption, bigger surface region/volume proportion, and lower propelling power. This article is a review of test results on the synthesis and strength of graphene nanofluids. Various studies manufacturing graphene nanofluids have been also reviewed.

Asadi et al. [4] presented a review of preparation methods and discussed different thermophysical properties of oil-based nanoparticles. Thermal oils, because of their high capacity in cooling and lubrication, have numerous applications in innovative businesses, such as the car industry. Scattering nanoparticles in a thermal oil, named a nanofluid, can change the cooling and lubrication efficiencies. This paper intends to introduce the new advances in the preparation and thermophysical properties of oil-based nanofluids. The impacts of different boundaries, for example, nanoparticle focus, size, and temperature on the upsides of properties including thermal conductivity, thickness, thickness were evaluated. The relationships accessible for the properties of oil-based nanofluids are accumulated for the individuals who expect to deal with oil-based nanofluids. Kumar et al. [5] studied and discussed different nanofluids for heat transfer in a heat exchanger. Different recent advancements in the preparation of nanofluids were also presented in this study. The heat exchanger is a critical segment of numerous mechanical frameworks in industry. To improve the heat transfer value nonfluids were widely utilized in heat exchangers. Finally, this paper sums up the recent investigations and advancements in the field of heat exchangers by utilizing nanofluids and the synthesis of nanofluids. Fuskele et al. [6] studied the recent advancements in the stability, preparation, and amalgamation of nanofluids. Nanofluids are weakened colloidal suspensions of nanoparticles in a base liquid that show astounding improvement in heat transfer execution in different applications. For the last decade various innovative advancement works have been done on the synthesis and stability of such materials. This chapter reports on the amalgamation of nanoparticles and formation of fixed nanofluids. Different kinds of nanoparticles with various base liquids were examined.

17.2 Preparation of hybrid nanofluids

Hybrid nanofluids can be prepared by one-step or two-step methods. The hybrid nanofluids are more superior in terms of properties than single nanofluids. The single NP exhibits lower performance than the combination of two NPs. The major processes for the preparation of hybrid nanofluids are discussed below (Fig. 17.1).

17.2.1 Two-step method

This is the most commonly used technique for the preparation of hybrid nanofluids. In this method, the process of NPs preparation and dispersion in base fluid take place separately. By using the two-step method, the hybrid nanofluids can be prepared in large quantities unlike in the one-step method. Fig. 17.2 shows the preparation of hybrid nanofluids using the two-step method.

Arshad et al. [8] presented a review article about graphene nanofluids preparations, applications, characterization, and preparation. Efficient heat

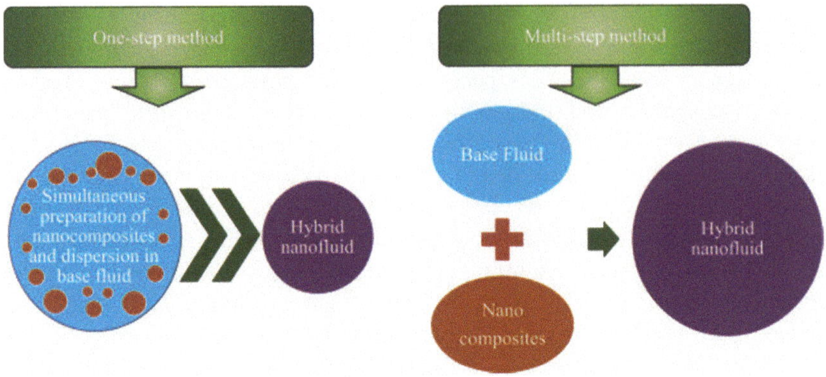

FIGURE 17.1 Preparation of hybrid nanofluid [7].

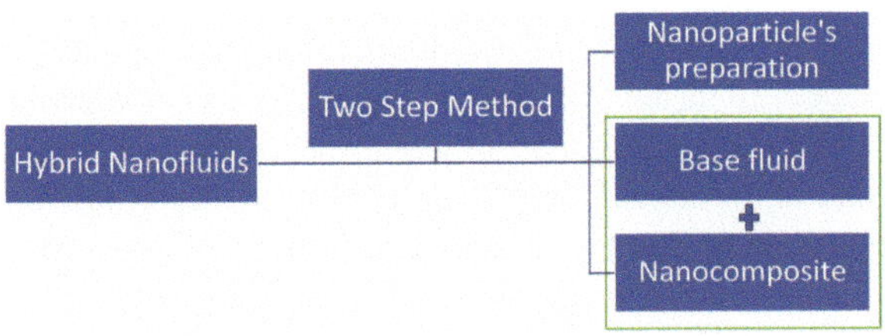

FIGURE 17.2 Two Step method for preparation of hybrid nanofluid.

transfer devices are required for a wide range of heat transfer applications. Graphene-based nanofluids showed the best performance in heat transfer applications, due to their high thermal conductivity, less erosion, and more stability. Babar et al. [9] reviewed the applications of hybrid nanofluids in various fields like solar energy, heat pipes, auto industry, processing industry, heat exchangers, and thermal management of electronics devices. Nanofluids are essential for the developments that have the capacity of advancing the field of heat transfer. The latest work has improved the thermophysical properties of nanofluids by presenting hybrid nanofluids, dealing with preparation methods, and by adding surfactants to improve their dependability. However, at the same time further consideration is required for the development of relationships/models that can foresee the thermophysical properties and heat transfer qualities.

17.2.2 One-step method

The one-step technique is also used for the preparation of hybrid nanofluids. In this method, the preparation of NPs and dispersion take place in single step. The one-step method for preparation of hybrid nanofluids is represented in Fig. 17.3. In one step, the metal source is evaporated in a vacuum, and

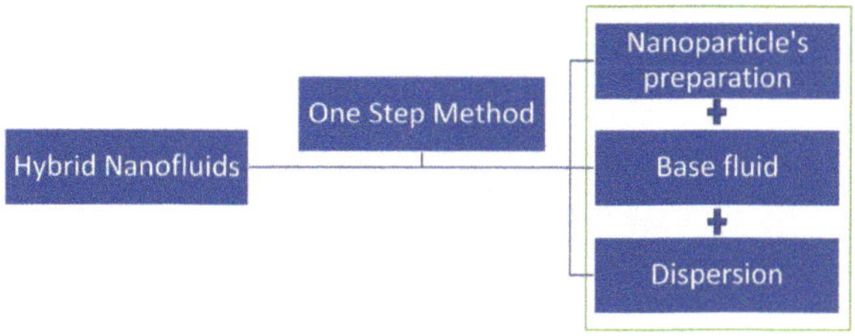

FIGURE 17.3 One-step method for preparation of hybrid nanofluid.

after condensation from vapor phase the hybrid nanofluids are produced [10]. The major advantage of using the one-step method is that the accumulation of NPs is low [11].

Munkhbayar et al. [12] studied a one-step pulsated method for the preparation of hybrid nanofluids with silver/multiwall carbon nanotubes in a base fluid of water. Silver wire was heated and evaporated using a high voltage pulse, and then the evaporated particles were condensed and directly mixed with the base fluid and multiwall carbon nanotubes.

17.3 Various hybrid nanofluids used in different heat sinks

With the rapid progression of technology in the numerous science fields many new innovative means were discovered. The heat transfer techniques were also improved. Many smaller electronic devices are being developed that are more reliable and faster. Such rapid development also leads to more heat generation. So, the cooling techniques of these compact devices need to be improved. Many efforts have been made by various researchers in the last few decades to increase the rate heat flux. Many optimized geometries of micro- and minichannel heat sinks were developed. Cooling techniques of these geometries were also improved by using different types as well as different concentrations of nanofluids.

Tuckerman and Pease [13] first worked on the heat sinks. They proposed that there is a reverse relationship between the channel width and the heat transfer. Ali and Arshad [14] investigated the thermal performance of inline and staggered heat sinks with pin fins using rutile and anatase base TiO_2 nanofluids. They have recorded the minimum base temperature for staggered heat sink using rutile base TiO_2 nano fluid (Fig. 17.4).

Khaleduzzaman et al. [15] experimentally analyzed the friction factor for TiO_2 nanofluids. They concluded that by using nanofluid at varying flow rates they achieved higher efficiency than water. Gwali et al. [16] performed experimental and theoretical studies on the heat transfer performance of a rectangular heat sink. They concluded that a heat sink gives better experimental and theoretical results at a Reynolds number between 400–450.

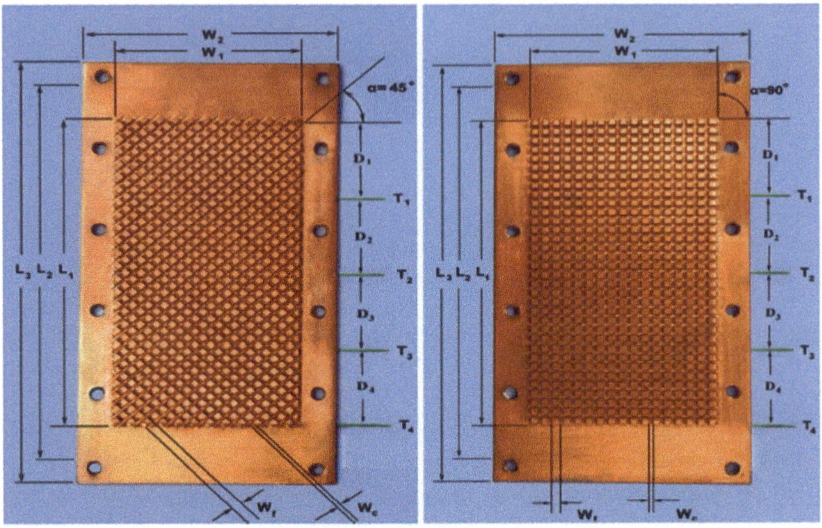

FIGURE 17.4 3D pictures of staggered and inline fin heat sinks [14].

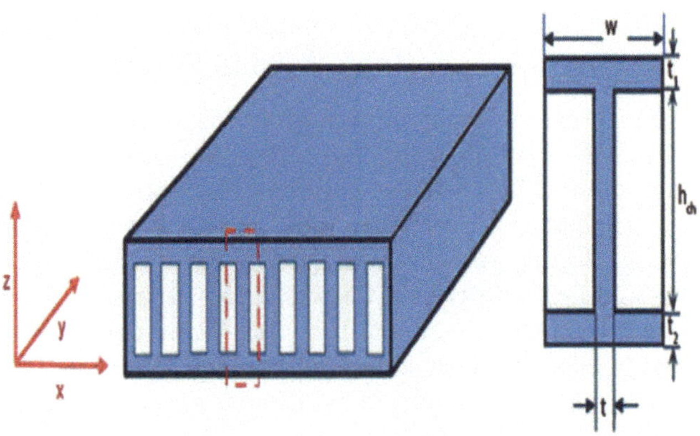

FIGURE 17.5 Mini channel heat sink [17].

Kumar and Sarkar [17] performed numerical simulation on two-phase hybrid nanofluid in minichannel. They recorded maximum heat transfer using volumetric concentration of 0.01 (Al_2O_3 + MWCNT) (Fig. 17.5).

Nimmagadda and Venkatasubbaiah [18] performed multiphase and experimental analysis on microchannels using smaller values of Reynolds number. At Re = 50 they recorded an increase in heat transfer of 35.37% compared with deionized water.

Nimmagadda and Venkatasubbaiah [19] numerically investigated the microchannel heat sink performance using hybrid nanofluid (water, Cu + Al, and methanol). Saeed and Kim [20] experimentally investigated the performance of

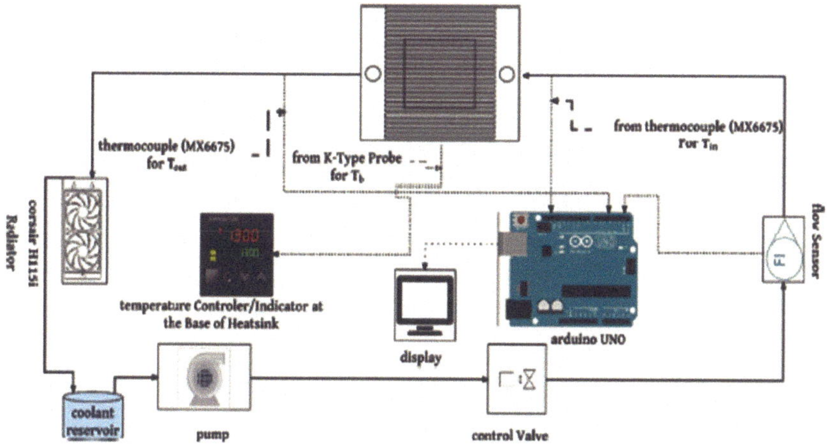

FIGURE 17.6 Schematic diagram for experimental setup [20].

a minichannel heat sink using (Al_2O_3-H_2O) nanofluid; the schematic diagram is shown in Fig. 17.6. They observed that thermal performance improved significantly using alumina nanofluid.

Tariq et al. [21] numerically and experimentally investigated the thermal performance of a facile minichannel heat sink using TiO_2-H_2O mixture. They observed that thermal performance was improved by using titanium oxide nanofluid.

Ali et al. [22] numerically and experimentally investigated the header design of a minichannel heat sink with different fin spacings using CuO-H_2O and Al_2O_3-H_2O nanofluid. They used different volumetric concentrations (0.4%, 0.67%) of copper oxide and alumina nanoparticles. They recorded the minimum base temperature using Al_2O_3-H_2O nanofluids.

Bahiraei et al. [23] studied a microchannel heat sink with ribs and a secondary channel using hybrid nanofluid (Fig. 17.7). They observed a decrease in pressure drop due to the presence of ribs. The convective heat transfer increases by increasing the Reynolds number and this results in an increase of pumping power.

Chandravadhana [24] reviewed hybrid as well as mono nanofluids inside microchannel heat sinks. Selvakumar and Singh [25] experimentally studied Al_2O_3-CuO/ H_2O hybrid nanofluid. They investigated the pressure drop and heat transfer characteristics of a thin channel heat sink. The increase in heat transfer recorded using a hybrid nanofluid was 24.35%. Sukhor et al. [26] numerically investigated the performance of a microchannel heat sink using a hybrid nanofluid, Al_2O_3-CuO/H_2O. They observed that the Nusselt number was improved as the transverse pitch was decreased from 3.81 to 1.81 mm. Kumar and Sarkar [27] numerically and experimentally investigated a minichannel heat sink (velocity profile shown in Fig. 17.8) using a

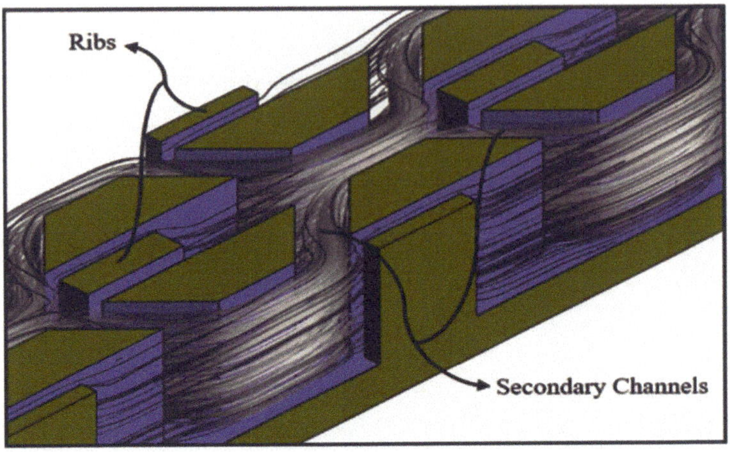

FIGURE 17.7 Ribs and secondary channels in the microchannel heat sink [23].

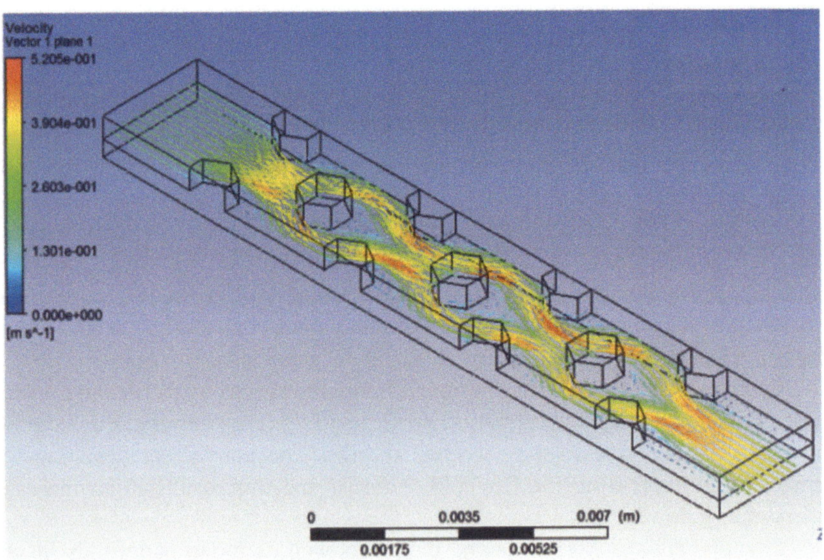

FIGURE 17.8 Velocity profile of hybrid nanofluid [26].

hybrid nanofluid (Al_2O_3-TiO_2/ H_2O) at different mixture ratios. The increase in heat transfer was recorded by increasing the Al_2O_3 particles ratio inside the heat sink.

Ardeh et al. [28] explored the performance of a double-layered microchannel heat sink under constant Reynolds number and constant fluid flow rate. The Al_2O_3-SiO_2/water and Al_2O_3-Cu/water hybrid nanofluids at various volume concentrations were used as working fluid. The results showed that Al_2O_3-SiO_2/water hybrid nanofluid lowered the wall temperature by 9.2°C, offered the lowest thermal resistance at higher pumping powers, and presented the highest performance factor at a volume concentration of 5% and Reynolds number of 50.

17.3.1 Thermal management of electronics using other fluids

The demand for storage of data has increased in recent years and is increasing day by day. It is challenging for electronic thermal management [29]. The result is the storage of data in compact and portable devices and this leads to the heating of that device due to improper thermal management of electronic components [30]. The conventional techniques used for thermal management of electronic devices are not suitable. The traditional techniques include fan cooling with large equipment, which increases the cost and power consumption [31,32]. Fan cooling was widely used in the past for the cooling of computer electronics and several numerical studies were carried out using natural and forced convection in electronic components [33–35].

During the last few years, phase change materials have been investigated in order to allow proper function of electronic components by storing and releasing heat from electronic components as latent heat [36,37]. The traditional electronic device consists of von Neumann Architecture which separated the central processing unit from the memory unit, whereas neuro-inspired computing is being adopted nowadays using PCM [38]. When PCM is used in electronic components, the temperature will increase via sensible heating in the case of heating of a heat sink. The paraffins are widely used as they have a temperature range of 40°C–45°C [39].

Parkrouh et al. [40] conducted a numerical study on the geometry optimization of a pin-fin heat sink filled with PCM. The optimization parameters were number, height, thickness of fin, heat sink, and critical temperature. Faraji et al. [41] carried out a numerical work using NEPCM for efficient cooling of a microprocessor. The PCM considered was n-eicosane and copper NPs. The equations that governed the problem were solved using the finite volume method. Simulations were done to check the numerical effects of the mass fraction of NPs for the thermal cooling of the heat sink and on the Nusselt number, and the liquid fraction was measured. The results concluded that the inclusion of NPs has a better cooling effect by decreasing the temperature. Mjallal et al. [42] numerically compared the thermal distribution of heat sinks by using PCM and without PCM. The two various types of PCM, namely salt hydrate and wax, were considered with changing heat flux. Experiments were also conducted to verify the results and the results concluded that there was a delay in peak temperature at low power, which is useful for mobile applications.

Ren et al. [43] used microencapsulated PCM (MEPCM)/expanded graphite (EG) along with pin-fins for thermal management of electronic components. The numerical study was conducted using the 3D Boltzmann constant and the results indicated that such a combination enhanced the HT and decreased the latent heat of microencapsulated phase change material. Joseph and Sajith [44] determined the effectiveness of composite PCM for cooling of electronic components. The comparison was done by using a paraffin wax/graphene composite and simple paraffin wax for cooling of the heat sink. The graphene was prepared using liquid-phase exfoliation and composite PCM was synthesized at various concentrations. The results concluded that heat transfer was superior using composite PCM and a maximum of 6°C lower steady-state temperature was obtained. An 11%–23% increased energy saving was observed using the composite PCM compared to a simple heat sink. Arshad et al. [45] used mono and hybrid PCM for thermal management of electronic components. The nanocomposites used were graphene nanoplatelets, multiwall carbon nanotubes, alumina, and copper oxide that were dispersed in RT-28HC PCM to form nanoPCM with 1wt.% of mass fraction for each nanoparticle. The various characterizations of prepared nanoPCM were also done to measure the thermophysical properties. The results indicated that graphene-based nanoPCM showed the highest thermal conductivity enhancement of 96% compared to pure PCM, and hybrid PCMs were better for the efficient thermal management of electronics.

Tariq et al. [46] investigated graphene-based RT-44HC and RT64HC nanoPCM in a heat sink for a decrease of baseline temperature. The NPs with various concentration, for example, 0.002 wt.%, 0.005 wt.%, and 0.008 wt.% were mixed in base PCM. Different power was used to measure the effect of nanoPCM in a heat sink. The results showed that a maximum reduction of 25% was observed for baseline temperature at a power of 0.86 KW/m^2. The RT-64HC composite was best for higher heating load, while RT-34HC was best for low heating loads. Leong et al. [47] studied the thermal performance of different cross fin heat sinks by using different PCMs. The PCMs considered were paraffin wax and 1-hexadeconal. The base temperature of the cross fin heat sink was 46.9°C compared to 51.6°C using the same heat sink without PCM. The heat sink with paraffin wax performed better than that with 1-hexadeconal.

Kim et al. [48] performed experiments as well as numerical simulations for cooling of electronic components using compact cascade PCMs. Numerical work was done using multiple PCMs to understand the fundamental mechanism and then experiments were conducted. The experimental results indicated that peak temperature inside the fin was reduced from 123.4°C to 107.2°C in one heat supply cycle and peak temperature was reduced to 13.1% using cascade PCM, compared to natural convection air. Also, the time required to attain peak temperature was decreased by 45% due to the larger heat capacity.

Amin Farzanehia [49] conducted experiments using a MWCNTs/paraffin-based heat sink for thermal control of an electronic chipset, and concluded that the heat sink incorporating NEPCM augmented the system performance by about 6%. Yasuke Tomizawa [50] synthesized microencapsulated PCM with thickness of 1.60–2.45 mm to regulate the temperature of mobile devices. Li et al. [51] investigated two types of flexible PCMs for the thermal control of compact electronic equipment. The latent heat of the two flexible PCMs was 160.2 J/g and 153.1 J/g and the experimental setup is shown in Fig. 17.9. According to the results, the temperature of the electronic chip decreased to 14.3°C. The graphene-based film of PCM was employed having a value "k" 1500 W/m which improved the latent heat capacity and rate of heat transfer. The diffusion of the heat area was improved by 32.4% at 3.2 W.

Alimohammadi et al. [52] carried out experiments by using simple PCM and NEPCM on electronic cooling, and compared the performance. The PCM that was used in the experiments was $Mn(NO_3)_2$, which is an inorganic salt, and iron oxide NPs were used. The NEPCM was prepared by an ultrasonication technique with 1% weight fraction of NPs. The heat flux was 1000–4000 W/m^2 and results showed that PCM and NEPCM decreased the temperature of electronic component by 14°C and 10°C, respectively. The NEPCM was suitable for temporary cooling and simple PCM was suitable for long-term cooling of electronic component. Colla et al. [53] investigated NEPCM for improving the cooling performance of electronic components. The alumina NPs were used in pristine paraffin to augment the value of "k" as well as the heat capacity. The NEPCM was prepared by mixing 0.5 wt.% and 1 wt.% of NPs in paraffin and the thermophysical properties were measured. The numerical work was also carried out by using ANSYS-15. The results concluded that the melting process was improved by using NEPCM compared to the reference case and more work was needed to understand the HT phenomena for NEPCM characterization.

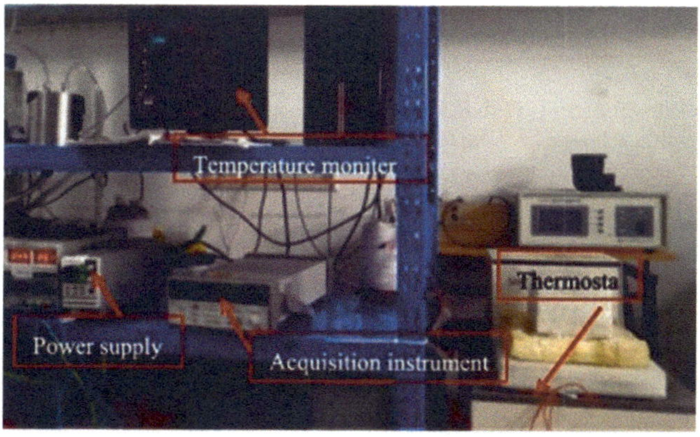

FIGURE 17.9 Experimental setup of electronic device cooling using PCM [51].

Zhu et al. [54] reported that the value of "k" was improved by silver-coated PCM nanocapsules in which n-octacosane was used as the core material and silica as the shell material. The value of "k" was improved from 0.246 to 1.346 W/m.K and the authors suggested applying this nanocapsule for cooling of electronic components. Haung et al. [55] carried out experiments of composite PCM for electronic thermal management using salt hydrate as PCM along with different carbon fillers. Such nanofillers enhanced the nucleating sites and increased the heat transfer. The different properties like particle size and specific surface area of nanofillers on thermal attributes of PCM were experimentally analyzed. The supercooling capacity of composite PCM was reduced to 97.24% compared to pure PCM. However, a slight change in phase transition temperature was observed after 200 cycle and there was no leakage after 2 wt.%. of carbon filler in PCM.

The thermal management of electronic chips using PCM has attracted considerable attention lately. Solid–liquid PCMs have unique cooling behavior, especially organic PCMs, except for their low thermal conductivity. The use of fins of 9% volume fraction of heat sink, NEPCMs with 0.2 to 1% mass fraction of nanoparticles, and metal foams with porosity of 0.95, in the heat sinks gave the best thermal management results, according to different studies [56]. Arshad et al. [57] dispersed TiO_2, Al_2O_3, and CuO nanoparticles in RT 35HC PCM for thermal management purposes. They prepared mono and hybrid nanoenhanced PCMs (NEPCMs) with uniform 1 wt.% concentration. Findings revealed that hybrid NEPCM of Al_2O_3 and CuO exhibited 53.7% increment in thermal conductivity and heat sink base temperature reduced by 13.95% using CuO based NEPCM. Similarly, Arshad et al. [58] prepared NEPCM by dispersing TiO_2 nanoparticles in RT-35HC PCM and performed extensive thermal analysis. The addition of TiO_2 nanoparticles enhanced the thermal conductivity by 59.5% and the optimal concentration of TiO_2 gave latent heat of 235.84 J/g and thermal conductivity of 0.284 W/m.K.

Wang et al. [59] studied a heat pipe coupled with a cavity containing paraffin enhanced with Al_2O_3, and CuO and MWCNTs nanoparticles for cooling several thermal sources. The concentrations of nanoparticles studied were 0.5, 1, 2, and 4 wt.%. The use of 1 wt.% Al_2O_3 nanoparticles enhanced PCM, the minimum thermal resistance was obtained, and maximum heat source temperature reduced by 17.4% at 8 W power level. Arshad et al. [45] prepared mono and hybrid NEPCMs using GNPs, MWCNTs, Al_2O_3, and CuO nanoparticles dispersed in RT-28HC PCM for cooling of microelectronics. Mass fraction of nanoparticles was constant and taken as 1 wt.%. Various thermal characterization techniques were used to investigate the chemical composition, structure morphology, and thermal behavior of NEPCMs. The hybrid NEPCM of GNPS/MWCNTs with mass percentage proportion of 75%/25% gave the highest thermal conductivity increase of 96% compared with pure PCM, with optimal phase transition enthalpy of 245.18 J/g, and showed uniform heating with enhanced phase change duration.

Arshad et al. [60] used carbon nanoadditives of graphene oxide, reduced graphene oxide, GNPs, and MWCNTs to prepare mono and hybrid NEPCM. The PCM used was RT-35HC, and mass concentration of nanoparticles was taken as 1 wt.%. Physical, chemical, and thermal characterization of mono and hybrid NEPCMs was carried out. The results revealed that hybrid NEPCM of GNPs and MWCNTs enhancement in thermal conductivity of 185.3% was observed and the enthalpy of fusion of 230.82 was obtained.

Hekmat and Molaeimanesh [61] investigated the hybrid thermal management of a lithium ion battery by integrating PCM and water cooling pipes. Experimentation was done for charging and discharging of the battery using active, passive, and hybrid cases of cooling. The passive cooling of the battery using PCM gave a maximum temperature of 32°C and a maximum difference among battery cells of 1.2°C, but the time required to reach 30°C was more than the active cooling case. The introduction of water-cooling pipes in PCM further improved the maximum temperature and temperature difference among battery cell.

Arshad et al. [62] numerically investigated the hybrid NEPCM for passive cooling of electronic equipment. Graphene oxide and silver nanoparticles were added in RT 28HC PCM in different proportions and different shape factors. The addition of optimum proportion and shape factor values of hybrid nanoparticles in PCM enhanced the thermal conductivity of PCM by 73.85%, reduced the melting time by 12.93%, and augmented the heat storage capacity by 8.65%. Kalbasi et al. [63] studied numerically the hybrid heat sink filled with PCM accompanied with air flow for electronic device cooling. The passive cooling using PCM was effective only before the inception of melting phase of PCM when the maximum heat was stored in PCM. While hybrid cooling resulted in better heat rejection performance before and after melting of PCM. Furthermore, the hybrid heat sink presented potential for competing with an air-cooled heat sink of double heat transfer coefficient.

Yang et al. [64] explored the thermal performance of finned heat sinks filled with PCM under uniform and intermittent thermal loads. The effect of PCM volume fraction, thermal loads, and forced convection was ascertained on heat sink thermal performance. The operational time of the PCM filled heat sink was extended considerably and the peak temperature was significantly lessened under intermittent heat load. While in case of forced convection, under intermittent load the baseline case was deemed as the better option than the PCM-based case, as the PCM positive effect of heat storage was outweighed by the negative effect of reduction of heat rejection rate.

Righetti et al. [65] studied the influence of ambient temperature, forced convection, and fast intermittent charging on the thermal performance of RT-70 PCM-filled aluminum heat sink with 3D periodic structures of pyramidal cells and compared it with a PCM-filled empty heat sink. The use of an enhanced heat sink lowered the charging–discharging time at all ambient temperatures. The introduction of forced convection drastically enhanced the heat dissipation by the heat sink and lowered the junction temperature, and the increase in air velocity reduced the discharging time. For fast intermittent charging case, the empty heat sink performed better than the enhanced heat sink. The authors recommended that the design of PCM-based cooling system should take into account the PCM's properties, ambient conditions, and system requirements for optimized thermal performance.

Mohammadi et al. [66] investigated the thermal management of electronic equipment using a nanofluid-enhanced mini channel heat sink. The experiment was carried out by varying parameters like heat flux, nanofluid mass fraction, and flow rate of coolants. The increase in flow rate from 100 to 200 mL/min lowered the electronic chip temperature by 2.3°C. The increase in silicon carbide/water nanofluid mass concentration up to 1 wt.% resulted in the maximum electronic chip temperature reduction of 3.7°C at 10 KW/m^2. In addition, for 1 wt.% nanofluid at 6 KW/m^2, maximum energy efficiency of 58.1% and thermal resistance of 0.58°C/W was attained.

Najafi et al. [67] assessed the cooling performance of a NEPCM-filled finned heat sink integrated with a copper minichannel heat absorber with water as coolant. The thermal characterization of the heat sink was done by variation of heat flux, heat transfer fluid flow rate, and nanoparticles. The nanoparticles used in the study were Fe_2O_3 and TiO_2 in the concentration of 2, 4, and 8 wt.%. The coolant flow rate of 50 mL/min was the preferred power consumption and temperature reduction. The maximum temperature drop of 17.9°C to 13.2°C and lowest thermal resistance was ascertained for various options of NEPCM-based heat sinks with uniform fluid flow.

Pourfattaha and Sabzpooshani [68] devised a novel cooling system that used jet impingement in conjunction with pin fins and vortex generators to reduce power electronics operating temperatures. The authors conducted a numerical research and then utilized a genetic algorithm and the response surface approach to optimize the cooling system's performance. The results revealed that the coolant flow rate and pin fin height had the greatest impact on the temperature of the power electronics. According to the findings of the genetic algorithm, the operating temperature of the insulated-gate bipolar transistor and the diode at the optimal point was 75°C and 59°C, respectively, and the pumping power was 3.9 W. The comparison of numerical and genetic algorithm results revealed that the genetic algorithm predicted thermal performance and pumping power with deviation less than 0.3% and 10%, respectively. Hayat et al. [69] combined the heat pipe, porous copper foam, and PCM-based heat sink as a hybrid cooling system, with and without a fan, for enhancing the thermal management of electronic equipment. RT 35HC PCM was employed in the study. The result showed that after 6000 seconds of charging at 3 kW/m^2, the hybrid cooling pipe with fan gave a maximum temperature reduction of 54%. Besides the hybrid cooling system improved the discharging rate of PCM.

TABLE 17.1 Summary of studies on thermal management of electronics components using advanced PCMs.

Type of heat sink	PCM Type	NPs	Results	References
–	Paraffin	Multi wall carbon nanotubes	Performance improved by 6%	[49]
Electronic chip		Graphene	HT area increased by 32.4%	[51]
–	RT-44HC, RT-64HC	Graphene NPs	Maximum decrement of base temperature of heat sink to 25% for RT-44/graphene NPs (0.008 wt.%)	[46]

Zeng et al. [70] employed a topology optimization technique to design a forced convection-based heat sink and analyzed its thermal behavior numerically and experimentally. According to the results, at 40 W heating load, the topology optimized heat sink saved 54.9% pumping power while maintaining the heat sink temperature at 44.5°C. Moradikazerouni et al. [35] numerically investigated a heat sink under forced convection for the thermal control of computer CPU. Results showed that the increase in Reynolds number from 3.80×10^3 to 2.28×10^4 lowered the thermal stress, thermal deformation, and surface temperature by 47.36%, 50%, and 28.57% respectively. Similarly, increasing the number and height of heat sink fins lowered the surface temperature by 25% and 20%, respectively.

Krishna et al. [29] investigated a heat pipe coupled with NEPCM for the thermal regulation of electronic devices. Al_2O_3 nanoparticles were added to PCM in concentrations of 0.1%, 1%, and 2%. According to the findings, the use of NEPCM lowered evaporator temperature by 25.75%, resulting in a 53% reduction in fan power consumption. Furthermore, the maximum energy stored by NEPCM with 2% nanoparticles was 30% of the input power. The use of 1% Al_2O_3 nanoparticles resulted in optimal heat transfer performance. A summary of studies on the thermal management of electronics components using advanced PCMs is presented in Table 17.1.

17.4 Conclusion

The hybrid nanofluids have extensive stable properties which have increased their demand for various cooling applications. The addition of hybrid nanofluids in a heat sink results in an increase in performance and better efficiency. The major conclusion points are:

- The performance of the base heat sink is very low. It can be enhanced by various methods, one of which is the use of hybrid nanofluids.
- By using hybrid nanofluids, the performance value of the base heat sink increased significantly and ultimately efficiency improved.
- Due to the extraordinary properties of hybrid nanofluids, they are extensively used in various applications, including heat sink systems.
- Most of the studies have concluded that the temperature decrease of heat sinks was better when using hybrid nanofluids compared with using ordinary fluids.

References

[1] S. Mukherjee, S. Paria, Preparation and stability of nanofluids-a review, J. Mech. Civil Eng. 9 (2013) 63–69.

[2] D.K. Devendiran, V.A. Amirtham, A review on preparation, characterization, properties and applications of nanofluids, Renew. Sustain. Energy Rev. 60 (2016) 21–40.

[3] T. Le, B. Omid, M. Somchai, I.M. Szilágyi, Review on the recent progress in the preparation and stability of graphene - based nanofluids Modified Hummers method, J. Therm. Anal. Calorim. 142 (3) (2020) 1145–1172.

[4] A. Asadi, et al., Recent advances in preparation methods and thermophysical properties of oil-based nanofluids: a state-of-the-art review, Powder Technol. 352 (2019) 209–226.

[5] P. Kumar, R.M. Sarviya, Recent developments in preparation of nanofluid for heat transfer enhancement in heat exchangers: a review, Mater. Today Proc. (2021).

[6] V. Fuskele, R.M. Sarviya, Recent developments in nanoparticles synthesis, preparation and stability of nanofluids, Mater. Today Proc. 4 (2) (2017) 4049–4060.

[7] M. Muneeshwaran, G. Srinivasan, P. Muthukumar, C.-C. Wang, Role of hybrid-nanofluid in heat transfer enhancement–s review, Int. Commun. Heat. Mass. Transf. 125 (2021) 105341.

[8] A. Arshad, M. Jabbal, Y. Yan, D. Reay, PT NU, J. Mol. Liq. (2019). p. #pagerange#.

[9] H. Babar, H.M. Ali, Towards hybrid nanofluids: preparation, thermophysical properties, applications, and challenges, J. Mol. Liq. (2019).

[10] H. Eshgarf, R. Kalbasi, A. Maleki, M.S. Shadloo, A review on the properties, preparation, models and stability of hybrid nanofluids to optimize energy consumption, J. Therm. Anal. Calorim. 144 (5) (2021) 1959–1983.

[11] L.S. Sundar, K.V. Sharma, M.T. Naik, M.K. Singh, Empirical and theoretical correlations on viscosity of nanofluids: a review, Renew. Sustain. Energy Rev. 25 (2013) 670–686.

[12] B. Munkhbayar, M.R. Tanshen, J. Jeoun, H. Chung, H. Jeong, Surfactant-free dispersion of silver nanoparticles into MWCNT-aqueous nanofluids prepared by one-step technique and their thermal characteristics, Ceram. Int. 39 (6) (2013) 6415–6425.

[13] D.B. Tuckerman, R.F.W. Pease, High Performance Heat Sinking for VLSI, IEEE Electron. Device Lett. 2 (5) (1981) 126–129. Available from: https://doi.org/10.1109/EDL.1981.25367.

[14] H.M. Ali, W. Arshad, Thermal performance investigation of staggered and inline pin fin heat sinks using water based rutile and anatase TiO2 nanofluids, Energy Convers. Manag. 106 (2015) 793–803.

[15] S.S. Khaleduzzaman, et al., Experimental analysis of energy and friction factor for titanium dioxide nanofluid in a water block heat sink, Int. J. Heat. Mass. Transf. 115 (2017) 77–85.

[16] D.B.S. Gawali, V.B. Swami, S.D. Thakre, Theoretical and experimental investigation of heat transfer characteristics through a rectangular microchannel heat sink, Int. J. Innov. Res. Sci. Eng. Technol. 03 (08) (2014) 15631–15640.

[17] V. Kumar, J. Sarkar, Two-phase numerical simulation of hybrid nanofluid heat transfer in minichannel heat sink and experimental validation, Int. Commun. Heat. Mass. Transf. 91 (2018) 239–247.

[18] R. Nimmagadda, K. Venkatasubbaiah, Experimental and multiphase analysis of nanofluids on the conjugate performance of micro-channel at low Reynolds numbers, Heat. Mass. Transf. 53 (6) (2017) 2099–2115.

[19] R. Nimmagadda, K. Venkatasubbaiah, Two-phase analysis on the conjugate heat transfer performance of microchannel with Cu, Al, SWCNT, and hybrid nanofluids, J. Therm. Sci. 9 (2017).

[20] M. Saeed, M.H. Kim, Heat transfer enhancement using nanofluids (Al_2O_3-H_2O) in minichannel heatsinks, Int. J. Heat. Mass. Transf. 120 (2018) 671–682.

[21] H.A. Tariq, M. Anwar, A. Malik, H.M. Ali, Hydro-thermal performance of normal-channel facile heat sink using TiO_2-H_2O mixture (Rutile–Anatase) nanofluids for microprocessor cooling, J. Therm. Anal. Calorim. (2020). no. 0123456789.

[22] M. Ali, A.A. Shoukat, H.A. Tariq, M. Anwar, H. Ali, Header design optimization of minichannel heat sinks using CuO–H_2O and Al_2O_3–H_2O nanofluids for thermal management, Arab. J. Sci. Eng. 44 (12) (2019) 10327–10338.

[23] M. Bahiraei, M. Jamshidmofid, M. Goodarzi, Efficacy of a hybrid nanofluid in a new microchannel heat sink equipped with both secondary channels and ribs, J. Mol. Liq. 273 (2019) 88–98.

[24] A. Chandravadhana, V. NandaKumar, K. Venkatramanan, Mono and hybrid nanofluid based heat sink technologies—a review, J. Phys. Conf. Ser. 1644 (1) (2020).

[25] P. Selvakumar, S. Suresh, Use of Al_2O_3 CuWater hybrid nanofluid in an electronic heat sink.pdf, IEEE Trans. Compond. Packag. Manuf. Technol. 2 (10) (2012) 1600–1607.

[26] N.B. Sukhor, A.S. Tijani, J. Kubenthiran, I.K. Muritala, Computational modeling of thermal characteristics of hybrid nanofluid in micro-pin fin heat sink for electronic cooling, Int. J. Green. Energy 00 (00) (2021) 1–19.

[27] V. Kumar, J. Sarkar, Numerical and experimental investigations on heat transfer and pressure drop characteristics of Al_2O_3-TiO_2 hybrid nanofluid in minichannel heat sink with different mixture ratio, Powder Technol. 345 (2019) 717–727.

[28] S. Sarvar-Ardeh, R. Rafee, S. Rashidi, Hybrid nanofluids with temperature-dependent properties for use in double-layered microchannel heat sink; hydrothermal investigation, J. Taiwan. Inst. Chem. Eng. 124 (2021) 53–62.

[29] J. Krishna, P.S. Kishore, A.B. Solomon, Heat pipe with nano enhanced-PCM for electronic cooling application, Exp. Therm. Fluid Sci. 81 (2017) 84–92.

[30] Y. Cai, Y. Wang, D. Liu, F.-Y. Zhao, Thermoelectric cooling technology applied in the field of electronic devices: updated review on the parametric investigations and model developments, Appl. Therm. Eng. 148 (2019) 238–255.

[31] H. So, A.P. Pisano, Micromachined passive phase-change cooler for thermal management of chip-level electronics, Int. J. Heat. Mass. Transf. 89 (2015) 1164–1171.

[32] S. Arora, Selection of thermal management system for modular battery packs of electric vehicles: a review of existing and emerging technologies, J. Power Sources 400 (2018) 621–640.

[33] R.C. Adhikari, D.H. Wood, M. Pahlevani, An experimental and numerical study of forced convection heat transfer from rectangular fins at low Reynolds numbers, Int. J. Heat. Mass. Transf. 163 (2020) 120418.

[34] S. Durgam, S.P. Venkateshan, T. Sundararajan, A novel concept of discrete heat source array with dummy components cooled by forced convection in a vertical channel, Appl. Therm. Eng. 129 (2018) 979–994.

[35] A. Moradikazerouni, M. Afrand, J. Alsarraf, S. Wongwises, A. Asadi, T.K. Nguyen, Investigation of a computer CPU heat sink under laminar forced convection using a structural stability method, Int. J. Heat. Mass. Transf. 134 (2019) 1218–1226.

[36] E.B. Jebasingh, V.A. Arasu, A comprehensive review on latent heat and thermal conductivity of nanoparticle dispersed phase change material for low-temperature applications, Energy Storage Mater. 24 (2020) 52–74.

[37] H.M. Ali, Experimental study on the thermal behavior of RT-35HC paraffin within copper and Iron-Nickel open cell foams: energy storage for thermal management of electronics, Int. J. Heat. Mass. Transf. 146 (2020) 118852.

[38] W. Zhang, R. Mazzarello, E. Ma, Phase-change materials in electronics and photonics, MRS Bull. 44 (9) (2019) 686–690.

[39] A.S. Fleischer, Thermal Energy Storage Using Phase Change Materials: Fundamentals and Applications, Springer, 2015.

[40] R. Pakrouh, M.J. Hosseini, A.A. Ranjbar, R. Bahrampoury, A numerical method for PCM-based pin fin heat sinks optimization, Energy Convers. Manag. 103 (2015) 542–552.

[41] H. Faraji, et al., Cooling of recent microprocessors by the fusion of nano-enhanced phase change materials, Mater. Today Proc. (2020).

[42] I. Mjallal, H. Farhat, M. Hammoud, S. Ali, I. Assi, Improving the cooling efficiency of heat sinks through the use of different types of phase change materials, Technologies 6 (1) (2018) 5.

[43] Q. Ren, P. Guo, J. Zhu, Thermal management of electronic devices using pin-fin based cascade microencapsulated PCM/expanded graphite composite, Int. J. Heat. Mass. Transf. 149 (2020) 119199.

[44] M. Joseph, V. Sajith, Graphene enhanced paraffin nanocomposite based hybrid cooling system for thermal management of electronics, Appl. Therm. Eng. 163 (2019) 114342.

[45] A. Arshad, M. Jabbal, Y. Yan, Preparation and characteristics evaluation of mono and hybrid nano-enhanced phase change materials (NePCMs) for thermal management of microelectronics, Energy Convers. Manag. 205 (2020) 112444.

[46] S.L. Tariq, H.M. Ali, M.A. Akram, M.M. Janjua, Experimental investigation on graphene based nanoparticles enhanced phase change materials (GbNePCMs) for thermal management of electronic equipment, J. Energy Storage 30 (2020) 101497.

[47] K.Y. Leong, S.P. Chew, B.A. Gurunathan, K.Z.K. Ahmad, H.C. Ong, An experimental approach to investigate thermal performance of paraffin wax and 1-hexadecanol based heat sinks for cooling of electronic system, Int. Commun. Heat. Mass. Transf. 109 (2019) 104365.

[48] S.-H. Kim, C.S. Heu, D.R. Kim, S.-W. Kang, Numerical modeling and experimental validation of a phase change material-based compact cascade cooling system for enhanced thermal management, Appl. Therm. Eng. 164 (2020) 114470.

[49] A. Farzanehnia, M. Khatibi, M. Sardarabadi, M. Passandideh-Fard, Experimental investigation of multiwall carbon nanotube/paraffin based heat sink for electronic device thermal management, Energy Convers. Manag. 179 (2019) 314–325.

[50] Y. Tomizawa, K. Sasaki, A. Kuroda, R. Takeda, Y. Kaito, Experimental and numerical study on phase change material (PCM) for thermal management of mobile devices, Appl. Therm. Eng. 98 (2016) 320–329.
[51] W. Li, F. Wang, W. Cheng, X. Chen, Q. Zhao, Study of using enhanced heat-transfer flexible phase change material film in thermal management of compact electronic device, Energy Convers. Manag. 210 (2020) 112680.
[52] M. Alimohammadi, Y. Aghli, E.S. Alavi, M. Sardarabadi, M. Passandideh-Fard, Experimental investigation of the effects of using nano/phase change materials (NPCM) as coolant of electronic chipsets, under free and forced convection, Appl. Therm. Eng. 111 (2017) 271–279.
[53] L. Colla, D. Ercole, L. Fedele, S. Mancin, O. Manca, S. Bobbo, Nano-phase change materials for electronics cooling applications, J. Heat. Transf. 139 (5) (2017).
[54] Y. Zhu, et al., Novel metal coated nanoencapsulated phase change materials with high thermal conductivity for thermal energy storage, Sol. Energy Mater. Sol. Cell 176 (2018) 212–221.
[55] J. Huang, J. Dai, S. Peng, T. Wang, S. Hong, Modification on hydrated salt-based phase change composites with carbon fillers for electronic thermal management, Int. J. Energy Res. 43 (8) (2019) 3550–3560.
[56] W. Hua, L. Zhang, X. Zhang, Research on passive cooling of electronic chips based on PCM: a review, J. Mol. Liq. 340 (2021) 117183.
[57] A. Arshad, M. Jabbal, Y. Yan, Thermophysical characteristics and application of metallic-oxide based mono and hybrid nanocomposite phase change materials for thermal management systems, Appl. Therm. Eng. 181 (2020) 115999.
[58] A. Arshad, M. Jabbal, L. Shi, J. Darkwa, N.J. Weston, Y. Yan, Development of TiO2/RT−35HC based nanocomposite phase change materials (NCPCMs) for thermal management applications, Sustain. Energy Technol. Assess. 43 (2021) 100865.
[59] J. Wang, K. Yu, R. Duan, G. Xie, B. Sundén, Enhanced thermal management by introducing nanoparticle composite phase change materials for cooling multiple heat sources systems, Energy 227 (2021) 120495.
[60] A. Arshad, M. Jabbal, L. Shi, Y. Yan, Thermophysical characteristics and enhancement analysis of carbon-additives phase change mono and hybrid materials for thermal management of electronic devices, J. Energy Storage 34 (2021) 102231.
[61] S. Hekmat, G.R. Molaeimanesh, Hybrid thermal management of a Li-ion battery module with phase change material and cooling water pipes: an experimental investigation, Appl. Therm. Eng. 166 (2020) 114759.
[62] A. Arshad, M. Jabbal, H. Faraji, M.A. Bashir, P. Talebizadehsardari, Y. Yan, Thermal process enhancement of HNCPCM filled heat sink: effect of hybrid nanoparticles ratio and shape, Int. Commun. Heat. Mass. Transf. 125 (2021) 105323.
[63] R. Kalbasi, Introducing a novel heat sink comprising PCM and air—adapted to electronic device thermal management, Int. J. Heat. Mass. Transf. 169 (2021) 120914.
[64] H. Yang, Y. Li, L. Zhang, Y. Zhu, Thermal performance enhancement of phase change material heat sinks for thermal management of electronic devices under constant and intermittent power loads, Int. J. Heat. Mass. Transf. 181 (2021) 121899.
[65] G. Righetti, C. Zilio, L. Doretti, G.A. Longo, S. Mancin, On the design of phase change materials based thermal management systems for electronics cooling, Appl. Therm. Eng. 196 (2021) 117276.
[66] M. Mohammadi, A. Taheri, M. Passandideh-Fard, M. Sardarabadi, Electronic chipset thermal management using a nanofluid-based mini-channel heat sink: an experimental study, Int. Commun. Heat. Mass. Transf. 118 (2020) 104836.

[67] F. Najafi, et al., Fabrication and experimental characterization of a modified heat-sink based on a semi-active/passive cooling strategy with fluid flow and nano-enhanced phase change material, Int. Commun. Heat. Mass. Transf. 126 (2021) 105371.

[68] F. Pourfattah, M. Sabzpooshani, On the thermal management of a power electronics system: optimization of the cooling system using genetic algorithm and response surface method, Energy 232 (2021) 120951.

[69] M.A. Hayat, H.M. Ali, M.M. Janjua, W. Pao, C. Li, M. Alizadeh, Phase change material/heat pipe and copper foam-based heat sinks for thermal management of electronic systems, J. Energy Storage 32 (2020) 101971.

[70] S. Zeng, B. Kanargi, P.S. Lee, Experimental and numerical investigation of a mini channel forced air heat sink designed by topology optimization, Int. J. Heat. Mass. Transf. 121 (2018) 663–679.

Index

Note: Page numbers followed by "*f*" and "*t*" refer to figures and tables, respectively.

A

Adaptive neuro fuzzy inference system, 207–208
Antibacterial activity via nanofluids
  on bacteria morphology, 294–296
    Staphylococcus aureus, 295*f*
  Escherichia coli, 284–285, 286*f*
    before and after treatment, 288*f*
    growth curve of, 295*f*
    and *Staphylococcus aureus*, 291*f*, 292*f*
  mechanisms of, 285–288
  parameters on, 289–294
    growth inhibition, 291
    growth slopes *versus* concentrations, 291*f*
    NPs size, 292–293
    NPs concentration, 289–291
    pH, temperature, and pressure, 294
    stabilizer, presence of, 293–294
  rate of NPs, 294
Apparent friction factor, 460
Aqueous ionic liquid, 137

B

Brownian motion, 159
Building energy systems
  heating, ventilation, and air conditioning systems, 487–489
  heat pump heating cycle power consumption, 488*f*
  HVAC systems, 490*t*
  nanofluids efficiency, 483–494
  phase change materials, 491*t*
  photovoltaic thermal systems, 483–487
  in PVT, 483*f*
    performance improvement, 484*t*
  thermal storage systems, 489–494, 493*t*
Building thermal energy systems
  barriers, 494–495
  of base fluid and nanoparticles, 480*f*
  domain knowledge, 480–483
  enhancement of, 483–494
Buongiorno model, 176–178

C

Cell centers, 188
Cell faces, 188
Computational fluid dynamics model
  fundamental relation for, 239–243
  Lattice Boltzmann method, 240–243
    for D2Q9, 241*f*
    for D3Q15, 242*f*
    for D3Q19, 243*f*
  macroscopic
    control equation, 239–240
    models, 239–240
    single-phase, 239
    two-phase, 239
Condenser, 350
Corrections for nanofliuids
  friction factor, 16–20
    of Fe_3O_4, 17*f*
    of Ni, 18*f*
    pressure drop of SiO_2, 19*f*
    of SiO_2, 19*f*
  numerical study, 12–13
  Nusselt number, 9–16
    of Al_2O_3, 9*f*
    of Fe_3O_4, 14*f*
    of MgO, 16*f*
    at TiO_2, 13*f*
Correlations, development of
  friction factor, 9
  Nusselt number, 8–9

D

Debye length, 157–158
Domain knowledge
  density of, 481–482

Domain knowledge (*Continued*)
Hamilton-Crosser model, 482
preparation methods, 481*f*
Reynold's number, 482

E

Einstein equation for viscosity, 118
Electrical double layer, 67–70
Electronics thermal management
design parameters, 405
heat pipes, 405–418
nanofluids
capability and models, 424–429
in heat transfer systems, 430–440
pulsating heat pipes, 418–424
Eulerian model, 180–181
Experimental methods
of forced convection apparatus, 4*f*
friction factor, 5
nondimensional numbers, 5–6
Nusselt number, 4

F

Flow of nanofluids in microchannels
continuous zigzag (CZ-HS), 463*f*
Hagenbach factor, 460
heat transfer for single-phase, 459
in microchannel heat sinks (MCHS), 462*t*
Navier-Stokes equations, 460
Fourier-transform infrared spectroscopy, 70–71
Four-season oils, 319
Friction factor correlations for single-phase fluids
laminar flow, 8
turbulent flow, 8

G

Geometry effect of microchannels
fluid path, 467–468, 471*t*
inclusions/exclusions in cross section, 467, 469*t*
in MCHS applications, 468*t*
shape of, 467

H

Harvesting solar energy by nanofluids
antibacterial activity, 282–296
economic performance of, 278–279
CPS achievement, 280*t*
evaporation efficiency of, 268*t*
evaporation flux, 267
on heat localization, 278
water and rGO, 279*f*
overview of, 266–267
parameters on light, 271–278
concentration of, 271
evaporation rate and vapor productivity, 272*f*
of GNP@MWCNT, 275*f*
GNP@MWCNT nanofluids over time, 273*f*
mass change *versus* time, 272*f*
nanoparticles size, 277–278
rGO NFs *versus* time, 277*f*
solar illumination intensity, 276
stability and optical property of, 272–276
UV-Vis and N-IR for several rGO, 274*f*
zeta potential of, 276*f*
ZnO NPs on evaporation efficiency, 278*f*
photocatalytic experiments by, 279–282
kinetic of, 282, 283*f*
mineralization by, 282
NFs under natural sunlight, 281*f*
UV-Vis spectra of RB5, 281*f*
steam generation experiments by, 267–271, 270*f*
Heat pipes
application of, 414–418
into aluminum case of PCB, 416*f*
axial groove testing, 418*f*
environmental test with PCB, 417*f*
miniature, 416*f*
performance testing, 415*f*
components and design requirements, 407*t*
designs, 412–413
axial grooves extrusions in aluminum alloy, 413*f*
sintered powder, 414*f*
thermal conductivity in, 414*t*
wrapped screen mesh, 413*f*
experimental compatibility testing, 408*t*
fluids merit, 412*f*
heat transport in, 411*f*
limitations, 409*t*
operation characteristics of, 406*f*
Heat transfer enhancement with nanofluids in automotive
for computational fluid dynamics model, 239–243

historical background, 229–234
applications of, 229–232
of nanofluid, 232–234
physical property, 234–238
Heat transfer mechanism
Boltzmann model of nanofluid, 253–256
density and heat capacity, 252
of Fe_3O_4-R134a(gas), 254*t*
of Fe_3O_4-R134a(liquid), 255*t*
diameter spheres, 157*f*
of different coolant
pressure contour of, 251*f*
temperature contours, 250*f*
dilute and dense flows, 156*f*
electrical double layer around a particle, 158*f*
external forces on particles, 161–162
fitting coefficients of viscosity, 253*t*
of Fe_3O_4-R13a, 253*t*
of Fe_3O_4-R13a (gas), 253*t*
of Fe_3O_4-R13a (liquid), 253*t*
Hamaker constant values, 157*t*
[HMIM][BF_4] graphene-based ionanofluids, 517*f*
ionanofluid
density, 519*f*
ionanofluids
viscosity variation against nanoparticle, 520*f*
ionic liquid atomic structure, 516*f*
of nanofluid parameters, 250–252
with nanofluids in automotive, 243–259
coefficient of gas in engine cylinder, 249*f*
computational fluid dynamics of water jackets, 245–250, 246*f*
computational models and boundary conditions, 245–248
density, specific heat capacity, and viscosity, 244–245
fluid dynamics calculation, 229–234
geometric model and mesh generation, 245
mesh model, 246*f*
nanofluid parameters, 243–245
parameters of engine, 244*t*
pressure contour of different coolant, 251*f*
test data of engine bench testing, 247*t*
utilization in engine coolant, 243–250
nanorefrigerant based on lattice Boltzmann method, 256–259
gas amount, 258*t*
liquid amount, 258*t*
model of flow in a tube, 257*f*
numerical simulation results, 257–259
physical model, 256–257
temperature contour of, 258*f*
particle-liquid interactions, 159–161
particle-particle interactions, 155–158, 156*f*
Ra on Nu of ionanofluids, 514*f*
real ionanofluid volume, 522
in refrigerant of automotive air conditioning, 250–259
Ru-ionanofluids density, 523*f*
specific heat against temperature, 525*f*
thermal conductivity, 250–251
of Fe_3O_4-R13a (gas), 250–251, 252*t*
of Fe_3O_4-R13a (liquid), 252*t*
of ionic liquids, 524*f*
variation, 518*f*
thermophoresis, 160*f*
uncharged suspended particles, 159*f*
viscosity, 251
of ionanofluids against temperature, 518*f*
ionanofluids, variation of, 521*f*
Heat transfer nanofluids by MXene
methodology, 134–140
aqueous ionic liquid, 137
aqueous poly (ethylene) glycol (PEG)-based, 136–137
diethylene glycol, 138
ethylene glycol (EG)-based MXene heat transfer nanofluid, 138
ionic liquid mixture-based MXene nanofluid, 138
MXene nanomaterial as an additive, 134–135, 134*t*
new class of silicone oil heat transfer nanofluid, 135–136
olein palm oil with MXene, 136
palm oil methyl ester-based MXene heat transfer nanofluid, 137
Ti_3C_2 nanomaterial, 135*f*
water-based MXene nanofluid, 137
physical property, 138–140
thermal conductivity measurement, 138–139
thermal stability testing, 139–140
viscosity measurement of, 139
results and discussion, 140–148
of formulated olein palm oil, 141*f*
in loading concentration and temperature, 143*f*

Heat transfer nanofluids by MXene (*Continued*)
of silicone oil-based, 141*f*
in temperature over base solution, 143*f*
thermal conductivity of, 140–144
thermal stability of, 146–148
formulated soybean oil, 147*f*
viscosity analysis of, 144–146
shear rate, 145*f*
temperature, 145*f*
soybean oil, 145*f*
Heat transfer of combustion engines
coefficient, 311–313
average transfer methods, 311*t*
gas motion patterns in cylinder, 312*f*
conduction coefficient, 313–316
coolant temperature, 309
engine
load, 308
size and dimension, 307
speed, 307–308
friction coefficient, 310
of engine parts, 310*f*
inlet air temperature, 309
materials, 309
aluminum piston, 309*f*
overhead and flathead engines, 308*f*
parameters affecting the, 307–316
thermal conductivity, 314*t*
temperature variation, 314*f*
through piston cylinder wall, 315*f*
History of solar energy collectors
highlight points, 343*t*
renewable and sustainable energy sources, 342–343
19th century, 342
Hybrid nanofluids, 95
in different heat sinks, 541–551
3D pictures of staggered and inline fin, 542*f*
electronic device cooling using PCM, 547*f*
experimental setup, 543*f*
mini channel, 542*f*
ribs and secondary channels in, 544*f*
preparation of, 539–541
thermal management of electronics, 545–551
using PCM components, 551*t*
velocity profile of, 544*f*
Hybrid nanofluids preparation
one-step method, 540–541, 541*f*
two-step method, 539–540, 540*f*

I

Ionanofluids
applications, 527*t*
characteristics
density, 512
factors affecting, 511*f*
specific heat, 512
thermal conductivity, 511–512
viscosity, 512
preparation methods
gold-based, 508*f*
light absorption through, 508*f*
nanoparticle stability techniques, 510*f*
postpreparation analysis of, 510*f*
single-step and two-step, 506*f*, 507–509
via one-phase process, 507*f*
of ZnO-[HMIM][PF_6], 509*f*
Ionic nanofluids
as base preparation, 505*t*
challenges of, 526–531
characteristics, 511–512
heat transfer mechanism, 513–526
preparation methods, 506–510
thermal applications for, 526

K

Key characteristics of nanofluids
base fluid, 457
estimating viscosity of, 451*t*
mathematical models for estimating thermal conductivity of, 453*t*
nanoparticle
concentration, 454–455
shape, 456
size, 455–456
thermal conductivity, 456–457
new model with experimental data, 454*f*
preparation technique, 458
temperature, 457–458
thermal conductivity of, 455*f*
thermophysical property of, 449–453
viscosity of Al_2O_3, 452*f*

L

Laminar flow, 6
Laser ablation method, 61–62
Least square support vector machine, 209–210
Lorentz force, 161–162

Lubricants type, 315*f*
basic engine oils and challenges, 319–320
nanotechnology application, 320*f*
process improvement, 319–320
of combustion engines, 316–319
multigrade engine oils, 318–319
temperature range, 318*t*
outdoor temperature, 318*f*
single-grade, 317
standardization references for, 316*f*
automotive engineering viscosity grade, 318*f*
temperature range
for summer single grade oils, 317*t*
for winter single grade oils, 317*t*

M

Machine learning
techniques
adaptive neuro fuzzy inference system, 207–208
basic model of a single node, 206*f*
kernel in regression errors, 209
least square support vector machine, 209–210
MLP-ANN, structure of, 207*f*
multilayer perception artificial neural network, 206–207
outputs' mean squared error (MSE), 210
radial basis function network, 208–209, 209*f*
structure of, 208*f*
thermophysical property
artificial neural network, 220*f*
experimental and predicted values of, 218*f*
experimental data and artificial neural network, 219*f*, 220*f*
experimental thermal conductivity data and ANN modeling, 218*f*
physical property of, 221*t*
and thermal performance, 217–223
Magnetic stirring, 67
Magnus lift force, 160–161
Mathematical models to simulate nanofluids
Eulerian, 180–181
continuity equation, 180
energy equation, 180–181
momentum equation, 180
multiphase Eulerian-Eulerian, 178–180
continuity equation, 179
energy equation, 179
mixture model, 178–180
momentum equation, 179
multiphase Eulerian-Lagrangian, 187–189
continuity equation, 183
energy equation, 183–184
momentum equation, 183
Nusselt number values, 185*f*
temperature-dependent thermophysical and constant, 185*f*
single-phase, 173–178
Buongiorno, 176–178
continuity equation, 174
energy equation, 174
homogenous, 174
momentum equation, 174
thermal dispersion model, 175–176
volume of fluid model, 181–183
continuity equation, 182
energy equation, 182
momentum equation, 182
Mesh, 187–188
Milling process, 65

N

Nanofluid application
challenges in solar collectors
environmental issues, 368
reusability of, 367
stability of, 367
in combustion engines
of heat transfer, 306–307
invented engine by Barnett, 306*f*
lubricants type, 316–320
parameters affecting, 307–316
pioneers and developers, 305*f*
threats and opportunities of, 333
using nanoparticles in, 320–333
in solar stills
conventional performance, 362*f*
direct use of, 362, 363*f*
indirect use of, 363, 364*f*
with nanofluids, 362–363
without nanofluids, 361–362
Nanofluid-based microchannel heat sinks
entropy analysis of, 464–467
fouling in, 466*f*
Nanofluid, characterization of
of CeO_2-MWCNT/water, 70*f*
Di-COOH, Di-Ag and Ag nanocomposites, 75*f*

Nanofluid, characterization of (*Continued*)
  Fourier-transform infrared spectroscopy, 70–71, 72*f*
  Sonication Time effect on the Absolute Zeta, 71*f*
  transmission electron microscopy, 71–74, 73*f*
  UV-Vis technique, 74–75, 76*f*
  X-ray crystallography, 74
  zeta potential, 67–70
Nanofluids capability and models
  thermal conductivity applied to, 426*t*
  thermal enhancement ratio for CuO and water, 427*t*
  viscosity models for, 429*t*
Nanofluids heat transfer systems
  after operation cycles, 437–438
  applications, 438–440
  evaluation stability, 436–437
    absorbance spectrometry, 437
    dynamic light scattering (DLS) method, 436–437
    electron microscopy, 437
    equipment used, 436*f*
    omega method, 437
    on surfactant percentage, 437*f*
    visual inspection, 436
  and machine learning, 210–216
    artificial neural network, 211*f*
    average Nusselt number and pressure drop using ANFIS, 214*f*
    correlation coefficient, 212*f*
    experimental against predicted q, 212*f*
    isotherm plot for Reynolds number, 213*f*
    by MLP modeling, 215*f*
    nanofluids' thermal performance using, 216*t*
    with Nusselt number and friction factor, 212*f*
    shell-and-tube heat exchanger, 211*f*
  nanoparticles types, 430–432
    carbon-based, 431
    examples based, 431*f*
    for formulation, 431*f*
    hybrid-based, 431–432
    metal-based, 430–431
    metal oxide-based, 431
  PHP operation with CuO-water, 439*f*
  preparation, 432–433
  in renewable energy
    challenges, 223
    data collection and representation, 204
    machine learning, 203–210
    model selection and validation, 204
    optimization model, 205
    thermophysical property and thermal performance, 217–223
  stability mechanisms, 435–436
    electrostatic, 435
    steric, 435
  stability strategy, 433–434
    carbon-based, 435*f*
    dispersion method, 433, 434*f*
    pH control, 434
    surface modification of nanoparticles, 434
    surfactants, 433–434
    of ultrasonic time, 434*f*
    with and without surfactant, 435*f*
  thermal conductances for, 439*f*
Nanofluids in solar energy collectors
  application of, 361–363
  challenges of, 364–368
  classification of, 344–356
  effective parameters on, 356–361
  history of, 341–343
  nanoparticles in solar stills, 363–364
Nanofluids preparation
  one- step method, 3
  one-step method, 29–30, 29*f*
  synthesis methods of, 28*f*
  two-step method, 3, 27–29, 28*f*
Nanofluids role in microchannel heat sinks
  entropy analysis of, 464–467
  flow of, 459–461
  geometry effect of, 467–468
  key characteristics of, 449–459
  thermal performance of, 461–464
Nanoparticles in internal combustion engines
  adding nanoparticles to engine oils, 322–326
    Al_2O_3 enriched nanolubricants as, 321*f*
    CO reduction, 325*f*
    graphene-enriched, 322*f*
    nanoparticle additives based on BTE and BSFC criterions, 325*f*
    nanoparticle function in fuel, 324*f*
    nanoparticles as fuel additives, 324*f*
    NOx reduction, 325*f*
    SAE40 engine oil additives, 321*f*
    soot reduction, 325*f*
    thermal conductivity and viscosity, 322*f*
    UHC reduction, 325*f*
  adding nanoparticles to fuel, 327

adding to pure oils, 328*t*
additives on cooling cycle of, 331–333
best EG-water-based nanocoolants, 331*f*
best water-based nanocoolant, 330*f*
conventional working fluids of, 332*t*
for exhaust gas recirculation systems, 331*f*
nanocoolant benefits, 330*f*
vehicle parts, 330*f*
additives on engines power, 329–330
friction effect, 329*t*
effects on friction, 327–329
Nanoparticles in solar stills
effect of using, 365*t*
frequency of, 367*f*
Nodes, 187–188
Numerical modeling of nanofluids' flow and heat transfer
mathematical simulation, 173–186
mechanism of, 154–162
numerical techniques to, 186–192
thermophysical property of, 162–173
Numerical techniques to simulate nanofluid
fluid dynamic, 186*f*
macroscale, 187–189
control volume finite element method, 189
2D computational domain and mesh, 187*f*, 188*f*
finite difference method, 187–188
finite element method, 189
finite volume method, 188–189
triangular element and control volume, 190*f*
mesoscale techniques, 191–192
dissipative particle dynamics, 192
Lattice Boltzmann method, 191–192
microscale techniques, 189–190
molecular dynamics principle, 191*f*
Nusselt number
correlations for single-phase fluid
Gnelinski equation, 7
laminar flow, 6–7
turbulent flow, 7
and friction factor of nanofluids
correlations, development of, 8–9
corrections for, 9–20
experimental methods, 3–6
preparation of, 2–3
for single-phase fluid, 6–8

O

Olcin palm oil, 136

P

Palm oil methyl ester-based MXene heat transfer nanofluid, 137
Parabolic trough solar collector utilization
optical efficiency of, 376*f*
property of, 379*f*
review of, 381*t*
Parameters on solar still performance
affecting basin type, 356*f*
climatic parameters, 359–361
ambient temperature, 360
solar radiation, 361
wind velocity, 359
design, 356–359
external condenser in, 360*f*
with heat exchanger, 359*f*
internal condenser in, 360*f*
Photographic technique, 32
Physical property of nanofluids
thermal conductivity, 234
classic, 235*t*
experimental based models for, 237*t*
theoretical models for, 236*t*
viscosity, 234–238
experimental based models for, 238*t*
Precipitation processes, 65
Pulsating heat pipes
application of, 421–424
for electronics cooling, 424*f*
evaporation section temperature, 423*f*
open-loop operation, 422*f*
thermal conductances for, 423*f*
closed-loop, 419*f*
with flow direction valve, 419*f*
design parameters for, 420*t*
open-loop, 419*f*

R

Radial basis function network, 208–209
Reduced gravity force, 161

S

Saffman lift force, 160–161
Solar energy collectors classification
application, 344–345
on concentration level ability, 345, 350*f*
passive and active, 346–347
benefits and drawbacks of, 346–347
on tracking or nontracking, 345–346, 346*f*
types of, 347–356
concentrating, 352–356, 357*f*
evacuated tube collectors, 349–352, 350*f*

Solar energy collectors classification (*Continued*)
flat plate, 347–348, 348*f*, 353*t*
heat pipe evacuated tube, 351*f*
and heat transfer mechanism, 348*f*
liquid and air heater type of flat plate, 349*f*
parabolic dish collector, 355*f*
parabolic trough, 354*f*
pipe and fin of flat plate, 349*f*
power tower receiver, 355*f*, 356*f*
thermosyphon evacuated tube, 352*f*
tube collectors, 349–352
U-pipe evacuated tube, 351*f*
Sol-gel method, 65–66
Specific heat capacity, 104–112
Stability mechanisms of nanofluids
electrostatic, 41–42, 41*f*
electrosteric, 44
steric, 42–44, 43*f*
of zeta potential concept, 42*f*
Stabilization techniques of nanofluid
chemical, 38–40
pH control of nanofluids, 39–40
surface modification, 39
surfactant, 38–39
optimization, 37*f*
physical method, 37–38
ball milling, 38
mechanical stirring, 37
Stable nanofluids for heat transfer application
evaluation of, 30*f*
at different zeta potentials, 31
electron microscopy and light scattering, 34–36
nanocluster sizes by DLS measurement, 36*f*
by observation method, 32*f*
pH measurement, 36
sedimentation and centrifugation methods, 32–33
spectral analysis method, 33
TEM measurement of Fe_3O_4 nanoparticles, 35*f*
transmittance measurement of Cu-Al LDH, 34*t*
3w method, 34
zeta potential analysis, 31, 31*t*
in heat transfer
evaluation of, 30–36
impact on thermophysical properties, 45–52
mechanisms, 40–44
preparation, 27–30
techniques, 36–40
on thermophysical property
aggregation degree on, 50*f*
density, 45
experimental and numerical models of, 46, 47*f*
Hamilton-Crosser model, 49
heat capacity of, 48, 48*f*
impact of, 45–52
Maxwell model, 49
nanoparticle aggregate structures, 51*f*
pH values on, 51*f*
by radius of gyration, 52*f*
thermal conductivity, 49–52
viscosity, 45–48
water-based Cu, 51*f*
Stern layer, 67–70
Summer oils, 317

T

Thermal conductivity, 96–104
measurement, 138–139
Thermal energy applications, 526
Thermal equilibrium model, 48
Thermal performance of nanofluids in microchannels
in heat and mass transfer, 464*t*
Nusselt number correlations and enhancement for, 465*t*
Thermodynamic modeling
energy and exergy analysis, 383–385
energy efficiencies, 391*f*
exergetic efficiency, 391*f*
heat capacity and thermal conductivity of, 386–387
heat transfer analysis, 385–386
heat transfer coefficient, 392*f*
HTFs thermophysical property, 390*f*
hybrid and mono nanofluids, 386–388, 387*f*
mathematical relations, 386
nanofluids dynamic viscosity, 387
Nusselt number, 392*f*
parabolic trough collector model, 389
performance evaluation criteria, 388
present and reference model, validation of, 389*f*
of PTC
design input parameters, 382*t*
model, 382*f*
and system description, 380–389

temperature impact on
criteria of examined HTFs, 396*f*
mean enhancement comparison of, 395*f*
Nusselt number enhancement of, 394*f*
pressure drop, 393*f*
pumping work, 393*f*
thermal efficiency enhancement of, 394*f*
thermophysical property of tested nanoparticles, 388*t*
volumetric concentration of nanoparticles, 386
Thermophysical property of nanofluids
characterization of, 67–75
classification of, 60*f*
density, 173
empirical correlations for, 174*t*
measurement techniques of, 75–84
nanofluid preparation, 66–67
magnetic stirring, 67
ultrasonication method, 67
nanoparticles preparations, 64–66
milling process, 65
precipitation processes, 65
sol-gel method, 65–66, 66*f*
one-step method, 59–64
chemical, 62–64
laser ablation, 61–62, 63*f*
submerged arc, 61, 62*f*
vapor deposition, 60–61
specific heat, 172–173
empirical correlations for, 173*t*
synthesis method and characterization techniques of, 59–67, 68*t*
thermal conductivity, 162–165
of different shapes, 165*f*
empirical correlations estimation, 166*t*, 168*t*
factors determining, 162*f*
with particle volume fraction, 163*f*
variation with particle size, 164*f*
two-step method, 64–67
viscosity, 165–171
experimental correlations for, 171*t*
on shear rate dependent, 170*f*
theoretical correlations for, 172*t*
Thermophysical property measurement techniques
conductivity, 76–79
constant for fluid, 77
constants analyzing, 78
temperature distribution of hot wire, 77–78
transient hot-wire method, 76–78, 77*f*
3w, 78
density, 83–84
specific heat, 81–83
C-thermal apparatus, 81*f*
viscosity of, 79–81, 80*t*
Thermophysical property of fluid
conductivity
of Al_2O_3, 101*t*
of f-ZnO, 102*f*
Maxwell model prediction and experimental data, 102*f*
of SiO_2, 101*t*
of unitary and hybrid, 98*t*
density, 112–116
with temperature, 117*f*
of unitary and hybrid, 113*t*
specific heat capacity, 104–112
of unitary and hybrid, 106*t*, 108*t*
variation of density with temperature, 111*f*
viscosity, 116–122
alteration, 121*f*
ratio alteration of TiO_2, 121*f*
of unitary and hybrid, 119*t*
Transient hot-wire method, 76–78
Transmission electron microscopy, 71–74
Transport property, 76
Turbulent flow, 6

U

Ultrasonication method, 67
Unitary and hybrid nanofluids
thermophysical property, 96–122
UV-Vis technique, 74–75

V

Vacuum evaporation, 29
Van der Waals forces, 155–156

W

Water-based MXene nanofluid, 137
Winter grade oils, 317

X

X-ray crystallography, 74

FSC
www.fsc.org
MIX
Papier | Fördert gute Waldnutzung
FSC® C083411

Kolibri 360 GmbH
Ferdinand-Jühlke-Straße 7
99095 Erfurt, Deutschland
produktsicherheit@kolibri360.de